Practical Reservoir Engineering

Practical Reservoir Engineering

PART I

**Methods for improving accuracy
or input into equations
and computer programs**

E. H. Timmerman

PennWell Books
PennWell Publishing Company
Tulsa, Oklahoma

Copyright © 1982 by
PennWell Publishing Company
1421 South Sheridan Road/P. O. Box 1260
Tulsa, Oklahoma 74101

Printed in the United States of America

1 2 3 4 5   85 84 83 82 81

Library of Congress Cataloging in Publication Data

Timmerman, E. H.
   Methods for improving accuracy or input into
equations and computer programs.

   (Practical reservoir engineering/E. H. Timmerman; pt. 1–   )
   Includes bibliographical references and index.
   1. Oil reservoir engineering—Mathematical models.
2. Oil reservoir engineering—Data processing.
I. Title. II. Series: Timmerman, E. H. Practical
reservoir engineering; pt. 1.
TN864.T55 pt. 1 [TN871] 622'.3382s 81–10591
ISBN 0–878–14–168–5 (set)   [622'.3382'0724] AACR 2

Contents

Chapter 4   Determination of Various Engineering Parameters 182

Chapter 5   Performance of Flowing and Shut-in Wells 227

Foreword

The experience of engineers and managers who have been using copies of parts of this book indicates that the practical engineer and manager interested in finding reliable answers to field-type problems will use this book daily. The content represents a condensation of a notebook occupying many feet on a book shelf. I am indebted to the management of Shell Oil Company in 1950 for an assignment that resulted in a reservoir engineering text for Shell. I received a set of notes and an expandable filing system permitted addition of useful items as they developed and were published during the following 30 years.

I wish to thank all of the individuals who published their thoughts and work. Special thanks for an education are due the staff of Shell Oil Company, including A.F. Van Everdingen, G.E. Archie, C.S. Matthews, M. Prats, and many others. No married engineer can be successful without a patient, understanding family; thank you Erline, Nancy, and Patsy. Special thanks are also given to the organizations that permitted me to use parts of their publications.

I hope that you will find these notes as useful as they have been to me during the past 30 years. The modern computers have taken the drudgery out of engineering, but they should not be substituted for sound thought and work. The first volume of this set is designed to help you analyze input information. The second volume should be helpful in checking the accuracy of output. In many cases, the information contained in this text can eliminate the use of computers since previous studies can be used directly in finding the solution of problems. Use of models without proper equations and accurate input can often result in poor forecasts, which may be less reliable than results from shortcut procedures and analogy. A curve match with past history is desired when data are available, but the match does not ensure that a forecast based on such a match will be accurate.

E. H. Timmerman

1 Introduction

The prime objective of reservoir or subsurface engineering is to understand the secrets of nature that are applicable to a specific oil and gas reservoir. Such information is essential when future performance and recoverable reserves are to be reliably forecast. Fields of expertise such as (1) geophysics-seismos, (2) petrophysics-logging, (3) geology-structure, deposition, rock solution-precipitation, oil source, and accumulation, (4) engineering-recovery mechanisms, flow of fluids in both rocks and tubing strings and EOR processes, and (5) economics-project profits from investments are involved. The successful petroleum engineer must blend these fields of knowledge to obtain an accurate understanding of the past, present, and future behavior of oil and gas reservoirs and must forecast the economic recovery required to prepare sound plans for development and operation of the oil and gas reservoirs. Proper use of this information enables the owner or operator to invest capital wisely and the operation of the reservoir can be made efficient. Product recovery also is improved.

The oil and gas industry, including subsurface engineering, is constantly changing with time to reflect new knowledge, economic conditions, and the location of the oil and gas reservoirs being discovered, drilled, and produced. Engineers during the 1930s devoted much time and energy to developing and understanding the basic concepts of fluid flow in rocks and hydrocarbon characteristics. Back then, knowledge regarding heat and electric flow was transferred to flow of oil and gas using mathematics, the slide rule, and hand-cranked calculators. The convenience of the 1980 hand-held computer was not even a dream. Engineers in early 1940s also investigated nature's secrets to obtain an understanding of the recovery and displacement processes using the new concepts such as relative permeability, measurement techniques, and the many other ideas routinely accepted by the 1980 industry. By about 1950, a general

consensus had been established as to the origin or source of oil and gas, the useable flow equations, the recovery mechanisms, and the major sound operating practices.

The electronic computer has displaced the hand-cranked calculator and slide rule. In 1980 most equations were programmed for hand-held computers. Large, relatively expensive computers are being used to model past performance of reservoirs and to forecast their future performance. Practical subsurface engineering has changed from a search and application of the basic theory-mathematical equations controlling the production of oil and gas to placing numbers into forms used to input data into the electronic computers. It is easy to fill out forms and run the programmed computer. Finding accurate values for each space in the computer input forms is much more difficult. "Garbage in results in garbage out" applies to both the computer output and management and engineering decision making. All engineers should be grateful for removal of the drudgery, but the computer should not be substituted for careful analysis of input data and equations, checking of output, and sound thinking. Good reliable or sound engineering demands that the best possible or most accurate and representative numerical values for all terms in the proper equations be used by the computer. Computer output must also be analysed in detail to be certain that results are realistic when compared with past experience, analogy, results observed in adjacent, or comparable reservoirs having similar characteristics and general concepts regarding reservoir behavior and production practices. A dash of common sense can be most valuable, and the computer should not be used blindly.

The concepts that permit reliable understanding of reservoir behavior have been assembled in a manner conducive to practical engineering. Scaling groups and other combinations of terms in the equations have been used so that field data and laboratory data can be compared and so that data from a reservoir can be easily compared with that of another. Graphs are based on scaled laboratory data, field results, and other data sources such as theoretical calculations using applicable equations. Some graphs may be applicable only to a specified area, but they serve as examples for preparation of similar graphs for other areas of interest.

The proper use of checklists, proven methods of organizing and assembling information, and a detailed knowledge of all assumptions used in the derivation of equations and programming is essential if sound results are to be obtained from a study. Some of these techniques are described in following sections of this introduction; however, equations are not derived.

References

Today the data bank is very large, and it is impossible to include even the better references in this text. As an illustration, the index for articles published by the Society of Petroleum Engineers in the USA requires four books. Many other organizations also have published a large number of papers. Refer to the following references when study in detail of any subject is to be undertaken.

Computer listings
- Listings maintained by oil companies and universities
- University of Tulsa Abstracts, Sidney Born Technical Library, 600 South College, Tulsa, Oklahoma 74104
- Kwic Indexing (IBM), distributed by SUNOCO, 4151 Southwest Freeway, Houston, Texas 77027

Publications by the Society of Petroleum Engineers of AIME 6200 North Central Expressway, Dallas, Texas 75206
- Indexes for each year in December issues of journals
- Books summarizing years 1921–52; 1953–66; 1967–74; 1975–
- Indexes contained in papers, books, monographs and reprints

References to papers published in trade magazines
- *Oil and Gas Journal,* 1421 S. Sheridan Road, Tulsa, Ok. 74101
- *Petroleum Engineer,* 800 Davis Building, Dallas, Texas 75202
- *World Oil,* 3301 Allen Parkway, Houston, Texas 77019

Papers published by U.S. Bureau of Mines
4800 Forbes Avenue, Pittsburgh, Pa. 15213

- One- and five-year indexes

Industry periodicals
- *Journal of Canadian Pet. Tech.* 400 Sherbrooke W., Montreal, Quebec
- *American Association of Petroleum Geologists,* Box 979, Tulsa, Okla. 74101
- *The Log Analyst,* 806 Main Street, Suite 1017, Houston, Texas 77002
- *Geophysics,* SEG, 3707 E. 51 St., Tulsa, Oklahoma 74135

Some of the many texts often quoted by authors are included in the back of the text.

Conversion Factors

Conversion factors are useful for fast conversion between units. These can either be performed by using a constant multiplier for each case or from graphs (Fig. 1–1). It's often useful to have a desk text handy for conversions, such as Steven Gerolde's *Universal Conversion Factors* (PennWell).

Conversion factors may be presented as graphs (Fig. 1–1).

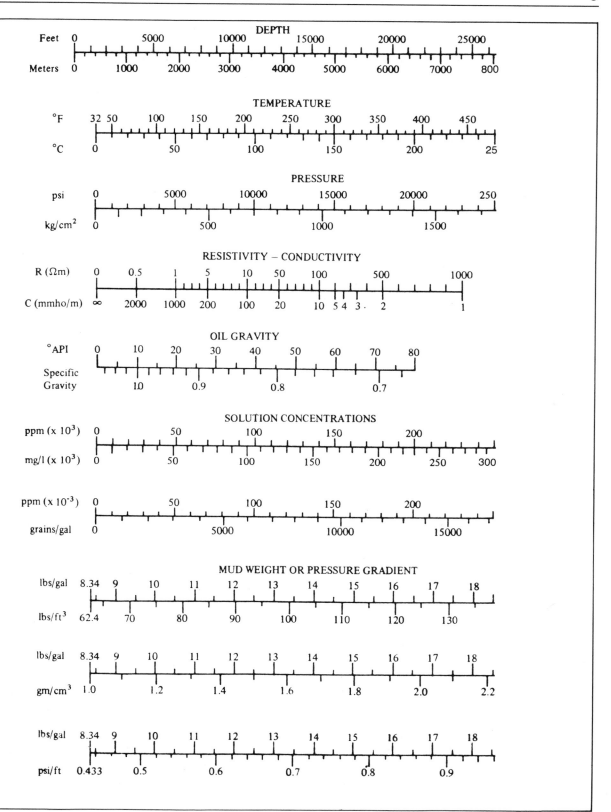

Figure 1-1. *Standard unit conversion.*

Equations and Mathematical Techniques

For details, consult texts related to mathematics for computers. Many equations are included in *Standard Mathematical Tables*, Chemical Rubber Co., 2310 Superior Ave., Cleveland 44114.

Method of Least Squares

When values are taken at equal intervals, the general equation is:

$$X = A + BY + CY^2$$

Values of X and Y are placed in as many equations as allowed by the data points. Each equation is thereafter multiplied by the coefficient of A and the equations are added to obtain one equation. The process is repeated, using the coefficients of B and C. Three summary equations are thereby made available. The average values for A, B, and C can be obtained by using any of the methods available for solving for three unknowns when three equations are available.

Averaging a Curve

	Decline in reservoir, psi	
Date	2 mo. drop	Effective
2–1–42	6	3
4–1–42	12	9
6–1–42	4	8
8–1–42	10	7
10–1–42	58	34
12–1–42	12	35

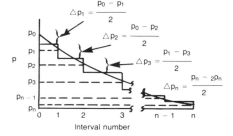

The Use of Logarithms

Any positive number can be represented by another positive number, called a base, raised to an appropriate power, an exponent, $x = b^y$. The exponent to which the base must be raised is called the logarithm of the number, x, for that specific base, b. Or, $y = \log_b x$.

Two bases normally used are 10 and $e = 2.718281828$. The relationship between these bases can be expressed as:

$$\log_{10} x = \frac{\log_e x}{\log_e 10}$$

$$= \frac{\ln x}{\ln 10}$$

$$= \frac{\ln x}{2.302585093}$$

Base 10 logarithms are called common logarithms, and base e logarithms are called natural logarithms. Logarithms for negative numbers are undefined.

The laws for exponents are as follows:

$$a^x \times a^y = a^{x+y} \qquad (ab)^w = a^x \times b^x$$

$$a^{-x} = \frac{1}{a^x} \qquad \frac{a^x}{a^y} = a^{x-y}$$

$$(a^w)^y = a^{xy} \qquad a^0 = 1 \quad a \neq 0$$

Logarithmic paper for plotting graphs is useful in several cases:

- When a gradually increasing or decreasing scale is desired (long range)
- The data fall along a straight line so that equations apply:
  A. $y = ab^{cx}$ (exponential curve); $\log y = cx \log b + \log a$
  Since this is a linear equation of $\log y$ and x, the curve becomes a straight line if the scale for y is logarithmic and the scale for x is linear. Semilog paper is used for plotting (Fig. 1–2).
  B. $y^2 = ax$ (parabolic curve) or $2 \log y = \log a + \log x$ (Fig. 1–3).
  This is a linear equation of $\log y$ and $\log x$ and becomes a straight line when log-log paper is used.
  C. $(x + a)(y + b) = k$ (hyperbolic curve equation)
  Since $\log (x + a) + \log (y + b) = \log k$ is a linear equation of $(x + a)$ and $(y + b)$, the curve becomes a straight line if $(x + a)$ is plotted against $(y + b)$ on log-log paper.
  D. $y^a = b \sqrt[c]{x}$ or a $\log y = \log b + (1/c)(\log x)$
  This too is a linear equation if a log y is plotted vs. $(1/c)(\log x)$ using log-log paper or log x is plotted vs. log y.
  E. If $Y = k_o/k_w$ and $X = S_w$, the curve is used in relative permeability work and illustrates

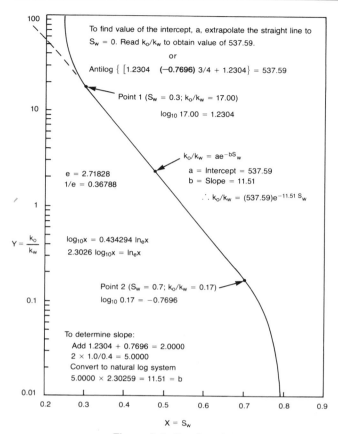

To find value of the intercept, a, extrapolate the straight line to $S_w = 0$. Read k_o/k_w to obtain value of 537.59.

or

Antilog $\{ [1.2304 \quad (-0.7696)\ 3/4 + 1.2304 \} = 537.59$

Point 1 ($S_w = 0.3$; $k_o/k_w = 17.00$)

$\log_{10} 17.00 = 1.2304$

$k_o/k_w = ae^{-bS_w}$
a = Intercept = 537.59
b = Slope = 11.51
∴ $k_o/k_w = (537.59)e^{-11.51\,S_w}$

e = 2.71828
1/e = 0.36788

$Y = \dfrac{k_o}{k_w}$

$\log_{10}x = 0.434294\ \ln_e x$
$2.3026\ \log_{10}x = \ln_e x$

Point 2 ($S_w = 0.7$; $k_o/k_w = 0.17$)
$\log_{10} 0.17 = -0.7696$

To determine slope:
Add 1.2304 + 0.7696 = 2.0000
2 × 1.0/0.4 = 5.0000
Convert to natural log system
5.0000 × 2.30259 = 11.51 = b

$X = S_w$

Figure 1–2. *Semilog plot.*

the method used to determine the slope of the curve. Similarly, the slope of a log-log curve can be obtained as shown in Fig. 1–3.

Quadratic Equation

If $ax^2 + bx + C = 0 \qquad a \neq 0$

then $\qquad x = \dfrac{-b \pm \sqrt{b^2 - 4ac}}{2a}$

Coordinate System

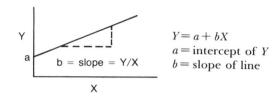

$Y = a + bX$
$a =$ intercept of Y
$b =$ slope of line

b = slope = Y/X

Trigonometric Functions

Trigonometric functions can be defined geometrically in terms of a right triangle.

$C^2 = A^2 + B^2$

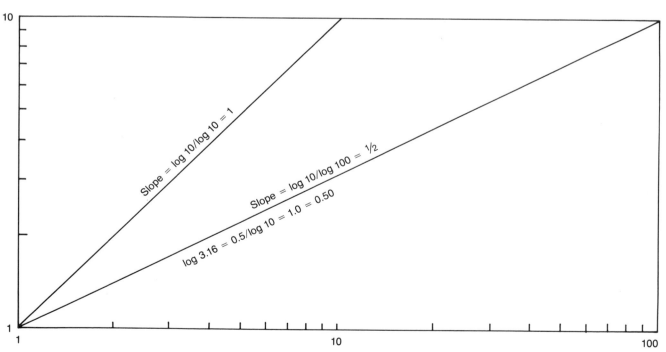

Slope = log 10/log 10 = 1

Slope = log 10/log 100 = ½

log 3.16 = 0.5/log 10 = 1.0 = 0.50

Figure 1–3. *Log-log plot.*

If the angle a is opposite side A, b is opposite B, and c opposite C, then

$$\sin a = A/C, \cos a = B/C, \tan a = A/B$$

Example: If you are building a roof on a tool shed 24 feet wide, and want to have an angle of 30° to provide drainage then $b = 12$ feet and $a = 30°$

$A = B \tan a$
$A = 12 \times \tan 30$
$= 12 \times .5773502692$
$= 6.92820323$ feet

Basic relations for the trigonometric functions are:

$$\sin a = \frac{1}{\csc a}, \cos a = \frac{1}{\sec a}, \tan a = \frac{1}{\cotan a}$$
$$\sin^2 a + \cos^2 a = 1$$

Valid also for any plane triangle

$$A/\sin a = B/\sin b = C/\sin c$$
$$C^2 = A^2 + B^2 - 2 AB \cos c$$

From calculus the functions can be defined as a series expansion

$$\sin a = a - \frac{a^3}{3!} + \frac{a^5}{5!} - \frac{a^7}{7!} + \ldots$$

$$\cos a = 1 - \frac{a^2}{2!} + \frac{a^4}{4!} - \frac{a^6}{6!} + \ldots$$

$$\tan a = a + \frac{a^3}{3} + \frac{2a^5}{15} + \frac{17a^7}{315}$$
$$+ \frac{62a^9}{2835} \ldots (a^2 < \pi^2/4)$$

where the angle a is expressed in radians.

Inverse Trigonometric Functions

Each function returns the value of the angle if the ratio for those two sides of the triangle is known.

$$a = \text{arc sin } (A/C) = \text{arc cos } (B/C) = \text{arc tan } (A/B)$$

The value of the argument for sin and cos functions must be in the interval $-1 \leq R \leq 1$ for arcsin and arccos to be defined, and the value of function will always be between $-\pi/2 \leq a \leq \pi/2$, or $-90° \leq a \leq 90°$. The value for the tan function is the interval $-\infty < R < \infty$.

Exponential Functions

Exponential functions occur frequently in the mathematical problems of biology, physics, chemistry, and engineering. The value of e^x given by the series expansion is:

$$e^x = 1 + X + \frac{X^2}{2!} + \frac{X^3}{3!} + \frac{X^4}{4!} + \ldots$$

The value of e can be evaluated by allowing $X = 1$; $e = 2.718281828$. Trigonometric functions can be expressed as functions of e^x.

$$\sin x = \frac{1}{2i} (e^{ix} - e^{-ix}), \cos x = \frac{1}{2} (e^{ix} + e^{-ix}),$$
$$\tan x = \frac{(e^{ix} - e^{-ix})}{i(e^{ix} + e^{-ix})}$$

Hyperbolic Functions

Hyperbolic functions may be defined as functions of exponentials

$$y = \sinh x = \frac{e^x - e^{-x}}{2}, y = \cosh x = \frac{e^x + e^{-x}}{2}$$
$$y = \tanh x = \frac{e^x - e^{-x}}{e^x + e^{-x}}$$

The function $y = a \cosh (x/b)$ is known as a catenary and describes the way a power cable, chain, or clothes line supported only by the ends will hang.

Basic relations for the hyperbolic functions are:

$\text{csch } x = 1/\sinh x, \text{sech } x = 1/\cosh x,$
$\tanh x = \sinh x/\cosh x = 1/\cosh x$
$\cosh^2 x - \sinh^2 x = 1$
$\sinh x = -\sinh -x, \cosh x = \cosh -x,$
$\tanh x = -\tanh -x$

The power series definition of the hyperbolic functions are:

$\sinh x = x + x^3/3! + x^5/5! + \ldots$
$\cosh x = 1 + x^2/2! + x^4/4! + x^6/6! + \ldots$
$\tanh x = x - x^3/3 + 2x^5/15 - 17x^7/315$
$+ \ldots (x^2 < \pi^2/4)$

Concepts Used in the Field

■ The pressure distribution in a well with skin:

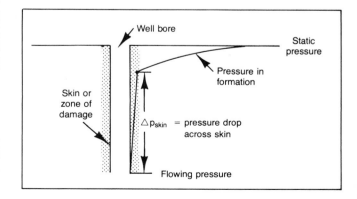

- The area drained by a well is proportional to the relative rates of flow of the wells (Fig. 1–4).

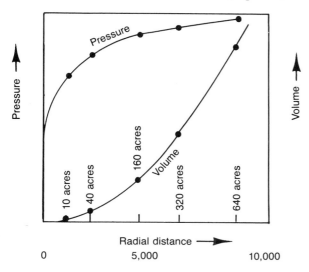

- Area varies as πR^2. The circumference of a circle varies as $2 \pi R$.

- Pressure varies as log distance.

Distance	log_{10}	ln
1	0	0
10	1	2.3026
100	2	4.6052
1,000	3	6.9078
10,000	4	9.210

- At the drainage boundary in a depletion- or solution-gas drive reservoir, $dp/dx = 0$

Drainage area		½ side of a square	Radius of circle
Acre	Sq ft		
40	1,742,000	660 ft	745 ft
80	3,485,000	933	1054
160	6,969,000	1320	1490
320	13,939,000	1867	2160
640	27,878,000	2640	2978

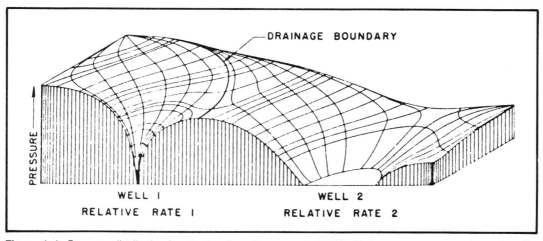

Figure 1–4. *Pressure distribution in a 2:1 rectangular reservoir. NOTE: Well pressures are not shown for either well. These pressures are far below the bottom of the diagram shown (after Matthews & Russell, SPE Monograph 1, © SPE-AIME).*

Mensuration Formulae for Some Areas, Surfaces, and Volumes

TRIANGLE

$S = \frac{1}{2} hc = \sqrt{s(s-a)(s-b)(s-c)}$

where S is the area and $s = \frac{1}{2}(a + b + c)$

$S = \frac{1}{2} ab \sin C$

area $= \frac{1}{2} c h$ where c = base

h = altitude

Rectangle

area = bh

Parallelogram (opposite sides parallel)

area = bh

$= ab \sin \theta$ a & b: sides

θ: angle between sides

TRAPEZOID

$S = \frac{1}{2}(a + b) h$

REGULAR POLYGON

$a_3 = r\sqrt{3} = 1{,}7321\, r$

$a_4 = r\sqrt{2} = 1{,}4142\, r$

$a_5 = \frac{1}{2} r \sqrt{10 - 2\sqrt{5}} = 1{,}1756\, r$

$a_6 = r$

$a_8 = r\sqrt{2 - \sqrt{2}} = 0{,}7654\, r$

$a_{10} = \frac{1}{2} r (\sqrt{5} - 1) = 0{,}61805\, r$

CIRCLE

$P = 2 \pi r = \pi d$

where P is perimeter

$S = \pi r^2 = \frac{1}{4} \pi d^2 = 0{,}7854\, d^2$

SECTOR OF A CIRCLE

$S = \frac{1}{2} r b = \frac{\pi r^2 \beta}{360} = 0{,}00873\, r^2 \beta$

$b = \frac{\pi r \beta}{180}$

SEGMENT OF A CIRCLE

$S = \frac{1}{2}\left\{ b r - c (r - h) \right\} = \frac{1}{2}\left(\frac{\beta \pi}{180} - \sin\beta \right) r^2$

$c = 2\sqrt{h(2r - h)} = 2 r \sin\frac{\beta}{2}$

ANNULARS

$S = (R^2 - r^2)\pi = (R + r)(R - r)\pi = (2r + b)\pi\, b$

where $b = R - r$

ELLIPSE

$S = \pi a b$

P (perimeter approximately) $= \pi \sqrt{2(a^2 + b^2)}$

SPHERE

$V = \frac{4}{3} r^3 \pi = 4{,}189\, r^3 = \frac{1}{6} d^3 \pi = 0{,}5236\, d^3$

$S = 4 \pi r^2 = 12{,}566\, r^2 = \pi d^2$

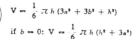

SPHERICAL SEGMENT OR FRUSTRUM OF SEGMENT

$V = \frac{1}{6} \pi h (3a^2 + 3b^2 + h^2)$

if $b = 0$: $V = \frac{1}{6} \pi h (h^2 + 3a^2)$

$S = 2 \pi r h$

if $a = r$ then: $V = \pi h (r^2 - \frac{1}{3} h^2)$

CUP

When the bases are ellipses

$V = \frac{1}{6} \pi h \left\{ (2a + a_1) b + (2a_1 + a) b_1 \right\}$

DRUM

$V = 1{,}0453\, h (0{,}4\, D^2 + 0{,}2\, Dd + 0{,}15\, d^2)$

In case when the limited arcs are those of a circle

$V = \frac{1}{12} \pi h (2 D^2 + d^2)$

CIRCLE

$(x - x_0)^2 + (y - y_0)^2 = R^2$

When $x_0 = y_0 = 0$: $x^2 + y^2 = R^2$

$R^2 = r^2 - 2 r r_0 \cos(\varphi - \varphi_0) + r_0^2$ (Polar equation)

PARABOLA

$y^2 = 2 p x$

$r = \frac{p}{1 + \cos\varphi}$

ELLIPSE AND HYPERBOLA

Ellipse: $\frac{x^2}{a^2} + \frac{y^2}{b^2} = 1$

Hyperbola: $\frac{x^2}{a^2} - \frac{y^2}{b^2} = 1$: $r = \frac{p}{1 + \varepsilon \cos\varphi}$

Where ε is the eccentricity $\frac{\sqrt{a^2 \mp b^2}}{a}$

IRREGULAR PRISM

Q is area of cross section at right angle to base

$V \text{ (volume)} = \frac{1}{3} Q (a + b + c)$

PRISM

S = lateral surface

V = volume

P = perimeter of normal section

Q = area of normal section

G = area of base

$V = G h = Q l$

$S = P l$

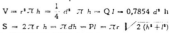

CYLINDER

P = approximate perimeter of normal section

Q = area of normal section, semiaxes of which are r and $\frac{h}{l}$

$V = r^2 \pi h = \frac{1}{4} d^2 \pi h = Q l = 0{,}7854\, d^2 h$

$S = 2 \pi r h = \pi d h = P l = \pi r \sqrt{2(h^2 + l^2)}$

PYRAMID

G = area of base

$V = \frac{1}{3} G h$

CONE

For a right cone the curved surface is:

$S = \pi r l = \pi r \sqrt{r^2 + h^2}$

where $l = \sqrt{r^2 + h^2}$

$V = \frac{1}{3} r^2 \pi h = 1{,}0472\, r^2 h = \frac{1}{12} d^2 \pi h = 0{,}2618\, d^2 h$

FRUSTRUM OF PYRAMID

G and g areas of lower and upper bases

$V = \frac{1}{3} h \left(G + g + \sqrt{G g} \right)$

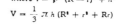

FRUSTRUM OF CONE

Surface of frustrum of regular right cone:

$S = \pi l (R + r)$

where $l = \sqrt{(R - r)^2 + h^2}$

$V = \frac{1}{3} \pi h (R^2 + r^2 + Rr)$

Regression Analysis

Statistics are often used when the computer is involved. Books are available, and the following is a very simple application used to obtain the equation for a straight line representing a data set.*

Data

$$n = 3 \quad \begin{array}{cc} 1.0 & 3.0 \\ 2.0 & 5.8 \\ 3.0 & 9.2 \end{array}$$

Plot

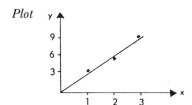

Model

$$y = a + b(x - \bar{x})$$

Sums

x_i	y_i	$x_i - \bar{x}$	$(x_i - \bar{x})^2$	$y_i - \bar{y}$	$(y_i - \bar{y})^2$	$(x_i - \bar{x})(y_i - \bar{y})$
1.0	3.0	−1.0	1.0	−3.0	9.00	3.0
2.0	5.8	0.0	0.0	−0.2	0.04	0.0
3.0	9.2	1.0	1.0	3.2	10.24	3.2
sums: $\sum_{i=1}^{3}$ 6.0	18.0	0.0	2.0	0.0	19.28	6.2

averages: $\bar{x} = 2.0$; $\bar{y} = 6.0$

Coefficients

$$a = \frac{1}{3} \sum_{i=1}^{3} (y_i) = \frac{1}{3} 18.0 = 6.0$$

$$b = \frac{\sum_{i=1}^{3} [(x_i - \bar{x})(y_i - \bar{y})]}{\sum_{i=1}^{3} (x_i - \bar{x})^2} = \frac{6.2}{2.0} = 3.1$$

Line

$$y = 6.0 + 3.1 \ (x - 2.0)$$

The original $y - x$ pairs do not satisfy this equation exactly. There is a small error or residual for each pair.

$$y_i = 6.0 + 3.1(x_i - 2.0) + \textit{small error}$$
$$3.0 = 6.0 + 3.1(1.0 - 2.0) + 0.1$$
$$5.8 = 6.0 + 3.1(2.0 - 2.0) - 0.2$$
$$9.2 = 6.0 + 3.1(3.0 - 2.0) + 0.1$$

* This section from *Concepts and Applications of Regression Analysis,* IBM.

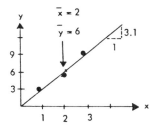

Two questions can be asked (and answered) here: (1) how good, or reliable, is the straight-line fit? and (2) how much confidence do you have in a prediction using the straight-line fit?

Adequacy

To answer the first question we develop a measure of the adequacy of the fit, and we call it r^2 *(r squared)*. In terms of the sums generated from n values of x_i and y_i:

$$r^2 = \frac{\left\{ \sum_{i=1}^{n} [(x_i - \bar{x})(y_i - \bar{y})] \right\}^2}{\sum_{i=1}^{n} (x_i - \bar{x})^2 \times \sum_{i=1}^{n} (y_i - \bar{y})^2}$$

This is how r^2 is used: if, for example, $r^2 = 0.92$, then 92% of the variation of y (about mean, $\bar{y}$) is accounted for by the straight-line fit (the other 8% is residual variation about the line of fit). By variation of y about $\bar{y}$ is meant the sum of the squares of $y_1 - \bar{y}, y_2 - \bar{y}, \ldots, y_n - \bar{y}$.

Some practitioners say the fit is reliable if the estimates for a and b of the line-of-fit are close to their "true" values. You know, for example, that a given slope b is good if it is 10 ± 1 ("true b" lies between 9 and 11) with a 95% probability. But if b is 10 ± 8, even with the same or lower probability, the slope is not too trustworthy. Thus many people test fit by looking at r^2, and also at these *confidence bounds* on a and on b. Confidence bounds on true a and true b:

$$a - t\sqrt{\frac{s^2}{n}} \leq \text{true } a \leq a + t\sqrt{\frac{s^2}{n}}$$

$$b - t\sqrt{\frac{s^2}{\sum_{i=1}^{n} (x_i - \bar{x})^2}} \leq \text{true } b \leq b + t\sqrt{\frac{s^2}{\sum_{i=1}^{n} (x_i - \bar{x})^2}}$$

where $s^2 = \frac{1}{n-2} \left[\sum_{i=1}^{n} (y_i - \bar{y})^2 - b^2 \times \sum_{i=1}^{n} (x_i - \bar{x})^2 \right]$

and t comes from the so-called t-table, under the headings 95% and $(n - 2)$.

Stipulate a level of assurance (75%, 90%, 95%,

etc.), and the corresponding t that you need in the formulas is found in a standard t-table under entries $(n-2)$ and %. Narrower bounds around true values come from more data (n is larger). Looser confidence needs (75%, rather than 95%, is stipulated) also cause true values to be in narrower ranges.

The miniature t-table shown below can be used in the example to follow. The confidence bounds on a and b (and other estimates in regression) cannot be calculated by using the 95%; the strict mathematics requires that you go via the table. Extensive t-tables can be found in almost every book on statistics.

Miniature t-table

% $n-2$	25%	5%	2.5%	0.5%
1	1.0000	6.314	12.706	63.657
10	0.7000	1.812	2.228	3.169
20	0.687	1.725	2.086	2.845
30	0.683	1.697	2.042	2.750
60	0.679	1.691	2.000	2.660

(This parameter is known as the degrees of freedom)

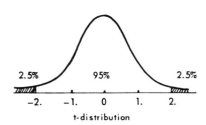

t-distribution

To get 95% confidence for $n = 62$, look up $n - 2 = 60$ and $1/2(100\% - 95\%) = 2.5\%$. Result: $t = 2.000$.

Take an example and review the points in question. Regression gives the line $y = 300 + 20 \times (x - 60)$; $a = 300$, and $b = 20$. How good are a and b? This is determined by the formulas for confidence bounds; these formulas need n, s^2, $\text{SUM}_{i=1}^{n} (x - \bar{x})^2$, a, b, and t (the t for some confidence percent). Assume $n = 62$, $s^2 = 6200$, $\text{SUM} (x - \bar{x})^2 = 10,000$, $a = 300$, $b = 20$, and the desired confidence level = 95% (95% with $(n-2) = 60$ gives a t-table value of 2.000). Substituting in the formulas, with a 95% probability

$$300 - 2\sqrt{\frac{6,200}{62}} \leq \text{true } a \leq 300 + 2\sqrt{\frac{6,200}{62}}$$

or $280 \leq \text{true } a \leq 320$

Similarly,

$$(20 - 1.57) \leq \text{true } b \leq (20 + 1.57)$$
or $$18.43 \leq \text{true } b \leq 21.57$$

Confidence

The second question was: "how confident are you in a prediction made with the regression equation?" The predicted $\dot{y}$ is 2300 when the new $\dot{x}$ is 160, by virtue of the regression formula $y = 300 + 20 (x - 60)$. There are confidence or reliability bounds on the true $\dot{y}$ (taking into account the natural randomness that y is supposed to have, as well as error from prediction).

Confidence bounds on predicted $\dot{y}$ for new $\dot{x}$:

$$(a + b(\dot{x} - \bar{x}) - p) \leq \text{true } \dot{y} \leq (a + b(\dot{x} - \bar{x}) + p)$$

where $$p = t \times s \sqrt{1 + \frac{1}{n} + \frac{(\dot{x} - \bar{x})^2}{\text{SUM}_{i=1}^{n} (x_i - \bar{x})^2}}$$

and t comes from a t-table under the assigned % and $(n-2)$.

As an example, let $n = 62$, $s^2 = 6200$, $\text{SUM}_{i=1}^{n} (x_i - \bar{x})^2 = 10,000$, $a = 300$, $b = 20$, $\bar{x} = 60$, $\dot{x} = 160$, and the desired confidence = 99% (99% and $(n - 2) = 60$ indicate a t-table value of 2.660). You can say about the predicted true $\dot{y}$ value that, with 99% probability, it lies between

$$300 + 20 (160 - 60) - p \text{ and } 300 + 20 (160 - 60) + p$$

where

$$p = (2.66 \times \sqrt{6200}) \sqrt{1 + \frac{1}{62} + \frac{(160 - 60)^2}{10000}}$$

$$= 2.66 \times 78.7 \times 2.02 = 422$$

The data and your desire for 99% certainty in this prediction lead to:

$$1878 \leq \text{true } \dot{y} \leq 2722$$

The confidence bounds get wider (that is, worse) the farther in either direction $\dot{x}$ is from the mean $\bar{x}$, of the data.

A great variety of new directions are available at this point. Staying with a simple approach, we go on now to find a general relation or equation between one variable and many others (multiple regression). For various applications of straight-line fits, see the section entitled "Applications of Regression" and the literature cited.

Summary Chart for Straight-line Regression

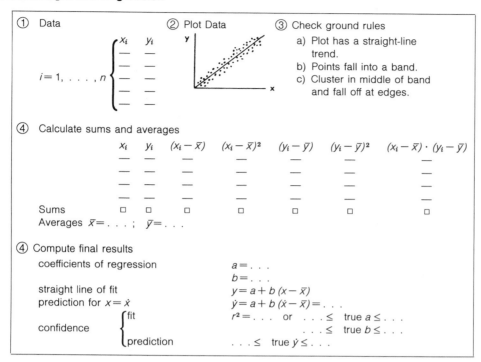

Definitions of Mean, Variance, and Standard Deviation

The *mean* or expected value of a distribution of a discrete random variable x where p_i is the probability that x takes on the value of x_i is

$$u = E(x) = \sum_i p_i x_i$$

For a continuous random variable, the mean is

$$u = \int_{-\infty}^{\infty} x f(x)\, dx$$

The mean is a measure of the central tendency of a distribution. The *arithmetic mean* of a set of N numbers $X_1, X_2, X_3 \ldots X_N$ is

$$\bar{X} = (X_1, X_2, X_3, \ldots X_N)/N = \left(\sum_{i=1}^{N} X_i\right)/N$$

The *geometric mean* is

$$G = \sqrt[N]{X_1, X_2, X_3 \ldots X_N}$$

The *harmonic mean* is

$$H = \frac{1}{(1/N)\sum_{i=1}^{N}(1/X_i)} \quad \frac{N}{\sum_{i=1}^{N}(1/X_i)}$$

The *variance* is a measure of the variability or degree of spread of the distribution and may be defined as

$$\sigma^2 = E(x-u)^2 = E(x^2) - [E(x)]^2$$

The *standard deviation* is the square root of the variance and is an indication of the variability of the distribution. It is applicable to both discrete and continuous random variables.

The *median* of a distribution is the point in the distribution of a random variable at which one-half of the values are greater than the value and one-half are less that the value.

The *mode* of a distribution is that value of the variable for which $p(x)$ or $f(x)$ is a maximum.

The engineer and geologist will frequently encounter distributions of specific forms such as the normal, the log-normal, the binomial, and the hypergeometric. The first two will be discussed in some detail.

The normal distribution of the continuous random variable x is given by

$$f(x) = (1/\sqrt{2\pi\sigma^2})\, e^{-(x-u)^2/2\sigma^2} \quad \infty < X < \infty$$
$$\text{Mean} = M = (\sum n_i x_i)/\sum n_i$$
$$\text{Variance} = \sigma^2 = \sum n_i (x_i - u)^2/\sum n_i$$

The mean, median, and mode of a normal distribution are coincident. This function has a single peak at the mean and is symmetrical about that point. The distribution is bell-shaped with inflection points at u and

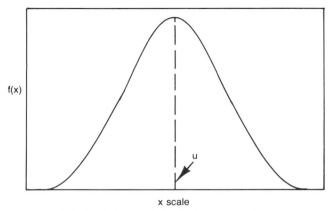

Figure 1–5. *Typical normal distribution on coordinate paper.*

σ (Fig.1–5). As the two parameters u and σ take on different values, the shape or position of the curve changes. As the mean changes, the curve is merely shifted laterally with no change in its shape. As the standard deviation changes, the shape changes but the mean remains the same. The total area under the curve is one, and the curve must be integrated to find the area between any two points on the x axis. Tables are available for this purpose. Most engineers solve the problem by using normal-probability paper. One axis of this paper is a probability scale, while the other is the random variable plotted on a coordinate scale. The two points needed to define the straight line are u and σ. The method should be clear after study of the example given under the log-normal distribution.

Log-normal Distribution

The log-normal distribution of a continuous random variable is similar to a normal distribution except that it is skewed on one side, as in Fig. 1–6.

If a random variable is log-normally distributed, the logarithms of the values of the random variable are normally distributed. The mathematical aspects of the log-normal density function are more complex, so it is usually convenient to take the logarithms of

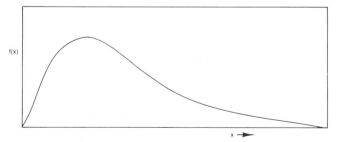

Figure 1–6. *A log-normal distribution.*

the data being studied and use the tables for normal distributions to determine the desired probabilities.

The log-normal distribution is of considerable importance in petroleum exploration. Many field-size (reserve) distributions can be described reasonably well with a log-normal distribution. Other parameters usually considered to be log-normally distributed are core permeability and the thickness of sedimentary beds.

The mean value of a log-normal distribution corresponds to the statistical notion of geometric mean. The mean value of a log-normal distribution skewed to the right will always be less than the arithmetic mean of the set of data. This is an important point that is often ignored in analyzing permeability data.

The cumulative log-normal distribution can be treated graphically in much the same manner that a normal distribution is treated. However, the graph must be log-probability paper on which the random variable axis is logarithmic. The following example illustrates how a set of per-well reserves are analyzed. Note that the equivalent analysis would be applicable to study normally distributed data, except that different graph paper would be used.

Recoverable Reserves Per Well (thousands of barrels)

Well no.	Recoverable reserves	Well no.	Recoverable reserves
1	11	29	127
2	149	30	29
3	48	31	32
4	50	32	77
5	51	33	118
6	41	34	13
7	82	35	152
8	121	36	32
9	63	37	157
10	231	38	74
11	61	39	39
12	215	40	95
13	86	41	48
14	200	42	57
15	192	43	139
16	151	44	26
17	44	45	22
18	111	46	116
19	100	47	110
20	32	48	17
21	95	49	40
22	98	50	55
23	42	51	111
24	51	52	110
25	120	53	20
26	161	54	288
27	69	55	83
28	24	56	140

The first step is to determine the frequency of reserves in given intervals, such as 10,000 barrels.

Cumulative Frequency Determination (thousands of barrels)

Random variable, X, recoverable reserves	Freq. in each interval	Cum. freq. less than or equal to X	Cum. freq. expressed as a fraction
$0 < x \le 10$	0	0	0
$10 < x \le 20$	4	4	0.071
$20 < x \le 30$	4	8	0.143
$30 < x \le 40$	5	13	0.232
$40 < x \le 50$	6	19	0.339
$50 < x \le 60$	4	23	0.411
$60 < x \le 70$	3	26	0.464
$70 < x \le 80$	2	28	0.500
$80 < x \le 90$	3	31	0.553
$90 < x \le 100$	4	35	0.624
$100 < x \le 110$	2	37	0.660
$110 < x \le 120$	5	42	0.750
$120 < x \le 130$	2	44	0.785
$130 < x \le 140$	2	46	0.821
$140 < x \le 150$	1	47	0.840
$150 < x \le 160$	3	50	0.893
$160 < x \le 170$	1	51	0.911
$170 < x \le 180$	0	51	0.911
$180 < x \le 190$	0	51	0.911
$190 < x \le 200$	2	53	0.947
$200 < x \le 210$	0	53	0.947
$210 < x \le 220$	1	54	0.964
$220 < x \le 230$	0	54	0.964
$230 < x \le 240$	1	55	0.982
$x > 240$	1	56	1.000
	56		

The random variable *x* is plotted versus the cumulative frequency less than or equal to the upper value of each interval. The graph paper used is called log-probability paper.

As in the case with most actual field data (as opposed to textbook examples) the data do not fall in a straight line. The causes are perhaps threefold: 1) insufficient data, 2) the data cannot be described adequately by a log-normal distribution, or 3) the data has bias, i.e., it may contain data from two or more different sample spaces. Probably, causes 1 and 3 are contributing most significantly, with the likelihood that the chief reason for the variation is the insufficient sample size. However, the fit appears to be of sufficient accuracy to permit estimation of the values desired.

The mean value is read from Fig. 1–7 at a cumulative frequency equal to 50% as 72,000 bbl. The reserves corresponding to $\pm\sigma$ can be read using the right ordinate scale of standard deviation. The reserves corresponding to $\mu + \sigma$ are 140,000 bbl, and the reserves corresponding to $\mu - \sigma$ are 37,000 bbl. The probability that reserves will exceed 45,000 bbl is the complement of the probability that reserves will be less than or equal to 45,000. The latter data is read from Fig. 1–7 as 0.24. Therefore, the probability of reserves being greater than 45,000 bbl is $1.0 - 0.24 = 0.76$.

Probability and statistics will be discussed in greater detail in a later section involving economics.

The areas under the normal distribution curve can be obtained from tables published in *Engineering Statistics and Quality Control*, by Irving W. Burr, McGraw-Hill, 1953. Tables of the binomial probabilities are available from "Tables of Cumulative Binomial Probabilities," Ordnance Corps Pamphlet ORDP 20–1, September 1952.

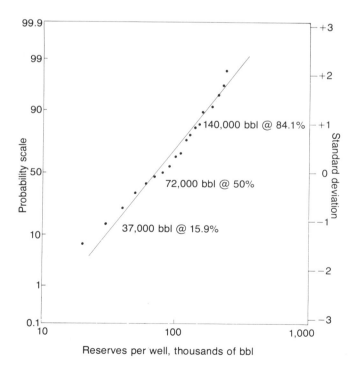

Figure 1–7. *Graphical representation of data of tables on log-normal probability paper.*

Symbols Used in Equations

The definitions shown below are taken from Arps, Chapter 37 of Frick, *Petroleum Production Handbook.* The symbols are defined by the Society of Petroleum Engineers and are published in several texts of SPE. Frick was copyrighted in 1962 by SPE-AIME.

The letter symbols and units used and enumerated alphabetically below follow the standards recommended by the Society of Petroleum Engineers of the American Institute of Mining, Metallurgical and Petroleum Engineers. *Initial conditions* are indicated throughout by the subscript *i, conditions at abandonment time* by the subscript *a,* and *conditions at the bubble point* by the subscript *b.*

A = *area;* in square feet; in acres

α = *angle* of formation dip, deg

B_g = *gas formation volume factor;* a dimensionless factor, representing the volume of free gas at a reservoir temperature $T°F$ and a pressure of p psia per unit volume of free gas under standard conditions of 60°F and 14.7 psia

B_o = *oil formation volume factor;* a dimensionless factor, representing the volume of oil saturated with solution gas at reservoir temperature T and pressure p per unit volume of stock-tank oil. It may be determined by *PVT* analysis of a bottom-hole or recombined sample or obtained from appropriate correlation charts. A typical oil-formation volume factor relationship against gas solubility R_s would be of the type

$$B_o = 1.05 + 0.0005 R_s$$

B_t = *two-phase formation volume factor for oil;* a dimensionless factor representing the volume of oil and its original complement of dissolved gas at reservoir temperature T and pressure p per unit volume of stock-tank oil. This two-phase formation factor for oil B_t is related to the oil-formation volume factor B_o, the gas-formation volume factor B_g, the gas-solubility factor R_s, and the gas-solubility factor at the bubble point R_{sb} by

$$B_t = B_o + 0.1781 B_g (R_{sb} - R_s)$$

b = *constant* (in decline-curve analysis)

C_f = *compressibility factor for reservoir rock;* expressed as change in pore volume per unit pore volume per psi. The compressibility factor C_f appears to vary inversely with rock porosity

from 10×10^{-6} for 2 percent porosity, to 4.8×10^{-6} for 10 percent porosity, and to 3.4×10^{-6} for 25 percent porosity

C_o = *compressibility factor for reservoir oil;* in volume per volume per psi for undersaturated oil above the bubble point. Typical values for C_o range from 5×10^{-6} for low gravity-oils to 25×10^{-6} for higher-gravity oils, with 10×10^{-6} being a good average

C_w = *compressibility factor for interstitial water;* in volume per volume per psi. Although the water compressibility C_w varies somewhat with pressure, temperature, and the amount of salt or gas in solution, 3×10^{-6} represents a good average value

D = *nominal decline rate;* the negative slope of the curve representing the natural logarithm of the production rate q against time $t;$ also the instantaneous rate of change of the production rate vs. time, divided by the instantaneous production rate $q;$ expressed as a decimal fraction with time in months or years

D_c = *effective decline rate;* the drop in production rate per unit of time (month or year) divided by the production rate at the beginning of the period; expressed as a decimal fraction

f_w = *water fraction of flow stream in reservoir which consists of oil and water*

G = *free reservoir gas in place,* scf

g_o = *gravity of stock-tank liquid* (oil or condensate), °API

G_p = *cumulative gas produced,* scf

G_s = *solution gas in place,* scf

H = *gross pay thickness,* ft

h = *net pay thickness,* ft

k = *absolute permeability,* millidarcies

k_o = *effective permeability to oil,* millidarcies

k_{rg} = *relative permeability to gas* as a fraction of absolute permeability

k_{ro} = *relative permeability to oil* as a fraction of absolute permeability

k_{rw} = *relative permeability to water* as a fraction of absolute permeability

L = *length,* ft

$\ln$ = *natural logarithm* to the base e

$\log$ = *common logarithm* to the base 10

m = *ratio between initial reservoir free gas volume and initial reservoir oil volume;* related to the amount of free initial reservoir gas G, the initial gas-formation volume factor B_{gi}, the amount of initial reservoir oil in place N, and the initial oil-formation volume factor B_{oi} by

$$m = \frac{GB_{gi}}{5.615 N B_{oi}}$$

N = *reservoir oil in place*, stb

N_p = *cumulative oil produced*, stb

n = *exponent*, in decline-curve analysis

p = *reservoir pressure*, psia; generally measured by bottom-hole pressure bomb at a depth representative of the entire reservoir, e.g., the midpoint of the oil or gas column. Although the vertical pressure gradient in oil fields may range from as low as 20 or 30 psi per 100 ft to as high as 90 or 100 psi per 100 ft of depth, typical hydrostatic gradients are usually in the 44 to 52 psi per 100 ft range.

p_c = *critical pressure*, psia

q_o = *rate of oil production*, b/d

q_g = *rate of gas production*, scf/d

q_t = *rate of total fluid production*, b/d; q_t designates the total flow rate of oil and free gas, the total flow rate of oil and water; both expressed in cu ft per day under reservoir conditions

R = *instantaneous producing gas-oil ratio*, scf/stb

R_p = *cumulative gas-oil ratio*, scf/stb, related to cumulative gas produced G_p and cumulative oil produced N_p by

$$R_p = \frac{G_p}{N_p}$$

R_s = *gas-solubility factor*; the number of standard cubic feet of gas, liberated under specified separator conditions, which are in solution in 1 bbl of stock-tank oil at reservoir temperature T and pressure p. It may be determined by *PVT* analysis of a bottom-hole or recombined sample or obtained from appropriate correlation charts. A typical gas-solubility relationship against pressure for medium-gravity crude would be of the type

$$R_s = 135 + 0.25p \text{ cu ft/bbl}$$

r = *ratio* of initial to final production rate q_i/q_a in decline-curve analysis

ρ_o = *density of reservoir oil*, g/cc

ρ_g = *density of reservoir gas*, g/cc

S_g = *free-gas saturation* under reservoir conditions, as a fraction of pore space

S_g' = *free-gas saturation* under reservoir conditions, as a fraction of hydrocarbon-filled pore space

$$S_g' = \frac{S_g}{1 - S_w}$$

S_{gc} = *equilibrium* (or critical) *free-gas saturation*, which is the maximum free-gas saturation reached when lowering the pressure below the bubble point, before the relative permeability to gas becomes measurable; expressed as a fraction of pore space under reservoir conditions

S_{gr} = *residual free-gas saturation* under reservoir conditions at abandonment time, as a fraction of pore space

S_o = *oil or condensate saturation* under reservoir conditions, as a fraction of pore space

S_{or} = *residual-oil saturation* under reservoir conditions, as a fraction of pore space, generally determined by multiplying the residual-oil saturation from core analysis with B_o

S_t = *total liquid saturation* under reservoir conditions, as a fraction of pore space: $S_t = 1 - S_g = S_o + S_w$

S_w = *interstitial water saturation as a fraction* of pore space; generally determined by (1) analysis of water content of cores taken with a non-aqueous drilling fluid, (2) capillary-pressure measurement on cores, or (3) quantitative analysis of electrical logs

T = *reservoir temperature*, °F, measured at a depth representative of the entire reservoir; e.g., at the midpoint of the oil or gas column. Vertical temperature gradients in oil fields range from 0.5°F to 3°F per 100 ft of depth with 1.5°F per 100 ft being a good average

t = *time*, days or months

μ_g = *reservoir gas viscosity*, cp, ranging from 0.01 cp at low temperatures and pressures to 0.06 cp for high gas gravities at very high temperatures and pressures, with 0.02 cp being a good average

μ_o = *reservoir oil viscosity*, cp, ranging from under 0.1 cp for volatile oils under very high temperatures and pressures, to very high values for low-gravity oils which will barely flow at all. Most reservoir oils, however, fall in the 1- to 2-cp range.

μ_w = *reservoir water viscosity*, cp, ranging from 0.2 cp at high temperatures to 1.5 cp at lower temperatures, with 0.5 cp being a good average

ϕ =   *effective porosity*, as a fraction of bulk pay volume; generally determined by laboratory analysis of cores, side-wall samples, or cuttings; quantitative analysis of electrical, radioactivity, or sonic logs. Typical values for ϕ range from as low as 0.03 in tight limestones, and from 0.10 to 0.20 in cemented and consolidated sandstones, to as high as 0.35 in unconsolidated sands.

ϕ' =   *effective hydrocarbon-bearing porosity*, as a fraction of bulk pay volumes

$$\phi' = \phi(1 - S_w)$$

V =   *gross pay volume*, acre-ft

V_g =   *net pay volume of the free-gas-bearing portion of a reservoir*, acre-ft

V_o =   *net pay volume of the oil-bearing portion of a reservoir*, acre-ft

W_c =   *cumulative water influx*, bbl

W_p =   *cumulative water produced*, bbl

Z =   *compressibility factor*, for the free gas in the reservoir; a dimensionless factor, which, when multiplied with the reservoir volume of gas as computed by the ideal-gas laws, yields the true reservoir volume

Electric Logging

English		Dimensions
d	hole diameter	L
D_i	electrically equivalent diameter of the invaded zone	L
e	individual bed thickness	L
E	electromotive force	mL^2/t^2Q
E_c	electrochemical component of the spontaneous electromotive force, equals the electrochemical component of the spontaneous potential (self potential) when the bed is sufficiently thick	mL^2/t^2Q
E_k	electrokinetic component of the spontaneous electromotive force, equals the electrokinetic component of the spontaneous potential (self potential) when the bed is sufficiently thick	mL^2/t^2Q
F	formation resistivity factor in general—equals R_o/R_w. (Numerical subscript indicates value of R_w. For example, $F_{0.3}$ represents the value	

of the formation factor measured with $R_w = 0.3$.)

F_{lim}	limiting value of formation resistivity factor when R_w approaches zero	
h	net pay thickness	L
H	gross pay thickness	L
I	resistivity index—equals R_t/R_o	
k	permeability	L^3
K	coefficient in the equation of the electrochemical component of the spontaneous electromotive force	mL^2/t^2Q
m	formation resistivity factor exponent	
n	saturation exponent	
p	pressure	m/Lt^2
Q	charge	Q
r	resistance	mL^2/tQ^2
R	resistivity	mL^3/tQ^2
R_a	apparent resistivity	mL^3/tQ^2
R_i	average resistivity of invaded zone	mL^3/tQ^2
R_m	mud resistivity	mL^3/tQ^2
R_{mc}	mud cake resistivity	mL^3/tQ^2
R_{mf}	mud filtrate resistivity	mL^3/tQ^2
R_o	resistivity of a formation 100 percent saturated with water of resistivity R_w	mL^3/tQ^2
R_s	resistivity of surrounding formations	mL^3/tQ^2
R_t	true formation resistivity	mL^3/tQ^2
R_w	in the laboratory: resistivity of an electrolytic solution; in the field: resistivity of interstitial water	mL^3/tQ^2
R_{xo}	resistivity of invaded zone close to the wall of the hole, where flushing has been maximum	mL^3/tQ^2
S	saturation	
S_g	gas saturation	
S_o	oil saturation	
S_w	water saturation	
t_{mc}	mud cake thickness	L
T	temperature	T

Greek

α alpha	spontaneous electromotive force reduction factor, equals the spontaneous potential (self potential) reduction factor when the bed is sufficiently thick: ratio of the value of E_c for a given shaly sand to the value of E_c for a clean sand having the same interstitial water	
σ sigma	conductivity	tQ^2/mL^3
ϕ phi	porosity	

Subscripts

a	apparent	*lim*	limiting value	*s*	surrounding formations	*w*	water
c	electrochemical	*o*	formation 100 percent sat-	*m*	mud	*xo*	invaded zone close to the
g	gas		urated with water (used	*mc*	mud cake		wall of the hole where
i	invaded zone		in R_o only)	*mf*	mud filtrate		flushing has been maxi-
k	electrokinetic	*o*	oil (except in R_o)	*t*	true		mum

A Proven Method to Solve Problems

1. Determine the objective of the desired study.
   First observe the forest, then study the tree. Is the study worth the time and money?
2. Obtain and organize properly all facts and ideas.
   Find all of the available data and applicable ideas. Review existing reports and available records. Then organize and prepare a preliminary outline. Determine the relative importance and the accuracy of the data. Outline and plan the study in detail. Develop a plan and theory for solving the problem, then find and derive equations and ideas. Ask the necessary pertinent questions and insist upon true, honest, reliable answers.

   Next, construct flow/minimum path diagrams for the study and construction phases so you can budget both your time and the available capital. Determine the best true value for each facet and parameter in relations, equations, and ideas.

   Finally, upgrade your information using statistics, analogy, computers, literature, and experience. Express or combine all data and evaluate everything. Be sure to understand the other fields of expertise involved. Don't be proud; consult with others when you need to.

3. Evaluate, think, and apply judgment professionally.
   Analyze the data, using computers, analogy, and graphs. Interrelate all of the data using equations. Plot the computer output using scaling groupings. Compare the results with your experience and the literature. Obtain an overall picture that provides the best fit of all the facts.
4. Make a sound, reliable, unbiased decision.
   Reduce the decision-supporting facts to writing. Sell the decision to management and obtain their approval. Also sell the public officials.
5. Act, implement the decision, and construct the facility.
   Supervise the activity carefully and review operational results.
6. Change the facility as indicated by its performance standards.

The reliability of any work study depends upon spending energy, time, effort, and money wisely; making sound decisions based on accurate data and ideas; hiring people with integrity, intelligence, and ability; and using reliable, unbiased, professional judgment.

Sources of Data Available

The manager, geologist, and engineer should study all available data before reaching conclusions that involve spending large sums of money. The adage "Garbage in equals garbage out" applies to management and engineering decisions. Sound evaluation of all data to arrive at the best possible facts is essential. The relative importance and value of each data source will vary with each field situation and how data are used. Some of these data sources are shown in Table 1–1.

Each data source may give a value for a specific parameter in the decision-making process. Selecting the most reliable value often is an art since sampling of the reservoir is usually inadequate. It is relatively simple to obtain information from a computer or from the human brain, our knowledge storage center. However, the process of selecting factual input and soundly analyzing the output often is the most difficult. This decision-making process must be done with the care and attention of those involved.

TABLE 1–1
Data Sources (Courtesy *Petroleum Engineer International*, May 1980)

Operation	Predrilling							During drilling (Well bore operations)													Post development						
	Gravity	Seismic				Geology—Eng. study							Logs						Wire line		Production					Special studies	
	Gravity	Time	Velocity	Amplitude	Character	Analogy, Regional knowledge and maps	Depositional environment	Drill rate	Mud log	Cuttings	Cores	Drillstem	Electric	SP	Acoustic	Density	Gamma ray	Neutron	Test	Cores	Flow test	Pressure	Water cut	GOR	History	Analogy	Engineering and geology
Depth markers		2	2			2	2	3	3	2	1	1	1	1	1	1	1	1	2	1	2						1
Structure and area	2	2	1	3	3	2						4									2	2	3	3	1		1
Hydrodynamics						1															2	1		3		1	1
Gross thickness			2		3	2	2	2	2	3	2	4	1	1	1	1	1	1			2						1
Net thickness			2	2		2	2	3	3	4	1	4	1	1	1	1	1				2						1
Lithology			2	2	3	2	2	3		2	1		3	3	2	2	2	3		1						2	1
Mechanical properties			2	2	3	2	2	3		2	1				2	2	2	3		2						2	1
Contacts			2	2	2	4			3	2	2	2	1	2	1	1		1	1	2	1		2	2	2		1
Pressure			2	3		1			3			1							2		1	1				1	1
Porosity			2	2	3	2	2	4		3	1	3			1	1		1		2	4					2	1
Permeability				4		2	2	4	1	1	4				3	3		3	2	3	2	1				2	1
Relative permeability												1									1	2	2	2	2	2	1
Fluid saturation			3	3	3	4		3	3	2	1	1	3	2	2	2		2	2	3	1		1	1	2		1
Pore sizes						2				2	1	4	4	4	4	4	4	4		3						2	
Producing mechanism	4	3	3	3		2	3																1	1	1	1	1
Hydrocarbon properties		4	4			2			3	4	3	1			4	4		4	2	4	1	2		1		2	1
Water properties						1						4	1						2			1	1			2	1
Production rate						2	2				2	2			4	4		4	2	3	1	1			1	2	1
Fluids produced												1									1		1	1	1	2	1
Well damage																					1	1			1		1
Recovery efficiency																							2	2	1	2	1

Code: 1. Best source.   2. Good data source.   3. Average data source.   4. Poor data source
Also, see World Oil, November 1978, p. 57.

Natural and Man-made Recovery Processes

This flow chart outlines recovery processes. Also see
Geffen (*Oil and Gas Journal,* May 1973) and Iyoho
(*World Oil,* Nov. 1978).

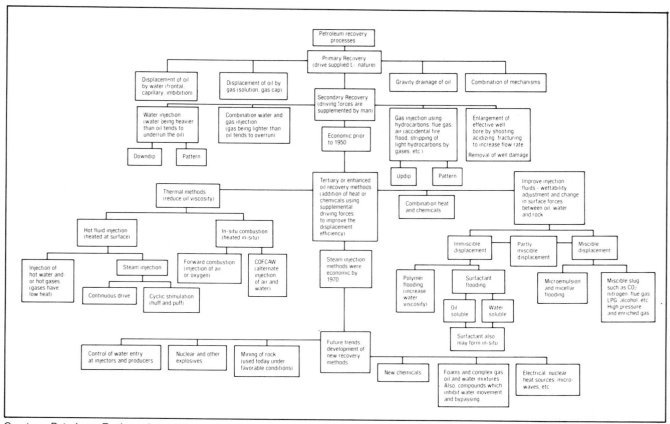

Courtesy *Petroleum Engineer International,* May 1980

Flow and Gradual Depletion of a Gas Well

Initial flow is transient as the drainage boundary increases with time. Flow later is semisteady state and pressure drops rather uniformly throughout the drainage area. When pressures are less than 3,000 in tight rocks or flow is at high rates, the $m(p)$ curves must be used. The p^2 approach may be used when all pressures are less than 1,000 psi. The flow history and related pressure behavior for oil wells is similar after adjustment for two or more phase flow using relative permeability concepts, since gas-oil ratio is almost constant with distance at any time. Well damage is equivalent to a reduced well radius, and fractures may be considered as an enlarged well radius.

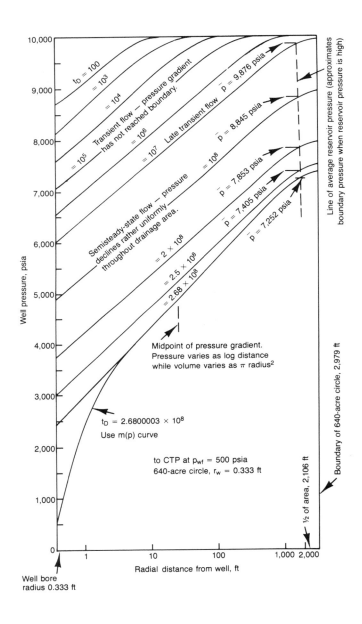

Figure 1–8. *Pressure profiles with time for constant-rate case. When $P_i = 10,000$ psia, $r_B = r_e/r_w = 2,979/0.333 = 8,936$ and $q_R = 500$ or flow rate is 191 Mscfd/md-ft of pay.*

Natural Oil Recovery Mechanisms

Gas Displacement Process–Gas Drive or Solution Gas Drive

Stock tank oil recovered as a result of gas displacement of the oil often is in the range of 5 to 25 percent of the stock tank oil originally in place. Ten to fifteen percent is a likely number. Many variables discussed in later chapters influence recovery. The gas may evolve from that in solution within the in-place oil, may in part be present in situ as a gas cap lying above the oil, or may be injected by man either into the crest of the reservoir or into wells located according to some pattern. The added gas usually improves oil recovery above that obtained from pure solution-gas drive. When gas is injected into the crest, the process is called pressure maintenance and when injected throughout the reservoir the mechanism is described as dispersed injection (Fig. 1–9).

The gas is usually a light hydrocarbon such as methane, but other gases such as CO_2, H_2S, and N_2 are often present naturally. Injected gas also may be air, with possibility of explosions. CO_2 is used as a tertiary recovery process. The large amounts of light hydrocarbons recovered in field gasoline plants suggest that the injected hydrocarbon gas stripped many light products from the remaining in-place oils. In a few cases, firefloods or in-situ combustion probably occurred when air was used as the injected gas. Hydrocarbon gases are expensive and the gas displacement efficiency is relatively low, so that water has been injected instead of gas since around 1950. When in-situ water approaches 50%, gas displacement may have an advantage over water. Gravity drainage effects may occur during the gas-drive process, and second-

ary gas may be observed during the time in which the oil is produced.

Water Displacement Process and Inhibition

Edge water may advance into the oil-bearing pore space and move the oil to the well. Also, the operator may supplement the edge-water drive by injecting water into downdip locations (pressure maintenance) or the water may be injected into oil-bearing rocks using a suitable pattern (pattern displacement). Water may enter from the bottom and through leaks in formation seals and may be injected at the base of the oil or below (Fig. 1–10).

Gas evolves from the oil if pressure declines and if the oil is below its bubble point so that the displacement process may be a combination of both gas and water. Oil recovery often is enhanced when gas in free and dissolved state is present in the oil-bearing rock at the time of abandonment of the reservoir.

The displacement efficiency of water may exceed 75% under favorable conditions such as the East Texas field. Efficiency may be low (less than 25% when in-situ water at the time of reservoir discovery is in range of 50%) when rocks are tight, and when the rock is oil wet. In some cases, recovery by gas displacement may exceed that of water displacement. The viscosity of gas is lower than water by a factor of 10 or more, and gas naturally tends to bypass the oil. In many fields, the water displacement efficiency ranges from 30 to 50%.

Water may be difficult to inject, particularly in shallow reservoirs and in low-permeability rocks. Small or short-length fractures seem to be beneficial, but operation above the fracturing pressure is detrimental. Water injection into abandoned reservoirs operated under gas displacement when permeability is from 20 to 200 md often recovers 0.5 to 1.5 times the oil previously recovered during solution-gas drive.

Capillary forces are important in the water displacement process when rocks are water wet. Capillary forces become relatively strong near the water-oil contacts. Oil banks form during most successful secondary recovery waterfloods, even if the water injection follows lengthy periods of gas injection. When rocks have been extensively fractured in all directions (Warren-Root Model), water seems to enter the fractures and the water is imbibed into the small matrix blocks to displace oil into the fracture system where it can

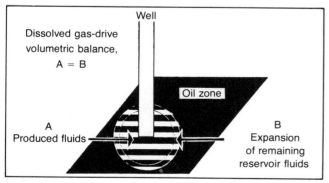

Figure 1–9. *Dissolved gas-drive reservoir. Dissolved gas comes out of solution and expands to force oil to the well as the pressure is reduced (after Norman Clark).*

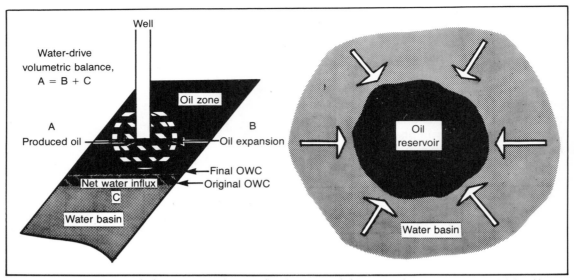

Figure 1–10. *Water from the aquifer enters the oil reservoir to help dissolved gas force oil to the well (after Norman Clark).*

move toward the wells. Edge water and water from surrounding shales also are imbibed during long periods of marginal oil production so that such areas, particularly when the accumulations are areal thin, are naturally flooded without being observed by the production staff. When fractures are few and long (Pollard type), bypassing of water often is a major problem.

Gravity Drainage

Gravity forces move oil downdip into the well. Gas often also moves updip to form a gas cap. Recovery can exceed 75% of the in-place oil present initially when pay is thick and vertical permeability is very high so that flow into the well is at economical rates. Injected gas tends to overrun the oil and water flows under the oil because of gravity forces. In many fields having good pay or high permeability, the vertical permeability approaches the horizontal permeability and steep dip of beds is not essential for gravity drainage.

Combination Displacement Mechanisms

Many oil reservoirs are subject to more than one oil-drive mechanism during both natural operations and during injection of fluids. The mechanism involves solution-gas drive, natural expansion of the gas cap, and water entry due to expansion of water in the aquifer (Fig. 1–11).

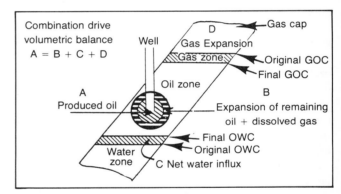

Figure 1–11. *Combination-type reservoir (after Norman Clark).*

General Conclusions

There are exceptions to every rule, and the following remarks should be used with intelligence.

1. Low viscosity of gas favors bypassing of oils by gas. Gas gravity favors overrunning of oils by gas. When water saturation at initial conditions is 20 to 40% of pore space, the gas displacement mechanism usually results in recovery of 5 to 30% of in-place oil compared with a recovery of 25 to 80% by water-displacement mechanism. When the reservoir is tighter and in-situ water initially is about 50% or above, recovery by gas displacement often exceeds recovery by water displacement. There is an intermediate water saturation where recovery by both processes may be equal. When oil gravity is low and viscosity is high, heat may be required before a favorable

recovery can be attained since fluid flow relates with both permeability and viscosity. Recovery may be very low when viscosity is high and when permeability is low.

2. A high water-oil ratio or a high gas-oil ratio in individual wells at final abandonment usually results in maximum recovery. Control of water and gas production is required to maintain the driving force. When edge-water drive is present, downdip wells are often shut in or recompleted in higher perforations when water production is relatively small to conserve natural energy. Production at high rates is desired when energy is unlimited.

3. Production of gas wells by pressure depletion usually is more efficient than the production of gas by water displacement. A water drive in a gas reservoir is not desired, and gas wells having a natural water drive are often produced at a very high rate to reduce the time when water is moving into the reservoir.

4. A gas cap should not be wetted with oil. The size of the gas cap should not be allowed to decrease when the oil rim is being recovered.

5. Recovery from unheated water displacement is often inefficient and may be uneconomical when oil viscosity exceeds about 20 cp and when $(\sqrt{k})/\mu$, expressed as md/cp, is less than about one. Also, water movement and injection when permeability is less then about 10 md is difficult and may be uneconomical. A high oil rate often is desired.

6. A natural water drive often is effective only when permeability is high and rate of fluid production is relatively low. Narrow reservoirs are more apt to have a natural water drive than wide reservoirs since the entire oil reservoir must be flooded with water at abandonment. Water from surrounding shales and the edges may flood an oil reservoir but may not be observed if the primary production period is long at slow rates. Injection of supplemental water during early life is often desired to maintain reservoir pressure and to make water available for the water displacement mechanism in oil reservoirs.

7. Operating a water-drive reservoir at high pressure increases costs but should increase recovery. Since the nonrecoverable oil contains gas in solution, less tank oil remains within the reservoir. Unitization problems in the US often delay installation of water injection facilities.

8. Gravity drainage may occur in reservoirs having high permeability or steep dip, particularly when pays are thick. The process is slow and may require restriction of production rate and gas injection into a gas cap. Recovery under gravity drainage may approach 75% of oil in place. Gas coning may be a problem.

9. Fracturing tends to improve the rate of production and injection. Small fractures do not seem to hurt operations nor reduce recovery in injection projects. Some waterfloods are operated at pressures that permit fracture extension when existing fractures plug; but long, natural, or man-made fractures of the Pollard model type are not favorable. If the fracture blocks are small, capillary-gravity forces combine to result in a good recovery efficiency from Warren-Root type fractures. Water enters through the fractures, and the other forces cause water to imbibe into the matrix blocks to force oil to the fractures and to the producing wells. The water may be injected or made available naturally from the edges and surrounding shales, which compress with reduction of reservoir pressure. Closely spaced fractures of the Warren-Root model type may be tight and may not make a substantial contribution to either fluid movement or bypassing when sealed by compression or precipitation.

10. The recovery processes are complicated, but today they are understood and are adaptable to mathematics and computer models. Relatively good forecasts, which are required when planning a project, are possible if available field and laboratory data, together with theoretical considerations, field analogy, etc., are properly evaluated and used. Graphs and monographs can be most useful and may eliminate many calculations and computer runs while reducing evaluation time and expense without sacrificing accuracy. It is usually more economical to recognize that a description of the reservoir is inadequate; revisions will later be required in both equipment and subsurface flood plan. In the past, there has been a tendency to overdesign. Later, changes were required as additional information was made available by the injection operation. The engineer and manager must recognize that only a small part of the reservoir is sampled by logging, coring, flow and buildup surveys, primary operations, etc., and that the information obtained cannot be entirely representative when applied to individual well-injection performance. Adjustments to include geological concepts usually do not overcome the problems.

11. Secondary and tertiary recovery methods should be combined with primary recovery operations when possible. Costs are thereby reduced and recovery usually is improved.

When the oil is displaced by only natural forces and in-situ fluids, recovery usually is defined as primary. When gas or water is injected by the lease operator, recovery is defined as secondary. The addition of heat and chemicals, solvents, special gases, and other exotic methods is commonly called tertiary recovery. The term has been broadened and is being changed to enhanced recovery methods, or EOR.

Screening Guide for EOR or Tertiary-recovery Projects

Thermal Methods

The application of heat, which reduces viscosity of oils, can be economic when oils are viscous so that unheated water displacement is inefficient. Two steam application methods are available.

The *cyclic or huff-and-puff* steam method uses the same well for both injection and production cycles. Heat loss is less when pay is thick, and a thick formation increases production and injection rates. The reservoir may be either a sandstone or carbonate, porosity usually is above 25%, permeability is high, often above 1,000 md, and viscosity is 50 to 10,000 cp.

The continuous *steam drive* often uses an initial huff and puff at the producing wells. A high residual oil at the start of the operation is favorable since oil recovery after normal water drive is used to justify the added cost of a steam flood. Steam can distill oils, and added oil recovered is not entirely limited to change in viscosity. Steam tends to overrun the oil, but it changes to hot water at a distance from the injector. Operating pressures are often less than 2,500 psi. Pay thickness in excess of 25 feet reduces heat loss to nonoil-bearing rock. A large gas cap, a strong water drive, shale, and a high permeability contrast are all detrimental.

The heat can be created by burning oil in situ after injection of air or oxygen, which is known as a *fire flood*. Fuel is required to run compressors, and some in-situ oil is burned to create heat. The combustion may be in the direction of air movement or in a reverse direction, The process is more efficient when slugs of water are combined or the COFCAW method is used. The fire distills some oil and reduces viscosity so that products can move to the well with steam and water. Some natural permeability to air must be present in either the oil or in-situ water. Good porosity and permeability are desired. The hot gases created by combustion tend to overrun the oil, and severe corrosion at producers and equipment failures due to heat are encountered. Gas caps, zones of high permeability contrast, fractures, shale, and rocks that deform with heat are detrimental. Emulsions that are hard to control often occur.

Chemical flooding includes *immiscible* and *miscible* techniques. Polymers may be used to increase the viscosity of injected water, and surfactant flooding may be accomplished by injection of anionic, cationic, and nonionic chemicals with injected water or oil slugs to reduce surface forces between water, oil, and rock in

an immiscible displacement process. Miscible displacement may be possible when high-pressure gas, enriched gas, CO_2, and microemulsion-micellar compounds are injected. Complete displacement of the in-situ oil within the affected volume is possible. The displacement of in-situ water by injected water and mud filtrates are good examples of the miscible process.

Polymers change the viscosity of injected fluids, usually water, so that the viscosity contrast and oil recovery are increased. To date, the process is limited to sandstone reservoirs, and some chemicals in the water prevent the use of polymers. Polymers may be used in normal waterfloods and may be used in cases in water which has been treated with chemicals designed to improve recovery.

Surfactants change the surface forces between water, oil, and rocks. The chemicals are expensive and are lost at a distance from injector because they adhere to the large rock surface area. The chemicals may be water soluble, oil soluble, and some methods create the surfactant in situ at the oil-water interface. Sandstone rocks are preferred, and the in-situ water can be displaced when required by injection of a slug of fresh water. Surfactant flooding is a variation of normal waterfloods.

Gas injection at high pressure at times can result in a miscible displacement. Oil gravity is usually above 40°API, and operating pressure in the formation is above 3,000 psi. A high concentration of C_2 to C_5 compounds is necessary. A high dip reduces overrunning problems. Risks are increased by the presence of high permeability contrast in pay stringers, high vertical permeability, and shale. High gas prices often preclude use of this method.

LPG slugs and enriched gas increase the presence of C_2 to C_5 compounds and favor operation at lower pressure. This is a variation of the gas injection process with similar problems. Operation at lower pressure enables its use in shallower reservoirs.

CO_2 and other gases such as N_2 are liquids when compressed that dissolve into the oil, similar to natural gas, and increase oil volume so that less oil remains in the reservoir at the conclusion of the displacement process. These compressed gas liquids are often miscible with the in-place oil so that displacement efficiency approaches 100%. A low oil viscosity (less than 10 cp), sufficient in-place oil to make the process eco-

nomical, moderate oil gravity (30–40°API), reasonable permeability (25–400 md), and a depth sufficient to keep the gas in liquid state (2,500 psi for CO_2) are desired. The process is applicable to both sandstones and carbonates. It has been successful at SACROC, West Texas. Companies have experimental projects in field-test stage and are making plans to move CO_2 from Colorado and New Mexico to West Texas. The natural CO_2 supply is limited, and large manufacturing plants are not located near the oil fields to supply additional CO_2 at economical cost.

Microemulsion and micellar floods are being tested with some economic success in sandstone reservoirs such as in Illinois. The Maraflood process probably is economic when conditions are favorable. Temperatures should be less than 250°F., and a bank of fresh water must often be injected to remove chemicals in the in-situ water. The hydrocarbon-base chemicals are in short supply, and a large capital investment and a time lag to construct additional plants at refinery locations will be necessary before large-scale use of the floods can occur. Polymers are used.

Other Remarks about EOR

Oil and gas production in the US has been declining since early 1970s, and many economists believe that the trend will not be reversed. Conservation is reducing the demand for oil in the US, which has peaked at almost 20 MM b/d with US production supplying about one-half of total demand. Worldwide demand is expected to grow, but production capability is expected to decline during the 1980s. In the US, drilling rigs in operation have increased rapidly during the last half of the 1970s, but reserve estimates for oil have continued to decline—although at a slower rate. About one-half of US production is considered as secondary oil, mostly waterflood mechanism. EOR potential recovery in the US from natural oil fields is placed at 20 to 40 billion barrels, depending upon the optimism of the estimator. Such a goal is relatively small compared to annual consumption.

The need for EOR is immediate, and the oil remaining in known reservoirs is over 250 billion barrels in the US alone. Unfortunately, this known in-place oil cannot be recovered by EOR methods today in most fields. Oil originally in place in fields discovered in the US during the past 140 years has been estimated to exceed 425 billion barrels, and recovery to date is only 30–40% of this amount. It is obvious that the prize available for the EOR operator is large, but development of useful processes has been difficult and slow. Overall primary-secondary recovery may approach 50%.

Crude prices have increased rapidly, but costs to install EOR methods have advanced, too. EOR projects underway are limited, as shown in an *Oil and Gas Journal* review appearing in the 3/28/80 issue. Except for thermal installations in heavy oil fields of California, most projects are experimental.

US Production during 1980 from EOR projects

Method	b/d production
Steam	295,000
In-situ combustion	12,000
Micellar-polymer	1,000
Polymer	1,000
Caustic	1,000
CO_2 miscible	22,000
Other gases	53,000
Total	385,000

The CO_2 miscible method appears to have good possibilities, and major companies intend to lay 500-mile pipelines to obtain natural CO_2 if the field tests prove economical in West Texas. The EOR chemicals are expensive, and slugs followed by water are often used.

Alternate sources of hydrocarbon energy will compete with EOR projects in both the US and Canada. These alternate resources include the Canadian tar sands, oil shale (the kerogen can be converted to gas and to oil), coal (can be converted to oil and gas), solar energy including alcohol, and nuclear energy. Most of these methods are operational, and the future for EOR depends upon relative cost and profitable operation and acceptance in the marketplace. At the present time, governments are helping test all alternates, and EOR is expected to play an important role in the future US energy picture. Economic cost today includes environmental considerations and public acceptance of real or imagined risks.

EOR projects can expect strong competition from tars, oil shale, coal and nuclear energy. The reserves of these sources far exceed those of the EOR projects. Environmental concerns regarding the hothouse effect from CO_2, which is formed by burning all hydrocarbons, is a risk of using EOR, tars, oil shale, and coals. Fear of accidents is delaying use of nuclear energy in the US, although other nations are installing required plants. Also, coal and oils contain sulphur, which must be removed to prevent acid rains. Many groups talk of conservation and solar energy, both having limited application today, and possibly would

TABLE 1–2
Comparative Costs of Emerging Fuel Technologies (Courtesy OGJ, June 23, 1980)

Fuel technology	Size (product)	Year of operation	Total capital investment (million $)	Unit capital investment ($/MMBTU/year)	$/Equivalent bbl Break-even	$/Equivalent bbl 15% return on investment after taxes	$/MMBTU Break-even	$/MMBTU 15% return on investment after taxes
Tar sands*	125,000 b/sd	1979	1,800	7	9	†17	1.50	2.80
Oil sands*	125,000 b/sd	1979	2,500	10	12	23	2.00	3.80
Oil shale*	50,000 b/sd	1990	1,080	11	10	22	1.70	3.70
Coal gasification‡								
Commercially proven								
High BTU gas	250 billion BTU/sd	1979	1,900	23	31	56	5.10	9.30
Medium BTU gas	250 billion BTU/sd	1979	1,600	20	26	48	4.40	8.00
Advanced gasifier								
High BTU gas	250 billion BTU/sd	1990	1,500	18	28	47	4.60	7.90
Coal liquefaction‡	50,000 b/sd	1990	1,300	13	28	44	4.70	7.30
Biomass-conversion								
Wood to methanol	3,700 b/sd	2000	90	32	14–20	28–42	5.7	10–15
Wood to ethanol	2,500 b/sd	2000	50	20	16–22	32–48	5.7	10–15
Wood to high BTU gas	37 billion BTU/sd	2000	240	20	30–42	60–90	5.7	10–15
Manure to high BTU gas	2 billion BTU/sd	2000	14	20	30–42	60–90	5.7	10–15

*Primary upgrading to synthetic crude oil. †Current OPEC price range for light Arabian crude is $25–36/bbl. ‡Based on coal with heat combustion of 10,000 BTU/lb priced at $1.25/MMBTU delivered.
Source: Bechtel Inc.

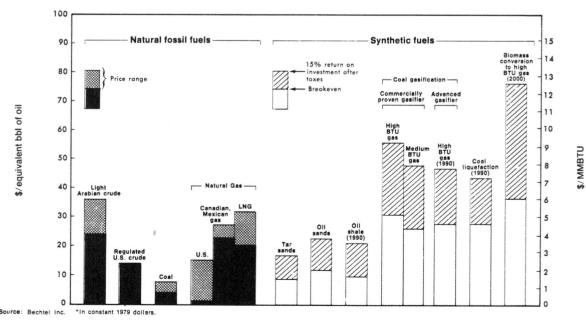

Figure 1–12. How Bechtel compares fuel production costs (courtesy OGJ).

prefer a no-growth environment rather than face the practical risks involved in the use of hydrocarbon and nuclear energy. Such extreme ideas are not expected to prevail during the long term, but they do distort the normal economic choices in the short-term situations (Table 1–2, Fig. 1–12).

Fusion with possibly some solar techniques, depending upon economics, is the long-term energy source of the world. The developing nations cannot attain economic success without an abundant and relatively inexpensive source of energy. They have the natural resources required by the more developed nations, whose standard of living will decline unless energy is supplied to less-developed peoples.

Basic Equations Used in Subsurface Engineering

The equations will not be derived in this text, which is directed toward the practical aspects of engineering. Derivations are included in many texts and papers, including Muskat's *Flow of Homogeneous Fluids through Porous Media*. Derivations of some equations are summarized by Ralph Nielson in *Petroleum Production Handbook*. The basic equations for ideal semisteady state flow as used in theory, the laboratory and the field follow:

Linear Flow

Define p_1 at $x = 0$ and p_2 at $x = L$; $\Delta p = p_2 - p_1$

$p = p_1 + x\Delta p/L$ for liquids
$p^2 = p_1^2 + x\Delta(p^2)/L$ for gas
$q = kA\Delta p/\mu L$ for liquids in laboratory
$q = 1.13 kA\Delta p/\mu L$ for liquids in field
$q = kA\Delta(p^2)/2\mu L$ for gas in laboratory
$q = 111\ kA\Delta(p^2)/\mu LT$ for gas in field

Radial Flow

Define p_w at $r = r_w$ and p_e at $r = r_e$
$\Delta p = p_e - p_w$

$p = p_w + \Delta p \ln(r/r_w)/\ln(r_e/r_w)$ for liquids
$p = 2\pi kh\Delta p/\mu \ln(r_e/r_w)$ for liquids in laboratory
$dp/dr = \Delta p/r \ln(r_e/r_w)$ for liquids
$p^2 = p_w^2 + \Delta(p^2) \ln(r/r_w)/\ln(r_e/r_w)$ for gas
$dp/dr = \Delta(p^2)/2p_r \ln(r_e/r_w)$ for gas
$\bar{p}_{av} = p_e - \Delta p/2 \ln(r_e/r_w)$ when r_e is much greater than r_w
$q = 7.07\ kh\Delta p/\mu \ln(r_e/r_w)$ for liquids in field
$q = \pi kh\Delta(p^2)/\mu \ln(r_e/r_w)$ for gas in laboratory
$q = 700\ kh\Delta(p^2)/T\mu \ln(r_e/r_w)$ for gas in field

Flooding Patterns with Unit Mobility Ratio

5 spot when d = distance between input and producer
$q = 1.13\pi\ kh\Delta p/\mu\ (\ln(d/r_w) - 0.619)$ for liquid in field

7 spot when d = distance between wells
$q = 4.52\pi\ kh\Delta p/\mu\ (3 \ln(d/r_w) - 1.707)$ for liquid in field

Direct line drive when d = distance from row of inputs to producers and a = distance between wells in a row; d is greater than a.

$q = 2.26\pi\ kh\Delta p/\mu(\pi d/a - 2 \ln(2 \sin h\pi r_w/a))$
for liquid in field.

Note: Sin $hz = z$ when z is small.

Where: k = permeability, darcys
μ = viscosity, cp
T = temperature, °R
q = liquids in lab., cc/sec
q = liquids in field, b/d
q = gas in lab., cc/sec at flow temp. and 1 atm
q = gas in field, scf/d
p in lab., atm abs
p in field, psia
r, L, and h in lab., cm
r, L, and h in field, ft
A in lab, sq cm
A in field, sq ft
$\pi = 3.1416$

Engineers must devote considerable effort to understanding the pressure and flow characteristics within the reservoir of interest of a study. However, the recoverable oil must always be of prime interest. The oil produced from an underground porous medium or reservoir is displaced and forced out of the pore medium by expansion of oil, gas, and water. When oil is above its bubble point, rock expansion also is important. Gravity-capillary forces can be important, and at times the in-situ subsurface water may be supplemented by rainfall, rivers that flow across the pay zone at the surface, lakes, and oceans. Injection of water and gas can also maintain reservoir pressure and drive oil to the producing wells. The material balance equation relates with oil and gas in place, and a rearrangement of the equation can be used to determine the contribution allocated to various types of drives as shown on p. 28 when units are consistent:

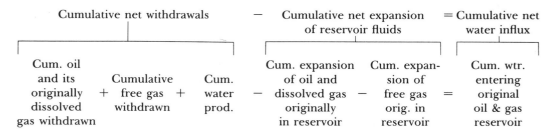

The rearranged material balance equation is:

$$N_p = \frac{N\{mB_{oi}[(B_g/B_{oi}) - 1] + B_g(R_{si} - R_s) - (B_{oi} - B_o)\} + W_e - W_p + W_i}{B_o + B_g(R_p - R_s)}$$

The above equation is divided by N_p to obtain three fractions, which give the fractional part of the production due to solution gas drive, segregation (gravity) drive, and water drive. The drive indexes are:

$$\text{Depletion drive index} = \frac{N[B_g(R_{si} - R_s) - (B_{oi} - B_o)]}{N_p[B_o + B_g(R_p - R_s)]}$$

$$\text{Segregation drive index} = \frac{N(mB_{oi}[(B_g/B_{oi}) - 1])}{N_p[B_o + B_g(R_p - R_s)]}$$

$$\text{Water drive index} = \frac{W_e - W_p + W_i}{N_p[B_o + B_g(R_p - R_s)]}$$

Where: N_p = cumulative oil production
N = oil initially in place
B = formation volume factors for oil and gas
R = gas-oil ratios
W = values for water

Compressibilities of oil, in-situ water, and rock are important when the oil is above the bubble point. Rock compressibility can be large when porosity is small.

The calculation of recoverable oil involves many concepts such as relative permeability and the Buckley-Leverett and Welge equations, some of which are summarized below.

1. Fractional flow:

$$f_w = \frac{1}{1 + \dfrac{k_o \mu_w}{k_w \mu_o}}$$

2. Relative permeability ratio:

$$\frac{k_o}{k_w} = ae^{-bS_w}$$

3. Fractional flow derivative:

$$\frac{\delta f_w}{\delta S_w} = \frac{b\dfrac{k_o \mu_w}{k_w \mu_o}}{\left[1 + \dfrac{k_o \mu_w}{k_w \mu_o}\right]^2}$$

4. Frontal advance:

$$x = \frac{5.615 \, qt}{\phi h w}\left(\frac{\delta f_w}{\delta S_w}\right) S_w$$

5. Welge saturation relations:

a. $\bar{S}_w - S_{w_2} = Q_i f_{o_2}$

b. $Q_i = \dfrac{1}{\left(\dfrac{\delta f_w}{\delta S_w}\right) S_{w_2}}$

Where: k_o = relative permeability to oil
k_w = relative permeability to water
μ_o = oil viscosity, cp
μ_w = water viscosity, cp
a = intercept of straight line portion of relative permeability ratio curve
b = slope of straight line portion of relative permeability ratio curve
q = water injection rate, bpd
Q_i = cumulative water injected, pore volumes
t = time, days
ϕ = formation porosity, fraction
h = formation thickness, feet
w = width of front, feet
x = frontal advance, feet
S_{w_2} = water saturation at producing end of linear system, fraction
$\bar{S}_w$ = average water saturation behind front, fraction
f_w = fraction of water in producing stream
f_o = fraction of oil in producing stream

The following ratios may be used to evaluate the validity of using assumptions made in the derivation or simplification of the equations. As an example, the Buckley-Leverett method should be used when capillary and gravity effects are zero, or N_{VE} is zero.

1. Capillary/viscous:

$$N_{Pc/\mu} = \frac{k_v \Delta P_c A}{887 \mu q h} \text{ or } N_{Pc/\mu} = \frac{k_v \Delta P_c L}{k_h \Delta p_{\text{datum}} h}$$

2. Gravity/viscous:

$$N_{\rho/\mu} = \frac{k_v \Delta \rho g \cos \alpha A}{887 \mu q} \text{ or } N_{\rho/\mu} = \frac{k_v \Delta \rho g \cos \alpha L}{k_h \Delta p_{datum}}$$

3. Capillary/gravity:

$$N_{Pc/\rho} = \frac{\Delta P_c}{\Delta \rho g h \cos \alpha}$$

4. Flow rate:

$$u_h \approx q/A \text{ or } u_h \approx k_h \Delta p_{datum}/\mu L$$

The vertical equilibrium number can be expressed

$$N_{VE} = N_{Pc/\mu} + N_{\rho/\mu}$$

$$N_{VE} = \frac{k_v \Delta P_c A}{\mu q h} + \frac{k_v \Delta \rho g \cos \alpha A}{\mu q}$$

$$N_{VE} = \frac{k_v \Delta P_c L}{k_h \Delta p_{datum} h} + \frac{k_v \Delta \rho g \cos \alpha L}{k_h \Delta p_{datum}}$$

Where:
- k_v = vertical permeability, md
- k_h = horizontal permeability, md
- ΔP_c = capillary pressure increment, psi
- A = characteristic cross-sectional area, ft^2
- μ = viscosity of the more viscous fluid, cp
- q = gross fluid rate, rb/d
- h = characteristic reservoir thickness, ft
- L = characteristic reservoir length, ft
- Δp_{datum} = pressure change, relative to a datum, across the characteristic length, psi
- $\Delta \rho g$ = density difference, psi/ft
- $\cos \alpha$ = cosine of dip angle
- μ = areal rate of fluid movement

Many assumptions were made when deriving the equations, and they should be adjusted together with data input when the assumptions are not applicable to the reservoir being studied. The usual assumptions include:

- The reservoir hydrocarbon pore volume is constant.
- The reservoir temperature and related parameters are constant.
- Equilibrium conditions exist throughout volume of interest at all times.
- The pay thickness, porosity, and permeability are uniform throughout the volume of interest in the flow equations.
- The other rock properties and relative permeability are uniform.
- Differential PVT properties best represent subsurface fluids.
- Recovery in practice is independent of rate.
- The well is open throughout the pay interval or fluids flow in radial, linear, or spherical system as used in equations.
- The porous system is uniform, homogeneous, and isotropic.
- Gravity and capillary forces are small or properly represented.
- Laboratory data is representative of field conditions when adjusted in the manner used by the engineer and computer.
- A constant gas-oil ratio, viscosity, etc., are at times assumed.

Boundary conditions at the well-bore sand face may assume flow at a constant rate or alternately flow at constant pressure. At the drainage boundary, flow may be zero for depletion or solution-gas drive reservoirs so that dp/dx at the boundary is zero. An infinite water influx or variations thereof may occur for a strong water-drive reservoir.

Also, computer programmers like to enter input information as straight lines, which may cause data input errors that must be recognized by the analyst or engineer. The computer programmer may have taken other liberties with both equations and input of information, which at times can distort output. The engineer must carefully review the details of the program to be certain that assumptions do not cause intolerable errors in computer output. Input data at times can be adjusted to overcome the problems.

General Organization of the Oil Industry

Many groups and individuals are involved in the oil industry. Smaller companies may hire consultants to perform some of the functions, and the importance of the respective groups naturally varies with the type and objectives of a company. It is normal for an individual working with a company to overemphasize his contribution to company success. It is also easy to fail to recognize that a successful company uses many fields of expertise. A very general outline of this interrelationship follows:

Management

This group gives overall direction, sets policy, and allocates money for capital expenditures and operations. Support groups such as accounting and financial are included. Company size determines duties.

Lease and Legal Groups

No company can operate without prospects for drilling or purchase and sale of properties. Joint operations are essential to spread the exploratory risks. Legal problems are not limited to contracts, since governmental rules, safety and environmental regulations, and forms must be submitted and negotiated.

Exploration Group

Exploration or regional geology is the basis for purchase of leases by the land group. The location of traps and a determination of the probable fluids contained therein is essential for successful exploration. The origin of the hydrocarbons, whether oil or gas, the nature of their transport to the traps, the probability of abnormal pressure, and size of accumulation should be considered. The type of reservoir rocks, probable porosity and permeability, shale content, the shape and type of trap expected and the source of the rocks and the environment of deposit are important. The possibility of hydrodynamics, faults, breaks in reservoir seals, and the presence of tars and heavy oils are determined from analogy and regional geology.

Log Analyst and Laboratory

Logs and laboratory analysis of cores, cuttings, and oils are used to determine porosity, permeability, and water content of the pores in the well. These tools also give the possible recovery, capillarity, wettability, and relative permeability of the rocks. Rock type, stratification, shale content, and many other characteristics are made available. The PVT properties of the oils and gas are determined by the laboratory.

Geologists

The local geology is determined by careful study of the logs and other information made available by the drill, logs, and other data sources. The size of the reservoir, the spillover points, the separation into reservoirs by faults and other barriers to flow, and a combination of data from various sources can give an indication of reservoir shape as required for further development.

Subsurface Engineers

The subsurface engineer working with geologists and petrophysicists determines the expected drive mechanism, recovery efficiency, reservoir characteristics, applicable flow, and recovery equations by combining information from the various data sources. Analogy with adjacent fields, correlations appearing in the literature, model studies, and other techniques are used to make forecasts and develop a plan for additional drilling and production of the reservoir. Methods for completing wells to obtain maximum production rate and recovery are selected. Future production techniques are anticipated, together with the inherent problems. Benefits from fracturing, sand-control devices, coning control by perforations, and alternative completion procedures are considered. The well is tested properly so that flow is at the highest efficient rate consistent with economics and the well is not damaged by production practices.

Contractors and Engineers

Surface facilities are installed to handle the indicated production. The possible benefits from artificial lift, the need for later remedial workovers, and other operational procedures can determine the location of surface facilities.

Economists

Economics of the projects must be determined including alternate operating plans. Money has value, particularly when interest rates are high, and decisions should be made using probable risks, discount or interest on money, and availability of capital.

Graphs, methods of analogy, the hand-held computer, and computer modeling should be used as economically justified to maintain production at highest efficient rate, MER, and to plan the present and future operation of the reservoir to recover the maximum product in the shortest time. Usually, the maximum profit is desired when expressed as a present value. Time at which expenditures are made and income is received is very important. Wells must be drilled and operated as cheaply as possible consistent with safety, good operations, and public considerations.

The problems in analyzing a potential reservoir are many and nature does not disclose secrets without careful analysis. Each field study is different to some degree, and data for some reservoirs can be both complicated and inconsistent. A good analyst eliminates the bad from the good information to reach reliable conclusions that permit sound investments and operating plans.

Checklist of Basic Data

An API committee chaired by Paul D. Torry published a list of useful basic data in *Secondary Recovery of Oil in the United States* (API, 1950). Except for possible contributions related to EOR projects, it continues to be appropriate for collecting basic information.*

It is doubtful whether all of the items listed in the schedule can be obtained or will be required for the evaluation of the secondary-recovery possibilities of most oil properties or fields. It has been made as complete as possible to insure the collection of all desirable information in an investigation involving a large number of fields. It may be abridged conveniently to conform to the necessities of any specific study.

For a preliminary investigation of secondary-recovery possibilities, information on the following subjects generally will be found to be very useful in determining whether a more exhaustive study is justified.

I—General:
   A. Map showing well locations, ownership, etc.
II—Reservoir data:
   A. Sand name.
   B. Well records, showing sand thickness, depth, initial production, etc.
   C. Stratigraphy and structure; existence of gas cap and presence of water-bearing formations.
   D. Core analyses showing porosity, permeability, and fluid saturations.
   E. Oil gravity, viscosity, gas-oil ratio, and pressure-volume-temperature relations.
III—Exploitation history:
   A. Primary-production history of gas, oil, and water; reservoir-pressure decline.
   B. Past secondary-recovery history, if any; vacuum, gas, air, and water (including natural and accidental floods).
IV—Secondary-recovery requirements:
   A. Secondary-recovery media (air, gas, or water), source, availability, and price.

V—Economics:
   A. Lease and royalty requirements.
   B. Drilling, plant, and production costs.
   C. Unitization or cooperation of ownership for secondary development.

If the preliminary check list suggests that the project is profitable or may be economically feasible, the following more detailed set of data should be assembled.

1. General Data

A. Name and title of investigator submitting report.
B. Date investigation was commenced.
C. Date investigation was completed.
D. Date report was submitted.
E. Common name of field.
F. Location of field: county, township, range, and section; U.S. Geological Survey quadrangle; other recognized survey, warrant, grant, etc.
G. Property map of field showing lease ownership.
H. Source of all information: name, title, address, and telephone.
I. Possible sources of additional information.
J. Area of field, in acres, for each producing horizon.
K. For each producing horizon, individual field map showing location of all wells drilled—and indicating, by conventional symbols, abandoned wells due to depletion or water encroachment, producing oil wells, producing gas wells, and dry holes.

2. History

A. Discovery date for each producing horizon; name and location of discovery well.
B. Age of field for each producing horizon.
C. Brief history of development: List active operators during initial stages of development; rate of development; market and price paid for crude at time of initial development; any factors connected with early field development which would have an adverse effect on subsequent secondary-recovery operation.
D. Record of redrilling, if any.

* All material from this selection is taken exclusively from Torry's book, reprinted with permission from the API.

E. Effect and location of accidental floods, if any, from casing leaks or improperly abandoned holes.

F. Date vacuum was applied and discontinued; date secondary-recovery operations were commenced and discontinued.

G. Record of all abandonment, and area of field abandoned.

H. Data on abandoned wells: Method of plugging; amount of casing pulled and amount of casing left in holes; junk left in holes.

I. Primary-oil production of field for each producing horizon, by convenient time periods.

J. Total primary production of field for each producing horizon.

K. Average primary recovery per acre for each horizon.

L. Average primary life of wells.

M. Average primary production of wells at time of abandonment.

3. Petroleum Geology

A. All available well records, including electrical, radioactive, temperature, and other types of logs. Comment on completeness and reliability of records.

B. Stratigraphy: Typical well records and location of wells from which records have been obtained; well elevations; general information regarding various geologic formations encountered—their thickness, continuity, and lithology; convergence of formations in field, and any unconformities. Records of deep test wells are important.

C. Structure: Surface and subsurface structure maps; faults, fracture zones, crevices, etc.; relation of occurrence of gas, oil, and water to structural conditions.

D. Name of sands encountered: Note whether they carry oil, gas, or water, and the amounts thereof, or whether they were dry.

E. Average depth of each sand: Average interval between sands and horizon markers recognized by drillers and paleontologists.

F. Average thickness of sands: Amount of sand usually penetrated.

G. Variation in thickness of sands throughout field, and extent and location of any pronounced lenticularity.

H. Composition of sands: Whether coarse or fine-grained, or conglomeratic; mineral composition of sand; occurrence of nodules and concretions; whether sand is cross-bedded, blocky, massive, flaggy, or thin-bedded; and whether there are any indications of open-joint planes or crevices of any kind; type of cementing material and degree of cementation; presence of silt, clay, or bentonite; secondary cementation, and any evidence of metamorphism; shape of sand grains; presence of shale breaks; description of petrographic examination of thin sections; heavy-mineral studies; grain-size analyses.

I. Composition of limestone reservoirs: Note mode of occurrence of oil and gas—whether in interbedded sandstone lenses, solution cavities, coral reefs, oölitic zones, fractures or joints, or along bedding planes and formation contacts.

J. Nature of cap rock and of beds immediately underlying reservoir; hardness; presence of water; shale replacements in the top or bottom of sand; irregularities in contact of reservoir with overlying and underlying beds. Is there a suitable spot for casing seat or packer placement above the oil reservoir?

K. Lateral changes of reservoir throughout field: Note variations in thickness, porosity, permeability, saturation, cementing material, shale breaks, etc. Locate shaly phases and shaly areas. Are the limits of oil production defined by thinning out of the pay zone, by variations in sand porosity, by gas-productive areas, or by water in the sand? Note effect of unconformities and disconformities.

L. Position of gas, oil, and water contacts in reservoir: Pay zones and their thickness. Note initial gas production of oil wells and variations of same throughout field. Note any early production of water with either gas or oil. Determine source of water, and whether salt or brackish.

M. Data on dry holes: Did producing formation pinch out or carry water?

4. Core-Analyses Data

A. Type of core: Cable-tool, side-wall, or chip; rotary with either diamond, wire-line, or conventional core-cutter heads; coring fluid used.

B. By whom and dates cored, sampled, and analyzed.

C. Oil-saturation data and profile:
   1. Discussion of any differences between saturation of specific cores and generally recognized oil saturation of reservoir.
   2. Oil content, in barrels per acre, and average oil content per acre-foot.

D. Water-saturation data and profile:
   1. Determination of interstitial-water content: Extent of infiltration of water from drilling fluid; indications, if any, of presence of water from accidental or intentional flood.

2. Comparison of water content determined by core analyses and calculated from electric logs.
3. Determination of water content by capillary-displacement method.
E. Porosity data and profile.
F. Permeability data and profile.
G. Interpretation of core analyses should include:
1. Vertical uniformity and distribution.
2. Continuity along bedding planes.
3. Classification of sand in respect to saturation, porosity, permeability, and thickness.
4. Determination of the minimum permeability that will be affected by secondary-recovery operations at the economic limit of production.
H. Core-analyses data should aid in the determination of:
1. Best well-completion practice.
2. Proper placement of packers.
3. Adaptability of the formation to the use of plugging agents.
4. Recoverable oil.
5. Input rates.
6. Well spacing.
7. Plant capacity, line sizes, water-storage facilities, etc.
8. Rate of return on investment.
9. Capacity requirements of pumps or compressors.
10. Fill-up time of reservoir, and life of project.
11. Possibility of flowing or pumping production.
12. Economic limit of produced water-oil or gas-oil ratios.
I. Residual oil saturation after complete water drive, or after complete gas expansion, and rate of oil production as indicated by laboratory flooding and repressuring tests.
J. Relative permeability of reservoir to gas and to oil.

5. Reservoir Characteristics and Behavior

A. Initial and present bottom-hole pressure: Pressure-volume relations; characteristics of production-decline curves.
B. Gas-liberation and oil-shrinkage data at various pressures: Analysis of produced gas and of gas after treating; deviation factors, either calculated or experimental.
C. Initial gas-oil ratios and accumulated gas-oil ratios: Gas-oil ratios at time of primary depletion.
D. Existence and extent of initial and present gas-cap zones: Indications of gas-cap expansion.

E. Extent and effectiveness of natural water encroachment.
F. Substantiation of existence of gas-cap and water zone by electrical logs in case flushing of cores renders detection difficult.
G. Reservoir temperature.
H. Viscosity of oil under existing reservoir conditions of temperature, pressure, and gas in solution.
I. Character of oil: Color, gravity, paraffin point, fluidity at low temperatures; distillation test—oil samples for analysis should be identified as follows: sample No.; well No.; lease, company, section, township, range, county, state, field; date taken; point taken, depth of well; sand and depth of sand; and general remarks.
J. Chemical analysis of connate water: Water sample for analysis should be identified as follows: sample No., well No., lease, company, section, township, range, county, state, field, date taken, point taken, depth of well, source of water, and general remarks. The sample should be collected under conditions that preclude contact with air insofar as that is possible. Analysis for barium, sulfur compounds, and the bicarbonate radical should be made as soon as possible. Observation in the field should be made whether the water is corrosive to casing, tubing, and other production equipment, and whether an odor of hydrogen sulfide is present.
K. Tendency of crude to emulsify or oxidize.
L. Presence of depleted gas pays and zones of high permeability which might prevent effective and uniform injection of gas or water into oil-bearing strata.
M. Data pertaining to possible pressure parting or other rock deformation under applied pressure.
N. Productivity indices.

6. Production Data

A. Total present daily oil production: Number of wells, present productive acres; individual lease and well daily oil- and water-production figures.
B. Potential tests of oil and water production, and method of obtaining same; production records of typical large, medium, and small wells, including date drilled, initial production, pressure, present oil and water production, gas-oil ratios, and date of abandonment; influence of gas production, variations in sand conditions, and location near center or edge of field on initial production, total production, and life of wells; reasons for any dry holes in field.

C. Gas-production history of field from strictly gas-bearing formations:
1. Number of producing gas wells.
2. Total present open-flow capacity.
3. Average present rock pressure.
4. Average gathering-line pressure.
5. Average main-line pressure.
6. Average age of gas wells.
7. Original rock pressure.
8. Original open-flow capacity.
9. Total gas production by convenient time periods.
10. Cumulative gas production.
11. Estimated gas reserves, with particular attention to gas available for secondary-recovery operations.
D. Water production of field, individual leases, and wells: Methods employed for disposal of water; determination of source of water either from leaking casing, improperly abandoned wells, or from some horizon producing oil.
E. Response of field to vacuum: Amount of vacuum applied and period of application; operation of gasoline-recovery plants in connection with vacuum; effect of vacuum on gravity of oil; average gasoline recovery from casinghead gas; type of gasoline extraction used; disposition of stripped gas.
F. Results, if any, from secondary-recovery operations:
1. Water flooding: Well-completion practice, and types and sizes of equipment used; well spacing, intake pressure, volume of water input, production, recovery results, duration of operations, and area affected; accidental or intentional nature of water floods.
2. Gas and air repressuring: Well-completion practice, and types and sizes of equipment used; well spacing, intake pressures, intake volumes, production, ratio of injected gas to oil produced, produced gas-oil ratio-recovery results, duration of operations, area affected; effect of air on quality of oil.
G. Estimated remaining primary reserves of field.
H. Estimated secondary reserves of field, based on results from existing projects or from core-analyses data.

7. Primary Development Practice in Field

A. Drilling: Cable-tools or rotary, drilling time, drilling cost, size of hole drilled, and type of rig used; special difficulties encountered in drilling.

B. Average well spacing.
C. Complete casing record: Sizes used, average amount used, average amount pulled, average amount left in wells, and present condition. Note age and condition of reclaimed pipe.
D. Cementing practice.
E. Use of explosives in well completion: Quantity and type of explosive used, placement of shots in reference to top and bottom of sand and in relation to pay sections, size of shells used, and average cleanout time required. Were tamped shots used and, if so, what was the type of tamping? Number of wells shot and general results of shooting.
F. Acid treatment in well completion: Quantity and type of acid used; method of placing either through packer on tubing or in open hole; amount of oil used for tamp, and method of removal of spent acid; number of wells acidized, and general results of acidization.
G. Gun-perforation of casing: Average number of shots and number of shots per foot; section of pay zone generally perforated; comparison of initial productions from gun-perforated wells and wells completed with open hole below casing or with perforated liner and screen.
H. Amount and size of tubing used.
I. Liners and screens used.
J. Gravel packing.

8. Primary Operating Practice in Field

A. Subsurface equipment in pumping wells: Anchor, working barrel, flood nipple, cups, balls and seats, rods, and production packers.
B. Surface equipment for pumping wells: Pumping jacks, derricks or gin poles, receiving tanks, gas lines, oil lines, well-head connections, pull rods or pull lines, and central powers; oil-storage facilities; condition of equipment.
C. Pumping frequency: Time required to pump off; estimated lifting costs.
D. Gas-lift installations: Type, pressure employed, and cost of operation.
E. Turbine and hydraulic pumps: Comparison of results with conventional pumping equipment; cost of operation.
F. Method of separation of oil and water: Treatment of emulsion.
G. Average total cost of surface-well equipment.
H. Common operating troubles:
1. Paraffin and basic sediment.

2. Mineral deposition on sand, casing, and tubing, and any salting-up of wells.
3. Casing and tubing leaks.
4. Corrosion of equipment.
5. Excessive wear and parting of rods.
6. Abrasion of working barrel by floating sand.
7. Accumulation of cavings and muck in bottom of hole.

I. Cleanout procedure: Method employed, size of shot used, time required, average cost, and general results.

J. Estimated increase in production by general rehabilitation of wells and equipment.

K. Possibilities for rehabilitation of abandoned wells.

9. Water Supply for Waterflooding Operations

A. Surface supply: Name and location of streams; proximity to site of operations; approximate area of water shed; average annual rainfall, in inches; stream flow during various seasons of the year; turbidity of water during various seasons of year; pollution, if any, of stream; corrosive nature of water; availability of dam sites; prior use of water for domestic and industrial supply, for carrying sewage, for hydroelectric-power generation, or for irrigation.

B. Subsurface supply: Depth, thickness, and productive capacity of subsurface water-bearing formations; turbidity of water; production by artesian flow or by pumping; effect of previous withdrawal of water on artesian pressure in flowing wells and on level of water table in pumping wells. Note all water-bearing horizons encountered in drilling oil wells.

C. Water-sample data: Same as 5-J.

D. Water treatment required.

E. Mixing test of produced water and supply water.

F. Observation of rate of precipitation or increase in turbidity of subsurface sample taken without exposure to air.

G. Estimated cost of water-supply development.

10. Gas Supply for Gas Repressuring

A. Availability of gas in field: By production from lease, by purchase from adjacent leases, or by purchase from gas company; amount of gas presently sold from field, company selling gas, and selling price.

B. Availability of gas outside of field:
   1. Ownership of gas.
   2. Purchaser of gas.
   3. Price paid per thousand cubic feet.
   4. Pressure base.
   5. Adequacy of supply for repressuring.
   6. Availability of any distress gas.
   7. Proximity to pipe lines.
   8. Amount of line required to make connection.
   9. Gathering-line pressures.
   10. Main-line pressure.
   11. Average closed-in rock pressure.
   12. Date of completion of wells.
   13. Maximum, minimum, and average open-flow capacity of wells.
   14. Age of wells.
   15. Original rock pressure at all sources.
   16. Recent production from wells.
   17. Estimated reserves available.
   18. Compressor facilities required.

C. Analysis of gas available for repressuring.

D. Possibility of using air for repressuring if gas supply is not adequate or is too costly.

11. Conjoint Use of Gas and Water

A. Possibilities of simultaneous gas injection at high structural position or near to present gas-oil contact in reservoir, and water injection at or near to present water-oil contact.

12. Economic Considerations

A. Effect of topography on development cost, including effect on roads, pipe lines, pumping equipment, well depth, etc.

B. Accessibility: Nearness to railroads, navigable streams, and highways.

C. Availability and cost of electric power.

D. State and pipeline proration practices: Well allowables.

E. State laws and regulations governing secondary-recovery operations: Restrictions on federal or Indian lands; protection of coal beds and other mineral deposits.

F. Name and address of operators with lease lists and acreage; description of leases: Probable productive acreage available for secondary-recovery operations, and cost of acreage; separation of deep rights from shallow rights; possible damage claims; existing lease terms; adverse lease terms; list of royalty owners and possibilities for unitization of royalty interests; willingness of operators to unitize and cooperate in secondary-recovery development; method generally favored for field and ability of all operators to contribute proportionate share of cost.

G. Production, excise, and *ad valorem* taxes.
H. Availability of labor, and wage scales: Nearness to supply houses.
 I. Adaptability of existing wells to secondary-recovery operations: Adaptability of secondary-recovery well-spacing pattern to lease boundary lines.
 J. Markets for oil:
   1. Grade of oil.
   2. Pipe-line facilities: Condition and ownership of same.
   3. Condition of main storage tanks.
   4. Frequency with which oil is run.
   5. Ultimate purchaser of crude.
   6. Desirability of crude.
   7. Location and capacity of refinery connections.
   8. Outlook for future market for oil.

   9. Current price paid for crude and any premium.
  10. Transportation of oil by tank cars or tank trucks.
  11. Differences in transportation facilities during various seasons of year.
K. Variation in development cost based upon different volume and pressure requirements of repressuring media.

13. Summary

Investigator's comments regarding the most applicable secondary-recovery method, probable secondary reserves, and desirability of further investigation and core-testing.

Scaling Groups and Factors

Scaling factors must be used when data from two fields are compared and when laboratory data are used in equations to forecast future performance and to understand past performance. Unscaled laboratory tests can be useless. Graphical presentations often are useful only when scaling groups are used in the scales. Use of scaling groups substantially reduces computation time. Geertsma, Croes and Schwarz (*JPT,* June 1956, © SPE-AIME) presented the scaling groups on p. 37 for fluid flow.

SIMILARITY GROUPS

Types of drive	Significant boundary condition	I — Darcy equations	II — General flow equations	III — Thermal balance equations	IV — Eq. of state	V — T-dependence of μ	VI — Diffusion	VII — Interfacial forces	VIII — Initial and boundary conditions	IX — Important dependent groups	X — Dimensionless time group θ
Cold-water drive	injection rate given	$\frac{l}{h}$ α $\frac{\mu_o}{\mu_w}$ $\frac{\rho_o}{\rho_w}$ $\frac{k\rho_w g}{v\mu_w}$ $\frac{\sigma\cos\theta\sqrt{k\phi}}{v\mu_w l}$	$\frac{v\rho_w\sqrt{k}}{\mu_w}$ $\frac{l}{\sqrt{k}}$ X						S_{oi}	$\frac{\Delta p k}{v\mu_w l}$; $\frac{q_w}{q_o}$	$\frac{tv}{lf}$
Cold-water drive	injection pressure given	$\frac{l}{h}$ α $\frac{\mu_o}{\mu_w}$ $\frac{\rho_o}{\rho_w}$ $\frac{\rho_w g l}{\Delta p}$ $\frac{\sigma\cos\theta\sqrt{\phi}}{\Delta\rho\sqrt{k}}$	$\frac{\rho_w k\sqrt{k}\Delta p}{\mu_w^2 l}$ $\frac{l}{\sqrt{k}}$ X						S_{oi}	$\frac{q_o\mu_w}{h\Delta p k}$; $\frac{q_w}{q_o}$	$\frac{t\Delta p k}{\mu_w l^2 f}$
Hot-water drive	injection rate given	$\frac{l}{h}$ α $\frac{\mu_{oi}}{\mu_{wi}}$ $\frac{\rho_{oi}}{\rho_{wi}}$ $\frac{k\rho_{wi} g}{v\mu_{wi}}$ $\frac{(\sigma\cos\theta)_i\sqrt{k\phi}}{v\mu_{wi} l}$	$\frac{v\rho_{wi}\sqrt{k}}{\mu_{wi}}$ $\frac{l}{\sqrt{k}}$ X	$\frac{\rho_{wi}c_w}{\rho_{oi}c_o}$ $\frac{\rho_r c_r}{\rho_{oi}c_o}$ $\frac{\lambda_w}{\lambda_o}$ $\frac{\lambda_r}{\lambda_o}$ $\frac{\lambda_o}{\rho_{oi}c_o v l}$ ϕ	$\beta_o T_i$	$A_{\mu_o,T}$ $A_{\mu_w,T}$		$A_{\sigma,T}$	$\frac{\rho_r c_r}{\rho_{oi}c_o}$ $\frac{\lambda_c}{\lambda_o}$ $\frac{T_b}{T_i}$ S_{oi}	$\frac{\Delta p k}{v\mu_{wi} l}$; $\frac{q_w}{q_o}$; $\frac{T}{T_i}$	$\frac{tv}{lf}$
Hot-water drive	injection pressure given	$\frac{l}{h}$ α $\frac{\mu_{oi}}{\mu_{wi}}$ $\frac{\rho_{oi}}{\rho_{wi}}$ $\frac{\rho_{wi} g l}{\Delta p}$ $\frac{(\sigma\cos\theta)_i\sqrt{\phi}}{\Delta\rho\sqrt{k}}$	$\frac{\rho_{wi} k\sqrt{k}\Delta p}{\mu_{wi}^2 l}$ $\frac{l}{\sqrt{k}}$ X	$\frac{\rho_{wi}c_w}{\rho_{oi}c_o}$ $\frac{\rho_r c_r}{\rho_{oi}c_o}$ $\frac{\lambda_w}{\lambda_o}$ $\frac{\lambda_r}{\lambda_o}$ $\frac{\lambda_o\mu_{wi}}{\rho_{oi}c_o k\Delta p}$ ϕ	$\beta_o T_i$	$A_{\mu_o,T}$ $A_{\mu_w,T}$		$A_{\sigma,T}$	$\frac{\rho_r c_r}{\rho_{oi}c_o}$ $\frac{\lambda_c}{\lambda_o}$ $\frac{T_b}{T_i}$ S_{oi}	$\frac{q_o\mu_{wi}}{h\Delta p k}$; $\frac{q_w}{q_o}$; $\frac{T}{T_i}$	$\frac{t\Delta p k}{\mu_{wi} l^2 f}$
Solvent injection	injection rate given	$\frac{l}{h}$ α $\frac{\mu_o}{\mu_d}$ $\frac{\rho_o}{\rho_d}$ $\frac{k\rho_d g}{v\mu_d}$	$\frac{D}{vl}$ $\frac{v\rho_d\sqrt{k}}{\mu_d}$ $\frac{l}{\sqrt{k}}$ X				$A_{\mu,c}$		C_{oi}	$\frac{\Delta p k}{v\mu_d l}$; $\frac{C_d}{C_o}$	$\frac{tv}{lf}$
Solvent injection	injection pressure given	$\frac{l}{h}$ α $\frac{\mu_o}{\mu_d}$ $\frac{\rho_o}{\rho_d}$ $\frac{\rho_d g l}{\Delta p}$	$\frac{D\mu_d}{\Delta p k}$ $\frac{\rho_d k\sqrt{k}\Delta p}{\mu_d^2 l}$ $\frac{l}{\sqrt{k}}$ X				$A_{\mu,c}$		C_{oi}	$\frac{q_o\mu_d}{h\Delta p k}$; $\frac{C_d}{C_o}$	$\frac{t\Delta p k}{\mu_d l^2 f}$

Relations with dimensionless groups in other engineering sciences.

$$\frac{v\rho\sqrt{k}}{\mu}\cdot\frac{l}{\sqrt{k}} = \frac{v\rho l}{\mu} \quad = \text{Re (Reynolds' group)}$$

$$\frac{\lambda}{\rho c v l} \quad = \text{Pé (Péclet)}$$

$$\frac{D}{vl}\cdot\frac{v\rho\sqrt{k}}{\mu}\cdot\frac{l}{\sqrt{k}} = \frac{\rho D}{\mu} \quad = \text{Sc (Schmid)}$$

$$\frac{vl}{\sigma\cos\theta\sqrt{k\phi}}\cdot\frac{v\rho\sqrt{k}}{\mu}\sqrt{\phi}\sqrt{k} = \frac{v^2\rho l}{\sigma\cos\theta} \quad = \text{We (Weber)}$$

$$\frac{v\mu}{k\rho g}\cdot\frac{v\rho\sqrt{k}}{\mu}\cdot\frac{l}{\sqrt{k}} = \frac{v^2}{gl} \quad = \text{Fr (Froude)}$$

$$\frac{v\rho\sqrt{k}}{\mu}\cdot\frac{\lambda}{\sqrt{k}\,\rho c v l} = \frac{\lambda}{c\mu} \quad = \text{Pr}^{-1}\text{ (Prandtl)}^{-1}$$

$$\frac{vl}{D}\cdot\frac{\lambda}{\rho c v l} = \frac{\lambda}{\rho c D} \quad = \text{Le (Lewis)}$$

$$\frac{k\rho g}{v\mu}\cdot\frac{v\rho\sqrt{k}}{\mu}\cdot\left(\frac{l}{\sqrt{k}}\right)^3\cdot\beta T = \frac{\rho^2 g l^3\beta T}{\mu^2} \quad = \text{Gr (Grashof)}$$

Latin

A = symbolic representation of a set of thermodynamic properties.

$A_{a,b}$ stands for the scaling rule: "the graphs of dimensionless a against dimensionless b should be congruent for the prototype and for the model" (dimensionless);

b = distribution factor of the interfacial force (dimensionless);

c = heat capacity per unit mass (J/kg °K);

C = volume concentration (dimensionless);

D = diffusion constant (m²/sec);

F = functions describing the relation between the properties of fluids and dependent variables; they always contain $A_{x,y}$ as a variable and have corresponding subscripts (dimensionless);

g = acceleration of gravity (m/sec²);

h = thickness of the sand (m);

J = Leverett's function (dimensionless);

k = absolute permeability (m²);

k_r = relative permeability (dimensionless);

l = length of the sand (m)

L = lithologic factor (dimensionless);

p = pressure (N/m²);

Δp = pressure difference between inflow and outflow faces of the sand (N/m²);

q = production rate (m³/m².sec);

$\vec{r}$ = geometrical vector (m);

S = saturation (dimensionless);

z = coordinate in direction perpendicular to x and y (m).

Greek

α = angle of inclination of the sand with respect to the horizontal (dimensionless);

β = thermal cubic expansion coefficient (°K⁻¹);

γ = scaling down factor (dimensionless);

μ = viscosity (N.sec/m²);

Θ = wetting angle of fluid interface (dimensionless);

$\theta = \dfrac{t}{\tau\phi}$ (dimensionless)

λ = thermal conductivity (J/m.sec°K);

σ = interfacial tension (N/m);

ρ = density (kg/m³);

τ = characteristic time (sec);

ϕ = porosity (dimensionless);

χ = similarity group for pore size distribution (dimensionless).

Subscripts

b = injection water;

c = cap rock;

c = concentration;

d = diluent,

D = diffusion;

e = outflow end of the sand;

i = initial;

m = mixture;

o = oil;

r = rock (except for D_r and k_r);

T = temperature;

w = water;

x = coordinate in direction of l;

y = coordinate in direction of h;

μ = viscosity;

σ = interfacial tension.

t = time (sec);

T = temperature (°K);

v = flow or injection rate (m³/m².sec);

x = coordinate in direction of l(m);

$X = \dfrac{x}{l}$ (dimensionless);

y = coordinate in direction of h(m);

$Y = \dfrac{Y}{h}$ (dimensionless).

2 Determination of Best Values for Rock Parameters

Finding sound or accurate values for terms appearing in the respective equations is essential. Many assumptions were used in the derivation of equations, and the applicability of the assumptions to the reservoir being studied must be determined. Equations required to solve most engineering problems have been programmed for hand-held and/or large computers. Unfortunately, the output is no better than the quality of the data initially placed into the computers. Curve match and modeling can be used to make forecasts only when basic data and equations are sound, accurate, and reliable.

Also, some understanding as to what constitutes a reliable result is required before the engineer can evaluate the reliability of computer output. The accepted truism, "Garbage in equals garbage out," must always be recognized even when the computer program has been designed to upgrade basic data by use of statistics, curve match, and similar methods to obtain more reliable values for the calculations.

Sound knowledge as to the oil, gas, and water in place in each part of the reservoir and aquifer related therewith at the start of a project is of prime importance. In addition to porosity, the saturations, and probable recovery mechanism, effective permeability and its variation within the reservoir are required. The sampling is inadequate and field and laboratory measurements are subject to error. Results, even when reliable, vary widely and must be properly averaged. The engineer making a study must recognize that all information must be analyzed and weighed properly. He must use the combined knowledge from data sources such as the laboratory, the engineer, the petrophysicist, the geologist, and many others. Rocks are not homogeneous, but many equations include this assumption. Stringers having low permeability such as shales influence flow and recovery and may result in separate small reservoirs.

Shale stringers and their probable effect on flow and performance must be determined and evaluated. Data from logs, production history, core analysis, special tests using flow and build-up theory, seismic interpretations, local and regional geology, and many other sources must be coordinated, and the best possible values and conclusion must be recognized. Emperical methods, analogy, performance of adjacent reservoirs, and past experience are valuable when selecting the best value for each parameter appearing in the best applicable equations useable by the computer. Graphical methods and scaling techniques must be used in the analysis of both input and output. Engineering is a science only when properly used with good judgment and hard careful evaluation of the available information.

Core Analysis and Logging Devices

Coring fluids may move and displace the in-situ fluids. Both core and log results are affected. Special expensive coring techniques often are not successful in obtaining cores in an undisturbed state. As an example, water in the drilling mud is miscible with the connate water, and the mud filtrate moves the in-place water as a bank ahead of the injected fluids, thereby affecting salinities of the water within the area investigated by logs. The following table is helpful in determining the techniques to use.

39

Coring Fluids	Filtrate	Usual Effect on Core Saturations*	
		Water	Hydrocarbons
Water base	Water	Increased	Decreased
Oil base	Oil	No change	Replaced
Inverted oil emulsion	Oil	No change	Replaced
Oil emulsion	Water	Increased	Decreased
Gas (hydrocarbons)	Gas	No change	Replaced
Air	—	Uncertain	Decreased

* Air and other gases can evaporate both water and light ends from oils. Also, oil and water are moveable when present in quantities above residual satuation values. Natural forces drive fluids from cores as pressure is reduced by bringing cores to the surface. Evaporation and gravity forces change core saturations during handling at surface.

Representative results of core analysis for many fields were assembled by E. H. Koeff and published in Frick's *Petroleum Production Handbook*. Obtaining cores is expensive, and the data should be used to advantage. Crossplots are used to evaluate the reliability of the data. Routine core analysis include:

Type of analysis	Use of results
Porosity	A factor in volume and storage determinations
Permeability horizontal and vertical	Defines flow capacity, crossflow, gas and water coning and relative profile capacity of different zones, pay and nonpay zones
Saturations	Defines presence of hydrocarbons, probable fluid recovery by test, type of recovery, fluid contacts, completion interval
Lithology	Rock type, fractures, vugs, laminations, shale content used in log interpretation, recovery forecasts, capacity estimates
Core-gamma ray log	Relates core and log depth
Grain density	Used in log interpretation and lithology
Water salinity	Used in log interpretation, invasion problems
Oil gravity	Useful in determination of fluid characteristics

Special core analysis tests may include:

Type of test	Use of results
Capillary pressure	Defines irreducible fluid content, contacts
Rock compressibility	Volume change caused by pressure change
Permeability and porosity versus pressure	Corrects to reservoir conditions

Type of test	Use of results
Petrographic studies mineral diagenesis	Used in log interpretation Origin of oil and source bed studies
clay identification sieve analysis	Origin of oil and log analysis Selection of screens, sand grain size
Wettability	Used in capillary pressure interpretation and recovery analysis—relative permeability
Electrical formation factor resistivity index	Used in log interpretation
Acoustic velocity	Log and seismic interpretation
Visual inspection	Rock description and geological study
Thin sections, slabs	

In addition to the above static tests, the following dynamic tests may be undertaken in the laboratory:

Type of test	Use of results
Air, water, and other liquid permeability	Evaluates completion, workover, fracture and injection fluids; often combined with flood-pot test
Flood-pot tests and waterflood evaluation	Results in values for irreducible saturations, values for final recovery with special recovery fluids such as surfactants, water, and polymers
Relative permeability gas-oil gas-water water-oil oil-special fluids thermal	Relative permeability is used to obtain values for effective permeability to each fluid when two or more fluids flow simultaneously; relative permeability enables the calculation of recovery versus saturation and time while values from flood-pot test give only end-point results
Gas and oil seals	Permeability is determined to 1×10^{-6} md; threshold pressures are determined
Particle movement reverse flow	Data are often obtained by careful observation during other tests; possible plugging
Water composition	Precipitates from water to be expected

The interpretation of electric logs in most pays developed in the 1980s is difficult and may be considered an art rather than a true science. Computer programs investigate many alternatives and use the more appropriate equations and data. However, major companies and some smaller units employ many log engineers to interpret logs. The reservoir engineer must constantly consult with logging engineers and geologists so that logical conclusions are used in studies. All engineers should be familiar with information in texts and papers by Schlumberger, Dresser Atlas–Layne Wells, Welex–Halliburton, and others.

Logs usually are employed to obtain formation type and thickness, connate or in-situ water, and porosity. The equations used for analysis of clean sands follow:

Formation factor $= F = R_o/R_w =$ Resistivity of a non-shaley formation 100% saturated with a brine divided by the resistivity of the brine saturating the formation

and $$F = a/\phi^m$$

where $a =$ constant
$\phi =$ formation porosity
$m =$ cementation factor

For a clean sand: $F = 0.81/\phi^2$

For sucrosic rocks: $F = 0.62/\phi^{2.15}$

For chalky rocks: $F = 1.0/\phi^2$

For compact or ooliscastic rocks: $F = 1/\phi^{2.2-3.0}$

For water saturation of a clean formation:
$$S_w^n = F\,R_w/R_t$$

Where: $n =$ saturation exponent, which often has a value of 2
$R_w =$ resistivity of brine saturating the formation
$R_t =$ true resistivity of clean formation as read from logs (adjusted when beds are thin, sands are shaley, invasion occurs)
$FR_w =$ resistivity of the formation when 100% saturated with the brine having resistivity R_w (may be read from the logs when part of the pay is water bearing)

Various logs are used for various purposes.

Gamma-ray Log

The gamma-ray tool primarily describes qualitatively the presence of radioactive salts in the formation. Since shales have a high concentration of these radioactive salts and sandstones, dolomites, and limestones have a low concentration, the device can distinguish between these formations. Although the gamma-ray method is primarily qualitative, some quantitative use is possible when values for shales and other rock types are known. The proportion of each rock can then be determined.

Spontaneous Potential Log

The SP log depends upon mud salinity and in-situ water salinity. The migration of ions in the wellbore face between different concentrations of salts—primarily sodium and chloride—caused by filtration of mud fluids into permeable formations surrounded by shales creates electrochemical and electrokinetic potentials. The SP log consists of a surface electrode that is maintained at constant potential and a downhole potential.

The magnitude of the SP deflection relative to a base line generally indicates the presence of permeable beds. Also, the SP can be used to pick lithology such as shales and sands. The SP can be used to determine the resistivity of the in-situ formation water when invasion of mud filtrate is small. The SP also can make a qualitative determination of the shaliness in sand beds after adjustments are employed as necessary.

Sonic Log (Compressional Velocity)

Sonic logging records the time required for a sound wave to travel through a specific distance of a formation. Most records are the reciprocal velocity of the compressional wave (first arrival) as a continuous recording. The velocities of the recorded wave, the rock matrix, and the in-situ fluid can be related to give a value for the fractional pore space or porosity of the rock. The pore space slows down the speed of sound waves. In an identical formation filled with like fluid, a common linear relationship exists between porosity and the sound velocity. If secondary porosity such as vugs or fractures are encountered, the sound wave may not be able to cross the barrier if the angle of intersection is not proper. Sonic logs may then skip. Also, sonic logs essentially ignore porosity other than intergranular.

Sonic Log (Shear Velocity)

Combining the velocity of the shear wave with that of the compressional wave should result in values for the porosity without a need for data as to rock composition and type. The method has problems and has not been used routinely.

Neutron Log

The neutron log records the gamma-ray intensity resulting from the bombardment of the formation with neutrons. The hydrogen atom, having a mass nearly equal to that of neutrons, slows down the emitted neutron and porosity can be determined. A formation having a high gamma-ray response also may contain hydrocarbons when natural gamma-ray intensity is low. The size of the borehole, the type of mud, and amount of drilling mud invasion are important corrections. The gamma ray and density logs are usually run in conjunction with the neutron log so that correction can be made for natural radiation and secondary porosity caused by vugs and fractures.

The neutron log can also indicate the presence of gas since gas, particularly when pressure is low or moderate, contains less hydrogen than either oil or water. Unfortunately, invasion of drilling mud filtrate often is substantial in gas-bearing pay.

Density Log

The scatter of induced gamma rays by a formation is proportional to the bulk density of the rock. Pad-type devices are used to eliminate scatter within the drilling mud in the well. The density log combined with the sonic log may be used to determine shale content in laminated sand-shale sequences. The density log is affected by mud cake thickness, and suitable correction may be necessary. Large washouts in the well bore may cause pads to lose contact, and a caliper log may be necessary.

The density log in combination with the neutron log can be used to locate gas-bearing pays.

Sonic-Neutron, Sonic-Density and Neutron-Density Crossplots

The reading indicated by the neutron, sonic, and density logs relate to formation lithology, fluid content, and porosity. Up to four unknowns, including porosity, can be obtained by using the three porosity logs in combination as crossplots and overlays. The neutron and density logs reflect total porosity, but the sonic log often does not reflect porosity in vugs and fractures. Base curves showing porosity for the density-neutron crossplot are available for clean sandstones, dolomite, and limestone. Similar plots can be made for other log and formation combinations.

Micrologs, Microlaterologs, and Proximity Logs

The distance investigated by these logs is measured in inches, and the results reflect the area invaded by drilling-mud filtrate. The log is used to determine permeable zones when mud cakes are not too thick. The microlog in conjunction with SES (Standard Electrical Survey) can be used to determine R_{xo} or resistivity of the zone near the well bore, which has been flushed by drilling fluids to indicate residual oil after displacement. A caliper log is run to eliminate affects of well-bore enlargement.

Conventional Logs

These are some of the first logs to be used successfully. In early days, pay quality of the reservoirs was good, and relatively fresh-water drilling muds were in general use. The 16″ normal log was useful in defining bed boundaries and measuring the resistivity of invaded zones. The 64″ normal log investigates deeper and usually gives reasonable or true resistivity values when invasion is not severe. The 18′8″ lateral log is a deep investigation log and avoids invasion problems, but thin beds do not respond properly unless thicker than 18 feet.

The conventional logs are more useful in sand-shale sequences and have been gradually replaced by other devices. Older wells were logged with conventional devices, and the engineer must be familiar with their interpretation when considering secondary recovery and EOR. Many departure curves are available to help with interpretation.

Focused Logs

These logs are used in salt-saturated muds and are useful in thin porous beds surrounded by resistive beds. The logs investigate thin beds with apparent resistivity approaching R_t, true values. The Laterolog 3 and Laterolog 7 are deep investigating tools and often give R_t without correction. The Laterolog 8, dual Laterolog, and spherical-focused log are shorter investigating tools and are more subjected to invasion of filtrate. Focusing the current minimizes the potential drop within the borehole and mud cake. Drilling mud filtrate can be a problem unless R_{mf} approaches R_w. Electrodes may read on the high side when various portions of the tool are crossing thin, highly resistive beds.

Induction Logs (IES and DIL Tools)

The 6FF40 tool is a deeper investigation tool with 40-inch nominal spacing, a 16-inch normal, and an SP electrode. The 6FF28 tool is similar but is used in smaller pipe. The dual-induction Laterolog 8 (DIL) tool uses a deep-reading induction device (6FF40) with medium induction (ILm), a Laterolog 8, and an SP electrode. The DIL device with its three focused resistivity readings of differing depths of investigation is superior to IES. ISF/sonic log also is available. Logging manuals should be consulted.

Determination of Maximum Depth of Burial

Porosity and permeability decrease with depth of burial of a formation. Nature is constantly attempting to reduce all surface elevations to sea level, and many reservoirs today are at depths less than their initial maximum burial depth. The sonic log is useful, as illustrated by Fig. 2–1 for Oklahoma, California, Wyoming-Alberta, and Texas Gulf Coast conditions. Data for Oklahoma and the Texas Gulf Coast indicate that the depth versus log correlation is dependent upon the age of the formation and that a series of curves are required. Formations that have a lesser depth than maximum may retain some overpressure characteristics, which is of interest during drilling. Also, rock compressibility characteristics differ for pressures above and below maximum burial depth. Changes in the sonic log versus depth correlation can be helpful in geological interpretation.

A summary review of these characteristics is by W. H. Lang, *Oil and Gas Journal* (January 28, 1980), and the references below should be studied by those interested in this important subject.

References

Athy, L. F. "Density, Porosity, and Compaction of Sedimentary Rocks." *AAPG Bul.,* Vol. 14, No. 1, 1930.
Bradley, J. S. "Abnormal Formation Pressure." *AAPG Bull.,* Vol. 59, 1975.
Faust, L. Y. "Seismic Velocity as a Function of Depth and Geologic Time." *Geophysics,* Vol. XVI, No. 2, 1957.
————. "A Velocity Function Including Lithologic Variation." *Geophysics,* Vol. XVIII, No. 2, 1953.
Hedberg, H. D. "Gravitational Compaction of Clays and Shales." *Am. Jor. of Science,* Vol. 31, No. 184, 1936.
Heiland, C. A. *Geophysical Exploration.* Prentice-Hall, 1940.
Jankowsky, W. "Diagenesis and Oil Accumulation as Aids in the Analysis of the Structural History of the Northwestern German Basin." *Zeitschrift der Deutschen Geologischen Gesellschraft,* Vol. 114, 1962.

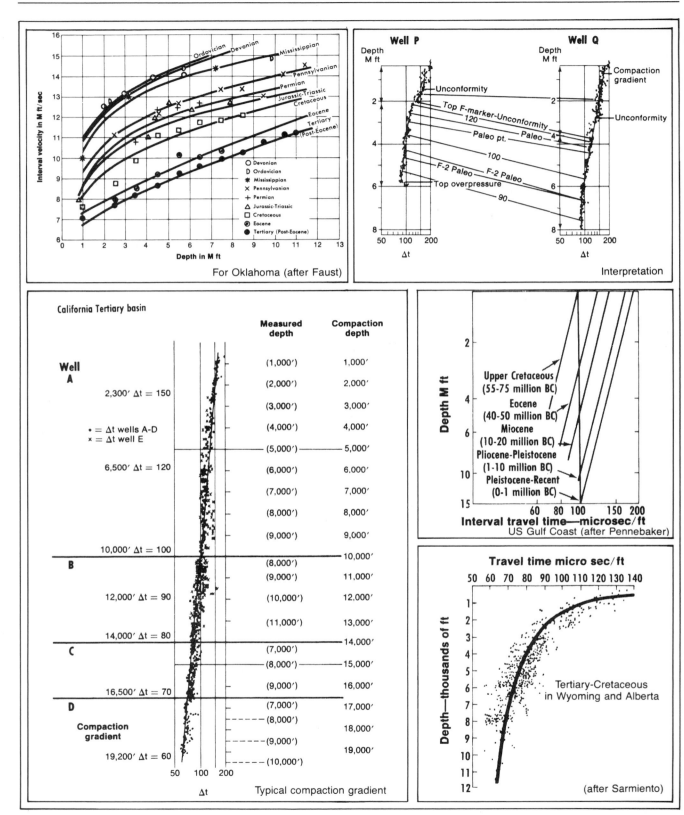

Figure 2–1. *Correlating sonic logs with maximum burial depth and geological age (after Lang, Oil & Gas Journal, 28 January 1980).*

Determination of Rock Characteristics

Examination of cores and cuttings by geologists, paleontologists, engineers and others normally determines the geological age of the rocks being penetrated by the drill. Obtaining the exact depth of cuttings collected from surface mud presents a problem that is solved by a number of methods, including drilling short intervals and circulating to the surface as an extreme limit. Cores naturally are more desirable than cuttings but usually are not cost effective. Drilling rate and similar special techniques furnished by gas logging trucks at the well site also are useful in defining rock characteristics. At some times, the study of cuttings under the microscope can give fair values for porosity as well as determine the marine life and fauna present at the time of initial deposition of the rock. Cuttings also often give knowledge as to the type and absence of hydrocarbons. Special study of shales obtained by cutting collection and coring may determine the source of the oil, migration paths, and similar facts.

Electric and radioactive logs show specific and distinct character so that rock type and features of the rock opposite the logging device can be determined. Some of the major characteristics of various types of rocks are shown in Figs. 2–2 and 3. Logs also are influenced by porosity and the fluids within the pore space. Analysis of logs can be complicated, and computer programs are available from service companies and consulting engineering firms to accomplish satisfactory analysis. As mentioned previously, the computer programs are useful, but major companies use many engineers plus their computers in analysis of logs. This fact indicates that the analysis of logs is not routine, and that the output from computer programs is not always satisfactory. Apparently, the ingenuity of the human mind can make a contribution not available from the best computer program.

Output from computer programs is presented in various formats, and that used by Schlumberger is shown as Figs. 2–4 and 5.

The log characteristics of fluids and "pure" rocks can be determined in the laboratory and by special use of logging devices at the surface. Values so determined are presented as the tables following. These values can be plotted in a number of ways, as shown on the remaining figures in this section, to obtain an indication of the lithology and composition of rocks. In general, the method involves the preparation of theoretical graphs based on the laboratory-derived information and then plotting the data from subsurface logging on such theoretical graphs to find satisfactory correlations. All graphs or crossplots shown are based on laboratory information and thus may serve as the basic or theoretical plots.

Fig. 2–6 is a plot of sonic versus neutron logs. Similar plots are available for neutron versus density and sonic versus density. These plots are often used to determine the porosity of an interval from logging devices. As shown on the illustrative plot, lithology and porosity are included on the theoretical plot. Fig. 2–7 shows the relation between density, gamma-ray logs, and percent of potassium salt. The theoretical relation for various minerals using the density versus neutron log is shown on Fig. 2–8. The proper neutron scales should be used. Clays may be present in sands in laminar, structural, and dispersed state as defined on Fig. 2–9. Fig. 2–10 presents a method for determining porosity in a sand filled with shale and water, which will be explained in detail in the discussion of porosity. The effect of the various shale types on the bulk density versus neutron plot is shown on Figure 2–11. The influence of gas in the pores on the neutron or density versus gamma-ray crossplot is shown by Figure 2–12. A plot of N versus M, as defined on Figure 2–13, shows a spread in lithology, which is therefore useful in making lithology determinations from logs. Similar data using scales of 1/N and M/N are shown by Fig. 2–14. Fig. 2–15 is interesting since a plot of MN versus density places quartz, limestone, dolomite, and anhydrate on a straight line. Many of these crossplots will be used in later sections, such as porosity.

	Mineral	Composition	Apparent log density	Average Δt	ϕ_N^* (GNT) p.u.	γ-Ray deflection (APIU)	Apparent $K_2O\%$
Sedimentary minerals	Calcite	$CaCO_2$	2.710	47.5	0	0	—
	Dolomite	$CaMg(CO_3)_2$	2.876	43.5	4	0	—
	Quartz	SIO_2	2.648	55.5	−4	0	—
Sedimentary formations	Limestone	(e.g., when $\phi = 10\%$)	2.540	62	10	5–10	0
	Dolomite	(e.g., when $\phi = 10\%$)	2.683	58	13.5	10–20	0
	Sandstone	(e.g., when $\phi = 10\%$)	2.485	65.3	3	10–30	0
	Shale		2.2–2.75	70–150	25–60	80–140	2–10
Evaporites — Non-radioactive	Halite	$NaCl$	2.032	67	0	0	—
	Anhydrite	$CaSO_4$	2.977	50	0	0	—
	Gypsum	$CaSO_4 \cdot 2H_2O$	2.351	52.5	49	0	—
	Trona	$Na_2CO_3 \cdot NaHCO_3 \cdot 2H_2O$	2.100	65	40	0	—
Evaporites — Radioactive	Sylvite	KCl	1.863	74	0	~500	63.0
	Carnallite	$KCl \cdot MgCl_2 \cdot 6H_2O$	1.570	78	65	200	17.0
	Langbeinite	$K_2SO_4 \cdot 2MgSO_4$	2.820	52	0	275	22.6
	Polyhalite	$K_2SO_4 \cdot MgSO_4 \cdot 2CaSO_4 \cdot 2H_2O$	2.790	57.5	15	180	15.5
	Kainite	$MgSO_4 \cdot KCl \cdot 3H_2O$	2.120	—	45	225	18.9
Other minerals	Sulfur**		2.030	122.0	<0(15.5″)	0	
	Lignite		0.7–1.5	140–170	←High→ Greater than 50%	0	
	Bituminous Coal		1.3–1.5	110–140		0	
	Anthracite Coal		1.4–1.8			0	
	Pyrite		5.0				
Shales	Glauconite		2.4–2.9				
	Illite		2.6–2.9				
	Kaolinite		2.6–2.7				
	Montmorillonite		2.0–3.0				

ϕ_N^* is apparent limestone porosity from neutron log
Compensated Neutron Porosity = CNP = SNP − (±3)

VALUES OF M AND N FOR COMMON MINERALS

Mineral	Fresh mud ($p_t = 1$)		Salt mud ($p_t = 1.1$)	
	M	N*	M	N*
Sandstone (1) $V_m = 18,000$	0.810	0.628	0.835	0.669
Sandstone (2) $V_m = 19,500$	0.835	0.628	0.862	0.669
Limestone	0.827	0.585	0.854	0.621
Dolomite (1) $\phi = 5.5$–30%	0.778	0.516	0.800	0.544
Dolomite (2) $\phi = 1.5$–5.5%	0.778	0.524	0.800	0.554
Dolomite (3) $\phi = 0$–1.5%	0.778	0.532	0.800	0.561
Anhydrite $P_{ms} = 2.98$	0.702	0.505	0.718	0.532
Gypsum	1.015	0.378	1.064	0.408
Salt			1.269	1.032

Fluids		Δt_f	P_f	$(\phi_N)_f$
PRIMARY POROSITY (Liquid-Filled):	Fresh Mud	189.0	1.00	1.00
	Salt Mud	185.0	1.10	
SECONDARY POROSITY (In Dolomite):	Fresh Mud	43.5	1.00	1.00
	Salt Mud		1.10	
(In Limestone):	Fresh Mud	47.5	1.00	1.00
	Salt Mud		1.10	
(In Sandstone):	Fresh Mud	55.5	1.00	1.00
	Salt Mud		1.10	

	$v_{ms}(h/sec)$	Δt_{ms} ($\mu sec/h$)
Sandstones	18,000–19,500	55.5–51.0
Limestones	21,000–23,000	47.6–43.5
Dolomites	23,000	43.5
Anhydrite	20,000	50.0
Salt	15,000	66.7
Casing (Iron)	17,500	57.0

* Values of N are computed for SNP Neutron Log.

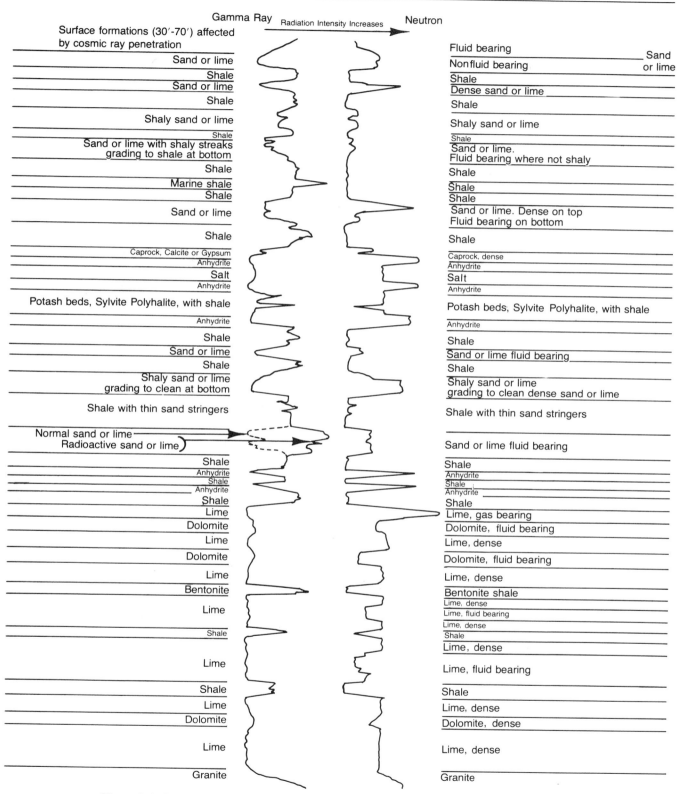

Figure 2–2. *Typical radioactive logs (from* Petroleum Production Handbook, *1962,* © *SPE-AIME).*

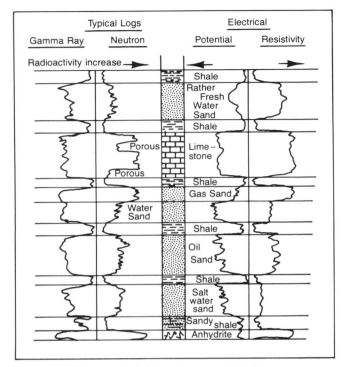

Figure 2–3. *Relations between typical logs.*

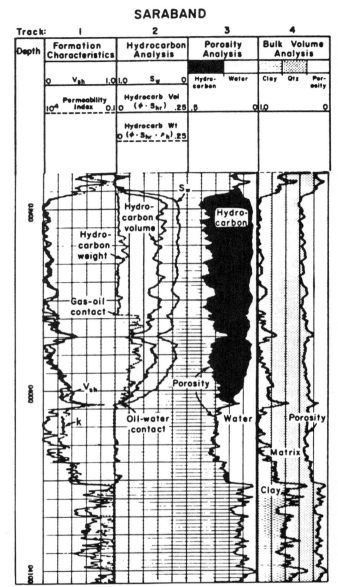

Figure 2–4. *Typical Saraband display (after Schlumberger).*

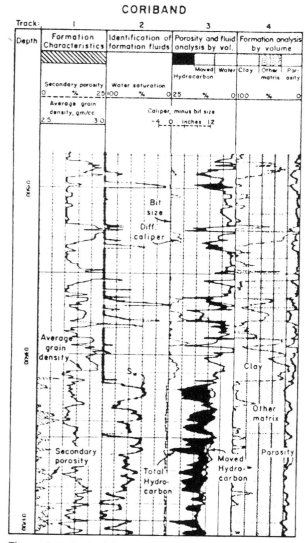

Figure 2–5. *Typical Coriband display (after Schlumberger).*

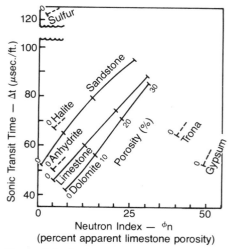

Figure 2–6. *Sonic vs. neutron.*
NOTE: Plots can be made using the tables for all figures in this section.

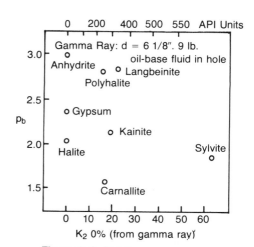

Figure 2–7. *Density vs. gamma ray.*

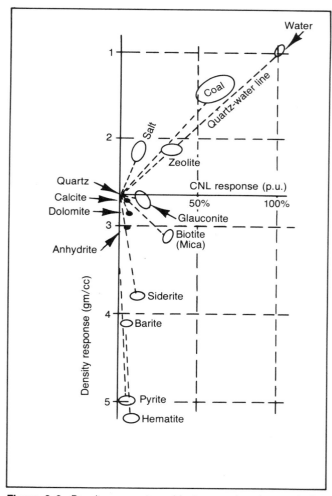

Figure 2–8. *Density vs. neutron. Most pure minerals contain little porosity and CNL deflection is small.*

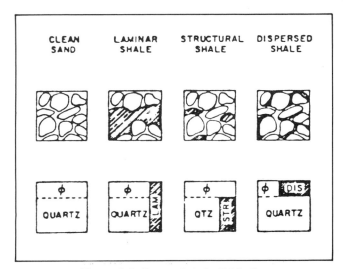

Figure 2–9. *Types of shale distribution.*

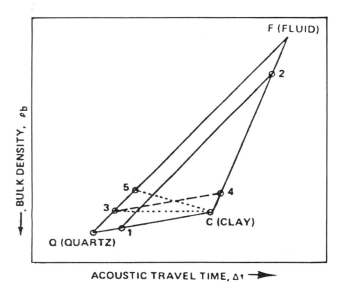

Figure 2–10. *Density vs. sonic plot.*

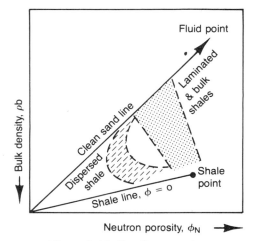

Figure 2–11. *Density vs. neutron.*

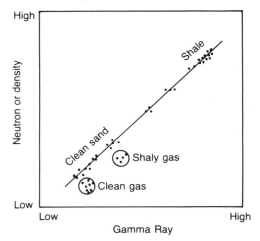

Figure 2–12. *Neutron or density plot vs. gamma ray. Shaly gas points will plot below the clean line and to the right of the clean gas sands. This type of plot is limited in that it requires just one lithology and shale (Fertl, OGJ, 25 September 1978).*

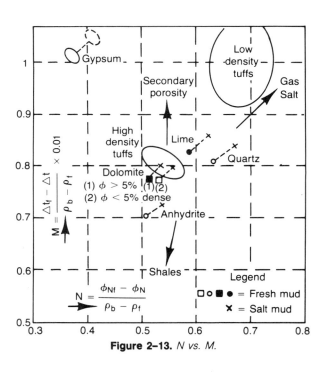

Figure 2–13. *N vs. M.*

These figures appear in papers by Fertl (Courtesy Dresser Atlas).

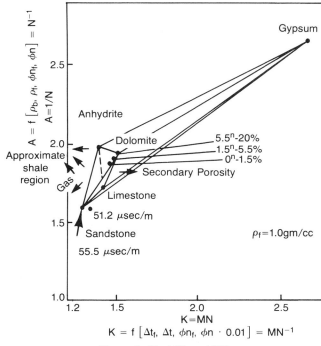

Figure 2–14. *1/N vs. M/N.*

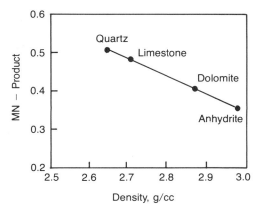

Figure 2–15. *MN vs. density crossplot.*

Porosity

The porosity of a rock usually is obtained from logs (neutron, sonic or density), core analysis, and various calculations. The logs tend to average an interval as required by the tool characteristics, while cores present values representative of a specific but small piece of rock. The core sample analyzed often has been upgraded by both field and laboratory since a tendency to pick the best sample cores for analysis is inherent in the process and in human nature.

Koepf lists the core characteristics of many rocks in Frick's *Petroleum Production Handbook*. These average values represent a useful data bank, particularly in wildcat areas.

Methods used to convert log data to porosity values are somewhat different for each tool used by the various logging companies. Manuals furnished by these companies should be studied in detail. The following discussion is based on publications of Schlumberger, when related to log determinations.

The porosity of cores is measured by using selected small plugs taken from cores. Full cores are analyzed when porosity is small, as found in calcareous materials. When the rock is not firmly cemented, the less consolidated and usually more permeable pay may

not be recovered during the coring operation. Obtaining reliable average values for porosity is not always easy, since many conditions including cutoff for producible pay are involved. Porosity of the cores is reported directly, usually as a percent of bulk volume. Results of core analysis should be compared with results of other methods used to obtain porosity, such as analogy, material balance, logs, volumetric-performance history, and a general knowledge of rocks such as presented by Koepf.

Examination of cores and cuttings using a microscope when calibration charts are available can give an estimate of porosity and an indication of permeability. The porosity values or porosity-depth logs for adjacent wells should be compared to obtain an indication of changes in pay characteristics and possible abnormalities. Capillary pressure measurements relate with porosity, as shown in a later section.

Porosity often decreases as the depth of burial of the formation increases, as shown in Fig. 2–16. The reduction for shales and shaley pays is greater than that for sands. Porosities decrease with the presence of shales and secondary deposits. Moving waters, often forced out of surrounding shales and sands as

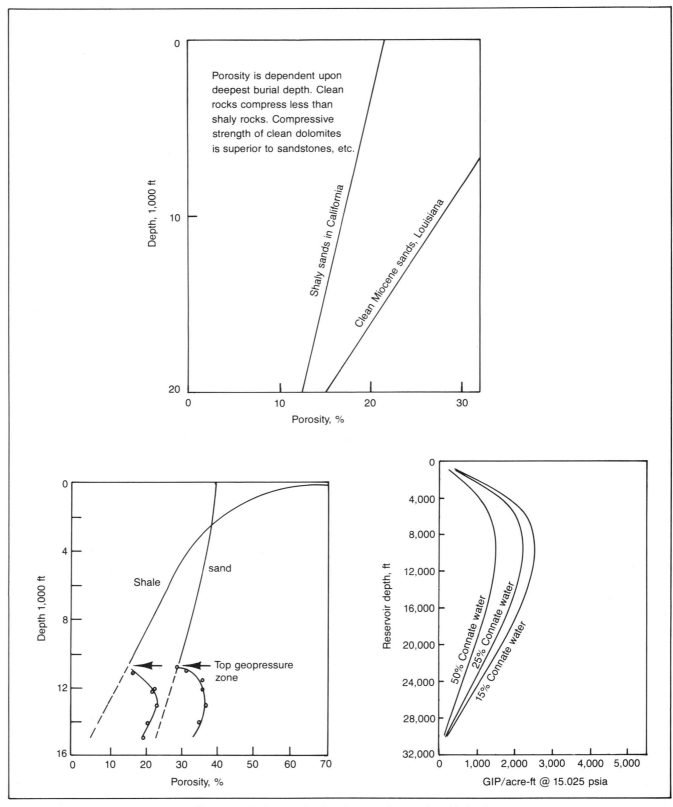

Figure 2–16. *Curves showing decrease in porosity with depth.*

compressive forces increase with depth of burial, both dissolve and precipitate minerals. Overburden weight can cause sand grains to imbed and dissolve at contact points. Shales can lose both mineral composition and interlayer water as overburden pressure increases, and many of these chemicals cause secondary precipitation within pore spaces. Fractures and vugs also may occur at both shallow and deep depth. Such fractures may not be observed properly by either core analysis or by the logging tools. The four-prong tool has been devised to log for fractures, and sonic tools sometimes skip when fractures are present. Vugs and fractures often are observed visually in cores and cuttings, but the fracture length, the degree of sealing, and the possibility of fracturing during coring must be evaluated. Fractures make a contribution to porosity that usually is small—zero to maybe 1.5% of bulk volume—but fractures and vugs if continuous can make a very substantial contribution to permeability. Fractures at depths greater than about 1,500 feet usually are vertical—70 to 90 degrees from horizontal.

The maximum porosity with cubic packing is 47.6% of bulk volume. Such sands must be uniform in size distribution. Uniform sands in the more likely rhombohedrel arrangement have a porosity of 26%. Nonuniform sands have lower porosity, and shales and other debris tend to reduce porosity. Some very tight gas-producible sands have porosities in the 5–10% range, and calcareous pays often have porosities in a range of 5–10%.

Porosities may be averaged in a manner similar to that discussed in the section on permeability. However, the following equations often are adequate.

Arithmetic average	$\phi = \Sigma\,\phi_i/\Sigma\,n_i$
Thickness weighted average	$\phi = \Sigma\,\phi_i h_i/\Sigma\,h_i$
Area weighted average	$\phi = \Sigma\,\phi_i A_i/\Sigma\,A_i$
Area thickness	maps of ϕh

Porosity appears in many engineering equations and is critical in the volumetric equation. A backward calculation usually is not effective since other terms in the equations also are not known with sufficient accuracy. Nevertheless, equations should always be used to be certain that the data are consistent. Accurate values for porosity are important in the volumetric equation, particularly when porosity is small. Unfortunately, accurate determination for these conditions is the most difficult, since both core analysis and the determination from logs is then also more difficult.

Porosity values for an area or basin such as presented by Koepf may be plotted versus depth. At times the scatter is not too severe and useful values are obtained, as shown in Fig. 2–16. Special geological conditions in rare cases can cause deviation from the areal curves. Production history of early wells in such fields should be carefully observed since the deviation probably is limited to a very small area of the field that relates with different geological conditions. Drilling may be delayed until production history is obtained.

Methods That may be Used to Determine Porosity Using Logs

Method	Log response relation	Optimum of range	Optimum formation type	Other variables
Sonic	$\Delta t = \Delta t_m + B\,\phi$	10%–20%	Consolidated	Lithology, ΔP, pore-size distribution
"Thermal" Neutron	$ND = C + D \log \phi$	1%–10%	Non-shaly liquid in pores	Hole size, shaliness, gas
"Epithermal" Neutron	$\phi \approx \phi_{snp}$	≤30%	Non-shaly liquid in pores	Borehole size, shaliness, gas
Density	$\rho_b = \rho_s(1 - \phi) + \rho_f(\phi)$	20%–40%	Unconsolidated sands	Grain density, borehole condition, *Gypsum*, *Pyrite* See SPWLA paper
NML	$\phi \approx FFI$	>6%–8%	Non-shaly	Shaliness, grain size in sandstone (?), magnetic content of mud and formation, gas
Cores	——	5%–30%	Consolidated	Sampling
R_t Tools	$R_t = I\,\phi^{-m}\,R_w$	10%–30%	Non-shaly	Shaliness, R_w, m, I
R_{xo} Tools	$R_{xo} = I_{xo}\,\phi^{-m}\,R_{mf}$	10%–30%	Non-shaly	Shaliness, R_{mf}, m, I

Data from Schlumberger Literature

Many of the tight gas sands being developed now do not conform with the criteria specified in the table. Interpretation of logs strains the available theory, and the selection of gas reservoirs as opposed to water-bearing intervals for perforation is not always possible. Water standing in such wells can be most damaging to production characteristics. Industry and the log service companies try to improve the log interpretations.

The curves presented as Figs. 2–17, 18 and 19 may be used to determine porosity when only quartz or sandstone, limestone, and dolomite are present. Anhydrite, gypsum, and shales complicate the porosity determination, as indicated by the curves. The method used for clean pays is simple, as illustrated by an example given with Fig. 2–17. Point P is a plot of field-measured values $FDC = 15$ and $CNL = 21$. The point P lies on the 18% porosity line. If the matrix is limestone and dolomite, which is indicated by the cutting description, proportioning the distance between the two curves, point P corresponds to about 40% dolomite and 60% limestone. As long as the formation does not contain shale-gypsum, the porosity value is not influenced by matrix lithology. If the

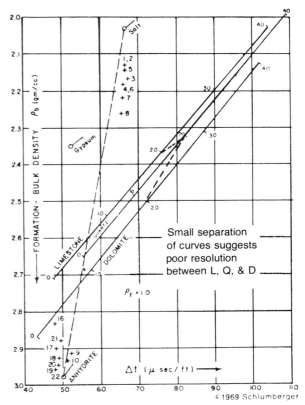

Figure 2–18. *Porosity and lithology determination from FDC density and sonic logs (base curve). Used to determine presence of salt, gypsum, and anhydrite (after Schlumberger).*

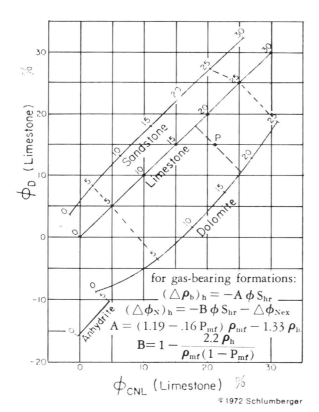

for gas-bearing formations:
$$(\triangle \rho_b)_h = -A \phi S_{hr}$$
$$(\triangle \phi_N)_h = -B \phi S_{hr} - \triangle \phi_{Nex}$$
$$A = (1.19 - .16\,P_{mf})\,\rho_{mf} - 1.33\,\rho_h$$
$$B = 1 - \frac{2.2\,\rho_h}{\rho_{mf}(1 - P_{mf})}$$

© 1972 Schlumberger

Figure 2–17. *Porosity and lithology determination from FDC density and CNL neutron in water-filled hole (base curve) (after Schlumberger).*

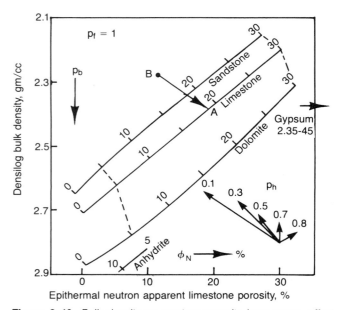

Figure 2–19. *Bulk density vs. neutron porosity base curve, effect of hydrocarbons. Arrow A-B represents correction of log point B for hydrocarbon effect for a gas case. The arrows at lower right represent approximate hydrocarbon shifts for various values of ρ_h when $OS_{hr} = 0.15$, $P_{mf} = 0$, and $\rho_{mf} = 1.0$ (courtesy Schlumberger).*

lithology were quartz and dolomite, the porosity would be 18.3% and mineral proportions would be 45% quartz and 55% dolomite. Lithology and porosity can both be determined with accuracy. Hydrocarbons also complicate the calculations, as illustrated in Fig. 2–19.

The curves are based on equations discussed in the following brief review of each log. Manuals and papers available from the individual service company logging the well should be reviewed before analyzing any log. The logging tools used may be different.

Porosity values can be obtained from an analysis of the neutron log, the density log, and the sonic log. Suitable corrections must be made for parameters other than porosity, including lithology, nature of fluids in the pore space, and shaliness. Two or more logs are often crossplotted to find values for these influences. The logging tools investigate an area near the well bore that often has been invaded by drilling fluids. The sonic log has the shallowest investigation of the area around the well bore.

The sonic tool measures the interval transit time, Δt, in microseconds for a sonic wave to travel through one foot of formation along a path parallel to the borehole. The basic method of analysis depends upon a relation:

$$\phi = (\Delta t - \Delta t_{ma})/(\Delta t_f - \Delta t_{ma})$$

where Δt_f is transit time in the pore fluid and Δt_{ma} is transit time in the rock matrix. These two values are available from the laboratory for single fluids and for clean matrix or pay formations. The value Δt is measured by the log. Compacted intergranular porosity is assumed and data may be corrected when formations are at shallow depth and not sufficiently compacted. The sonic tool tends to ignore secondary porosity and may skip as it attempts to cross fracture planes and vugs. Fractures and gas complicate the interpretation. It is useful in separating salt, gypsum, and anhydrate from more usual reservoir rocks. Basic values are shown in the following table:

Log-Derived Clay Content Indicators (after Ferti and Frost, *JPT*, September 1980, © SPE-AIME)

Logging curve	Mathematical relationship	Favorable conditions	Unfavorable conditions
SPONTANEOUS POTENTIAL (SP-Curve)	$V_{cl} = 1.0 - (PSP/SSP) + 1.0 - \alpha$ $V_{cl} = (PSP - SP_{min})/(SSP - SP_{min})$ $V_{cl} = 1.0 - C - \alpha$ $1.0 - \alpha = \log A/\log\{(A - V_{cl} \cdot B)/(1 - V_{cl} \cdot B)\}$ where $A = R_t/R_{xo}$, $B = R_t/R_{cl}$ $1.0 - \alpha = (K.V_{cl}.W)/(K.V_{cl}.W + \phi S_{xo})$	Waterbearing, laminated shaly sands ($<R_t$) $c < 1.0$ as function of clay type Knowledge of several parameters required including α, R_t, R_{xo}, R_{cl}. Similar limitations as for straight forward SP-equations K = log derived coefficient, W = clay porosity from bulk and matrix ρ_{cl}; S_{xo} = flushed zone water saturation; laboratory-derived, too many requirements.	R_{mt}/R_w approaches 1.0. Thin, $\gg R_t$ zones Hydrocarbon bearing. Large-electrokinetic and/or invasion effects.
GAMMA RAY	$V_{cl} = (GR - GR_{min})/(GR_{max} - GR_{min})$ $V_{cl} = C(GR - GR_{min})/(GR_{max} - GR_{min})$ $V_{cl} = (GR - W)/Z$ $V_{cl} = 0.33(2^{2\,VCL} - 1.0)*$ $V_{cl} = 0.083(2^{3.7\,VCL} - 1.0)*$ *where $VCL = (GR - GR_{min})/(GR_{max} - GR_{min})$	Only clay minerals are radioactive. $C < 1.0$, frequently approximately 0.5 when $V_{cl} < 40\%$ W, Z = geologic area coefficient Highly consolidated and Mesozoic rocks Tertiary clastics	Radioactive minerals other than clays (mica, feldspar, silt). Only potassium-deficient kaolinite present. Uranium enrichment in permeable fractured zones. Radiobarite scales on casing. Severe washouts ($\ll GR$) Younger, unconsolidated rocks Older, consolidated rocks
SPECTRALOG Gamma ray spectral logging provides individual measurements of potassium (K, %) and thorium (Th, ppm) content	$V_{cl} = (A - A_{min})/(A_{max} - A_{min})$ $V_{cl} = C(A - A_{min})/(A_{max} - A_{min})$ $V_{cl} = 0.33(2^{2\,VCL} - 1.0)*$ $V_{cl} = 0.083(2^{3.7\,VCL} - 1.0)*$ *where $VCL = (A - A_{min})/(A_{max} - A_{min})$	Conditions similar to gamma ray discussion A = Spectralog readings (K in %, Th in ppm). A_{min} = minimum value (K or Th) in clean zones. A_{max} = maximum values (K, Th) in essentially pure shales.	Similar to gamma ray discussion. However, uranium enrichment in permeable, fractured zones and radiobarite build up are no limitations. If Th-curve is used, localized bentonite streaks should be ignored.

Log-Derived Clay Content Indicators (Continued)

Logging curve	Mathematical relationship	Favorable conditions	Unfavorable conditions
RESISTIVITY If several resistivity logs are available, use the one which exhibits highest resistivity values in subject well.	$V_{cl} = (R_{cl}/R_t)^{1/b}$ where b = 1.0 b = 2.0 $V_{cl} = \{R_{cl}(R_{max} - R_t)/(T_t(R_{max} - R_{cl}))\}^{1/b}$ $V_{cl} =$ same as above, where $(1/b) = 1.0$ when $R_{cl}/R_t \geq 0.5$ $(1/b) = 0.5/(1 - R_{cl}/R_t)$ when $R_{cl}/R_t < 0.5$	Low porosity zones (carbonate, marls), pay zones with low $(S_w - S_{wir})$. R_{cl}/R_t from 0.5 to 1.0 R_{cl} approaches R_t In clean hydrocarbon bearing zones one calculates $V_{cl} = 0$	High porosity water sand, high R_{cl} — values.
NEUTRON PULSED NEUTRON	$V_{cl} = \phi_N/\phi_{Ncl}$ $V_{cl} = (\phi_N - \phi_{MIN})/(\phi_{Ncl} - \phi_{MIN})$ $V_{cl} = (\Sigma - \Sigma_{min})/(\Sigma_{max} - \Sigma_{min})$ $V_{cl} = (\Sigma_{cl}/\Sigma)(\Sigma - \Sigma_{min})/(\Sigma_{max} - \Sigma_{min})$	High gas saturation or very low reservoir porosity ϕ_{min} can be varied. Fresh water environment low porosity and gas bearing zones. V_{cl} calculates zero in clean zones.	ϕ_{Ncl} is low.
DENSITY-NEUTRON	$V_{cl} = \dfrac{\rho_B{}^*(\phi_{Nma} - 1.0) - \phi_N{}^*(\rho_{ma} - \rho_f) - \rho_f{}^*\phi_{Nma} + \rho_{ma}}{(\rho_{sh} - \rho_t)^*(\phi_{Nma} - 1.0) - (\phi_{Nsh} - 1.0)^* (\rho_{ma} - \rho_f)}$		Too low V_{cl} in prolific gas zones. Don't use with severe hole conditions. Lithology affected.
DENSITY-ACOUSTIC	$V_{cl} = \dfrac{\rho_B{}^*(\Delta t_{ma} - \Delta t_f) - \Delta t^*(\rho_{ma} - \rho_f) - \rho_f{}^*\Delta t_{ma} + \rho_{ma}{}^*\Delta t_f}{(\Delta t_{ma} - \Delta t_f)^*(\rho_{sh} - \rho_f) - (\rho_{ma} - \rho_f)^*(\Delta t_{sh} - \Delta t_f)}$	Less dependent on lithology and fluid conditions than DEN-NEU crossplot Use in gauge boreholes.	Badly washed out, wellbores Highly undercompacted formations (shallow, overpressures).
NEUTRON-ACOUSTIC	$V_{cl} = \dfrac{\phi_N{}^*(\Delta t_{ma} - \Delta t_f) - \Delta t^*(\phi_{Nma} - 1.0) - \Delta t_{ma} + \phi_{Nma}{}^*\Delta t_f}{(\Delta t_{ma} - \Delta t_f)^*(\phi_{Nsh} - 1.0) - (\phi_{Nma} - 1.0)(\Delta t_{sh} - \Delta t_f)}$	Use only in gas bearing zones with low S_w.	Similar effects due to shaliness on both logs.

The density tool relates with the electron density of the formation. For common formation materials, the electron density is proportional to actual density. The porosity is derived from the bulk density of clean liquid-filled formations by the following equations:

$$\phi = (\phi_{ma} - \phi_b)/(\phi_{ma} - \phi_{liq})$$

where ϕ_{ma} is the matrix density, ϕ_{liq} is the liquid density, and ϕ_b is from the log. The determination is complicated by shale and gas in the formation, and a combination of porosity logs is necessary to resolve the problems.

The neutron tool responds chiefly to the presence of hydrogen atoms. The response is basically a measure of porosity when the pore space is filled with liquids. The log is usually scaled on the basis of either a limestone or a sandstone matrix by suitable use of proper equipment in the logging truck. A different lithology may require the use of corrections. Shale and gas effect the porosity recorded, and corrections must be made as necessary.

The resistivity tool can be used to estimate porosity in clean, water-bearing sands by using the following equation:

$$F = R_o/R_w \quad \text{or} \quad F = R_{xo}/R_{mf}$$

Archie also showed that $F = a/\phi^m$. When circumstances are proper, the flushed zone readings in the hydrocarbon bearing formation can be used with the formula:

$$F = S_{xo}^2 \ (R_{xo}/R_{mf})$$

Empirical data indicate that $S_{xo} = S_w^{1/5}$ when S_{xo} is the average residual oil saturation. Also, it is possible to plot any of the porosity tool results or actual porosity versus R_t on log-log paper and obtain porosity by calibrating the porosity scale at a value of one or log = 0 at aR_w, as shown in Fig. 2–20.

The presence of clay within the pores or intermixed as thin beds among the sand can be determined from crossplots such as the neutron-density crossplot, the sonic-density crossplot, the gamma-ray log, the SP log, and the resistivity logs. The density-neutron crossplot is used as a total shaliness indicator when the formation matrix lithology is constant, preferably known. The density-sonic crossplot is useful when sands are compacted and sands combine with shale laminae. The gamma-ray log usually results in an upper limit for shaliness when no nonshale radioactive minerals are present. The method is used to illustrate the qualitative application. Values of gamma ray for adjacent shales and clean sands are read from the respective logs. Care must be taken to be certain that no unconformity is present in the interval of interest. Then:

$$V_{clay} = (GR_{log} - GR_{sand})/(GR_{shale} - GR_{sand})$$

The SP log, if read properly and representative of the section, may be used in a similar manner in water-bearing sands. In hydrocarbon-bearing sands, the SP will give an upper limit for shale content. The resistivity logs are also influenced by shale in a more complicated manner, as discussed later.

In nature, shale and sand sequences are often found as separate beds. Thickness of the layers of shale, sand, and limestone vary, depending upon the geolog-

ical conditions existing at time of deposition. Since most beds were at that time subject to wind and wave action, substantial intermixing, as presently found in cores and by logging, is to be expected.

The porosity logs in shaly formations may be analyzed using methods devised by the respective logging companies. The following discussion, consolidated from various instructions by Schlumberger, is presented as an illustration of the general method.

The Porosity Logs in Shaly Formations*

The responses of density and neutron logs to laminar and structural shales are taken to be the same as to the nearby bedded shales. Their responses to dispersed shale may also be taken to be the same, if we associate the same amounts of bound water and impurities with the dispersed shale as with the bedded shales.

On the other hand, sonic log response to interstitial dispersed-shale is quite different from its response to laminar or structural shale.

The Neutron Log

The neutron log responds to all the hydrogen present in a shaly formation (including the hydrogen of water bound in the shale). If hydrocarbons are present, and assuming the sonde was calibrated in fresh-water-bearing formations, the Neutron porosity reading, ϕ_N, is given by:

$$\phi_N = \phi S_{xo}\phi_{Nmf} + V_{sh}\phi_{Nsh} + \phi S_{hr}\phi_{Nh} - \Delta\phi_{Nex} + (1 - \phi - V_{sh})\phi_{Nma}$$

Where:

ϕ is the actual porosity of the shaly formation (excluding dispersed shale and associated bound water)

S_{xo} is flushed-zone water saturation

S_{hr} is the residual hydrocarbon saturation in the region investigated by the tool

ϕ_{Nmf} is the response of the neutron log to the water in the formation investigated, presumed to be invasion filtrate. If the water is fresh, $\phi_{Nmf} \approx 1$

ϕ_{Nsh} is the neutron log response to the shale (determined, for example, from log readings in the adjacent shales). The value of ϕ_{Nsh} will vary with locality and formation

ϕ_{Nh} is the neutron log response to the hydrocar-

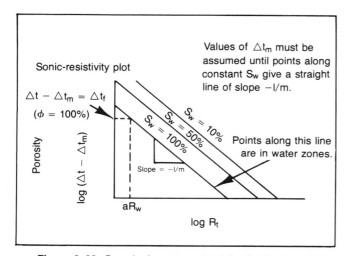

Figure 2–20. *Porosity from crossplot (after Schlumberger).*

* This section taken from Schlumberger.

bon. For usual oil densities the hydrogen index of oil will be close to that of water. $\Delta\phi_{Nex}$ is the excavation effect on the neutron log ϕ_{Nma} is the neutron response to the matrix (lithology effect). If the sonde is calibrated in limestone formations, lithology effect in limestone will be zero, and this term disappears. Similarly, sandstone calibration makes lithology effect in sandstone essentially zero

In terms of the clay parameters

$$\phi_N = \phi S_{xo}\phi_{Nmf} + V_{clay}\phi_{Nclay} + \phi S_{hr}\phi_{Nh} - \Delta\phi_{Nex} + (1 - \phi - V_{clay})\phi_{Nma}*$$

where $\phi_{Nma}*$ is the neutron response to the nonclay solids.

The Density Log

The relation for bulk density is:

$$\rho_b = \phi(S_{xo}\rho_{mf} + S_{hr}\rho_{ha}) + V_{sh}\rho_{sh} + (1 - \phi - V_{sh})\rho_{ma}$$

where ρ_{ma} is the density of the shale-free fraction of the matrix and ρ_{ha} is the apparent density of the hydrocarbon. This equation is sometimes written as:

$$\rho_b = \phi\rho_{mf} - \phi S_{hr}(\rho_{mf} - \rho_{ha}) + V_{sh}(\rho_{sh} - \rho_{ma}) + (1 - \phi)\rho_{ma}$$

where $- \phi S_{hr}(\rho_{mf} - \rho_{ha})$ is the hydrocarbon effect on the Density Log, and $V_{sh}(\rho_{sh} - \rho_{ma})$ is the shaliness effect.

In terms of V_{clay},

$$\rho_b = \phi(S_{xo}\rho_{mf} + S_{hr}\rho_{ha}) + V_{clay}\rho_{clay} + (1 - \phi - V_{clay})\rho_{ma}$$

A relation for ϕ_D, the apparent porosity seen by the density log, can be derived

$$\phi_D = \phi(S_{xo} + S_{hr} \cdot \phi_{Dh}) + V_{sh}\phi_{Dsh}$$

where ϕ_{Dh} and ϕ_{Dsh} are the apparent density-log porosities of the hydrocarbon and the adjacent shales:

$$\phi_{Dh} = \frac{\rho_{ma} - \rho_{ha}}{\rho_{ma} - \rho_{mf}}$$

$$\rho_{Dsh} = \frac{\rho_{ma} - \rho_{sh}}{\rho_{ma} - \rho_{mf}}$$

For shaly sands, ρ_{ma} is taken to be 2.65. The value ϕ_{Dsh} is generally smaller than 0.2 and may sometimes be very small or negative. In some areas field experi-

ence shows that it is permissible in water sands and filtrate-invaded sands to take $\phi = \phi_D$. (This is equivalent to taking $\rho_{sh} = \rho_{ma}$.)

Sonic Log

In *compacted* formations hydrocarbon saturation is usually seen by the sonic as if it were water, so

$$\Delta t = (1 - \phi - V_{sh}) \Delta t_{ma} + (V_{lam} + V_{str}) \Delta t_{lam} + (\phi + V_{dis}) \Delta t_f$$

and

$$\phi_S = \frac{\Delta t - \Delta t_{ma}}{\Delta t_f - \Delta t_{ma}} = \phi + V_{dis} + (V_{lam} + V_{str})(\phi_S)_{lam}$$

In *uncompacted sands*,

$$\phi_S = C_p(\phi + V_{dis}) + (V_{lam} + V_{str})(\phi_S)_{lam}$$

where C_p is a correction factor which is greater than 1 in uncompacted formations. Its value is obtained by comparing ϕ_S and ϕ_N values in clean, liquid-filled sands.

In uncompacted gas sands ϕ_S may be appreciably larger than predicted.

In these equations, Δt_{lam} and $(\phi_S)_{lam}$ refer to the sonic response to laminar and structural shales, and

$$(\phi_S)_{lam} = \frac{\Delta t_{lam} - \Delta t_{ma}}{\Delta t_f - \Delta t_{ma}}$$

Δt_{lam} is read from the adjacent shales: the value of $(\phi_S)_{lam}$ ranges between 0.2 and 0.4 in most cases, but may be as low as 0.1 for carbonate shales.

For shaly sandstone Δt_{ma} ranges from 55.5 to 51.2 μsec/ft ($v =$ 18,000 to 19,500 ft/sec).

Dispersed-Shale Simplified Model

The sonic response to dispersed shale is taken equal to that of water ($\Delta t_{dis} = \Delta t_w$ and $(\phi_S)_{dis} = 1$). Laboratory data and field experience show this approximation to be usually valid for values of q ($= V_{dis}/\phi_{im}$) up to 40 or 50%. (For higher values of q the formation is considered to have too low permeability to be a commercial producer.)

When only dispersed shale is present the sonic response is described by:

$$\phi_S = \phi + V_{dis} = \phi_{im}$$
(in compacted formations)

$$\phi_S = C_p(\phi + V_{dis})$$
(in uncompacted formations)

Again in uncompacted gas sands ϕ_S may be appreciably larger than predicted.

In areas where, for water-bearing and filtrate-invaded formations, $\phi_D = \phi$ (i.e., $\rho_{sh} = \rho_{ms}$),

$$\phi_S - \phi_D \approx V_{dis} = q\phi_{im}$$

$$\frac{\phi_S - \phi_D}{\phi_S} \approx \frac{V_{dis}}{\phi_{im}} = q$$

This is the basis of the "Q" log ($Q = (\phi_S - \phi_D)/\phi_S$), which is useful in many areas as an indicator of permeability as affected by dispersed shale.

Neutron-Density Crossplot

When the lithology is known and uniform (and the neutron log is calibrated accordingly), and the formations investigated by the logs are water saturated, the responses of neutron and density logs become

$$\phi_N = \phi + V_{sh}\phi_{Nsh}$$

$$\rho_b = \rho_{ms}(1 - \phi - V_{sh}) + \rho_{sh}V_{sh} + \rho_f\phi$$

or

$$\phi_D = \phi + V_{sh}\phi_{Dsh}$$

These equations are written assuming the neutron and density responses to laminated shale and dispersed shale are the same. This means V_{sh} is a bulk-volume total shale measurement regardless of the form of the shale. The porosity equations can be solved graphically for porosity, ϕ, and bulk-volume total shale fraction, V_{sh}, by means of a crossplot.

Direct use of the crossplot assumes 100% water saturation in the zone investigated by the tools. Introduction of gas or light hydrocarbons decreases ϕ_N and decreases ρ_b (increases ϕ_D). This would cause the point to shift in a northwesterly direction. When gas or light hydrocarbons are present an additional shaliness indicator, such as gamma ray or SP, is needed in order to evaluate the amount of this shift.

When the shale is composed of wet clay and silt, in terms of bulk-volume fractions:

$$V_{sh} = V_{silt} + V_{clay}$$

Furthermore, we may characterize the shale by a silt index, I_{silt}, given by:

$$I_{silt} = V_{silt}/V_{sh}$$
$$V_{clay} = V_{sh}(1 - I_{silt})$$
$$\phi_{Nsh} = \phi_{Nclay}(1 - I_{silt})$$
$$\phi_{Dsh} = \phi_{Dclay}(1 - I_{silt})$$
$$\phi_{Ssh} = \phi_{Sclay}(1 - I_{silt})\phi_{Ssilt} \times I_{silt}$$

The neutron-density crossplot can be used, as illustrated in Figs. 2–20 through 2–23. The density of

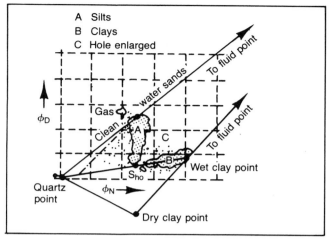

Figure 2–21. *Plot of field data from logs for a silty-shaly sand using density and neutron porosity. Field values designated as A are silty shales, while B are wet clays (after Schlumberger).*

sand or quartz is known to be 2.65 and its porosity is zero. Likewise, fresh water is known to have a density near 1.0, possibly as high as 1.1 if very salty, and this density may be related with a completely filled pore space of 100% for the condition when no matrix is present. A clean water-sand line can be drawn between the points, and this line can be directly calibrated in porosity units by simple proportion. The wet clay points are found below this line and are designated as *B* on Fig. 2–21. A line is drawn from the quartz point through the average of the wet-clay values. The limit or end of this line is the wet-clay point and is taken to be near the end of the wet-clay values, possibly except for a few points, which are believed to be nonrepresentative. The wet-clay point is connected to the fluid point. If neither the porosity or the density of the true dry clay is determined in the laboratory, the dry-clay point can be established on the line. It is obvious that the porosity measured in the laboratory, based on the use of cores, relates with porosity lines parallel to the line connecting the quartz and dry shale point. The porosity of the wet clay, whether measured in the laboratory or by logs, must be less than the dry porosity since shales are known to accept water and swell. Figs. 2–22 and 23 show the use of the method for determining porosity and clay content, respectively.

Shale also retards flow of fluids, and Fig. 2–24, which is similar to that used by Fertl of Dresser Atlas in various publications, is a good example. Here, shale is based on the use of the Q factor. Fertl has also assembled shale equations and shale characteristics in the following table:

Clay Mineral Characteristics (after Fertl and Frost, *JPT*, September 1980, © SPE-AIME)

Clay minerals	Composition	Remarks	Density $\rho cl(g/cm^3)$	Hydrogen index, HI	Cation exchange capacity C_{ec} (meq/100g)	Spectral gamma-ray distribution (avg.) Potassium K(%)	Uranium U(ppm)	Thorium Th(ppm)
Chlorite	$(Mg,Al_4,Fe)_{12}[(Si,Al)_8O_{20}](OH)_{16}$	Low water absorptive properties. As coating and/or pore bridging in reservoir pore space. Small effect on resistivity measurement due to moderate surface area.	2.60–2.96	0.34	10–40			
Illite	$K_{1-1.5}Al_4[Si_{7-6.5}Al_{1-1.5}O_{20}](OH)_4$	No absorbed water. As coating (pore-lining) and/or pore bridging in reservoir, pore space. Reduce resistivity measurements. Moderate surface area.	2.64–2.69	0.12	10–40	4.5	1.5	<2.0
	$\begin{cases} K_2(Mg,Fe^{+2})_{6-4}(Fe^{+3},Al,Ti)_{0-2} \\ [Si_{6-5}Al_{2-3}O_{20}]O_{0-2}(OH,F)_{4-2} \end{cases}$	Biotite As pore bridging and/or thin mica seams as laminae in reservoir rock.	2.7–3.2	0.12		6.7–8.3		<0.01
	$K_2Al_4[Si_6Al_2O_{20}](OH,F)_4$	Muscovite Drastic effect on vertical permeability.	2.76–3.0	0.13		7.9–9.8		<0.01
Kaolinite	$Al_4[Si_4O_{10}](OH)_8$	"Patchy" Kaolinite as discrete particles in reservoir pore space. As migrating fines creating internal formation damage. Small effect on resistivity measurements. Low surface area.	2.61 (Theoretical density); 2.60–2.68 (Extensive Literature); 2.63 (Most frequently quoted)	0.36	3–15	0.42	1.5–3.0	6–19
Smectites	$(\tfrac{1}{2}Ca,Na)_{0.7}(Al,Mg,Fe)_4(Si,Al)_8O_{20}(OH)_4$	Montmorillonite, montronite. As coating and/or pore bridging in reservoir pore space. Critical to physical and chemical formation damage. Large reducing effect on resistivity measurements. High surface area.	2.20–2.70	0.13	80–150	0.16	2.0–5.0	14–24
		Low-iron smectite	2.53					
		3.6% iron content	2.74					
		Bentonite				<0.5	1–20	6–50

*Data in this table compiled from References 7 and 19–21.

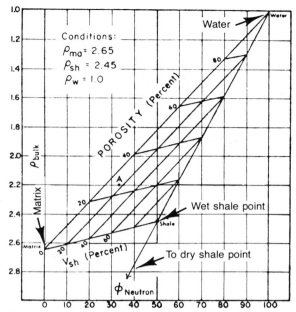

Figure 2–22. *Use of bulk density and neutron logs to obtain porosity in a sand-shale environment. Wet shale porosity is not comparable to core porosity (after Schlumberger).*

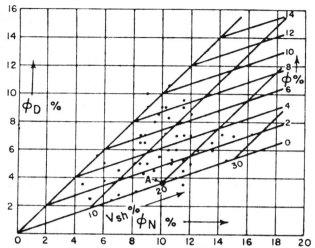

Figure 2–23. *Use of density and neutron logs to determine shale content (courtesy Schlumberger).*

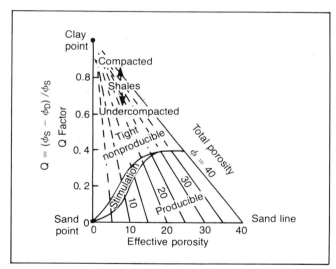

Figure 2–24. *Use of Q factor and effective porosity to determine producible pay intervals as proposed by Fertl of Dresser Atlas (Oil and Gas Journal, 3 July 1978).*

Water Saturation in Porous Rocks

Water saturation usually is obtained from resistivity logs and capillary pressure information. Routine core analysis determines the fluids remaining in the cores, and special coring techniques may be used to determine in-situ water using oil-base drilling muds when the in-situ water is immobile. Also, the production history includes determination of water produced that can be related to in-place water by use of relative permeability and analogy.

Porous formations contain water when buried even at shallow depths. Many formations were deposited in a marine environment and all rocks are saturated with water mixed with air (gradient 0.3–0.5 psi/ft).

Oil and gas from shales and other possible source rocks migrated with water as mobile fluids were displaced from the ever-compressing rocks as depth of burial increased. Knowledge as to source rocks and fluid migration known to regional geologists is useful. Generally, oil and water migrated from shales as a result of the reduction is porosity associated with the compression of rocks as burial depth increased. Gravity and other forces caused the oil to become small droplets, which gradually increased in size and collected on structure and other traps to form oil accumulations. Oil stains are frequently found near the top of sand stringers at downdip locations.

In most reservoirs, the oil has displaced the in-situ water from the pore space so the competition between gravity and capillary forces is in equilibrium. The water saturation is highest near the water-hydrocarbon contact and gradually decreases with height above such contact in water-wet pays. Formation rocks may also be oil-wet, and most formations containing oil are believed to have relatively neutral wetting characteristics. The capillary-gravity forces have adjusted during millions of years, and many oils contain surfactants that tend to change wettability in favor of the oils. Capillary forces cause water content to be higher at the same elevation in tighter rocks than in the porous and permeable formations.

Gas can displace both water and oil from traps. Oil can displace water but usually not gas. A trap once filled with gas will continue to be gas filled even if oil later migrates under the gas column in water-wet formations. The moving water—possibly oil—can dissolve gas from gas and oils so that tars and heavy oils are at times found near the water-oil contact. Core-handling procedures are capable of changing wettability of rocks, and laboratory wettability tests should not be accepted without careful analysis. The water content of some gas reservoirs is less than 10%, which might suggest some movement of in-situ water

by convection of the in-place gas. In oil reservoirs, free gas usually is found as a cap on top of the oil when excess gas is available. The gas-oil ratios may either increase or decrease with depth of the oil accumulation as required to satisfy gravity, capillary, and migration conditions. Shale stringers can separate oil accumulations so that oil characteristics may differ rapidly with depth, particularly when two or more sources of the oil are contributing to the oil accumulations. In areas where the depth of burial is decreasing, capillary and gravity forces are probably in rather good equilibrium. If burial depth is increasing today, migration conditions also must be evaluated.

The oil, gas, and water content of cores is routinely determined during laboratory analysis. Since in-situ fluids are usually partly displaced by the drilling fluids and fluid expansion as the core is brought to the surface further changes the fluid content remaining in the cores, empirical methods must be used to correct the laboratory data to obtain some indication of in-situ conditions. Examples of these methods are presented as Figs. 2–24a, 25. The mud fluid invasion is more substantial when gas is present and in semidepleted reservoirs. A large pressure differential between the drilling column and the reservoir will distort the correction techniques. A high GOR expels more fluids.

Geological conditions can change with time so that traps can later leak and leave only a residual oil with little or no gas content. Earth movements and moving water can tilt the traps so that equilibrium conditions are disturbed and the in-situ water level found at discovery does not satisfy the laws of gravity. The engineer and geologist working together can usually develop a consistent and logical explanation for the available facts encountered in each area.

Coring techniques using oil-base mud can accurately determine in-situ water content of formations when the in-place water is low and immobile. Low loss mud is desired, and samples analyzed should be taken at the center of the core to reduce invasion effects. Oil-base muds usually are not miscible with the in-place water, but the mud fluids at times do displace water, particularly near the water-oil contacts. Careful planning is necessary when special coring techniques and pressure core barrels are to be used. Water content at the edge of the core also should be determined.

Resistivity logs are the most frequently used method for determining in-place water at any time. Manuals prepared by the respective logging companies should be consulted and their engineers are avail-

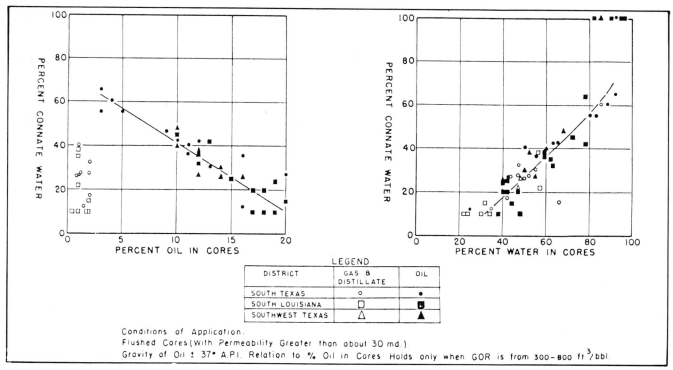

Conditions of Application:
Flushed Cores (With Permeability Greater than about 30 md.)
Gravity of Oil ± 37° A.P.I. Relation to % Oil in Cores Holds only when GOR is from 300–800 ft³/bbl.

Figure 2–24a. *Conversion of water from core analysis to in-situ water in formation (Archie).*

able to help answer questions. For convenience, some of the methods used by Schlumberger are presented as examples in the following pages.

The water saturation from resistivity tools method is based on the experimental fact that the resistivity of a shale/clay-free formation is proportional to the

Water Saturation from Resistivity Tools (courtesy Schlumberger)

Log	Application	Optimum R_a range	Optimum environment	Limitations	Optimum geologic environment
ES	R_t, R_i	1–500	$t > 10'$	Interbedded thin beds, invasion, low R_m's	thick sand-shale sequences
LL-7	R_t	1–10,000	resistive beds $t > 2' - 3'$	calibration deep invasion	carbonates salt muds
LL-3 (guard)	R_t (conductivity)	1–200	thin beds, $t > 2'$ moderate R_t's	high resistivities calibration deep invasion	carbonates salt muds
IL (6FF40, 5FF40, ILM)	R_t, R_i	0.1–100	low R_t's intermediate thickness ($t > 5'$, preferably $t > 10'$) deep investigation	high resistivities thin-bed effects complexity of interpretation	sand-shale sequence $R_t < 100$ ohm-m
ML	permeability indicator	0.5–100	invaded beds smooth borehole w/mud cake	qualitative only	sand-shale sequence
MLL	R_{xo}	0.5–100	invaded beds mudcake $< \frac{1}{4}''$	mudcake effects rough borehole effects	medium porosity invaded carbonates
Proximity	R_{xo}	0.5–100	invaded beds	rough borehole calibration $R_a > R_{xo}$	medium porosity invaded sands or carbonates

resistivity of the brine with which it is fully saturated. This constant of proportionality is known as the formation resistivity factor, F. If $R_t = R_o$ is the logged resistivity of a non-shale formation that is 100% saturated with formation brine having a resistivity of R_w, $F = R_t/R_w$. For a given porosity, this ratio remains nearly constant for all values of R_w below about one ohmmeter. In relatively nonsaline waters, the value of F is reduced as R_w increases and as the grain size of the sand decreases.

The porosity, ϕ, of a rock is the fraction of the total volume occupied by the voids or pores. The formation factor is a function of porosity, pore structure, and pore size distribution. Archie proposed that

$$F = a/\phi^m$$

where a is a constant and m is the cementation factor, both determined empirically or by laboratory measurements. Representative values for m are given as follows:

Typical *m* and *n* Values as Defined by Archie

Pay	Lithology	Ave. m	Ave. n
Wilcox, US Gulf Coast	ss	1.9	1.8
Sparta, S. La. (Opelussea)	ss	1.9	1.6
Cockfield, S. La.	ss	1.8	2.1
Government wells, S. Texas	ss	1.7	1.9
Frio, S. Texas	ss	1.8	1.6
Miocene, S. Texas	Cons. ss	1.95	2.1
	Uncon. ss	1.6	2.1
Travis Peak and Cotton Valley	ss	1.8	1.7
Rodessa, E. Texas	ls	2.0	1.6
Edwards, S. Texas	ls	2.0	2.8
Woodbine, E. Texas	ss	2.0	2.5
Annona, N. La.	Chalk	2.0	1.5
Nacatoch, Arkansas	ss	1.9	1.3
Ellenburger, W. Texas	ls and Dol.	2.0	3.8
Ordovician Simpson, W. Texas and New Mexico	ss	1.6	1.6
Pennsylvanian, W. Texas	ls	1.9	1.8
Permian, W. Texas	ss	1.8	1.9
Simpson, Kansas	ss	1.75	1.3
Pennsylvanian, Okla.	ss	1.8	1.8
Bartlesville, Kansas	ss	2.0	1.9
Mississippian, Illinois	ls	1.9	2.0
	ss	1.8	1.9
Pennsylvanian, Illinois	ss	1.8	2.0
Madison, No. Dakota	ls	1.9	1.7
Muddy, Nebraska	ss	1.7	2.0
Cretaceous, Saskatchewan, Canada	ss	1.6	1.6
Bradford, Pennsylvania	ss	2.0	1.6
Frio, Chocolate Bayou, La.	ss	1.55–1.94	1.73–2.22
Agua Dulce, S. Texas	ss	1.71	1.66
Edinburgh, S. Texas	ss	1.82	1.47–1.52
Hollow Tree, S. Texas	ss	1.80–1.87	1.64–1.69
Jackson, Cole Sd., S. Texas	ss	2.01	1.66
Navarro, Olmos & Delmonte, So. Texas	ss	1.89	1.49
Edwards Lime, Darst Creek, So. Texas	ls	1.91–2.02	2.01–2.08
Viola, Bowie Field, N. Texas	ls	1.77	1.15
Lakota, Crook Co., Wyoming	ss	1.52	1.28

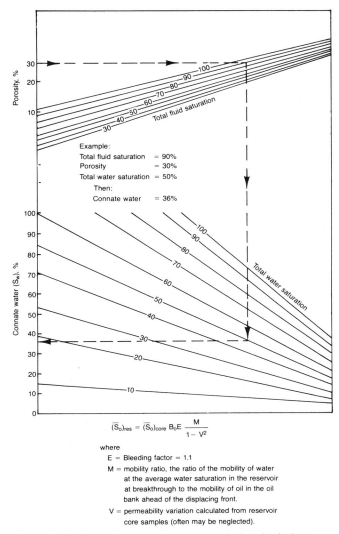

Figure 2–25. *Converting water from core analysis to in situ (courtesy Petroleum Engineer).*

$$(\overline{S}_o)_{res} = (\overline{S}_o)_{core}\, B_o E\, \frac{M}{1 - V^2}$$

where

E = Bleeding factor = 1.1

M = mobility ratio, the ratio of the mobility of water at the average water saturation in the reservoir at breakthrough to the mobility of oil in the oil bank ahead of the displacing front.

V = permeability variation calculated from reservoir core samples (often may be neglected).

As a first approximation, values for m and n are both taken as 2 with a equal to one, as proposed by Archie in early 1940s. The Humble values use 0.62 for a and 2.15 for m so that values for F are the same when porosity is 5% but differ so that the Archie value is 1.35 times the Humble values when porosity is 30%. Laboratory measurements indicate that m has a value of about 1.3 for unconsolidated sands and oolitic limestones, a value of 1.4–1.5 when sands are very slightly consolidated, a value of 1.6–1.7 for friable or slightly cemented sands, and a value of 1.8–1.9 plus for firm and consolidated sands with porosity less than 15%. Lower porosity sands, intergranular porosity-type limestones, dolomites, and chalks have values of m in the range on 2.0 to 2.2.

The values of n are in a range of 1.5 to 3.0 with typical values from 1.8 to 2.0. Oil-wet rocks often have values from 3 to 4.

The presence of a water-oil contact or adjacent water-bearing sands having comparable characteristics as to both pay and water can be used to calculate both m and a using the approach in Fig. 2–28.

If S_w is the fraction of the pore volume occupied by formation water, $1 - S_w$ must be the fraction occupied by either gas or oil, both being electrical insulators. Experiments by Archie using clean formations showed that the logged resistivity of formations containing hydrocarbons and in-situ water, R_t, may be related with the in-situ water by use of the following equation:

$$S_w^n = (FR_w)/R_t$$

where n is the saturation exponent and is generally near 2. $FR_w = R_t$ when the formation is 100% saturated with water having a resistivity of R_w. The Archie equation usually is rearranged in the form

$$S_w = \sqrt{(FR_w)/R_t}$$

Representative values for F are in Fig. 2–26.

The equation can easily be solved with today's hand-held computer, but in 1940 Archie was using the two curves shown as Fig. 2–27. The various service logging companies also published slide rules for quick solution when conditions were ideal so that the simple equation was applicable. Methods for determining R_w will be discussed later in this section. The Archie curves quickly evaluate a possible range of values.

When water-based muds are employed while drilling an oil-gas-bearing formation, the drilling fluids often displace the in-situ water by a miscible process so that the water in the pores near the drill bore is changed to that of the drilling mud filtrate. The drilling mud filtrate also moves some oil, which may at times equal the oil displaced during waterflood operations. If the oil remaining after this displacement process is defined as S_{xo},

$$S_{xo} = \sqrt{FR_{mf}/R_{xo}}$$

where R_{mf} is the resistivity of the drilling mud filtrate at reservoir conditions and $R_{xo} = R_t$ as measured for the invaded zone by use of a short logging device. Use of logging tools of differing degrees of investigation permits one to evaluate the depth of invasion. The difference between S_w and S_{xo} gives a value for the oil displaced by the mud filtrate in the area around the well bore. Based on field and laboratory data,

Key equations

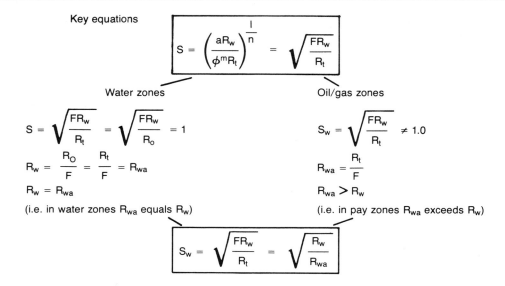

$$S = \left(\frac{aR_w}{\phi^m R_t}\right)^{\frac{1}{n}} = \sqrt{\frac{FR_w}{R_t}}$$

Water zones

$$S = \sqrt{\frac{FR_w}{R_t}} = \sqrt{\frac{FR_w}{R_o}} = 1$$

$$R_w = \frac{R_O}{F} = \frac{R_t}{F} = R_{wa}$$

$$R_w = R_{wa}$$

(i.e. in water zones R_{wa} equals R_w)

Oil/gas zones

$$S_w = \sqrt{\frac{FR_w}{R_t}} \neq 1.0$$

$$R_{wa} = \frac{R_t}{F}$$

$$R_{wa} > R_w$$

(i.e. in pay zones R_{wa} exceeds R_w)

$$S_w = \sqrt{\frac{FR_w}{R_t}} = \sqrt{\frac{R_w}{R_{wa}}}$$

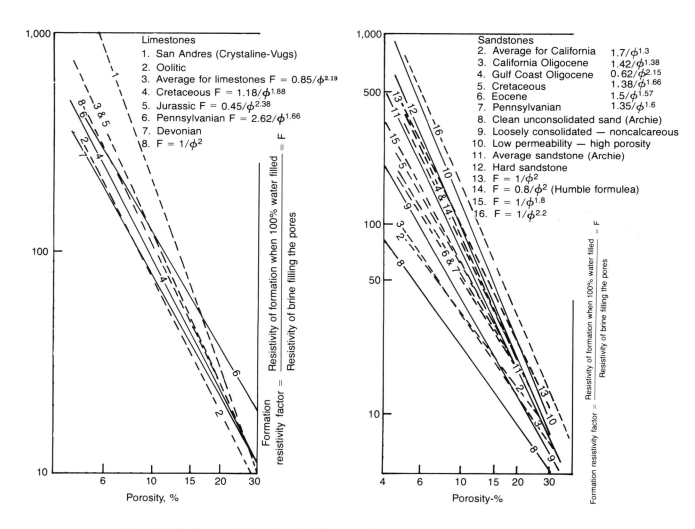

Figure 2–26. *Representative values for F (courtesy* Continuous Tables, Petroleum Engineer 1979).

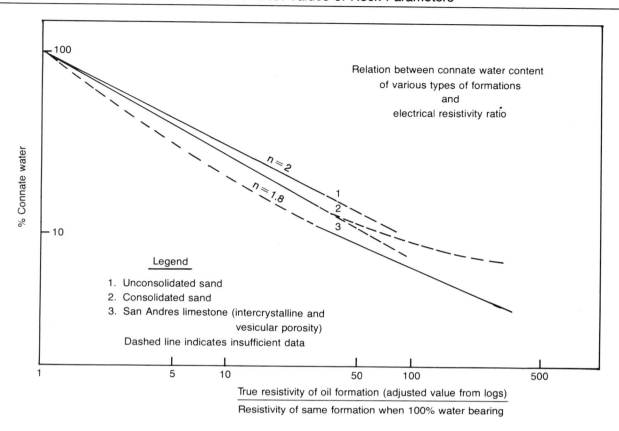

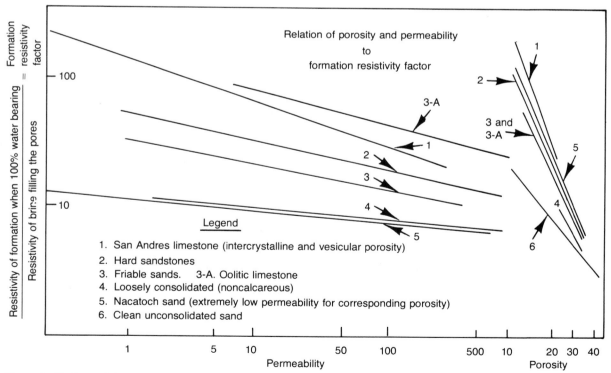

Figure 2–27. *Determining water saturation. Curves prepared by Archie. (courtesy Shell Oil Co.) 1) Determine porosity of zone of interest; 2) Select rock type; 3) Read value for formation resistivity factor; 4) Read value for water resistivity; 5) Read log for true resistivity of zone; 6) Read water saturation from curve. If water-wet sand is present, values for 100% water-filled sand may be read from logs.*

the following empirical relation is often used $S_{xo} = (S_w)^{1/5}$ and $S_w = \left(\dfrac{R_{xo}/R_t}{R_{mf}/R_w}\right)^{5/8}$. Study of the section entitled "Oil Saturation" suggests that the displacement of oil by water is more complicated.

Several types of physical displacements are involved, as discussed in later sections of this text. The displacement under these conditions is fast and pressure controlled so that the data represent conditions near an injection well and may not be applicable in most of the volume located at a distance from the well where capillary forces enter the displacement process. Sections dealing with capillary pressure and recovery mechanisms should be reviewed.

Pickett, *JPT* (November 1966), reviewed the methods for determining water saturation from logs. Equations for porosity from logs can be combined with the resistivity equations and thereby eliminate the determination of values for many of the terms in the respective equations. The general approach is presented as Fig. 2–28 where porosity is plotted versus field logged values of R_t to determine both a and m values when values include a water-saturated zone.

The thickness of the oil- or gas-bearing beds rela-

Porosity — resistivity plots.

Given
1. Resistivity device to determine R_t
2. Porosity device

Equations for porosity-resistivity

$$S_w{}^n = \frac{aR_w}{\phi^m R_t} \quad \text{(Archie's equation)}$$

Taking the log of both sides
n log S_w = log(aR_w) − mlogϕ − log R_t
Rewriting the equation:
log ϕ = l/m log aR_w − n/m log S_w − l/m log R_t

Porosity-resistivity plots

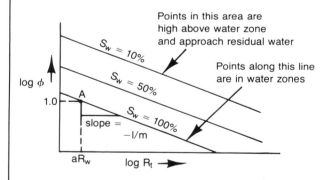

Points in this area are high above water zone and approach residual water

Points along this line are in water zones

Discussion
(1) Letting S_w = 100% and ϕ = 100% then point A represents an R_t which is equal to aR_w. See above equation.
(2) For a constant S_w the line through the points has a slope of −(l/m).
(3) When porosity is not known but porosity response values are known, the equation can be rewritten.

The porosity logs may be substituted for porosity as illustrated using the sonic log.

Equations for sonic versus resistivity
Example (Sonic log):

$$\phi = \frac{\triangle t - \triangle t_m}{\triangle t_f - \triangle t_m}$$

$$\log (\triangle t - \triangle t_m) = \log (\triangle t_f - \triangle t_m) + \frac{l}{m} \log aR_w$$
$$- \, n/m \log S_w - l/m \log R_t$$

Advantages
(1) No knowledge of aR_w is required if water zone is present
(2) No knowledge of cementation exponent, m, is required
(3) Exact porosity response equation need not be known. Data accuracy, however, does not usually warrant slight variation in porosity response equations.
(4) Lithology may be apparent in slope changes.

Disadvantages
If water zone is not available, this method requires estimates of an R_w for calculation of water saturation.

Comments
(1) If a water saturation other than 100% S_w is known (i.e., connate water saturation) then a value for the saturation exponent, n, can be calculated independently, see plot and equations.
(2) This graphical technique can also be used in evaluatory S_{xo} to movable oil saturation. Plot log ϕ vs log R_{xo}. The intercept at ϕ − 100% would be equal to aR_{mf}.

Figure 2–28. *Use of porosity and resistivity logs to determine parameters used to calculate water saturation (courtesy Petroleum Engineer).*

tive to the logging tool required a suitable correction in the R_t values. Modern tools have been designed to eliminate this correction, but older logs being reviewed in secondary or EOR projects must be properly adjusted. Some minerals present in the sands conduct electricity and the short circuit caused by pyrite and iron must be recognized. Fortunately, these minerals also can be recognized by the porosity logs. Shales complicate the analysis of all logs, and the Schlumberger method for considering the influence of shales and clays is summarized in Fig. 2–29.

The logging equations may be simplified as justified by the field data and desired accuracy. The most accurate value for each term in the equations is desired. The method of simulation may be used to calculate various theoretical curves that may be matched to obtain a best fit, based on the equations and data. For shaley sands, Miyairi et al. in "Water Saturation in Shaley Sands," SPWLA Seventeenth Annual Logging Symposium, June 1976, used the simulation technique to conclude that a plot of porosity versus Y in a 100% water-bearing zone gives a straight line, the formation factor line. The slope of this line and its intercept with 100% porosity give values for m and a, respectively. The Y term is

$$Y_1 = \tfrac{1}{2} m \log \phi - \tfrac{1}{2} \log a = \sqrt{R_w/R_t} - V_{cl}\sqrt{R_w/R_{cl}}$$
$$Y_2 = m \log \phi - \log a = R_w/R_t - V_{cl} R_w/R_t$$

The correct equation can be selected from the results obtained.

The water content of oil-bearing pore space when the height above the water-oil contact is substantial often relates with pore size. Then the capillary pressure curve shows a constant water saturation versus depth for each rock type or porosity. This water saturation also is considered to approximate irreducible values for the effective displacement process. For these conditions, $(\phi\, S_w)$ is frequently a constant. Then

$$Z = V_{cl}\sqrt{R_t/R_{cl}} + \phi^{n/2}\sqrt{R_t/aR_w}$$
$$\text{Log } Z = \tfrac{1}{2} n \log \phi - \tfrac{1}{2} n \log (\phi\, S_w)$$

A plot of ϕ versus Z on log-log paper results in a straight line, and n can be calculated from the slope of this line. The porosity must have been properly corrected and V_{cl} is obtained from the clay indicators discussed in Fig. 2–29. The Y and Z versus porosity plots are illustrated in Fig. 2–30.

The resistivity of the in-place water, R_w, appears in most equations. Local sections of SPWLA have often collected representative values for waters found in formations within their area of interest so that values are available by simply reading a tabulation and correcting for a desired temperature. Water salinity—

usually chloride content of a produced water—can be determined. Salinity is converted to resistivity, and temperature corrections are made by using Fig. 2–31. The salinity of ground waters can change with depth and areally, and values should be properly interpreted.

The SP curve can often be used to obtain an acceptable value for R_w. The SP in clean formations is related to the chemical activities of the formation water, a_w, and the mud filtrate, a_{mf}, by

$$SSP = -K \log (a_w/a_{mf})$$

for NaCl solutions, $K = 71$ at 77°F or 25°C, and K varies directly in proportion to the absolute temperature, $460 + °F$. For NaCl solutions that are not too concentrated, resistivities are inversely proportional to activities. This relation is not exact at higher concentrations, and an equivalent resistivity by definition defined by $R_{we} = 0.075/a_w$ at 77°F is used to obtain

$$SSP = -K \log (R_{mfe}/R_{we})$$

The value of SSP can be obtained directly from the SP curve if there are enough thick, clean, water-bearing beds in the vertical section of interest. A line is drawn through the SP (negative) maxima opposite thick permeable beds and another line called the shale base line is drawn through the SP opposite the intervening shale beds. The difference in millivolts between these lines is the value for SSP. SP anomalies, if present, should be noted. If the permeability is very low, the reservoir is partly depleted, and very heavy muds were used, the SP should probably not be used to obtain the values for R_w. Values for R_{mf} are shown in the heading section of the logs. Schlumberger makes the following recommendation:

1. For predominantly NaCl muds:
   a. If R_{mf} at 75°F is greater than 0.1 ohm-m, use $R_{mfe} = 0.85\ R_{mf}$ at formation temperature. This relation is based on measurements made on many typical muds.
   b. If R_{mf} at 75° is less than 0.1 ohm-m, use the NaCl (solid) R_w-R_{we} curves on SP-2 to derive a value of R_{mfe} from the R_{mf} value corrected to formation temperature by Chart.
2. For gyp muds the dashed curves drawn for "average" fresh formation waters on Chart SP-2 may be used to derive a value of R_{mfe} and R_{mf}.
3. Lime-base muds, despite their name, usually have a negligible amount of Ca in solution and may be treated as regular mud types.

These values may be placed in the SSP equation, and R_{we} can be calculated. R_w is read from Fig. 2–32.

RESISTIVITY OF SHALY FORMATIONS

LAMINATED SAND-SHALE SIMPLIFIED MODEL

R_t, the resistivity in the direction of the bedding planes, is related to R_{sh} (the resistivity of the shale laminae) and to R_{sd} (the resistivity of clean formation streaks) by the parallel-conductivity relationship

$$\frac{1}{R_t} = \frac{1 - V_{lam}}{R_{sd}} + \frac{V_{lam}}{R_{sh}}$$

where V_{lam} is the bulk-volume fraction of the shale, distributed in laminae, each of uniform thickness.

For the clean-sand laminae $R_{sd} = F_{sd} R_w / S_w^2$, where F_{sd} is the formation resistivity factor of the clean sand. Since $F_{sd} = a / \phi_{sd}^2$ (where ϕ_{sd} is the sand-streak porosity) and $\phi = (1 - V_{lam}) \phi_{sd}$ (where ϕ is the bulk-formation porosity),

$$\frac{1}{R_t} = \frac{\phi^2 S_w^2}{(1 - V_{lam}) a R_w} + \frac{V_{lam}}{R_{sh}}$$

R_{sh} is taken as the resistivity of nearby shale beds.

DISPERSED SHALE SIMPLIFIED MODEL

In this model the formation conducts electrical current through a network made up of the pore water and the dispersed shale. As suggested by L. de Witte , it seems acceptable to consider that the water and the dispersed shale conduct like a mixture of electrolytes. Development of this assumption yields

$$\frac{1}{R_t} = \frac{\phi_{im}^2 S_{im}}{a} \left[\frac{q}{R_{dis}} + \frac{S_{im} - q}{R_w} \right]$$

where

ϕ_{im} is the "intermatrix porosity", which includes all the space occupied by fluids and dispersed shale. (ϕ_{im} corresponds to an "intermatrix formation factor", $F = a / \phi_{im}^2$).

S_{im} is the fraction of the "intermatrix porosity", ϕ_{im}, occupied by the formation-water, dispersed-shale mixture.

q is the fraction of the "porosity", ϕ_{im}, occupied by the dispersed shale.

R_{dis} is the resistivity of dispersed shale.
Also, it can be shown that $S_w = (S_{im} - q) / (1 - q)$ where S_w is the water saturation in fraction of true porosity.

Combining these relations and solving for S_w we get

$$S_w = \frac{\sqrt{\frac{a R_w}{\phi_{im}^2 R_t} + \left(\frac{q (R_{dis} - R_w)}{2 R_{dis}} \right)^2} - \frac{q (R_{dis} + R_w)}{2 R_{dis}}}{1 - q}$$

Usually ϕ_{im} is obtained from the Sonic log
The value of q is obtained from Sonic and Density logs. In some areas, where $\rho_{dis} \approx \rho_{ma}$, it is acceptable to take ϕ equal to ϕ_t, in invaded sands or water

sands. The value of R_{dis} is difficult to evaluate, but since it is usually considered to be several times greater than R_w, its value is not too critical. When R_w is very small compared to R_{dis} and the sand is not very shaly, a form independent of R_{dis} can be used.

$$S_w = \frac{\sqrt{\frac{a R_w}{\phi_{im}^2 R_t} + \frac{q^2}{4}} - \frac{q}{2}}{1 - q}$$

TOTAL SHALE RELATION

Based on the above ideas, laboratory investigations and field experience   , it has been found that a simple relation of the following form applies well for many shaly formations independently of the distribution of the shale and over the range of S_w values encountered in practice:

$$\frac{1}{R_t} = \frac{\phi^2 S_w^2}{a R_w (1 - V_{sh})} + \frac{V_{sh} S_w}{R_{sh}}$$

In using this formula, R_{sh} is taken equal to the resistivity of the adjacent shale beds and V_{sh} is the shale fraction as determined from a total clay indicator

For a shale consisting of clay and silt, by analogy with the relation $R_o = a R_w / \phi^m$, R_{clay} is related to R_{sh} by:

$$R_{sh} = \frac{R_{clay}}{(1 - I_{silt})^x}$$

x is usually taken to be 2, but can be determined by statistical analysis of shale resistivities. Experience indicates x is generally between 1.4 and 2.4.

The relation for R_t in terms of the clay parameters is:

$$\frac{1}{R_t} = \frac{\phi^2 S_w^2}{a R_w (1 - V_{sh})} + \frac{V_{clay} (1 - I_{silt}) S_w}{R_{clay}}$$

Similar relations can be written for R_{shxo} and R_{xo}.

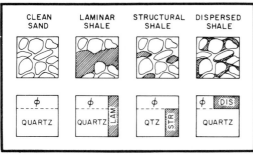

– Forms of shale classified by manner of distribution in formation. Pictorial representations above, volumetric representations below.

Figure 2–29. *Schlumberger correction for shales.*

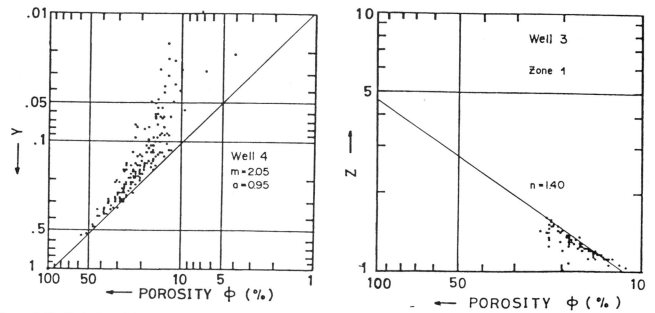

Figure 2–30. *Illustration of the use of Y-ϕ and Z-ϕ plots (Miyairi et al., "Water Saturation in Shaly Sands" SPWLA Seventeenth Annual Logging Symposium, June 1976).*

The R_{wa} log may be used to obtain R_w since $R_{wa} = R_t/F$. R_t is from a deep-investigating resistivity log, and F is computed from porosity logs. For clean, water-bearing zones, $R_t = R_o = FR_w$ and the R_{wa} log reaches a minimum value equal to the formation water resistivity or R_w.

The $R_{wa} - SP$ plot may be used to determine R_w at shallow to moderate depths where formation water resistivity often changes rapidly. R_{wa} is plotted on the log scale versus SP deflection on the linear scale. This plot may also discover hydrocarbon zones.

R_w may be determined from the R_{xo} and R_t logs by use of

$$R_{xo}/R_t = (S_w/S_{xo})^2 (R_{mf}/R_w)$$

The value of R_{xo}/R_t is a maximum in water zones where $S_w = S_{xo} = 1$. R_{xo}/R_t is computed over an interval containing clean, invaded, water-bearing sands to obtain the maximum value of $R_{xo}/R_t = R_{mf}/R_w$. R_{mf} measured at surface is converted to reservoir temperature, and R_w is calculated. Other more involved methods for determining R_w are included in the manuals of various logging companies.

R_{xo} is determined preferably from the microlaterolog or MSFL. The microlog and the proximity log may be useable. The proximity log is used directly as an R_{xo} tool if invasion is greater than about 40 inches. For shallower invasions, the proximity log is affected by the uninvaded zone and appropriate charts

must be used. An estimate of R_{xo} is available from the formula:

$$R_{xo} = 0.62 R_{mf}/\phi^{2.15} (1 - S_{or})^2$$

using ϕ for porosity logs and an assumed value for S_{or}. Pad devices determining R_{xo} are sensitive to mudcake effects and borehole vugosity but are usually insensitive to bed thickness.

The useable ranges of the R_t logs also must be recognized. Good R_t values are possible with unfocused electrical logs only in thick, homogeneous beds having porosities greater than 15%. Reliable R_t determinations in beds thinner than about 30 feet are possible only if invasion is shallow, resistivity is low to moderate, and when R_{xo} is much less than R_t. The laterolog suite is preferred for all porosity values when R_{mf}/R_w is less than 2.5. At higher values of this ratio and for high values of porosity, say above 5–15%, induction logs are useful. Both types of logs may be useful when porosity is less than 15% and the value of R_{mf}/R_w exceeds 2.5. Literature available from the logging companies often includes helpful charts that make it possible to convert questionable basic data into usable information. Schlumberger makes the following comments regarding the radius investigated by various logs:

Only in combinations of vertically focused devices with adequate depths of investigation can be depended on to permit a good R_t determination under a wide

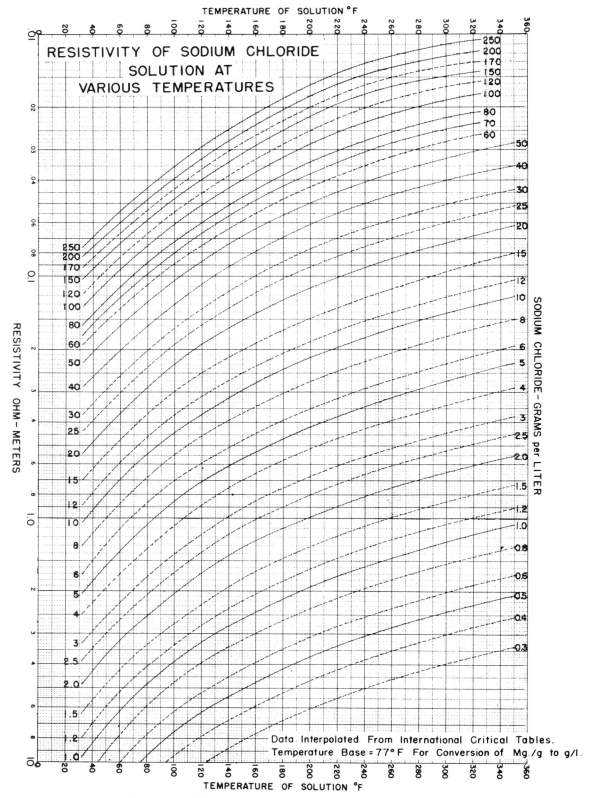

Figure 2–31. *Water resistivity related with salt content and temperature.*

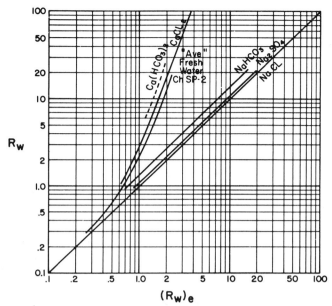

Figure 2-32. R_{we} related with R_w (courtesy Schlumberger).

range of conditions. Their readings are less affected by borehole effects and adjacent beds. The corrections required are generally small and often negligible. Induction logs and laterologs are such devices.

When invasion is not too deep, the deep induction logs (6FF40, ILd) and laterologs (LL3, LL7, LLd) may read fairly close to R_t. Thus in shallow-invasion territories, uncorrected values of R_{IL} or R_{LL} can often be used in place of R_t in preliminary log computations or cross-plots, or in plotting R_{wa} or movable oil curves. When invasion is very shallow, even a relatively shallow investigation tool may give a fairly good value of R_t.

Induction logs and laterologs differ in the way they respond to an invaded formation. To a first approximation the invaded zone and the uncontaminated zones are in parallel for the IL currents; they are in series for the LL currents. This fact is reflected in the forms of the IL geometrical-factor and LL pseudo-geometrical-factor equations. It also means that the IL reading is more influenced by the more conductive of the two zones, whereas the LL reading is more influenced by the more resistive zone. It follows that if $R_{xo} > R_t$, this is a factor tending to give the IL an advantage over the LL for determination of R_t. If $R_{xo} < R_t$, this is a factor tending to give an advantage to the LL.

At deeper and deeper invasions a larger proportion of the ILm geometrical factors falls within the invaded zone, whereas most of the ILd signal is still from the noninvaded zone. The relative positions of the three curves on the DIL-LL8 log is sufficient in many cases to indicate whether invasion is deep or shallow. This, of course, is not possible with only two resistivity logs. Thus, for $R_{xo} > R_t$:

Zero or very shallow invasion

$$R_{LL8} \approx R_{ILm} \approx R_{ILd} \approx R_t$$

Moderate invasion

$$R_{LL8} > R_{ILm} \approx R_{ILd} \approx R_t$$

Deeper invasion

$$R_{xo} \approx R_{LL8} > R_{ILm} > R_{ILd}$$

Very deep invasion

$$R_{xo} \approx R_{LL8} \approx R_{ILm} > R_{ILd}$$

The logarithmic-scale presentation permits a qualitative interpretation by inspection of the DIL-LL8 log. If there is enough invasion for the R_{xo} zone to influence the LL8 reading, and invasion is fairly uniform, the ratio R_{LL8}/R_{ILd} will be a qualitative indicator of the R_{xo}/R_t value, and hence of the S_w/S_{xo} value. The ratio R_{LL8}/R_{ILd} is seen directly from the separation of the two curves on the logarithmic scale. A decrease of this separation along with an increase in R_{ILd} value would signal the possibility of displaced hydrocarbons.

The presence of an annulus can sometimes be seen from the relative separation of the ILm-ILd curves on the DIL-LL8. This is interesting because existence of an annulus indicates that the formation is oil- or gas-bearing.

The reader must recognize that literature relative to logging is extensive, and many of the better articles appear in publications by SPWLA rather than by SPE. The logging companies can offer valuable suggestions and useful correction charts. The object of this brief summary is to make available or consolidate some of the more accepted techniques so that the reservoir engineer can communicate with the log analyst. The reader may also wish to refer to *Handbook of Well Log Analysis* by Pirson, which contains a set of calculation forms useful during an analysis of more routine logs.

Oil Saturation

The initial oil saturation at the time that a field is discovered usually is determined by simply substracting the average water saturation from the total effective porosity. In other words, the reservoir initially is filled with oil and water when no initial gas cap is present. The water content is determined from use of logs, capillary pressure, and core information. Porosity can be determined from cores, logs, and other available techniques. Later during the productive life, gas too must be present when oil is below its bubble point, and determination of separate values for gas and oil can be difficult. Special logs and techniques must then by employed, and manuals of service companies should be consulted. The Interstate Oil Compact Commission published a book evaluating various methods during June 1978 called "Determination of Residual Oil Saturation," which concludes, "All of the procedures discussed are in need of refinement to provide the accuracy in determination of residual oil saturation that is desirable."

Nature during geological time causes oil, gas, and water to segregate so that the fluids are located ac-

to find gas in a downdip trap in the line of migration, and updip traps may remain filled with water.

The term *residual oil saturation* is used in several contexts in the literature such as:

1. The irreducible oil saturation (lowest possible value)
2. An average or material-balance type saturation that exists at a designated time or set of conditions
3. Oil saturation remaining in a water-swept area or zone
4. Oil saturation in a gas-swept zone

Methods for measuring residual oil saturation include:

> Core analysis
> Specialized coring techniques
> Reservoir engineering studies
> Logging using special techniques
> Pressure transient measurements
> Chemical tracers with field injection and coring

The logging techniques often involve fluid injection as follows:

Logging tool	Technique	Used in casing	Field tested to date	Expected accuracy
Resistivity	Conventional	No	Yes	poor
	log-inj-log*	No	Yes	good–special
Pulsed neutron	Conventional	Yes	Yes	poor
	Log-inj-log*	Yes	Yes	good–excellent
	Log-inj-log with chem strip	Yes	Partially	poor–good
Nuclear magnetism	Inj-log	No	Yes	excellent
Carbon/oxygen	Conventional	Yes	Yes	poor
Gamma radiation	log-inj-log	Yes	No	unknown
Dielectric const.	Conventional	No	Partially	unknown

* After water displacement or after waterflood. All methods survey only the area around a well.

cording to their density and recognized natural laws. Density or gravitational forces place the gas at the top, the oil in the middle (except in rare cases where oil density is greater than that of water) and water at the bottom. Small traps located within the reservoir due to lenses and shale stringers might contain limited amounts of free gas within the oil reservoir. The time of migration and accumulation of gas and oil within the water-filled aquifer controls the location of the oil and gas reservoirs within traps in an area. Gas can displace an oil accumulation and an oil accumulation may leak from a faulty trap. It is not unusual

It is obvious that the residual oil saturation at the start of any process is critical in economics. Likewise, the probable saturation at the conclusion of a project based on field testing is equally important.

Residual oil remaining after use of a recovery mechanism relates with the oil originally in place, as illustrated in Fig. 2–33. The degree of scatter of data from various field test methods usually is substantial, as illustrated in Fig. 2–34.

Methods of analogy and study of the performance of adjacent fields using proper scaling techniques often can aid in forecasting recovery possibilities. Later sections dealing with the recovery efficiencies, water-flood or water displacement process, and solution-gas drive or gas displacement process should be reviewed. These sections contain methods for analyzing both field and laboratory data to determine recovery efficiencies and residual oil saturation following solution-gas and water displacement. An experienced engineer may use material balance and field history to obtain values for residual oil saturation. At times pilot floods or small-scale operations are useful. Data sets from the various sources must be coordinated using volumetric, material-balance, fractional flow with true effective permeability for saturations then existing and other concepts to obtain a reliable set of values applicable to existing field conditions.

Tracers may be used in coring fluids to measure flushing. No coring tool available today completely eliminates flushing of the core by mud filtrate during coring. The IOCC book concludes, "There is need for improvement in calibration and accuracy of most well logging systems." Core properties, such as poros-ity and formation resistivity factor, should be corrected to in-situ stress conditions. The center portions of large cores may represent in-situ conditions existing at the time of coring. An injection profile survey should be combined with tracer surveys to determine the fraction of the formation sampled.

Large areal variations in oil saturations remaining in reservoirs should be anticipated, particularly when partially depleted by gas, gravity, and water displacement mechanisms. All values for oil in place at the start of secondary or EOR operations must be coordinated properly, and average values for the area to be subjected to the new recovery process must be obtained. The residual oil that will remain after completion of the new recovery process is of equal importance in the determination of the economics of a project. If construction and planning time are substantial, costs must be increased to allow for inflation (discussed in a later section dealing with economics). Again, the need for experience and careful use of all possible data sources is essential.

Basic Tools and Techniques to Determine Residual Oil Saturation (after Fertl, *Petroleum Engineer*)

Basic tool	Technique	Can be used when hole is cased	Has been field tested	Expected accuracy*
Reservoir performance Well Tests	Volumetric determination	Yes	Yes	Poor
	Pressure transients		Yes	
	Fluid compressibility	Yes		Very poor
	Effective permeability	Yes	Yes	Poor
	Water-oil ratio	Yes	Yes	Poor
Cores				
Conventional	Saturation measurements from resh cores.	Core must be cut while drilling holes.	Yes	Poor
	Lab flooding techniques, inhibition, centrifuge, etc.	Core must be cut while drilling holes.	Yes	Poor to fair
Pressure	Core with specially designed mud.	Core must be cut while drilling holes.	Yes	Poor to excellent
Single-well tracer	Back-flow hydrolyzed tracers.	Yes	Yes	Good to excellent
Logging tools				
Resistivity	Conventional	No	Yes	Poor
	Log-inject-log	No	Yes	Good to excellent
Pulsed neutron capture	Conventional	Yes	Yes	Poor
	Log-inject-log with water-flood	Yes	Yes	Good to excellent
	Log-inject-log with chemical strip	Yes	Partially	Fair to good
Nuclear magnetism†	Inject-log	No	Yes	Excellent
Carbon/oxygen	Conventional	Yes	Yes	Poor
Gamma radiation	Log-inject-log	Yes	No	Unknown (but could be excellent)
Dielectric constant†	Conventional	No	Partially	Unknown

* Expected accuracy at 2 standard deviations-percent pore volume.
Excellent 0–4, Good 4–8, Fair 8–12, Poor-greater than 12.
† Subject to "well-bore" errors, because of shallow penetrations.

Basic Considerations for Log-Inject-Log ROS Determinations (after Fertl)

Reservoir	Logging devices	Well conditions	Injection
1. High porosity, high residual oil saturation and good permeability.	1. Properly functioning, calibrated instruments.	1. Enough rat hole so entire zone can be logged.	1. Nonuniform injection profiles suggest poor fluid displacement in stratified formations.
2. Select uniform reservoir.	2. Multiple repeat runs (6 to 10) at proper logging speed, time constant etc. to reduce statistics.	2. Evaluate a short single zone rather than a too long zone or multiple zones to facilitate control of proper injection procedures.	
3. Avoid fractured or fracturing of reservoir which is very detrimental to sweep efficiency.	3. Zones investigated by logs must be completely covered by the injection. Does not necessarily guarantee complete fluid replacement around the cased wellbore.	3. Newly perforated intervals rather than zones with old perforations to avoid formation slumping, sand production, and resulting drastic porosity changes	3. Proper control of injection pressure (vs fracture gradient) and rates.
4. Availability of reliable porosity information.		4. Avoid tests in old injection wells, since ROS may be drastically reduced due to "stripping effects."	4. Injection fluids prepared under controlled conditions (i.e. batch mixing, calculated and/or measured Σ_w-values).
5. Gas saturation is zero in subject reservoir.		5. Satisfactory well completion and zone isolation.	

NOTE: Do not be concerned about what at first appears to be conflicting ROS-data obtained from reservoir engineering concepts, single well tracer tests, core analyses, and log-derived tests. Closely study the valid reasons for apparent discrepancies, which are many. Keep in mind that results may be weighted by permeability, porosity, depth of investigation and vertical resolution of logs, etc. Also note that no single method alone gives totally meaningful results of both the amount and the distribution of residual oil saturation.

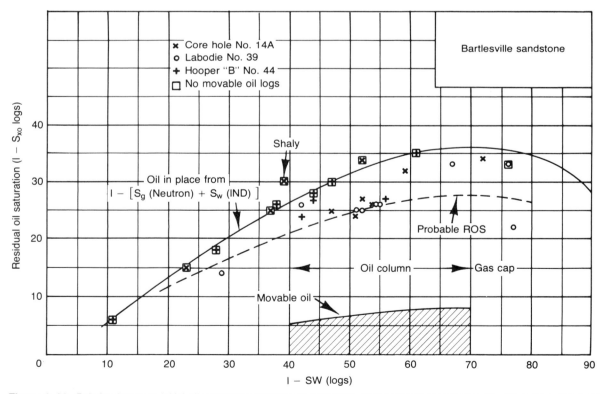

Figure 2–33. *Relation between initial oil saturation and final oil saturation after recovery by solution-gas drive. The difference between the two curves is expected moveable oil recoverable by waterflood operation (after GTN Roberts, Shell Oil).*

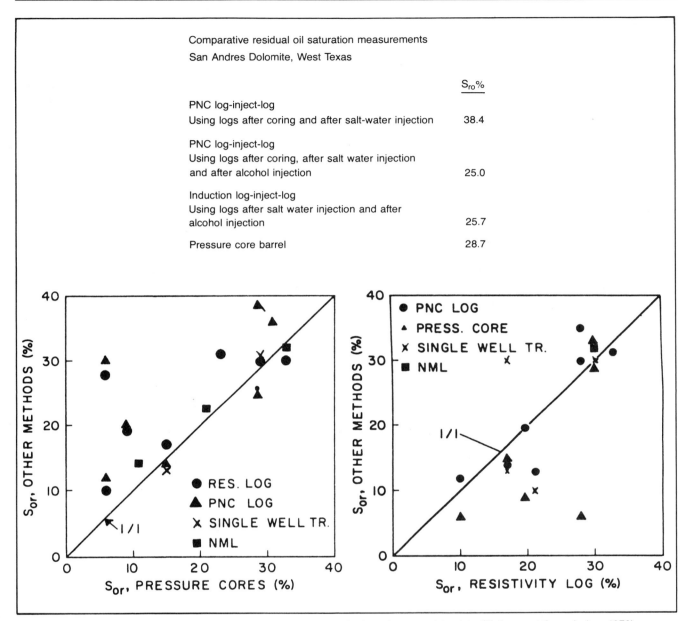

Comparative residual oil saturation measurements
San Andres Dolomite, West Texas

	$S_{ro}\%$
PNC log-inject-log Using logs after coring and after salt-water injection	38.4
PNC log-inject-log Using logs after coring, after salt water injection and after alcohol injection	25.0
Induction log-inject-log Using logs after salt water injection and after alcohol injection	25.7
Pressure core barrel	28.7

Figure 2–34. *Comparison of results from residual oil determinations (courtesy Interstate Oil Compact Commission, 1978).*

Gas Saturation

Drill-stem and production tests are often used to determine whether a zone contains oil, gas, or only water. Also, close observation of cores when first taken from the core barrel can indicate the presence of gas. Gas also may be observed in drilling mud. Gas zones are often invaded by drilling fluids since the viscosity of gas is less than that of oil and water. The qualitative determination of the presence of gas often is rather easy, but quantitative determination can be difficult, particularly when a reservoir has been partly depleted by production. The pay zone is completely filled with oil, gas, or water, and one can be determined if the other two are known.

At initial in-situ conditions—before the reservoir

is produced—any free gas present will be found at the top of the accumulation of oil or water. The gas-oil or gas-water contact should be at a rather uniform depth throughout the reservoir unless the reservoir is being subjected to a hydrodynamic pressure gradient. Leaking traps pressure differences across an accumulation can cause inclined contacts.

The gas content of a rock when no oil is present can be determined from logs, as illustrated by Fig. 2–35. The density of gas often differs from that of oil and water by a significant amount. Neutron and density logs respond to total porosity, but sonic logs ignore vug and fracture porosity. The density of gas depends upon reservoir pressure. Gas contains hydrogen in a different quantity than oil, and neutron logs reflect this fact.

Core analysis of a gas zone usually cannot be converted to a good estimate of the amount of gas present. Invasion of gas zones is substantial and displacement of liquids as the core is brought to the surface may not remove all injected water. At other times some in-situ water may also be removed.

Determination of the exact gas and oil content is very important in secondary and tertiary or EOR decisions and economic planning. Unfortunately, separation of the oil content from the gas content is most difficult. The cumulative production can be subtracted from the original content for each component to obtain an estimate that is as good as the production records and determination of the quantities present originally. Also, the fractional flow equation can be combined with the existing gas-oil ratio of production to obtain an estimate of subsurface content based on relative permeability concepts. Logging techniques also offer possibilities. Unfortunately, rather accurate values often are desired and the above methods do not supply them. The possibility of substantial changes in water content of the reservoir during the time of production also must be carefully investigated by use of production history and logging tools. Small structural traps, irregular shale stringers, etc., may offer possibility of free gas accumulations, which may exceed average gas content related with preferential displacement, gravity drainage, and other recovery mechanisms.

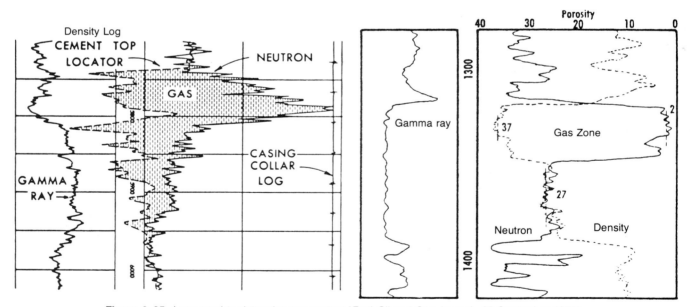

Figure 2–35. *Logs used to determine gas content (Fertl*, Oil and Gas Journal, *25 September 1978*).

Permeability

The permeability of rocks usually is measured in the laboratory using cores obtained with the inherent selection problems. Fractures and vugs contribute substantially to flow capacity but contain little porosity. Secondary precipitation often seals fractures and vugs during geological time, and at great depth pressure causes fractures to heal because of compressive forces. Permeability also can be generally related to logs.

The best values for permeability often are obtained from flow and build-up testing. If pay thickness can be determined, accuracy is good and averaging problems are not encountered with this method. Flow and build-up test data must be analyzed properly using techniques detailed in Chapter 5. The basic equations are:

Linear flow: Plot pressure versus $\sqrt{t}$ or $\sqrt{\Delta t}$

Slope $= m = 16.23 \ (q_{sc}B\mu/A) \sqrt{1/k\phi\mu c_t}$ for oil

Spherical flow: Plot $1/\sqrt{t}$ or $1/\sqrt{\Delta t}$

$k^{2/3} = 2453 \ (q_{sc}B\mu\sqrt{\phi\mu c_t})/m$ for oil

Radial flow: Plot pressure versus t or Δt

$kh = 162.6 \ q_{sc}B\mu/m$ for oil

Deviated holes, location of perforations, changes in fluid and rock properties, reservoir boundaries, and many other factors influence test results. The flow time must be realistic when permeability is large and the $m(p)$ function often is required when testing gas wells. The theory for flow and buildup testing is reviewed in detail in SPE Monographs 1 and 5.

Flow of fluids in rocks follows Darcy's law, but adjustments often must be made to allow for particle movement, the presence of two or more fluids, changing composition and characteristics of the fluids with distance and time, sand lenses, crossflow and zones of different permeability, local solution channels, and many other factors. Skins—both negative and positive—crossflow in the well bore during build-up or shut-in periods, lift efficiencies, location and number of perforations, and similar considerations must be recognized when applying measured permeability obtained from the laboratory, logs, and flow tests to prediction of well and reservoir behavior using sound equations over the life of field operations.

Permeability may be directly measured in the laboratory using representative cores that have been properly handled and cleaned so that in-situ conditions have not been disturbed. Values obtained from laboratory tests must be averaged properly using techniques such as detailed in the following section taken from *Petroleum Engineer*.

Various Methods Used to Average Permeability

Values for permeability are usually obtained from laboratory core analysis and field flow and buildup tests. Before comparing such values, adjustments should be made for effects such as connate water, relative permeability, deviated wells, overburden pressure.

When cross flow between zones is not suspected, the horizontal permeabilities of the parallel zones may be averaged by

$$\bar{k} = \frac{k_1 + k_2 + k_3 + k_4 + \ldots + k_n}{n}$$

when all values represent zones of equal thickness.

This implies that in stratified or laminated reservoir rocks with beds in parallel average permeability, $\bar{k}$ can be calculated for both linear and radial flow conditions by the equation:

$$\bar{k} = \left(\sum_{i=1}^{n} k_i h_i\right) / \left(\sum_{i=1}^{n} h_i\right)$$

where: $i =$ number of reservoir layers
$k_i =$ permeability of individual layer
$h_i =$ thickness of individual layer

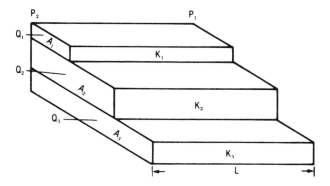

Average permeability of beds in series under a linear flow regime is defined by:

$$k_{avg} = \frac{L_t}{L_1/k_1 + L_2/k_2 + L_3/k_3} = \frac{\Sigma L_i}{\Sigma L_i/k_i}$$

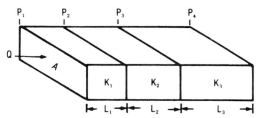

For radial flow system of constant thickness with a permeability of k_e between the drainage radius r_e and some lesser radius r_a, and an *altered* permeability k_a between the radius r_a and the wellbore radius r_w as shown:

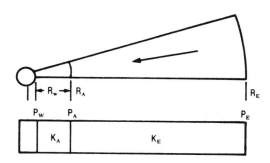

$$k_{avg} = \frac{k_a k_e \ln(r_e/r_w)}{k_a \ln(r_e/r_q) + k_e \ln(r_a/r_w)}$$

Other special techniques must be used when the drainage radius of the zones differ by substantial quantities as calculated from the radius of investigation equations.

In some of these methods for averaging permeability, the relative number of sequences and range of values in each sequence may influence results and sensitivity tests should always be made.

Reservoirs with cross flow tend to behave as a more or less uniform reservoir or pay but a continuous flow barrier such as an impermeable shale results in multiple zone behavior. When considering the possibility of cross flow, or lack thereof, one must realize that flow in the vertical direction is often through a large area related with πr^2 through two faces while the circumference of a circle is $2\pi r$ which is multiplied by pay thickness.

A method using logarithm is also sometimes of value.

$$\log \bar{k} = (\log k_1 + \log k_2 \ldots + \log k_n)/n$$

When two or more fluids are flowing, the average or total

$$\left[\frac{k}{\mu}\right]_t = \frac{k_o}{\mu_o} + \frac{k_g}{\mu_g} + \frac{k_{to}}{\mu_w}$$

$$\left[\frac{k}{\mu}\right]_t = \frac{162.6}{mh}[B_o q_o + B_g(q_{gt} - q_o R_s) + B_w q_w]$$

The permeability here is effective permeability or core permeability times the respective relative permeabilities. Flow and buildup tests usually are used to obtain effective permeabilities, as of conditions existing at time of test.

When permeability is known to vary with both distance and thickness, maps showing reservoir character, which may be planimetered, are useful as illustrated below.

Core permeability also should be compared with logs before averaging.

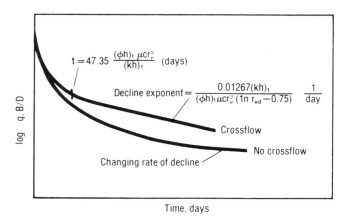

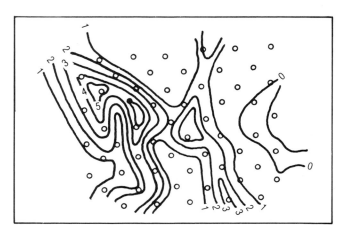

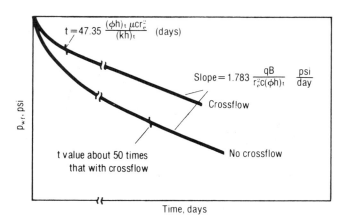

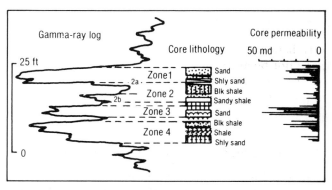

Statistical Methods of Averaging Permeability

The most probable behavior of a hetrogeneous system often approaches that of a uniform system where core permeability is averaged by the geometric mean technique.

$$k_{avg} = \bar{k} = \sqrt[n]{k_1 \times k_2 \times k_3 \times k_4 \times \ldots \times k_n}$$

This average relates with the log normal distribution. A plot of the number of samples in any permeability range versus the log permeability yields a bell shape configuration.

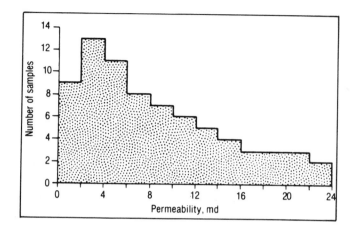

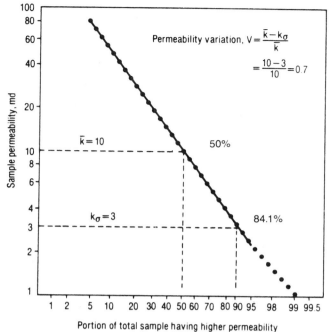

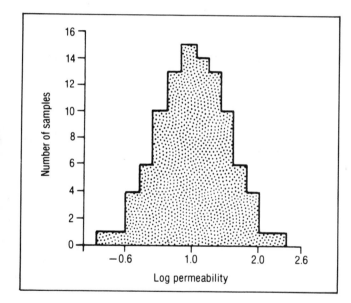

The data can also be plotted on logarithm probability paper.

The best straight line is drawn through the points, with the central points weighted more heavily than the more distant points. The permeability variation is then

$$V = (\bar{k} - k_\sigma)/\bar{k}$$
$$V = (\log \bar{k} - \log k_\sigma)/\log \bar{k}$$

where:

$\bar{k}$ is mean permeability or permeability value at 50% probability

k_σ is permeability at 84.1% of the cumulative sample

The possible values of the permeability variation range from zero to one, with a completely uniform system having a value of zero.

The permeability data may be arranged by layers (Stiles Method) or may be arranged by equally sized groups according to decreasing (or increasing permeability).

Craig presents a detailed example showing ten permeability intervals—also representing ten depth intervals with identical thickness of one foot in each of ten wells or a total of 100 permeability measurements. Results of an analysis are summarized in the following table:

Average layer permeability (md)

Layer	Permeability ordering	Positional approach
1	84.0	10.0
2	37.0	6.8
3	23.5	4.7
4	16.5	10.4
5	12.0	20.5
6	8.9	12.1
7	6.5	8.6
8	4.6	18.4
9	3.0	14.3
10	1.5	10.9

Arithmetic average permeability	28.2 md
Mean permeability	10.0 md
Ratio of maximum to minimum layer permeability:	
Permeability ordering	84.0/1.5 = 56.0
Positional approach	20.5/4.7 = 4.37

The permeability values are arranged into a list of declining values and separated into ten groups of ten values each to obtain the permeability ordering column. The values shown under positional approach are calculated using the geometric average equation previously presented for each depth interval for the ten wells and for the entire sample. All 100 samples were used to calculate the arithmetic average.

Permeability in series may be averaged using the following equation:

$$n/k_{avg} = 1/k_1 + 1/k_2 \ldots + 1/k_n$$

In a similar manner porosity thickness values can be obtained which correspond with each of the permeability values. Both groups of values can be expressed as fractions of the total values so that the following plot may be constructed.

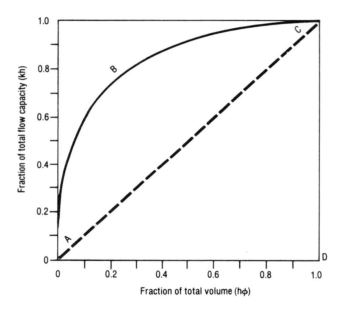

Area (ABCA)/(Area ADCA) = (Lorenz Coefficient
The values of this coefficient range from zero to one and a uniform reservoir has a value of zero. Several different permeability distributions can yield the same value for the Lorenz coefficient.

Vertical permeability is less than horizontal permeability even in relatively clean sands. The difference becomes substantial in poor quality sands.

Spherical flow permeabilities may relate horizontal and vertical flow stream lines so that

$$k_s = \sqrt[3]{k_v k_{h1} k_{h2}}$$

Value of spherical flow permeability

k_h	600	500	400	300	200	100	10
$k_v = .8k_h$	557	464	371	278	186	93	9
$k_v = .7k_h$	533	444	355	266	178	89	9
$k_v = .6k_h$	506	422	337	253	169	84	8
$k_v = .5k_h$	476	397	317	238	159	79	8
$k_v = .4k_h$	442	368	295	221	147	74	7
$k_v = .3k_h$	402	335	268	201	134	67	7
$k_v = .2k_h$	351	292	234	175	117	58	6
$k_v = .1k_h$	247	232	186	139	93	46	5

The objective when using the previously presented techniques is to find values which adequately describe the reservoir rock for use in applicable equations so that sound decisions can be made. Use of more than one technique often is necessary for proper understanding of the reservoir.

Several of these methods-equations appear in following books:

1. Muskat, *Physical Principles of Oil Production*, 1949
2. Pirson, *Elements of Oil Reservoir Engineering*, 1950
3. Calhoun, *Fundamentals of Reservoir Engineering*, 1960
4. Craft and Hawkins, *Applied Petroleum Reservoir Engineering*, 1959
5. Smith, *Mechanics of Secondary Oil Recovery*, 1966
6. Matthews-Russel, *Pressured Buildup and Flow Tests in Wells.*
7. Craig, Forest F., Jr., *The Engineering Aspects of Waterflooding*, SPE Monograph 3.
8. Testerman, *JPT*, Aug. 1962, p. 889–893.

The permeability measured using air adjusted for Klinkerberg effects may differ from values obtained using water. The value may be further lowered by the presence of two or more nonmoving fluids, and the correction for movement of multifluids is adjusted by use of the relative permeability concept. Permeability in the horizontal direction may be preferential and permeability in the vertical direction often is less than that in the horizontal direction, particularly in tighter rocks (Fig. 2–36). Each of these possibilities must be investigated and adjustment made when results are affected sufficiently to cause errors in final conclusions. The permeability of tight rocks with permeability less than 1 md depends severely upon confining pressure so that core data and field performance might require adjustments.

Permeability is based on Darcy's law and is defined:

1 darcy = 1,000 millidarcys; 1 millidarcy = 0.001 darcy

$$k = \frac{Q\mu}{A(P/L)}$$

1 darcy $= \dfrac{\text{cc/sec cp}}{\text{cm}^2 \text{ atm/cm}}$

$\qquad = 9.869 \times 10^{-7} \dfrac{\text{cc/sec cp}}{\text{cm}^2 \dfrac{\text{dyne/cm}^2}{\text{cm}}}$

$\qquad = 9.869 \times 10^{-9} \text{ cm}^2$

$\qquad = 1.062 \times 10^{-11} \text{ ft}^2$

$\qquad = 7.324 \times 10^{-5} \dfrac{\text{ft}^3/\text{sec cp}}{\text{ft}^2 \text{ psi/ft}}$

$\qquad = 9.679 \times 10^{-4} \dfrac{\text{ft}^3/\text{sec cp}}{\text{cm}^2(\text{cm water/cm})}$

$\qquad = 1.127 \dfrac{\text{B/D cp}}{\text{ft}^2(\text{psi/ft})}$

$\qquad = 1.424 \times 10^{-2} \dfrac{\text{gal/min cp}}{\text{ft}^2(\text{ft water/ft})}$

Determination in the laboratory analysis of cores is based on horizontal rectilinear semisteady-state flow using following equations:

Oil:
$$Q_o = \frac{kA(P_1 - P_2)}{\mu L}$$

Gas:
$$Q_b = \frac{kA}{2\mu L}\frac{P_1{}^2 - P_2{}^2}{P_b}$$

Define $\bar{P}$ as $(P_1 + P_2)/2$, and $\bar{Q}$ as the volume rate of flow at $\bar{P}$ such that $\bar{P}\bar{Q} = P_b Q_b$;

$$\bar{Q} = \frac{kA(P_1 - P_2)}{\mu L}$$

Area for a circle is $A = \pi r^2 = 0.785\, d^2$
for a square is $A = w^2$

The above equations define flow in porous medium and may be compared with the following equations applicable to flow in pipe and conduits:

Poiseuille's equation for viscous flow

$$v = \frac{d^2 \Delta P}{32 \mu L}$$

Fanning's equation for viscous and turbulent flow

$$v^2 = \frac{2d\,\Delta P}{f\rho L}$$

where v = fluid velocity, cm/sec
$\quad d$ = diameter of conductor, cm
$\quad \Delta P$ = pressure loss over length L, dynes/cm^2

L = length over which pressure loss is measured, cm
μ = fluid viscosity, poise
ρ = fluid density, g/cc
f = friction factor, dimensionless

A more convenient form of Poiseuille's equation is

$$Q = \frac{\pi r^4 \Delta P}{8\mu L}$$

where r is the radius of the conduit, cm, Q is the volume rate of flow, cc/sec, and other terms are as previously defined.

The rule for classifying permeability is as follows:

Classification	Permeability range, md
poor to fair	1–15
moderate	15–50
good	50–250
very good	250–1000
excellent	1000–plus

Vugs often may approximate circular openings and, the permeability of vugs relates with the radius so that

$$k = r^2/(8 \times 9.869 \times 10^{-9}) = 12.5 \times 10^6\, r^2$$

when r is in cm and k is in Darcies or 1000 md.

When r is in inches and k is in Darcies,

$$k = 80 \times 10^6 \times r^2 = 20 \times 10^6 \times d^2$$

A circular opening of 0.005 inches has a permeability of 2,000 Darcies.

The permeability of fractures is a direct function of the width of the fracture so that

$$\Delta p = 12\,\mu\,v\,L/W^2 \qquad q = W^2 A\,(p_1 - p_2)/12\,\mu\,L$$
$$k = 84.4 \times 10^5\, W^2$$

when W is in cm and k is in Darcies.

$$k = 54.4 \times 10^6\, W^2$$

when W is in inches and k is in Darcies.

The permeability of a fracture 0.01 inches wide is 5,440 Darcies.

Fractures and vugs increase flow capacity substantially but make only a small contribution to porosity in oil and gas reservoirs.

The presence of capillary fluid reduces the air permeability as measured in the laboratory, and relative permeability concepts as shown in Fig. 2–37 must be used when two or more fluids are flowing at the same time. Clays and shales often expand in the presence of fresh water so that permeability with fresh

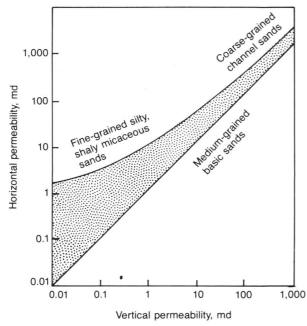

Figure 2–36. *General relation between horizontal and vertical permeability (after Fertl).*

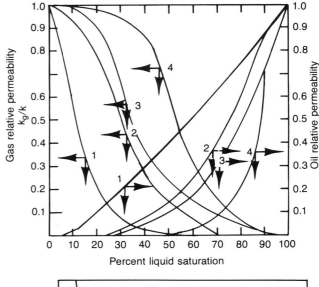

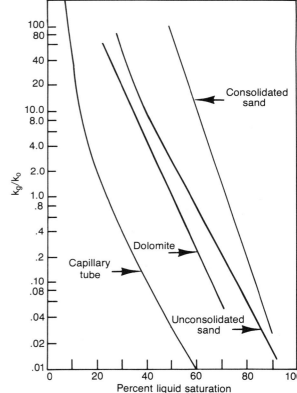

Figure 2–37. *Relative permeability ratios for various media: 1) capillary tube; 2) dolomite; 3) unconsolidated sand; 4) consolidated sand (after Calhoun,* Fundamentals of Reservoir Engineering).

water may be less than when salt water is present in the formation and cores.

Permeability usually becomes less with depth since weight of the overburden tends to compress the pore space. When the initial pressure is abnormally high, the pore space has not been completely compressed to adjust for overburden weight. Some further compression of pore space and reduction in permeability probably occurs in these reservoirs as fluid pressures are reduced during production. Fortunately, the process is slow and no reduction is noticed in field operations. Permeability varies in vertical and horizontal, and geology—type of deposition of the productive formation—should be studied in detail.

Viscosity of gas is 10 to 40 times more favorable to flow than that of most liquids at reservoir conditions. Economical flow is possible from gas reservoirs when pay is very tight (less than one md and at times less than 0.01 md).

The permeability of a pay may at times be related with pay characteristics determined from logs. Possible relations between porosity and permeability for a representative group of reservoirs and sand quality are shown in Fig. 2–38. The slopes of the lines on these semilog plots appears to depend upon pay quality, but straight lines are observed over substantial portions of the curves. The exponential curve may be reduced to an equation such as $y = ab^{cx}$. The core analysis data naturally show usual scatter, but a general relation between porosity and permeability usually can be established. Also, porosity can be obtained from logs so that log-derived porosity can be related with permeability and values can be approximated for intervals not cored. Pay quality shows a general

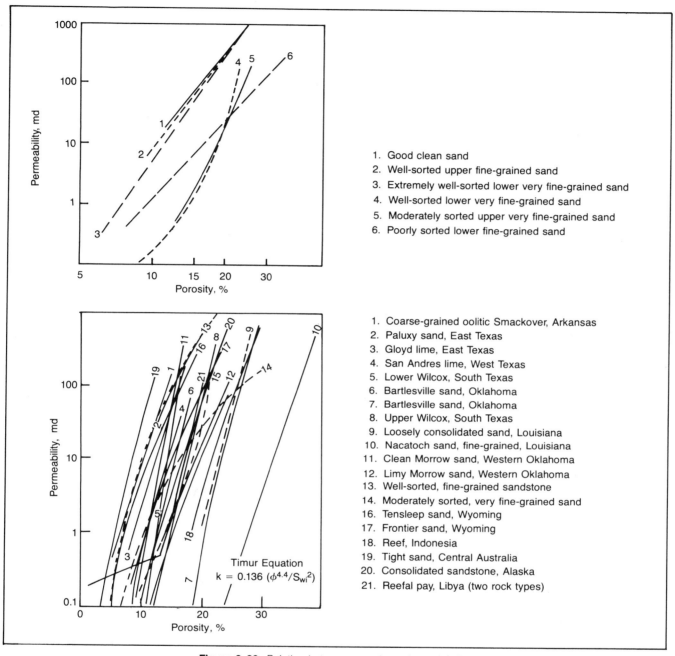

1. Good clean sand
2. Well-sorted upper fine-grained sand
3. Extremely well-sorted lower very fine-grained sand
4. Well-sorted lower very fine-grained sand
5. Moderately sorted upper very fine-grained sand
6. Poorly sorted lower fine-grained sand

1. Coarse-grained oolitic Smackover, Arkansas
2. Paluxy sand, East Texas
3. Gloyd lime, East Texas
4. San Andres lime, West Texas
5. Lower Wilcox, South Texas
6. Bartlesville sand, Oklahoma
7. Bartlesville sand, Oklahoma
8. Upper Wilcox, South Texas
9. Loosely consolidated sand, Louisiana
10. Nacatoch sand, fine-grained, Louisiana
11. Clean Morrow sand, Western Oklahoma
12. Limy Morrow sand, Western Oklahoma
13. Well-sorted, fine-grained sandstone
14. Moderately sorted, very fine-grained sand
16. Tensleep sand, Wyoming
17. Frontier sand, Wyoming
18. Reef, Indonesia
19. Tight sand, Central Australia
20. Consolidated sandstone, Alaska
21. Reefal pay, Libya (two rock types)

Figure 2–38. *Relation between porosity and permeability.*

relation with irreducible water saturation as well as with geological descriptions. Various equations have been proposed using log information, such as:
Timur equation

$$k_{(\text{Darcy})} = 0.136 \, (\phi^{4.4}/S_{wi}^2)$$

Sheffield equation

$$k = (1/F) \, [\phi/(1-\phi)]^2 \, (1/S_{wi})$$

where $F = a/\phi^m$ = Archie cementation factor so that m has a value of 1.3 for unconsolidated rocks and oolitic limestones; a value of 1.4 to 1.5 for very slightly cemented sands; a value of 1.6 to 1.7 for slightly cemented sands; 1.8 to 1.9 for highly consolidated sands with porosity less than 15%; and 2.0 to 2.2 for low-porosity sands, intergranular porosity-type limestones, and dolomites and chalks.

Tixier equation

$$(k/20)^{1/2} = [2.3/R_o(\rho_w - \rho_o)](\Delta R_t/\Delta h)$$

where ρ is the specific gravity of oil and water and $(\Delta R_t/\Delta h)$ is resistivity gradient in ohmmeters per foot of depth as determined from a deep investigation tool such as focused logs.

For rather ideal conditions such as water-wet sands that are shale free and of intergranular type, the equation reduces to:

$$k^{1/2} = 250 \ \phi^3/S_{wi} \text{ for medium-gravity oils}$$
$$k^{1/2} = 79 \ \phi^3/S_{wi} \text{ for dry gas sands}$$

Pirson equation

$$k = [(850,000/^\circ API \text{ gravity}) - 3.5 \text{ depth in ft}](R_w^2)/FR_oR_{ti})$$

where k is in Darcies and R_{ti} is the true resistivity of oil bearing zone above the transition zone. The equation is not applicable to high-gravity crudes and depths below 6,500 feet.

All of these equations must be used with care since they are approximations only. Values from these sources should be checked against values determined from core analysis. Results can be plotted on graphs relating various parameters, such as shown in Fig. 2–39.

Permeability is reduced by confining pressure so that correction to air permeability measured in the laboratory using cores is necessary for low-permeability rocks located at deep depth. The rock in a laboratory reacts rather fast to a loss of confining pressure, and hysteresis is observed. In the field, the reaction of rocks to a reduction in pressure is much slower and less noticeable in flow behavior. The measured core permeability and porosity must be reduced to allow for the change in confining pressure as illustrated by Fig. 2–40. Also, field production might decline with pressure drop and time.

Permeability values are normalized on the basis of permeability measured at 1,000 psi net confining pressure to allow direct comparison of the influence of confining pressure independent of permeability level. The equation for the relation is:

$$k = k_{1,000}\left(1 - S\log\frac{P_k}{1000}\right)^3$$

where S, the magnitude of the negative slope, is given by

$$S = \frac{1 - \left(\frac{k}{k_{1000}}\right)^{1/3}}{\log\frac{P_k}{1000}}$$

Use of the straight-line relation simplifies both testing and handling of data. Permeabilities need be measured at only two confining pressures to fix the slope parameter of the equation well enough for most engineering purposes. Measurements usually are made at 1,000 psi (6.90 mpa) and at the reservoir net overburden pressure.

Increasing values of S imply increasing effects of confining pressure. Moderate effects of stress, such as seen in testing higher permeability rocks, produce S factors in the vicinity of 0.1 to 0.2. Most tight gas sand yielded S factors in the range of 0.3 to 0.6 with factors over 0.7 indicating large reductions. A rock decreased tenfold in permeability below routine permeability by reservoir overburden pressure would have an S factor of approximately 0.4.

Connected fractures can substantially increase permeability, and detection of natural fractures is very important. This subject was summarized by McCoy, Kumar, and Pease in *World Oil* (December 1980). An abnormally high flow rate (kh from cores is less than kh from flow or build-up analysis) often suggests natural fractures. A skip on the conventional sonic log may be caused by fractures. The variable density log and the 3-D velocity log may show attenuation of the shear arrivals when natural fractures are present. Focused induction logs and microdevices may show anomalous spiking. The noise log, the flowmeter, and the temperature log all show anomalous character at fracture planes. The caliper log and four-prong log may locate fractures. Fractures may have a high radium content. The compensated density log may show positive RHO correction in a nonrugose environment. Some of the characteristics of these logs are given in the following table:

Characteristics of Measurements Used in the Detection of Fractures (after McCoy et al.)

Method	Spacing or vertical resolution		Approx. depth of investigation		Percent of circum. of 8¾-inch borehole surveyed	Effect of vugs associated with fractures	Can be recorded in gas or air-filled holes
	Inches	cm	Inches	cm			
Spontaneous potential	12	30.48	0	0	100%	None	No
Calipers 3-Arm Bow Spring Recorded with: Induction Electrog	18	45.72	0	0	25%	None	Yes
BHC Acoustilog	18	45.72	0	0	25%	None	Caliper Only
1-Arm Compensated Densilog	6	15.24	0	0	6%	None	Yes
Sidewell Epithermal Neutron	6	15.24	0	0	6%	None	Caliper Only
2-Arm Proximity-Minilog	12	30.48	0	0	36%	None	Caliper Only
Micro-Laterolog	12	30.48	0	0	36%	None	Caliper Only
4-Arm 4-Arm Dual Caliper	1	2.54	0	0	4%	None	Yes
High Resolution 4-Arm Diplog	12	30.48	0	0	50%	None	Caliper Only
Minilog Micro Inverse 1-inch × 1-inch	2	5.08	1.5	3.81	7%	Improves	No
Micro Normal 2 inch	4	10.16	4	10.16	7%	Improves	No
Spectralog	24	69.96	6	15.24	100%	None	Yes
Correction curve on compensated density curve	8	20.32	1.5	3.81	10%	Improves	Yes

Characteristics of Measurements Used in the Detection of Fractures (Continued)

Method	Spacing or vertical resolution		Approx. depth of investigation		Percent of circum. of 8¾-inch borehole surveyed	Effect of vugs associated with fractures	Can be recorded in gas or air-filled holes
	Inches	cm	Inches	cm			
Spontaneous potential	12	30.48	0	0	100%	None	No

Resistivity
  Dual Induction
    Focused Log:
    Shallow Focused
	14	35.56	30	76.20	100%	None	No

Medium Induction	60	152.40	70	177.80	100%	None	Yes
Deep Induction	72	182.88	120	304.80	100%	None	Yes
Proximity Log	12	30.48	10	24.40	7%	Improves	No
Dual Laterolog: Shallow Laterolog	24	60.96	30	76.20	100%	None	No
Deep Laterolog	24	60.96	120	304.80	100%	None	No
Micro Laterolog	8	20.32	6	15.24	7%	Improves	No

Comparison of porosity
  Measurements:
    Sidewall Acoustilog
    BHC Acoustilog
	6
	24
		15.24
		60.96
			0 to very
			shallow
					4%
					100%
						None
						None
							No
							No

Compensated Densilog	18	45.72	6	15.24	12%	Improves	Yes
Compensated Neutron	24	60.96	8	20.32	30%	Improves	No
High resolution 4-arm diplog	½	1.27	1.5	3.81	4%	Improves	No
Accoustical variable density frac log BHC Acoustilog	Several Available from 36 to 84 or 91.44 to 213.36		very shallow		100%	None	No
Sidewall Acoustilog	9 or 15 22.86 or 38.10		very shallow		4%	None	No

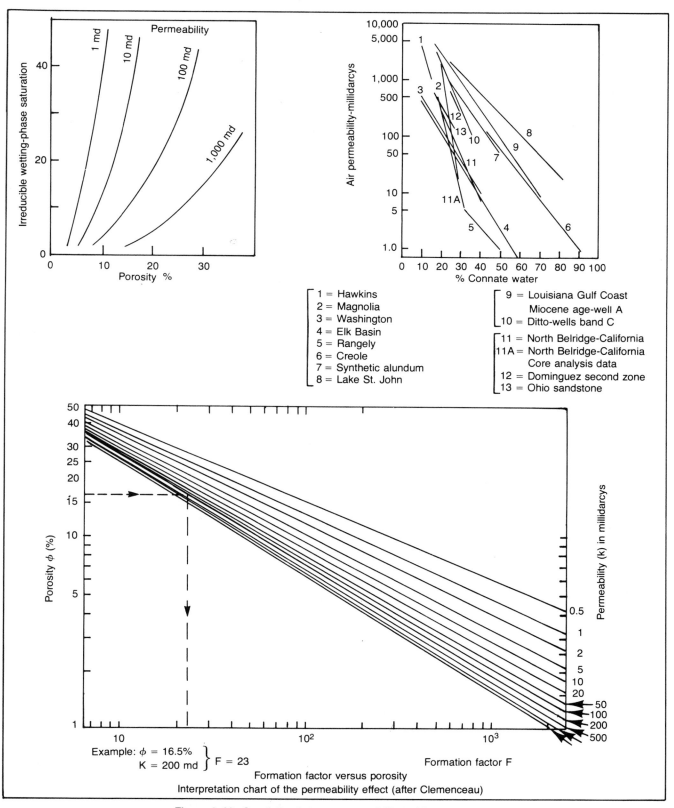

1 = Hawkins
2 = Magnolia
3 = Washington
4 = Elk Basin
5 = Rangely
6 = Creole
7 = Synthetic alundum
8 = Lake St. John

9 = Louisiana Gulf Coast
    Miocene age-well A
10 = Ditto-wells band C

11 = North Belridge-California
11A = North Belridge-California
      Core analysis data
12 = Dominguez second zone
13 = Ohio sandstone

Example: ϕ = 16.5%
         K = 200 md $\big\}$ F = 23

Formation factor F

Formation factor versus porosity
Interpretation chart of the permeability effect (after Clemenceau)

Figure 2–39. *Correlation between permeability and log parameters.*

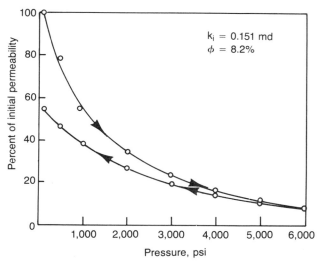

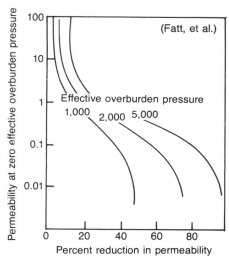

Figure 2–40. *Influence of confining pressure on core permeability (after Thomas and Ward, JPT, February 1972, © SPE-AIME).*

Wettability, Contact Angles, and Threshold Pressures

The capacity of a liquid to spread on a given surface is called wettability. Wettability is a function of the type of fluid and the solid surface. Oils contain chemicals capable at times of changing the initial wettability, and drilling fluids and fluids injected often contain surfactants that affect wettability and contact angles between oil and water. Water poured on a glass spreads out in a thin sheet and its surface makes a very small angle with glass. The glass is water wet or hydrophilic. Mercury on glass does not spread and the angle formed is greater than 90 degrees. The mercury is said to be nonwetting to glass. The interfacial tension of the mercury is about seven times that of water. The liquid with the lower interfacial tension with the solid is the wetting phase.

If oil and water are spread on a pure silica surface (sand), the water will displace the oil. Surface forces are relatively large at and very near an advancing water front in water-wet, oil-bearing formations. Since the contact angle, which is generally measured through the denser phase, can vary with different surfaces and fluids, there are varying degrees of wettability. Oil reservoirs at initial conditions often possess a rather neutral wettability. Some pores are water wet and others are oil wet. Since in most reservoirs the migrating oil entered water-bearing pays and the oil has remained within the pores for a very long time period, this observation is expected. Fig. 2–41 shows the probable location of oil and water in both oil-

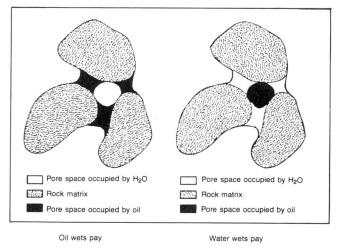

Figure 2–41. *Basic wettability concept (after Calhoun).*

wet and water-wet situations. Some laboratory studies indicate that most reservoirs are preferentially oil wet, but such results may not be sound since many factors in the extraction, drilling, and handling of cores would be expected to favor such a conclusion. Field performance history, results of logging, and a high recovery suggest the in-situ presence of a more water-wet condition than indicated by many core analyses.

As illustrated in Fig. 2–42, a contact angle near zero indicates a rather complete wetting by the water

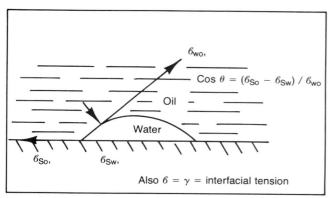

$$\cos \theta = (6_{So} - 6_{Sw}) / 6_{wo}$$

Oil

Water

Also $6 = \gamma$ = interfacial tension

Figure 2–42. *Definition of contact angles.*

phase. An angle of 90 degrees suggests a neutral condition, and an angle of 180 degrees indicates complete wetting by oil.

The distribution of fluids within the interstices of the rock relates with wettability at initial conditions and at a later time when the reservoir is partly depleted. Because of the capillary forces, the wetting fluid tends to occupy the smaller pores and throats of the flow channels. The nonwetting phase occupies in the more open channels.

The effects of wettability must be considered in all determinations of residual oil saturations, capillary pressure, and other similar tests. Important changes in the physical properties of core samples have been noted, which are apparently due to changes that occurred in the rock-wetting characteristics during laboratory extraction with solvents. Heat, air, and surfactants change wettability.

Some of the unexplained irregularities in capillary pressure and other laboratory core data could possibly be attributed to the uncertainties in the wetting properties of the rock samples and possible changes in these wetting properties due to aging and laboratory procedures.

Methods of calculating wettability include measurements of threshold pressures, determining areas under capillary pressure curves, and by direct measurement of contact angles using core surfaces and reservoir fluids. The major disadvantage of these methods is that measurements at surface conditions do not necessarily represent in-situ conditions. Mike Holmes and Doug Tippie (SPE paper 6856, October 1977) combined log data with capillary measurements to estimate wettability.

The Jamin effect is discussed in the following section, taken from Calhoun's *Fundamentals of Reservoir Engineering:* The pressure necessary to keep the interface from moving to the right at point B within the capillary, or, under static conditions in Fig. A is:

$$P_B - P_A = \frac{2\gamma \cos \theta}{r}$$

Now consider a discrete globule of one fluid within another fluid with which it is immiscible, as in Fig. B. There are two interfaces; the pressure drop across each interface is the same but opposite in direction to the other, and there is no net pressure necessary to prevent motion. The total pressure drop between the points A and B is zero, as seen by:

$$P_B - P_A = \left(\frac{2\gamma \cos \theta}{r}\right)_A - \left(\frac{2\gamma \cos \theta}{r}\right)_B = 0$$

Now, if either term of this equation were modified, the net pressure drop between points A and B would not be zero. This condition gives the Jamin effect, i.e., a resistance to flow. The difference may not be zero due to a change in any one of the three terms γ, $\cos \theta$, or r.

Consider first a variation in r as in Fig. C. The capillary is no longer considered to have a uniform radius. The difference in pressure between points A and B is now:

$$P_B - P_A = 2\gamma \cos \theta(1/r_A - 1/r_B)$$

Inasmuch as r_B is less than r_A a positive pressure is required at point A to retain the bubble in the position shown. If flow were to the right, a bubble of oil in the water stream could block such a channel until the pressure drop between points A and B was sufficiently great to push the bubble past the smallest constriction at the point where the channel again widened.

Consider a variation in the angle θ as shown in Fig. D. The situation occurs in many instances due to the existence of different advancing and receding contact angles. The angle θ at A is defined as an advancing contact angle and that at B as a receding contact angle. The former is always larger than the latter. Such a deformation of a bubble takes place when it is on the verge of movement, as toward the right in the figure shown. The resultant pressure between points A and B is:

$$P_B - P_A = \frac{2\gamma}{r}(\cos \theta_A - \cos \theta_B)$$

Inasmuch as θ_A is larger than θ_B, P_A will be larger than P_B. A pressure drop between A and B is necessary to initiate flow. A total of n such bubbles within a

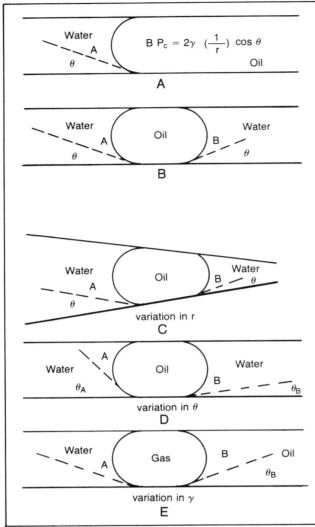

A

$$B\ P_c = 2\gamma \left(\frac{1}{r}\right) \cos \theta$$

variation in r
C

variation in θ
D

variation in γ
E

(after Calhoun, *Fund. of Reservoir Eng.*)

$$P_B - P_A = \frac{2}{r} (\gamma_A \cos \theta_A - \gamma_B \cos \theta_B)$$

and again if $\gamma_B \cos \theta_B$ were greater than $\gamma_A \cos \theta_A$, a positive pressure drop from A to B is necessary to initiate flow to the right.

During the 1940s, engineers worried unnecessarily about this effect. Today's tight sand development (permeability less than 1 md) is forcing engineers to review the older theory relative to the Jamin effect and threshold pressures. Wells in tight pays tend to flood out by water entry in very small amounts, and return of the wells to production following water fracture is difficult if not impossible. The combined effects at millions of pore restrictions can cause capillary pressure forces that are very large compared with the flowing pressure gradients available at the well. If flow is established, these forces act as a rubber band to restrict flow capacity and reduce effective permeability.

Figs. 2–43 and 44 show the effect of wettability on relative permeability, water displacement (water-drive recovery), and the movement of fine material present initially or released by operations within the formation. Some of these problems were noted by Hewitt, *JPT* (August, 1963) as follows:

The causes of damage that are related to rock properties are: 1. swelling of indigenous clays that constrict the pores; 2. dispersion of indigenous, nonswelling particles, rearrangement of the particles during fluid flow, and plugging of the pore system; and 3. a combination of swelling and dispersion; slight swelling promotes loosening and mobility of fine particles.

Causes of damage that originate from a specific production operation are: 1. invasion of solid particles from drilling mud or injected fluids; 2. invasion of solid particles from oil well cementing operations; 3. precipitation of salts from the reservoir brine in the zone of relatively lower pressure around the well bore; 4. water block or emulsion block; 5. direct or indirect effects of bacteria; and 6. completion practices not suitable for a specific reservoir.

A partial change in wettability often is associated with problems so discussed.

tube would require a pressure drop of $n(P_B - P_A)$ to move them.

A third situation giving the Jamin effect occurs when the interfacial tension is variable. For example, as shown in Fig. E, if the bubble of gas is bounded on one side by oil and on the other by water. The net effect between A and B is then:

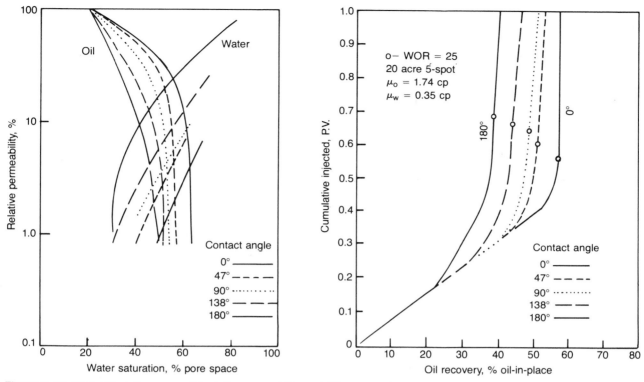

Figure 2–43. *Wettability influence on (A) relative permeability and (B) recovery by water displacement of oil (after Owen and Archer, JPT, July 1971, © SPE-AIME).*

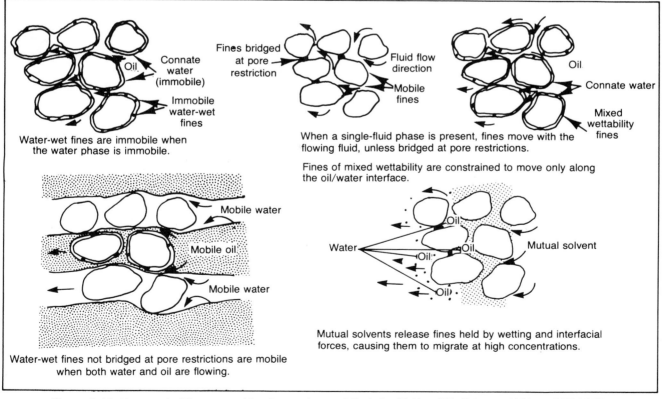

Figure 2–44. *Movement of fines caused by changes in wettability (after McKee, JPT, February 1979, © SPE-AIME).*

Capillary Pressure, Interfacial Tension, and Surface Tension

High school science classes often include experiments showing the capillary rise of water and mercury in capillary tubes of various sizes, such as illustrated in Fig. 2–45. The principles apply to water-wet and oil-wet pay rocks. Also, mercury often is used to measure the capillary pressure of cores in the laboratory. The application of these principles to oil reservoirs is illustrated by Fig. 2–46. Tight sands or pays usually contain more water than more permeable pays, and the capillary pressure curves differ considerably. Clean pays with 45–50% water often produce water-free oil on completion (before coning) and dirty pays with 60–65% water are capable of producing little water. Capillary pressure can be related with permeability and water saturation, as illustrated by Figs. 2–47 and 48. Typical relations between capillary pressure and water saturation for sandstone and calcareous cores are illustrated in Figs. 2–48 and 49.

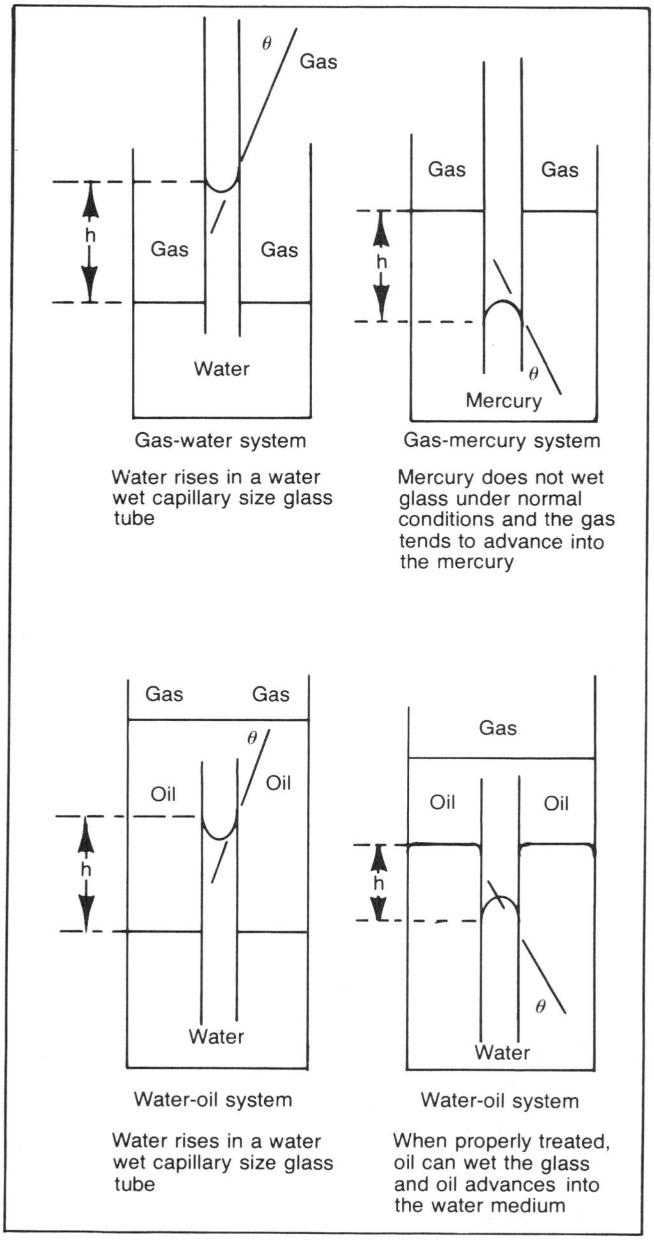

Capillary pressure measurements on cores can be made using the following procedures:

- Desaturation of displacement process through a porous diagram or membrane (restored state or Welge method)
- Centrifugal method
- Dynamic capillary pressure method
- Mercury injection method
- Evaporation method

The displacement method can require 10 to 40 days to analyze a sample. However, the fluids closely simulate those found in the reservoir. Also, the core must be representative and not affected by drilling fluids. The mercury injection method offers the advantages of speed and the range of pressure investigated is increased. Disadvantages are the difference in wetting properties and the permanent loss of the core since mercury cannot be completely removed.

The basic capillary pressure equation is:

$$P_{c_2} = \frac{P_{c_1} (\gamma \cos \theta)_2}{(\gamma \cos \theta)_1}$$

$$P_{c_2} = (\rho_w - \rho_o) h$$

$$h = \frac{P_{c_2}}{\rho_w - \rho_o}$$

subscript 1 refers to lab data or values
subscript 2 refers to reservoir data or values

P_c = capillary pressure, psi
γ = interfacial tension
θ = contact angle
ρ_w = water density, psi/ft
ρ_o = oil density, psi/ft
h = height above zero capillary pressure, ft

Technique: (1) Convert lab P_c to equivalent reservoir P_c
(2) Calculate height above free water zone

Figure 2–45. *Illustration of capillary forces (after Calhoun).*

$$\gamma_{o\text{-}w} = \frac{rh (\rho_w - \rho_o) g}{2 \cos \theta}$$

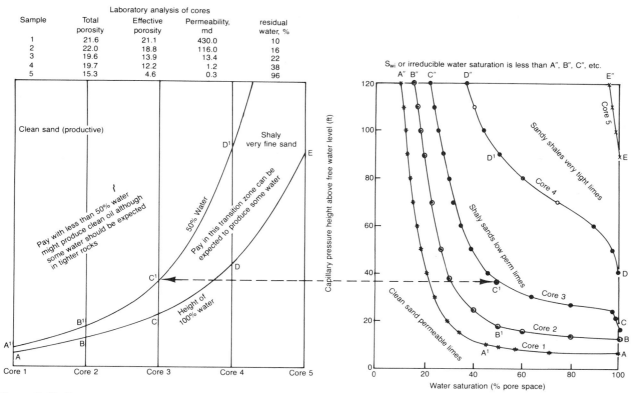

Sample	Total porosity	Effective porosity	Permeability, md	residual water, %
1	21.6	21.1	430.0	10
2	22.0	18.8	116.0	16
3	19.6	13.9	13.4	22
4	19.7	12.2	1.2	38
5	15.3	4.6	0.3	96

Laboratory analysis of cores

Figure 2–46. *Data illustrating relations between sand type, porosity, permeability, and capillary pressure (courtesy Shell Oil Company).*

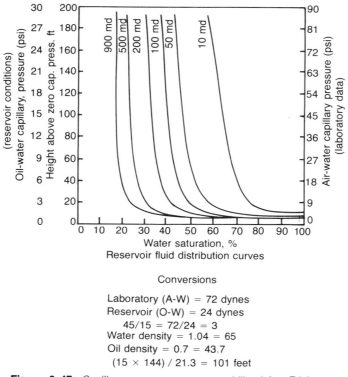

Reservoir fluid distribution curves

Conversions

Laboratory (A-W) = 72 dynes
Reservoir (O-W) = 24 dynes
45/15 = 72/24 = 3
Water density = 1.04 = 65
Oil density = 0.7 = 43.7
(15 × 144) / 21.3 = 101 feet

Figure 2–47. *Capillary pressure vs. permeability (after Frick, © SPE-AIME).*

Since interfacial tension enters as a ratio, pressure in any consistent units can be used with the interfacial tension in dynes per centimeter.

Procedures are available for laboratory measurement of interfacial tension and contact angles when reliable values are desired. Surface forces might be changed by chemicals, heat, and other factors and proper handling of cores and fluids is required. In the absence of laboratory data, the following values may be used as approximations:

surement is 72 dynes, the value for γ cosine θ at reservoir conditions is believed to approximate 26 dynes, and the measured densities of the reservoir water and oil are 68 and 53 lb/ft³

$$P_{c(reservoir)} = 18 \times 26/72 = 6.5 \text{ psi}$$

$$h = 6.5/\frac{(68-53)}{144} = 62.4 \text{ feet}$$

Using this method, construction of height scales for

System	θ Contact angle	Cosine of the contact angle	γ Interfacial tension	γ (Cosine θ)
Laboratory				
Air-water	0	1.0	72 @ 70°F	72
			60 @ 200°F	60
Air-brine	0	1.0	same	60–75
Oil-water	30	0.866	48	42
Air-mercury	140	0.765	480	367
Air-oil	0	1.0	24	24
Mercury-vacuum	140	0.765	468 @ 0°C	358
			480 @ 60°C	367
Reservoir				
Oil				0–40
Gas				zero
Water-oil	30	0.866	30*	26*
Water-gas	0	1.0	50*	50*

* Pressure and temperature dependent. Value shown reasonable to 5000 ft.

The free-water level is defined as the level at which capillary pressure is zero. As illustrated in Fig. 2–46, this water level is at the lowest possible height in clean, more permeable pay; this value often serves as the reference value. The free-water level often is not the level at which only water is produced from a well. The 100% water level (water-bearing zones) can be determined by use of logs, drill-stem tests, and production history when adjusted using relative permeability data.

The P_c value measured by mercury injection may be converted to a P_c value for an oil-water system (both at surface) as follows:

γ Cos θ for mercury-vacuum at 70°F   360
γ Cos θ for oil-water system at 70°F   42

$$\frac{P_{c(Hg/vac.)}}{P_{c(oil-water)}} = \frac{360}{42} = 8.57$$

When the laboratory measured $P_c = 18$ psi, which also relates with a water saturation, $S_w = 35\%$, γ cosine θ for water-oil system used in the laboratory mea-

capillary pressure presentation becomes easy. Other conversions follow:

γ (Cosine θ)	$P_c/P_{c(Hg.-vac)}$	=	γ (Cosine θ)/367
70	1/5.1		0.196
50	1/7.4		0.136
35	1/10.5		0.095
25	1/14.7		0.068
10	1/36.8		0.027
1			0.0027

The density of fresh water is 62.4 lb/cu ft = 0.433 psi/foot and is reflected in a water rise of 100 feet for each 43.3 psi. Densities of reservoir fluids usually are in the following ranges:

Liquid	Specific Gravity	lb/cu ft	lb/ft
water	1.0–1.1	62.4–68.6	0.433–0.476
oil	0.7–0.9	43.7–56	0.303–0.389
gas	0.1–0.25	6.2–20	0.043–0.139

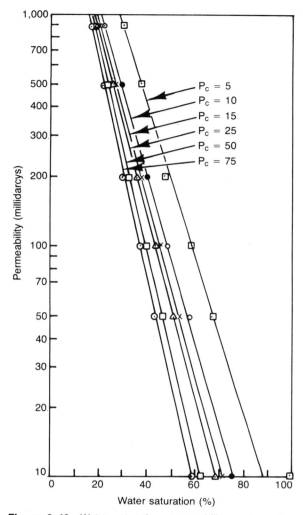

Figure 2–48. *Water saturation, permeability, and capillary pressure correlation (after Frick, © SPE-AIME).*

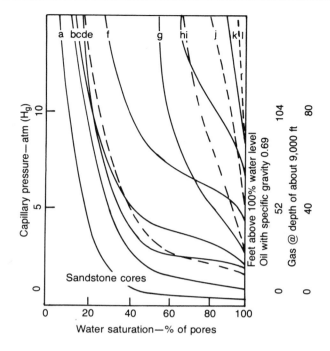

Sandstone cores

Curve letter	Total porosity	Permeability to air,
	%	md
a	17	285
b	12	8
c	19	13
d	14	3
e	32	30
f	20	1
g	12	0.5
h	11	0.3
i	28	2
j	25	0.4
k	15	0.3
l	25	0.1

Figure 2–49. *Illustrative capillary pressure curves for sandstone cores.*

Capillary pressure information differs for the imbibing and the desaturation conditions and the proper data must be used. The height of the water column calculated from laboratory capillary pressure data must agree with the height calculated from logs. Differences must be explained.

The capillary pressure may be related with radius of the pores as shown below:

Pore entry radii calculation

$$Ri = \frac{2\,\gamma \cdot \cos\theta \cdot C}{Pc}$$

Distribution function calculation

$$\frac{1}{D(Ri)} = \frac{Vi \cdot 10^4}{\Delta R}$$

Where:
Ri = Pore radii, microns (average value at midpoint of ΔR)
Pc = Capillary pressure in laboratory, psi
γ = Interfacial tension, dynes/cm
θ = Contact angle, degrees
$D(Ri)$ = Distribution function (as a function of Ri), cm²
Vi = Incremental volume of mercury injected (fraction of total pore volume)
C = Conversion constant equal to 145×10^{-3}
ΔR = Incremental change in radius over which Vi was determined

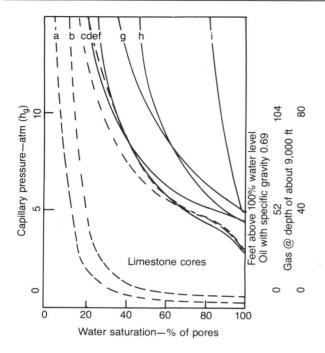

Limestone cores

Curve letter	Total porosity	Permeability to air,
	%	md
a	11	886
b	13	51
c	10	0.4
d	37	15
e	12	1.6
f	28	12
g	25	4
h	17	1
i	18	0.6

Figure 2–50. *Illustrative capillary pressure curves for calcareous cores.*

The rise of water in surface rocks (ground-water analysis) is much less than that encountered in tighter consolidated rocks found at deep depths. The surface rocks have porosities in range of 41%. The capillary rise in surface rocks is shown below:

Material	Grain size (mm)	Capillary rise (cm)
Fine gravel	5–2	2.5
Very coarse sand	2–1	6.5
Coarse sand	1–0.5	13.5
Medium sand	0.5 –0.2	24.6
Fine sand	0.2 –0.1	42.8
Silt	0.1 –0.05	105.5
Silt	0.05–0.02	200

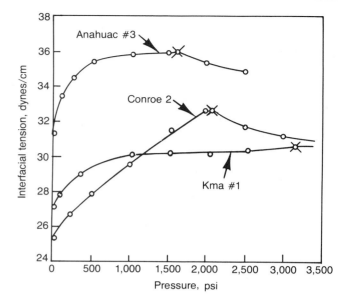

(1) 33.5° API crude liberating 640 ft³ of gas per barrel when flashed from 3,120 psi, 178° F. to atmospheric pressure and 78° F.

(2) 36.9° API crude liberating 550 ft³ of gas per barrel on flashing from 2,035 psi and 170° F. to 78° F. and atmospheric.

(3) 41.3° API crude liberating 650 ft³ gas per barrel when flashed from 1,570 psi and 130° F. to atmospheric.

Note: All experiments conducted at reservoir temperatures.

Figure 2–51. *Interfacial tension between water and saturated oils (after Holcott, Trans. AIME, 1939).*

Fluid saturation-capillary pressure relationships are affected by the physical properties of the rock samples (pore structure, porosity, and permeability). Leverett proposed a correlating function, called the *J*-function, as follows:

$$J(S_w) = \frac{P_c}{\sigma} \left(\frac{k}{\phi} \right)^{1/2}$$

where: P_c = capillary pressure, dynes/cm
σ = interfacial tension, dynes/cm
k = permeability, sq cm
ϕ = fractional porosity

Note that in using capillary pressure data to determine connate water saturation, practical oil-field units can be used. This will only shift the position of the $J(S_w)$ versus S_w curve.

Use of the *J*-curve should be limited to single reservoirs since the same values have been known to not apply for a group of closely related fields. The method is illustrated in the following table:

Air-Brine Capillary Pressure Data

Sample number	Depth feet	Perm md	Porosity percent	$\left(\frac{k}{\phi}\right)^{0.5}$	$P_{cL} = 1\ psi$		$P_{cL} = 2\ psi$		$P_{cL} = 4\ psi$		$P_{cL} = 8\ psi$		$P_{cL} = 15\ psi$		$P_{cL} = 35\ psi$		$P_{cL} = 100\ psi$*	
					J	S_w	J	S_w	J	S_w	J	S_w	J	S_w	J	S_w	J	S_w
1	10,097.5– 10,098	15	13.4	1.06	0.015	100	0.030	100	0.061	86.9	0.121	49.2	0.227	39.1	0.529	32.3	1.511	26.9
2	10,101.0– 10,101.5	73	15.1	2.20	0.031	100	0.063	84.9	0.126	44.5	0.251	32.6	0.471	26.3	1.100	21.8	3.141	17.4
3	10,370– 10,371	21	14.9	1.187	0.0170	100	0.0339	99.2	0.0678	76.2	0.136	51.9	0.254	41.1	0.594	31.7	1.696	25.2
4	10,379– 10,380	14	12.9	1.04	0.015	100	0.030	100	0.059	92.9	0.119	54.7	0.223	39.5	0.520	28.0	1.488	21.5
5	10,382– 10,383	28	14.4	1.39	0.020	100	0.040	99.1	0.079	68.7	0.159	46.3	0.298	34.7	0.695	25.4	1.992	19.6

* Equivalent pressure from centrifuge

J-Function Calculation

1) $$J = \frac{P_{cL} \, (k/\phi)^5}{\sigma_L \cos \theta_L}$$

2) $$P_{cR} = P_{cL} \frac{\sigma_r \cos \theta_r}{\sigma_L \cos \theta_L}$$

3) $$P_{cR} = \frac{h}{144} (\rho_{wr} - \rho_{or})$$

where:
$$\theta = \theta_r = \theta_L = 0°$$
$$\cos \theta = 1.0$$
$$\gamma = \sigma_L = 70 \text{ dynes/cm}$$
$$\gamma = \sigma_r = 28 \text{ dynes/cm}$$
$$\rho_{wr} = 58.5 \text{ lbs/ft}^3$$
$$\rho_{or} = 46 \text{ lbs/ft}^e$$

$$h = \frac{322.56 \, J}{(k/\phi)^{0.5}}$$

Note: ϕ in percent in above equation

Averaging of Capillary Pressure Data

1. The table presents the basic laboratory data on the five samples and the assumptions, equations, and calculations used to correct the data to reservoir conditions. The Leverett J-function was utilized as a convenient method of averaging the capillary data. Plot J versus k/ϕ, Sw, k, ϕ, etc. An average curve was constructed through the data points of J versus k/ϕ.
2. To denormalize the data, the following procedure can be used:
   a. Determine S_{wi} from the log and core data
   b. Determine k/ϕ from log, cores and transient data
   c. At various values of $(S_w - S_{wi})$, determine J value
   d. Calculate h using the following equation:

$$h = \frac{322.56 \, J}{(k/\phi)^{0.5}}$$

Relative Permeability Concepts

The absolute permeability of a rock to flow of a homogeneous fluid that does not react with the rock is defined as k. When two or more fluids flow at the same time, the relative permeability of each phase at a specific saturation then existing is the ratio of the effective permeability of the phase to the absolute permeability and

$$k_{ro} = k_o / k$$
$$k_{rg} = k_g / k$$
$$k_{rw} = k_w / k$$

where subscripts o, g and w represent oil, gas and water, respectively.

The sum of the relative permeability is less than

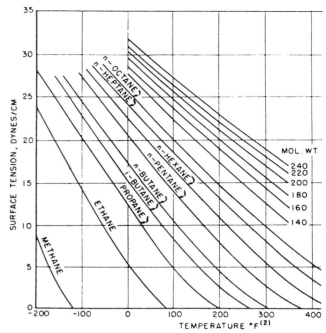

Figure 2–52. *Surface tension of oil components (after Katz, Monroe, and Trainer,* JPT, *September 1943,* © *SPE-AIME).*

one when two or more fluids are flowing simultaneously. Flow in laboratory thin sections indicates that the separate phases flow largely in separate channels. Critical or irreducible saturation values for each phase are required before flow in measurable quantities can start as illustrated by Figs. 2–53, 54. Water is often

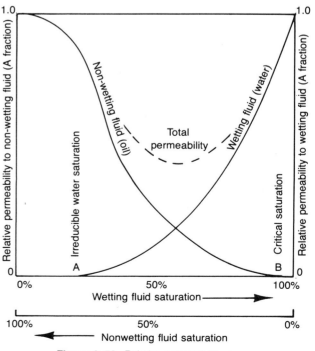

Figure 2–53. *Relative permeability curve.*

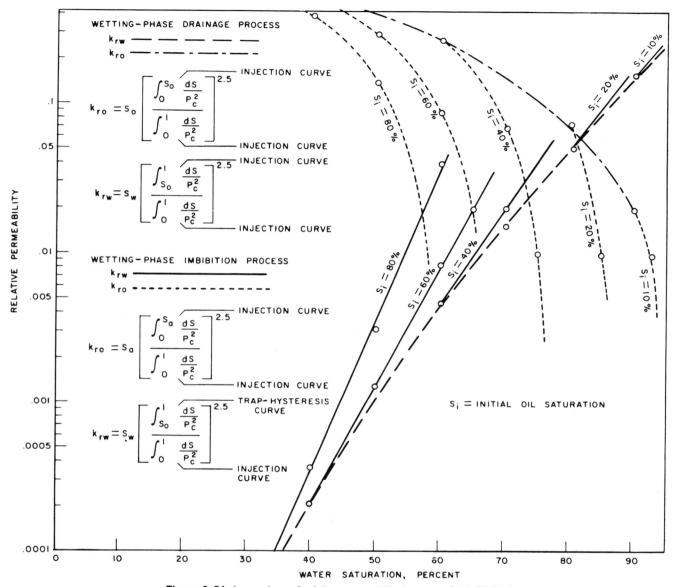

Figure 2-54. *Low values of relative permeability (courtesy Shell Oil Co.)*

the wetting fluid, but oil-wet pays exist. Oil and gas are often the nonwetting fluids. In solution-gas drive reservoirs, water saturation may be small and non-movable so that liquid saturation is substituted for the wetting fluid saturation. For this condition, B is the critical gas saturation (value 90–100%) and A is the irreducible water saturation, which is often the minimum water saturation indicated by logs and capillary pressure data at a height above the water-oil contact.

Recoveries calculated using the various recovery equations can be sensitive to values of relative permeability to oil at very low oil saturations. Conventional laboratory data can be extended to low oil saturations

by plotting measured values of effective permeability to oil on log-log paper using $(S_o - S_{or}^*)/(1 - S_{wi} - S_{or}^*)$ for the saturation in the presence of connate water. S_{or}^* should be small and near zero. The log-log plot will approximate a straight line, and the general equation should be useful:

$$K_{ro} = (S_o - S_{or}^*)/(1 - S_{or}^* - S_{wi})^n$$

The slope of line n should be approximately four but may be as small as three and as large as six. The data measured in the laboratory differ substantially for each core analyzed, and the relative permeability information is normalized to obtain relative permeability curves having common scales when relative permeability is plotted versus saturation, as illustrated on Fig. 2–55.

Using end values to which all data for each core analyzed are adjusted, the measured values can be adjusted or normalized using the following equations:

$$k_{ro} = A(1 - S_w)^n$$
$$k_{rw} = C(S_w)^m$$

n and m are *not* the Archie values. The use of these equations is illustrated in the following section.

Normalized value	Type of system		
	gas/oil	gas/water	water/oil
k_{ro}^*	k_{ro} @ S_L/k_{ro} @ S_{wi}	—	k_{ro} @ S_w/k_{ro} @ S_{wi}
k_{rw}^*	—	k_{rw} @ S_w/k_{rw} @ S_{gr}	k_{rw} @ S_w/k_{rw} @ S_{or}
k_{rg}^*	k_{rg} @ S_L/k_{rg} @ S_{gr}	k_{rg} @ S_w/k_{rg} @ S_{gr}	—
S_w^*	$\dfrac{(S_L - S_{wi})}{(1 - S_{wi} - S_{gr})}$	$\dfrac{(S_w - S_{wi})}{(1 - S_{wi} - S_{gr})}$	$\dfrac{(S_w - S_{wi})}{(1 - S_{wi} - S_{or})}$

S_{wi} = Water saturation at initial conditions, percent

k_o @ S_{wi} = Oil permeability at initial conditions (S_{wi}), md

S_{or} = Oil saturation at terminal conditions, percent

k_w @ S_{or} = Water permeability at terminal conditions (S_{or}), md

$R.F.$ = Recovery factor, percent oil in place

Also, all pore space is assumed to be filled with oil, gas, and water at all times. The absolute permeability may be less than the air permeability measured for cores in the laboratory if the pore space should be increased by handling and drying the core or the natural permeability of the core is altered by clay swelling or movement.

The shape of the relative permeability curves suggests that relative permeability curves may be approximated by the following general equations;

The relative permeability of a formation and a core is influenced by a number of factors that usually are related to water saturation and irreducible water saturation and wettability. Fertl has illustrated the relation with capillary pressure and the use of the fractional flow equation, as shown in Fig. 2–56. He also shows the direction of change with changing wettability with Fig. 2–57. Tighter, lower-quality rocks usually have poor relative permeability characteristics and higher in-situ or irreducible water saturation. Data presented by Wyllie in Frick's *Petroleum Production Handbook* and by Arps and Roberts in *JPT* (August 1955) are combined as shown in Fig. 2–58 for the water-oil system and Fig. 2–59 for the gas-oil system when production is from water-wet sandstone. The curves are presented as relative permeability ratios that are directly useable in the fractional flow equations. A fractional flow equation may be used directly to obtain end points or suggested recovery factors when viscosities are known. The middle range of the curves is usually involved in recovery calculations, and the use of log-log type equations often is permissible over the limited range of interest.

Laboratory values for relative permeability should be used with caution to forecast the future performance of an oil or gas reservoir. Methods described in a later part of this text should be used to determine relative permeability ratios based on field performance whenever sufficient field information or history is available. The slopes of the two data sets often are comparable, but starting points are usually different. In Louisiana clean Miocene sands, the irreducible water saturation may be in the range of 10 to 25% and little or no moveable water is producible until water saturation in a pay opposite the perforations exceeds about 45%. In tighter pays, the irreducible water may be above 60%, and first measureable water is noticed when water saturation opposite the perfora-

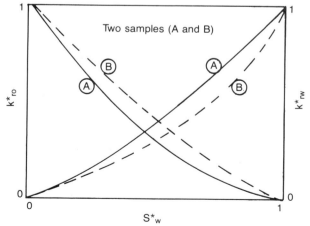

Figure 2–55. *Normalized relative permeability curve for two wells (from JPT, February 1976).*

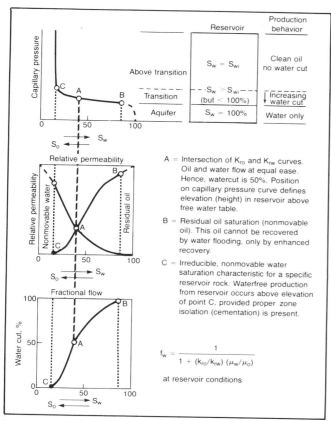

Figure 2–56. *Relation between capillary pressure, relative permeability, and fractional flow (after Fertl, OGJ, 22 May 1978).*

tion is only slightly higher. As discussed under log analysis, shales and small pores increase water saturation. In gas-oil systems or solution-gas drive recovery systems, a critical free gas must be present before gas can flow. The amount of free gas present before gas moves in measurable quantities often is 3% and may be greater than 10%. Temperature as well as water salinity can cause mica and shales to break up in water-sensitive rocks so that particles move to block pores and reduce effective permeability.

The relative permeability concept has received the attention of many authors during the past 40 years, and the publication indexes, particularly those in SPE publications, should be consulted. The relative permeability may be related to pore size distribution by the following equations, which are expressed in terms of capillary pressure:

$$K_{rw} = \left[\frac{S_L - S_{wi}}{1.0 - S_{wi}}\right]^2 \frac{\int_{S_{wi}}^{S_L} \frac{dS_w}{(Pc)^2}}{\int_{S_{wi}}^{1.0} \frac{dS_w}{(Pc)^2}}$$

$$K_{rnw} = \left[\frac{1.0 - S_L}{1.0 - S_{wi}}\right]^2 \frac{\int_{S_L}^{1.0} \frac{dS_w}{(Pc)^2}}{\int_{S_{wi}}^{1.0} \frac{dS_w}{(Pc)^2}}$$

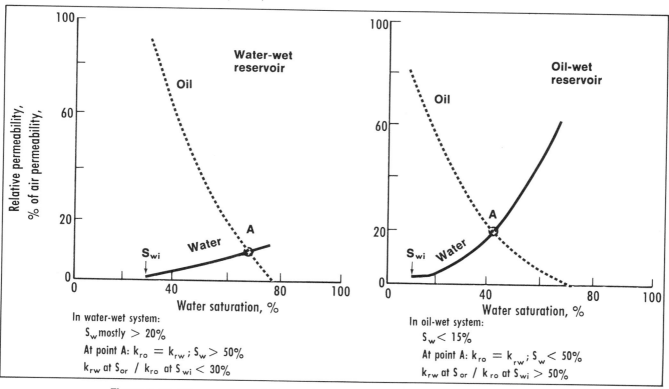

Figure 2–57. *Influence of wettability on relative permeability (after Fertl, OGJ, 22 May 1978).*

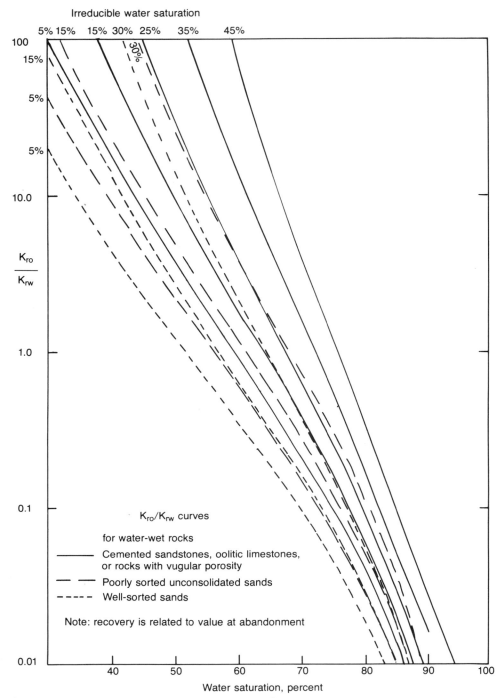

Figure 2–58. *Relative permeability ratio curves for the oil-water system* (*after Frick*, Petroleum Production Handbook, © *SPE-AIME*).

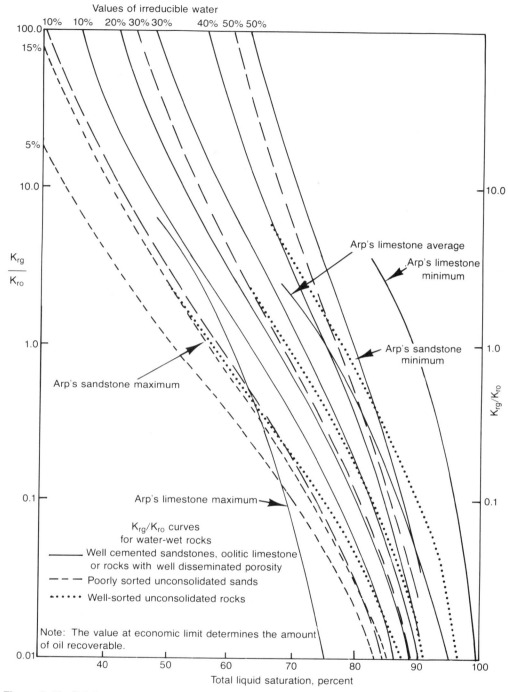

Figure 2–59. *Relative permeability ratio curves for the gas-oil system (after Frick,* Petroleum Production Handbook, © *SPE-AIME).*

$$\frac{K_g}{K_o} = \frac{K_{rg}}{K_{ro}} = \frac{K_{rnw}}{K_{rw}}$$

Where: K_{rw} = Relative permeability to wetting phase (K_{rw} & K_{ro})

K_{rnw} = Relative permeability to non-wetting phase (K_{rg})

S_{wi} = Minimum connate water saturation

S_L = Total liquid saturation = (1.0 − gas saturation)

P_c = Capillary pressure

S_w = Wetting phase saturation from Pc curve

Note: Equations as written apply to case where initial water in sample is at irreducible (S_{wi}).

A rough estimate of the relative permeability is available from logs such as

$$k_{rw} = \left[\frac{S_w - S_{wi}}{1 - S_{wi}}\right]^4$$

$$k_{ro} = \left[\frac{S_o}{1 - S_{wi}}\right]^4$$

$$k_{rg} = \left[1 - \frac{S_o}{S_m - S_{wi}}\right]^2 \cdot \left[1 - \left(\frac{S_o}{1 - S_{wi}}\right)^2\right]$$

Capillary pressure curves for the imbibition and drainage cycles are not identical. The above equations are based on a linear relationship between the reciprocal of the capillary pressure squared and effective saturation and apply to the drainage direction only where the wetting phase saturation is decreasing so that

$$k_{rw} = \left(\frac{S_w - S_{wi}}{1 - S_{wi}}\right)^2 \frac{\int_{S_{wi}}^{S_w} dS_w/p_c^2}{\int_{S_{wi}}^{1.0} dS_w/p_c^2}$$

Also, intergranular-type porous media with limited cementing material to complicate pore configuration are assumed.

Flow of oil and water in a water-wet, intergranular-type porous medium can often be expressed as:

$$k_{rw} = S_w^4 \left(\frac{S_w - S_{wi}}{1 - S_{wi}}\right)^{1/2}$$

$$k_{ro} = \left[1 - \frac{S_w - S_{wi}}{1 - S_{wi} - S_{nwt}}\right]^2$$

Gas flow in the drainage direction in the presence of a flowing liquid in intergranular rocks often may be represented:

$$k_{rw} = S_{we}^{3/2} S_w^3$$

$$k_{rg} = (1 - S_{we}) [1 - S_{we}^{1/4} S_w^{1/2}]^{1/2}$$

In the log-based equations, S_w is the water saturation and S_{wi} is the irreducible water saturation.

Triangular graph paper is usually used to present relative permeability when three phases flow simultaneously, as illustrated by Fig. 2–60. For water-wet systems, operating on a drainage cycle with respect to water and oil (gas saturation is increasing at the expense of water and oil saturations) in unconsolidated well-sorted sand grains, the following equations may be used:

$$k_{rg} = \frac{S_g^3}{(1 - S_{wi})^3}$$

$$k_{ro} = \frac{S_o^3}{(1 - S_{wi})^3}$$

$$k_{rw} = \left(\frac{S_w - S_{wi}}{1 - S_{wi}}\right)^3$$

If the rock is a cemented sandstone, vugular rock, or oolitic limestone, the equations shown below may be used:

$$k_{rg} = \frac{S_g^2 [(1 - S_{wi})^2 - (S_w + S_o - S_{wi})^2]}{(1 - S_{wi})^4}$$

$$k_{ro} = \frac{S_o^3 (2S_w + S_o - 2S_{wi})}{(1 - S_{wi})^4}$$

$$k_{rw} = \left(\frac{S_w - S_{wi}}{1 - S_{wi}}\right)^4$$

Field experience suggests that oil, water, and gas flow as banks of each fluid in primary and secondary operations that are successful, and the pore volume

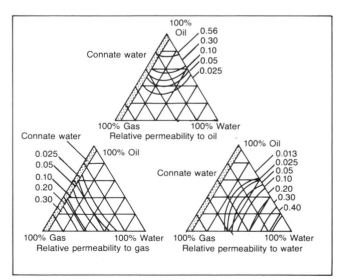

Figure 2–60. *Three-phase relative permeability (after Caudle, Slobad, and Brownscombe,* JPT, *May 1951,* © *SPE-AIME).*

subjected to simultaneous flow of three fluids is rather limited. Oil banks often are not present in the respective streamlines in tertiary operations, and three-phase relative permeability becomes more important.

Wyllie in Frick defines $S*$ as

$$S* = S_o/(1 - S_{wi})$$

for the oil-gas system to arrive at the following relations:

Type of formation	k_{ro}	k_{rg}
Unconsolidated well-sorted sand	$(S*)^3$	$(1 - S*)^3$
Unconsolidated poorly sorted sand	$(S*)^{3.5}$	$(1 - S*)^2(1 - S*^{1.5})$
Cemented sands, oolitic limestones, and some vugular porosity	$(S*)^4$	$(1 - S*)^2(1 - S*^2)$

$S_w^* = (S_w - S_{wi})/(1 - S_{wi})$ may be substituted for $S*$ for the oil-water system equations.

Wyllie also proposes the following equations for three-phase relative permeabilities when the pay is water wet for the drainage cycle with respect to water and oil:

	Unconsolidated well-sorted sand	Cemented sandstone, oolitic limestone, vugular rocks
K_{rw}	$(S_w^*)^3$	$(S_w^*)^4$
K_{ro}	$(S_o)^3/(1 - S_{wi})^3$	$(S_g)^3/(1 - S_{wi})^3$
K_{rg}	$(S_o)^3(2S_w + S_o - 2S_{wi})^4/(1 - S_{wi})^4$	$(S_g)^2[(1 - S_{wi})^2 - (S_L - S_{wi})^2]/(1 - S_{wi})^4$
where	$S_w^* = (S_w - S_{wi})/(1 - S_{wi})$	
	$S_L = (S_w + S_o)$	

Various crossplots can aid in selecting other values for these constants, as discussed in the next section of this text.

Relative permeability data for limestone reservoirs has been assembled as Fig. 2–61. Similar data for the western tight sands is shown as Fig. 2–62. Results of other laboratory determinations are presented as Fig. 2–63. Relative permeability ratio curves can be calculated from field performance (Fig. 2–64).

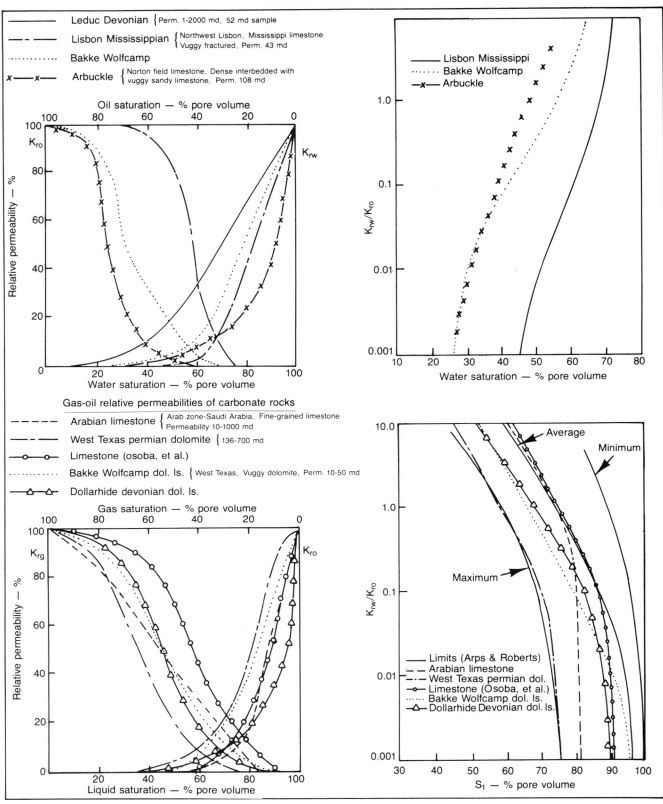

Figure 2–61. *Relative permeability of carbonates.*

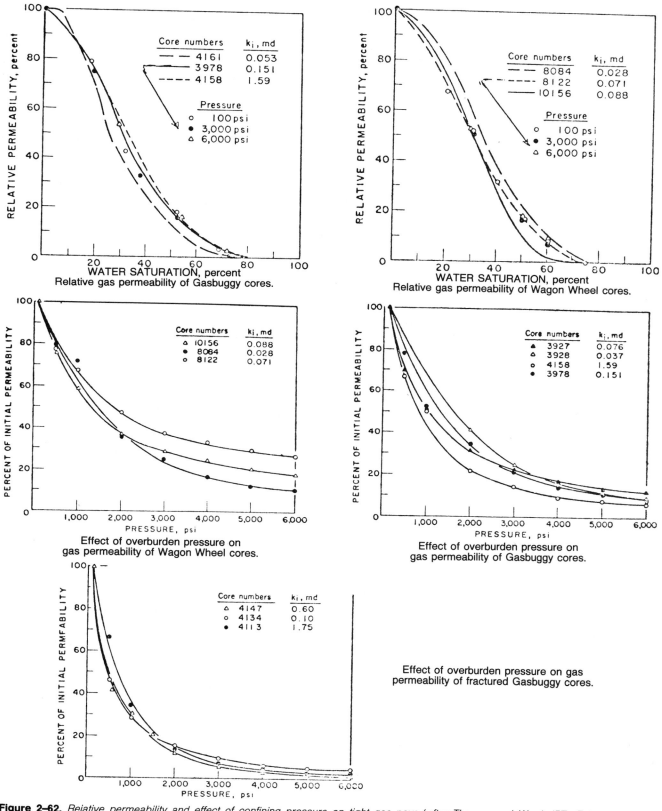

Relative gas permeability of Gasbuggy cores.

Relative gas permeability of Wagon Wheel cores.

Effect of overburden pressure on gas permeability of Wagon Wheel cores.

Effect of overburden pressure on gas permeability of Gasbuggy cores.

Effect of overburden pressure on gas permeability of fractured Gasbuggy cores.

Figure 2–62. *Relative permeability and effect of confining pressure on tight gas pays (after Thomas and Ward, JPT, February 1972). These sands have a very low permeability after adjustment to reservoir conditions. The in-situ water is high and the relative permeability further reduces the effective permeability so that rates are low even after large fracture operation. Separation of gas from water zones using logs is difficult and the water standing opposite the pay is detrimental (© SPE-AIME).*

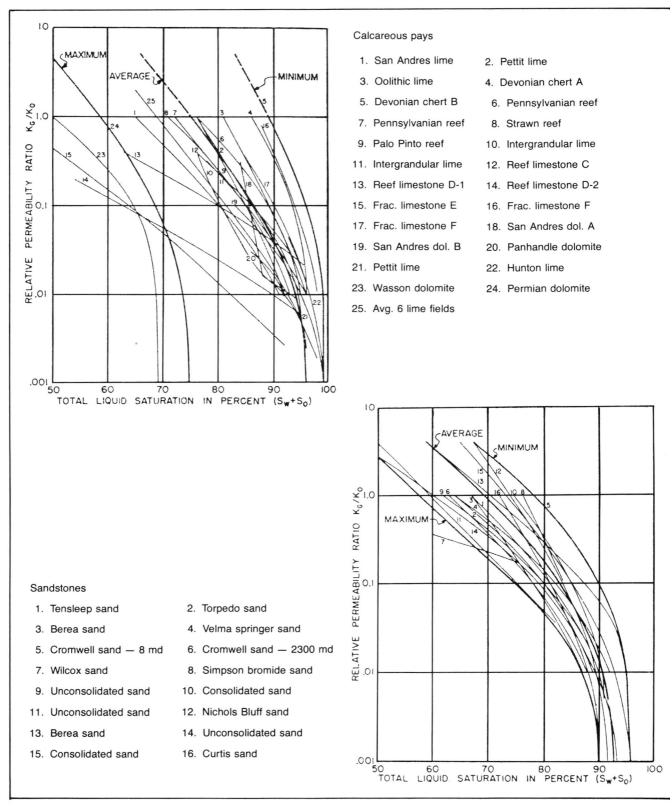

Figure 2–63. *Relative permeability ratio curves for dolomite, limestone, and chert cores and sand data (after Arps and Roberts, JPT, August 1955, © SPE-AIME).*

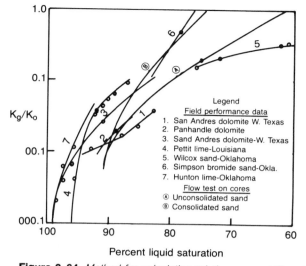

Average pressure	R	ΔN	ΔN/N	β	v	r	μ_o/μ_g	S_o	k_g/k_o
3.448	850	.476 × 10⁶	.0041	1.443	.000840	752	30.4	.710	.00187
3.303	920	1.743 × 10⁶	.0150	1.432	.000875	725	32.1	.696	.00321
3.153	990	2.818 × 10⁶	.0242	1.420	.000910	695	34.0	.684	.00556
2.938	1,020	4.652 × 10⁶	.0399	1.403	.000970	547	36.8	.664	.00682
2.813	1,000	6.030 × 10⁶	.0517	1.393	.001010	632	38.4	.652	.00694
2.678	1,180	7.360 × 10⁶	.0632	1.382	.001062	608	40.5	.638	.01085
2.533	1,420	8.751 × 10⁶	.0752	1.371	.001122	580	42.4	.625	.01620
2.453	1,510	9.873 × 10⁶	.0847	1.364	.001162	565	43.6	.615	.01848
2.318	1,660	11.259 × 10⁶	.0965	1.354	.001230	540	45.5	.609	.0224
2.153	1,920	12.619 × 10⁶	.1083	1.340	.001330	509	48.0	.589	.0292
1.978	2,220	13.998 × 10⁶	.1202	1.326	.001453	476	50.8	.575	.0377
1.818	2,480	15.321 × 10⁶	.1318	1.313	.001590	446	53.8	.563	.0456
1.658	2,710	16.552 × 10⁶	.1420	1.301	.001758	416	57.4	.550	.0567
1.625	2,800	16.929 × 10⁶	.1453	1.298	.001795	410	58.2	.546	

Legend
Field performance data
1. San Andres dolomite W. Texas
2. Panhandle dolomite
3. Sand Andres dolomite-W. Texas
4. Pettit lime-Louisiana
5. Wilcox sand-Oklahoma
6. Simpson bromide sand-Okla.
7. Hunton lime-Oklahoma
Flow test on cores
Ⓐ Unconsolidated sand
Ⓑ Consolidated sand

Figure 2–64. *Method for calculating relative permeability ratio curves (after H. H. Kaveler, AIME Trans., 1944, © SPE-AIME).*

Analysis of Core Information Combined with Log Data

Reservoir rocks are not uniform but vary both vertically and horizontally. Also, the rock sampled by cores and logs is a very small part of the total volume containing hydrocarbons. The problem involved with obtaining representative average values is very substantial. Statistical evaluation methods and equations are reliable, but reservoir engineering as applied to a specific reservoir remains an art rather than a science. Determining cutoff values—reservoir volumes that will not effectively contribute to recovery—is most difficult and depends upon the anticipated recovery mechanism applicable for the field. Also, determining sound equations representative of field characteristics of the rock or average values thereof is a major task.

The author is indebted for the following analysis techniques to Wayne Beeks, Keplinger and Associates, Denver (formerly with H. K. van Poollen and Associates). The methods used are included on the respective graphs that follow this discussion. The digital log analysis methods used by the respective logging companies, conventional core analysis, and the usual special core analysis techniques have been combined to select the following weighted average petrophysical values for use in recovery analysis work. Please study Figs. 2–65 through 2–81 and the basic information from cores and logs shown on Tables 1 & 2.

Weighted average porosity	17.1%
Weighted average water saturation	29.7%
Geometric mean permeability	30.0 md

The water saturation was calculated using the Archie equation since the pay sandstone contains only small amounts of clay and conductive minerals are absent:

$$S_w = \left(\frac{aR_w}{\phi^m R_t}\right)^{1/n}$$

Core analysis and logging tools indicate the following average values should be used in this equation:

	Core values	Log values	Fig. no.
R_w	0.018*	0.029 #	2–65, 66, 68
a	1.0	0.7 #	2–68
m	1.705	1.66 #	2–67–69
n	1.918	±2.00 #	2–71, 73

* Water sample believed contaminated by drilling fluids.
# Value used.

Core analysis indicates that matrix density of this sandstone is 2.65 gm/cc, but the log analysis (Fig. 2–66) indicates a value of 2.69 gm/cc. Values used in the determination of porosity using logs are fluid density, $\rho_f = 1.05$ gm/cc, $\Delta t_m = 53.5$ μsec/ft, and $\Delta t_f = 189$ μsec/ft. The porosity determined from cores was 17.3%, and that from logs was 17.0%. (Fig. 2–77). Using a statistical average from logs for the value of $R_t = 4.317$, the average water saturation from logs was 29.7%, which compares with a value of 27.6% obtained from core analysis (special and conventional analysis) after adjustment.

The number and type of graphs used depend upon the type and quality of available information. An engineer simply must use good judgment, and no engineer can be lazy if mistakes are to be avoided.

Relative permeability data for various core samples also show a degree of scatter. Data usually are normalized, but the normalized data also show the scatter, as illustrated by Figs. 2–82 and 85. The normalized values vary between the end points zero to one for both scales. Equations used to normalize the data follow:

$$k_{ro}^* = \text{normalized relative permeability to oil} = \frac{k_o \ @ \ S_w}{k_o \ @ \ S_{wi}}$$

$$k_{rw}^* = \text{normalized relative permeability to water} = \frac{k_w \ @ \ S_w}{k_w \ @ \ S_{or}}$$

$$S_w^* = \text{normalized water saturation} = \frac{S_w - S_{wi}}{1 - S_{wi} - S_{or}}$$

k_{rw}^* and k_{ro}^* usually are plotted versus S^* and such plots exhibit curvature, which suggests that the curves may be described by equations $k_{ro}^* = (1 - S_w^*)^n$ and $k_{rw}^* = B(S_w^*)^m$. Values for n and m are not the same as those in the Archie equation.

The constants in the equations can be fitted using methods presented in the previous discussion. Solution often requires the use of iterative procedures. The laboratory values for S_{or} cannot be used.

Wayne Beeks has suggested the possible use of a graphical technique using the basic data shown on Table 3. The first step involves comparison of values obtained for S_{wi} from logs (at height above water-oil contact), capillary pressure data, waterflood displacement data obtained from special laboratory analysis of cores and field performance, etc., with the data from relative permeability tests. Inconsistent data should be studied for possible cause and may be discarded.

The method is based upon the selection of the best value for S_{or} using trial-and-error procedures. The basic equations used follow:

$$k_{ro}' = \frac{k_o \ @ \ S_w}{k_o \ @ \ S_{wi}} = (1 - S_w')^n$$

$$k_{ro} = \frac{k_o \ @ \ S_w}{k_a} = A(1 - S_w')^n$$

$$k_{rw}' = \frac{k_w \ @ \ S_w}{k_w \ @ \ S_{or}'} = (S_w')^m$$

$$k_{rw}'' = \frac{k_w \ @ \ S_w}{k_o \ @ \ S_{wi}} = B(S_w')^m$$

$$k_{rw} = \frac{k_w \ @ \ S_w}{k_a} = C(S_w')^m$$

where:

$$S_w' = \frac{S_w - S_{wi}}{1 - S_{or}' - S_{wi}}$$

$$A = \frac{k_o \ @ \ S_{wi}}{k_a}$$

$$B = \frac{k_w \ @ \ S_{or}'}{k_o \ @ \ S_{wi}}$$

$$C = AB = \frac{k_w \ @ \ S_{or}'}{k_a}$$

$S_{or}' =$ value at $k_{ro} = 0$. Does not equal laboratory S_{or}.

End Points to Basic Equations

Parameter	End points at	
	S_{wi}	S_{or}'
S_w'	0.0	1.0
k_{ro}'	1.0	0.0
k_{ro}	A	0.0
k_{rw}'	0.0	1.0
k_{rw}''	0.0	B
k_{rw}	0.0	C

The method is used as shown in the following description.

1. Using coordinate graph paper, plot laboratory k_{ro}' values raised to the 0.25 power (y-axis) versus laboratory S_w values (x-axis). An analysis of the resulting plot will normally indicate either of the following two possibilities (basic data and sample curves for this step are not shown):

   a. Data plot in good straight line. This indicates that the exponent n in Equation 1 is equal to 4.0. Extrapolation of the $(k_{ro}')^{0.25}$ line to 0 gives the theoretical end-point S_{or}' value ($S_{or}' = 1 - S_w$).

   b. Data plot exhibits a curved relationship. This is the expected result since an exponent n value of 4.0 was derived under relatively ideal and theoretical conditions. Extrapolate the curved line to a value of $(k_{ro}')^{0.25}$ equal to 0. The estimated theoretical end-point S_{or}' value equals $(1 - S_w)$ when $(k_{ro}')^{0.25}$ equals 0.

2. Calculate values of S_w' using the S_{or}' values determined in Step 1 and the laboratory S_w values. Plot $(1 - S_w')$ on x-axis versus laboratory k_{ro}' values (y-axis) on log-log paper. Analysis of this plot has the following two possibilities:

   a. Data plot in good straight line. This indicates that the S_{or}' value used to determine S_w' is correct. Determine the slope of the line by either of the two below methods:

      1) Slope $n = \tan \theta$ where the angle θ is measured from the x-axis to the straight line.

      2) Slope $n = y/x$ where y and x represent distances measured along the k_{ro}' and $(1 - S_w')$ axis.

   b. Data plot exhibits a curved relationship; if the curve is concave upward, the S_{or}' value used to determine $(1 - S_w')$ must be reduced.

3. Estimate a new value of S_{or}' and recalculate $(1 - S_w')$. Return to Step 2 and repeat as many times as necessary to obtain a good straight line on the log k_{ro}' versus log $(1 - S_w')$ plot. Central data points should be given the most weight since k_{ro}' values at low $(S_w - S_{wi})$ and high S_w values might be questionable. If essential, change value of 0.25.

4. With the theoretical end-point S_{or}' value obtained, determine the slope n as shown above in Step 2a. Plot $(k_{ro}')^{1/n}$ versus S_w on coordinate paper. This step is essentially a final check and the data should be in a good straight line.

5. Plot laboratory k_{rw}'' values (y-axis) versus the final S_w' values (x-axis) determined in Step 3 above on log-log paper. Analysis of this plot has the following possibilities:

   a. Data plot in good straight line. This is the expected result. Central data points should again be given the most weight.

   b. Data do not plot in good straight line; if satisfied that there is no problem with the k_{rw}' data and good results were obtained, force a straight line through the k_w'' data points. The failure to obtain a good straight line can be considered to be caused by laboratory errors in most cases.

6. Determine the slope m required for Equation 2 as shown above in Step 2a. Extrapolate the line to a value of $S_w' = 1.0$. The intercept value is equal to the following:

$$\text{intercept} = \frac{k_w \ @ \ S_{or}}{k_o \ @ \ S_{wi}} = B$$

This value is indicative of the preferential wettability of the core sample.

7. Repeat Steps 1 through 6 for all samples. Make various crossplots to determine if there is a definite relationship between S_{or}', n, m, A, B, and C and various other petrophysical parameters.

8. Using the appropriate values for S_{wi}, S_{or}', n, m, A, B, and C, write the equations for k_{ro} and k_{rw}.

The curves used are presented as Figs. 2–82 through 102.

The various end points determined by use of the equations, method, and figures are summarized as follows:

(Also see Table 3 for initial rough values)

Parameter	Value	Figure
End point residual oil saturation (S_{or}')	0.219*	2–92
Slope n (k_{ro} equation)	6.35*	2–93
Slope n (k_{ro} equation)	6.20	2–95
Slope n (k_{ro} equation)	6.31	2–96
A	0.262*	2–91
Slope m (k_{rw} equation)	1.25*	2–94
Slope m (k_{rw} equation)	1.22	2–99
Slope m (k_{rw} equation)	1.29	2–97
C	0.26*	2–100
S_{wi}	0.16*	log analysis

$$k_{ro} = 0.262 \ [1 - 1.610(S_w - 0.16)]^{6.35}$$
$$S_w' = (S_w - 0.16)/(1.00 - 0.29 - 0.16) = 1.61(S_w - 0.16)$$
$$k_{rw} = 0.472 \ [(S_w - 0.16)]^{1.25}$$
$$(1.61)^{1.25} \times 0.26 = 0.472$$

The fractional flow curve shown in Fig. 2–102 is calculated using the equation and methods shown below: The basic equations are:

$$Q_o = \frac{(K_{ro} \cdot K_a)(A)(dP)}{(\mu_o)(L)}$$

$$Q_w = \frac{(K_{rw} \cdot K_a)(A)(dP)}{(\mu_w)(L)}$$

$$\text{Fractional flow of water} = \frac{\text{water}}{\text{water} + \text{oil}} = f_w$$

$$f_w = \frac{Q_w}{Q_o + Q_w}$$

$$f_w = \frac{\dfrac{(K_{rw} \cdot K_a)(A)(dP)}{(\mu_w)(L)}}{\dfrac{(K_{ro} \cdot K_a)(A)(dP)}{(\mu_o)(L)} + \dfrac{(K_{rw} \cdot K_a)(A)(dP)}{(\mu_w)(L)}}$$

Calculation of Fractional Flow

(1)	(2)	(3)	(4) $\frac{K_{rw}}{K_{ro}}$	(5) $\frac{K_{ro}}{K_{rw}}$	(6) $\frac{\mu_w}{\mu_o}$	(7) $\frac{K_{ro}}{K_{rw}} \cdot \frac{\mu_w}{\mu_o}$	(8) 1.0 + (7)	(9) f_w 1.0/(8)
S_w	K_{ro}	K_{rw}						
0.27 ·	0.60	0.0	0		0.333			0.0
0.30	0.34	0.002	0.0059	170.	0.333	56.5	57.5	0.017
0.40	0.10	0.01	0.10	10.	0.333	3.33	4.33	0.231
0.49	0.03	0.03	1.0	1.	0.333	0.333	1.33	0.751
0.60	0.009	0.09	10.0	0.1	0.333	0.033	1.03	0.968
0.65	0.0031	0.125	40.3	0.025	0.333	0.0083	1.008	0.993
0.70	0.001	0.17	230.0	0.0043	0.333	0.00143	1.001	0.996

Where: S_w = Water saturation, fractional pore volume

K_{ro} = Relative permeability to oil, fraction

K_{rw} = Relative permeability to water, fraction

μ_o = Oil viscosity, centipoise

μ_w = Water viscosity, centipoise

f_w = Fractional flow of water

$$f_w = \frac{1}{1 + \frac{K_o}{K_w} \cdot \frac{\mu_w}{\mu_o}}$$

Given: Oil viscosity   = 1.65 cp = μ_o

Water viscosity = 0.55 cp = μ_w

Relative permeability (water-oil)

Calculation of Waterflood Performance

(1) $S_w 1$ end face	(2) $S_w 2$ average	(3) f_w @ $S_w 1$	(4) Cumulative oil produced (fractional PV) $S_w 2 - C_w$	(5) $S_w 2 - S_w 1$	(6) f_o (1.0 − f_w)	(7) Q_i Water injected (pore volumes) (5)/(6)
0.270	0.270	0	0	0	1.0	0
0.505	0.554	0.829	0.284	0.049	0.171	0.286
0.550	0.607	0.925	0.337	0.057	0.075	0.760
0.600	0.646	0.965	0.376	0.046	0.035	1.31
0.650	0.677	0.994	0.407	0.027	0.006	4.5
0.700	0.700	1.00	0.430	0	0.0	∞

Basic equation: $S_w 1 = S_w 2 - f_o\, Q_i$
or $Q_i = (S_w 2 - S_w 1)/f_o$

Where: $S_w 1$ = Water saturation at producing well, fraction of pore volume

$S_w 2$ = Average water saturation between producing and injection well, fraction of pore volume

f_w = Fractional flow of water

C_w = Interstitial water saturation (initial water in reservoir at start of water injection)

f_o = Fractional flow of oil = (1.0 − f_w)

Q_i = Cumulative water injected, pore volumes

$$f_w = \frac{\frac{Krw}{\mu w}}{\frac{Kro}{\mu o} + \frac{Krw}{\mu w}}$$

$$f_w = \frac{1}{\frac{Kro}{\mu o} \cdot \frac{\mu w}{Krw} + 1}$$

$$f_w = \frac{1}{1 + \frac{Kro}{Krw} \cdot \frac{\mu w}{\mu o}}$$

The basic data and assumptions used in the calculation of the fractional flow curve shown on Fig. 2-102 follow:

PVT Data

B_o = 1.3912 RB/STB

B_w = 1.0467 RB/STB

μ_o = 0.82 cp

μ_w = 0.33 cp

Assumptions

Irreducible water saturation (S_{wi}) = 0.16

Sweep efficiency = 100%

No gravity or capillary effects

Horizontal and linear reservoir

The analysis of the fractional flow curve is as follows:

	At break-through	At 95% water cut
$(S_w)_{avg}$ (behind front)	0.352	0.441
S_w (at front)	0.302	0.395
f_w (surface water cut)	0.740	0.95
Recovery factor $= (S_w - S_{wi})/(1 - S_{wi})$	0.229	0.334

Overall, the special core data suggest poor to average relative permeability characteristics and neutral wettability. Rapid increase in water cuts with substantial recovery after breakthrough is anticipated. Moderate water sensitivity is indicated, but this appears to be more an effect of particle movement than clay swelling, as permeability reductions were observed for all water salinities tested.

Water sensitivity test results are shown on Fig. 2–103 data presented on Fig. 2–104 show that the compressibility of the reservoir rock exceeds that of an average formation (Hall's correlation). Figs. 2–103 through 108 show a usual influence for the effect of overburden.

The relative permeability data determined in the laboratory often do not represent actual field performance. The laboratory results should be compared with field performance using techniques presented in a later chapter discussing displacement efficiency.

Two additional data sets are often obtained during special core analysis testing in the laboratory. The reliability of such data can be determined by use of crossplots similar to those shown previously. Such tests are assumed to be relatively unimportant for purposes of the study analyzed here and Tables 4 and 5 are included simply to show the nature of the data obtained from gas-oil relative permeability testing and waterflood or flood-pot testing.

Again, the engineer should recognize that each reservoir is different and that each analysis should be based on good judgment. Simple cookbook methods serve as a checklist of procedures, which should be revised to meet the needs of a specific study.

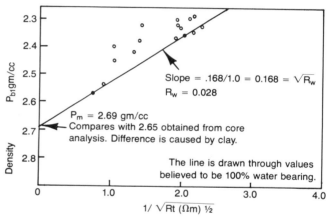

Figure 2–66. ρ_b vs. $1/\sqrt{Rt}$.

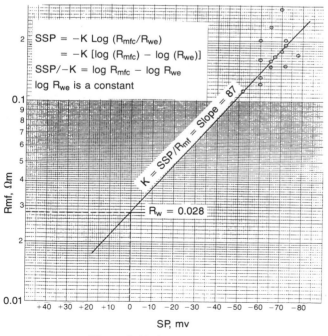

Figure 2–65. Log Rmf vs. SP.

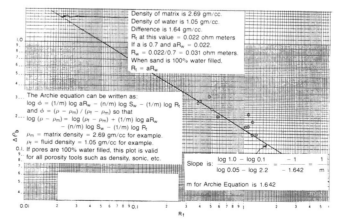

Figure 2–67. Log $\Delta\rho_b$ vs. log Rt.

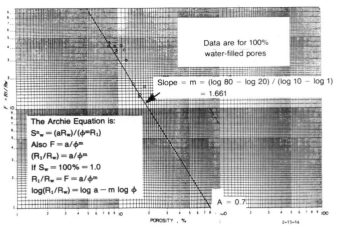

Data are for 100% water-filled pores

Slope = m = (log 80 − log 20) / (log 10 − log 1)
= 1.661

The Archie Equation is:
$S_w^n = (aR_w)/(\phi^m R_1)$
Also $F = a/\phi^m$
$(R_1/R_w) = a/\phi^m$
If $S_w = 100\% = 1.0$
$R_1/R_w = F = a/\phi^m$
$\log(R_1/R_w) = \log a - m \log \phi$

A = 0.7

Figure 2–68. *Log F vs. log ϕ.*

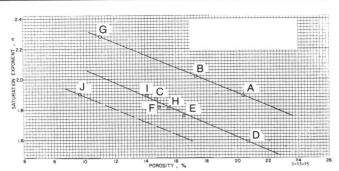

Figure 2–70. *Porosity vs. saturation exponent.*

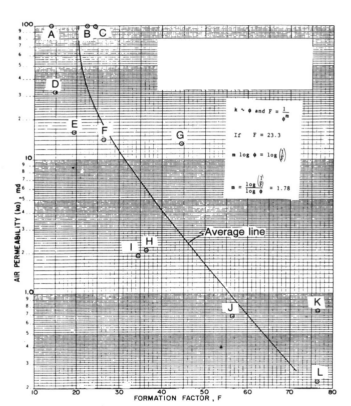

$k \sim \phi$ and $F = \dfrac{1}{\phi^m}$

If   F = 23.3

$m \log \phi = \log\left(\dfrac{1}{F}\right)$

$m = \dfrac{\log\left(\dfrac{1}{F}\right)}{\log \phi} = 1.78$

Average line

Figure 2–69. *Air permeability vs. formation factor.*

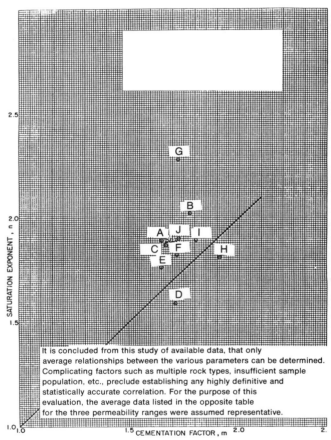

It is concluded from this study of available data, that only average relationships between the various parameters can be determined. Complicating factors such as multiple rock types, insufficient sample population, etc., preclude establishing any highly definitive and statistically accurate correlation. For the purpose of this evaluation, the average data listed in the opposite table for the three permeability ranges were assumed representative.

Figure 2–71. *Saturation exponent vs. cementation factor.*

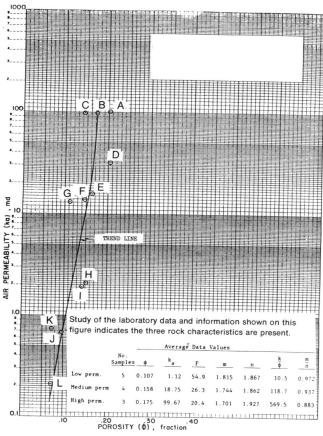

Figure 2–72. *Air permeability vs. porosity.*

Study of the laboratory data and information shown on this figure indicates the three rock characteristics are present.

Average Data Values

	No. Samples	ϕ	k_a	F	m	n	$\frac{k}{\phi}$	$\frac{m}{n}$
Low perm.	5	0.107	1.12	54.9	1.815	1.867	10.5	0.972
Medium perm	4	0.158	18.75	26.3	1.744	1.862	118.7	0.937
High perm.	3	0.175	99.67	20.4	1.701	1.927	569.5	0.883

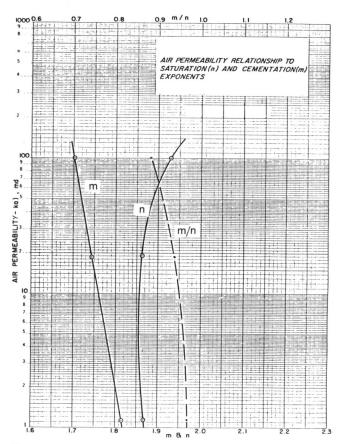

AIR PERMEABILITY RELATIONSHIP TO SATURATION (n) AND CEMENTATION (m) EXPONENTS

Figure 2–74. *Air permeability relationship to saturation (n) and cementation (m) exponents.*

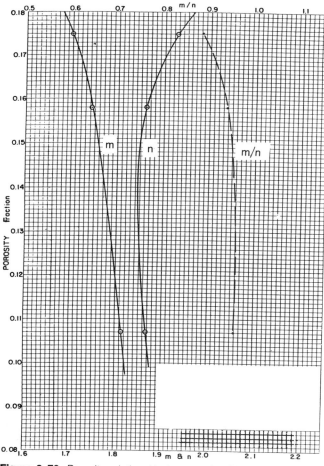

Figure 2–73. *Porosity relationship to saturation (n) and cementation (m) exponents.*

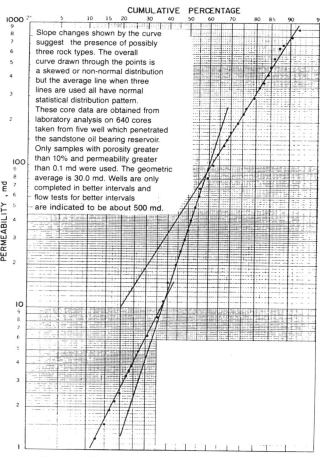

CUMULATIVE PERCENTAGE

Slope changes shown by the curve suggest the presence of possibly three rock types. The overall curve drawn through the points is a skewed or non-normal distribution but the average line when three lines are used all have normal statistical distribution pattern. These core data are obtained from laboratory analysis on 640 cores taken from five well which penetrated the sandstone oil bearing reservoir. Only samples with porosity greater than 10% and permeability greater than 0.1 md were used. The geometric average is 30.0 md. Wells are only completed in better intervals and flow tests for better intervals are indicated to be about 500 md.

Figure 2–75. *Permeability frequency distribution.*

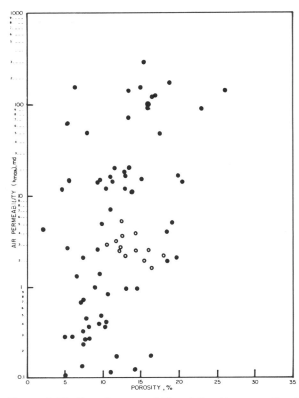

Figure 2-76. *Porosity-permeability relationship conventional core data; data scatter is great.*

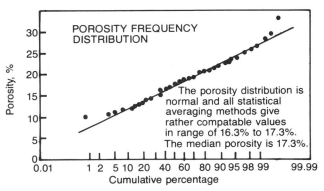

POROSITY FREQUENCY DISTRIBUTION

The porosity distribution is normal and all statistical averaging methods give rather compatable values in range of 16.3% to 17.3%. The median porosity is 17.3%.

Figure 2-77. *Porosity frequency distribution.*

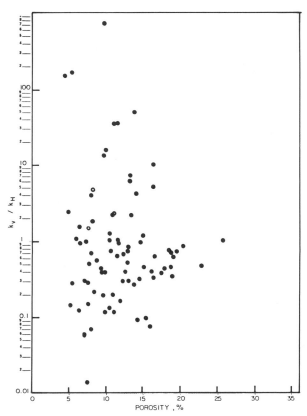

Figure 2-78. *Porosity vs. k_v/k_h conventional core data; data scatter is great.*

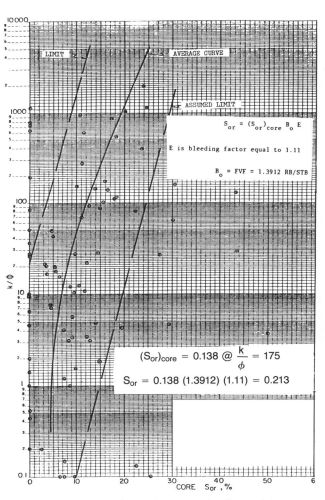

LIMIT

AVERAGE CURVE

ASSUMED LIMIT

$S_{or} = (S_{or})_{core} \ B_o \ E$

E is bleeding factor equal to 1.11

B_o = FVF = 1.3912 RB/STB

$(S_{or})_{core} = 0.138 @ \dfrac{k}{\phi} = 175$

$S_{or} = 0.138 \ (1.3912) \ (1.11) = 0.213$

Figure 2-79. *k/ϕ vs. S_{or} conventional core data.*

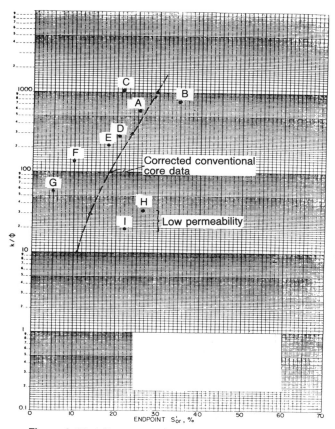

Figure 2–80. k/φ vs. endpoint s_or conventional core data.

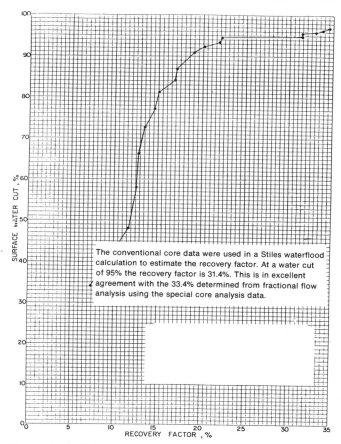

The conventional core data were used in a Stiles waterflood calculation to estimate the recovery factor. At a water cut of 95% the recovery factor is 31.4%. This is in excellent agreement with the 33.4% determined from fractional flow analysis using the special core analysis data.

Figure 2–81. Stiles waterflood prediction conventional core data; Water cut = water/(water + oil).

TABLE 2-1
Summary of Data from Logs and Associated Computer Calculations

Well no.	Top upper pay ft KB	ft ss	Top lower pay ft KB	ft ss	Net pay thickness Upper, ft	Lower, ft	Oil-water contact ft KB	ft ss	Average porosity Upper, %	Lower, %	Average water saturation Upper, %	Lower, %	Hydrocarbon pore volume upper, ft
A	8,764	8,437	Absent		144	—	8,986	8,659	17.9		32.5		17.4
B	8,803	8,489	Absent		150	—	8,994	8,679	18.7		38.2		17.3
C	8,656	8,334	8,913	8,591	192	0	S.O.	—	17.2		31.6		22.6
D	8,938	8,634	Absent		2	0	8,951	8,647	20.1		55.0		0.2
E	8,838	8,504	N.P.		71	—	N.L.	—	19.1		16.3		11.4
F	8,728	8,404	9,048	8,724	234	0	8,981	8,657	19.4		30.1		31.7
G	8,678	8,342	8,746	8,410	5	50	—	—	12.1	17.1	37.3	22.2	0.4
H	8,645	8,318	Absent		0	—	BSMT.	—	—		—		0
I	8,739	8,411	Absent		144	—	S.O.	—	18.7		29.1		19.1
J	8,861	8,543	Absent		145	—	9,013	8,695	18.2		33.0		17.7
K	8,700	8,365	9,025	8,690	246	0	S.O.	—	21.5		29.9		37.1
L	8,983	8,668	Absent		0	—	8,992	8,677	—		—		0
M	8,863	8,538	Absent		81	—	9,001	8,676	13.4		39.3		6.6
N	8,897	8,587	N.P.		89	—	8,992	8,682	19.2		32.6		11.5
O	8,756	8,426	N.P.		154	—	8,970	8,651	18.3		20.4		22.4
P	8,805	8,491	Absent		167	—	8,977	8,662	18.5		29.8		21.7
Q	8,750	8,418	N.P.		201	—	N.P.	—	18.3		35.6		23.7
R	8,686	8,354	8,827	8,495	88	102	N.P.	—	17.4	16.2	24.0	36.0	12.2
S	8,644	8,317	Absent		135	—	N.P.	—	14.5		34.1		12.9
T	8,762	8,433	N.P.		157	—	8,990	8,661	14.2		31.3		15.3
U	8,894	8,540	N.P.		77	—	9,026	8,672	18.1		27.5		10.1
V	8,593	8,264	Absent		113	—	N.P.	—	19.7		18.5		18.1
W	8,830	8,509	N.P.		132	—	9,012	8,691	16.2		27.9		15.4
X	8,682	8,358	Absent		91	—	N.P.	—	16.1		21.0		11.6
Y	8,702	8,373	Absent		100	—	N.P.	—	15.3		25.5		11.4
Z	8,733	8,410	Absent		153	—	N.P.	—	16.8		32.6		17.3
A-1	8,810	8,488	Absent		105	—	N.P.	—	16.6		26.8		12.8
A-2	8,711	8,363	Absent		252	—	N.P.	—	17.9		28.5		32.3
A-3	8,813	8,478	9,635	9,300	170	0	9,000	8,665	18.4		20.4		24.9
A-4	8,772	8,460	N.P.		102+	—	N.P.	—	16.5		30.9		11.6
B-1	8,720	8,401	Absent		23	—	N.P.	—	11.8		32.0		1.9
B-2	8,800	8,486	Absent		190	—	8,999	8,685	16.1		33.3		20.4
B-3	8,765	8,443	Absent		95	—	S.O.	—	17.1		27.7		11.8
C-1	8,733	8,408	Absent		85	—	N.P.	—	17.3		27.4		10.7
C-2	8,596	8,274	N.P.		136	—	N.P.	—	16.9		21.2		18.1
D-1	8,797	8,490	Absent		170	—	8,983	8,676	17.4		33.1		19.8
D-2	8,832	8,507	Absent		135	—	8,985	8,660	15.6		34.4		13.8
D-3	8,965	8,656	Absent		28	—	8,993	8,684	16.8		48.4		2.4
E-1	8,913	8,607	9,245	8,939	52	0	8,984	8,678	16.7		36.3		5.5
E-2	8,959	8,657	9,318	9,016	24	0	8,984	8,682	18.5		41.5		2.6
E-3	8,840	8,531	Absent		103	—	N.P.	—	14.2		33.3		9.8
F-1	8,914	8,611	Absent		66	—	8,993	8,691	18.2		37.7		7.5
F-2	8,860	8,561	Absent		37	—	8,944	8,645	14.3		41.7		3.1
F-4	8,865	8,561	Absent		0	—	N.P.	—	—		—		0
F-5	8,815	8,513	Absent		45	—	N.P.	—	14.6		32.8		4.4
G-1	8,940	8,639	Absent		43	—	8,992	8,691	13.5		31.1		4.0

TABLE 2-2
Summary of An Analysis of Conventional Cores

	Permeability,[1] md						Porosity,[2] %					
Well No.	1	3	9	37	9	All Wells	1	3	9	37	9	All Wells
No. of data values	227	88	95	41	69	520	227	88	95	41	69	520
Arithmetic mean	219.7	360.9	195.2	408.3	203.8	251.9	17.6	16.5	16.4	19.0	17.2	17.2
Geometric mean	18.5	54.2	16.4	123.8	68.5	30.0	17.1	16.2	16.0	18.7	16.7	16.8
Harmonic mean	2.1	7.6	1.6	23.8	2.2	2.4	16.5	15.8	15.6	18.4	16.3	16.3
Series mean	115.4	95.8	57.0	199.2	95.0	104.2	17.2	16.2	15.9	19.0	16.8	16.8
Median	14	95	17	147	139	34.7	17.9	16.3	16.0	19.6	16.9	17.3
Range	2,279.9	5,269	2,750.9	3,838.5	1,232.9	5,269.9	19.1	13.6	15.0	14.8	22.7	23.2
Standard deviation	462.0	848.4	456.7	708.9	237.8	550.3	4.23	3.14	3.63	3.21	4.17	3.92
Variance	213,441	719,834	208,580	502,572	56,548	302,828	17.79	9.83	13.16	10.30	17.43	15.39
Standard deviation (Biased)	461.0	843.6	454.3	700.2	236.1	549.8	4.21	3.12	3.61	3.17	4.15	3.91
Variance (Biased)	212,501	711,654	206,385	490,314	55,727.8	302,246	17.71	9.72	13.02	10.05	17.18	15.36
Moment of skewness	2.66	4.43	3.32	3.19	1.96	4.53	0.17	−0.23	0.34	−0.63	0.76	0.25
Moment of kurtosis	9.65	23.98	15.01	14.76	7.34	31.56	2.53	2.59	2.32	3.32	4.49	2.92

Note: [1] Permeability values less than 0.1 md excluded from analysis
      [2] Porosity values less than 10% excluded from analysis

TABLE 2-3
Summary of Laboratory Relative Permeability Data

Sample number	Depth KB, ft	k_a md	ϕ %	S_{wi} %	k_a ϕ	$k_o @ S_{si}$ md	$\dfrac{k_o @ S_{wi}}{k_a}$ fraction	S_{or} %	$k_w @ S_{or}$ md	$\dfrac{k_w @ S_{or}}{k_a}$ fraction	R.F. %
A	159	15.0	11.7	1,060	40	0.252	42.8	23	0.145	51.4
B	131	17.0	10.3	771	41	0.313	52.8	22	0.168	41.0
C	117	20.1	14.8	582	22	0.188	45.4	16	0.137	46.7
D	70	24.9	16.1	281	34	0.486	31.3	15	0.214	62.7
E	28	13.1	16.1	214	5.8	0.207	39.6	3.2 (4.9)*	0.114	52.8
F	20	14.7	16.0	136	13	0.650	43.5	5.8 (5.9)*	0.290	48.3
G	6.7	18.5	24.0	36.2	1.0	0.149	Large Permeability Reduction			
H	4.6	13.9	13.0	33.1	1.1	0.239	27.7	0.64 (1.2)*	0.139	67.8
I	2.6	13.1	15.4	19.8	0.97	0.373	43.7	0.70 (1.3)*	0.269	48.3

Sample number	Depth KB, ft	k_a md	ϕ %	S_{wi} %	S'_{or} %	$\dfrac{k_w @ S'_{or}}{k_o @ S_{wi}}$ fraction	$\dfrac{k_o @ S_{wi}}{k_a}$ fraction	Slope n	$\dfrac{k_{wo} @ S_{or}}{k_a}$ fraction	Slope m	S'_{or}/S_{wi} ratio
A	159	15.0	11.7	0.215	1.00	(0.252)	(6.11)	(0.252)	1.39	1.84
B	131	17.0	10.3	0.350	0.83	(0.313)	(6.69)	(0.260)	1.13	3.40
C	117	20.1	14.8	0.255	1.60	(0.188)	(6.31)	(0.301)	1.44	1.72
D	70	24.9	16.1	0.207	0.59	0.486	3.90	(0.287)	2.19	1.29
E	28	13.1	16.1	0.180	0.80	(0.207)	(6.14)	(0.166)	0.84	1.12
F	20	14.7	16.0	0.100	1.00	0.650	8.78	0.650	1.26	.62
G (omit)	6.7	18.5	24.0	—	—	0.149	—	—	—	—
H (omit)	4.6	13.9	13.0	0.265	0.59	(0.239)	2.42	(0.141)	0.87	1.10
I (omit)	2.6	13.1	15.4	0.220	1.00	(0.373)	(6.49)	0.373	0.66	1.43
										1.30	1.43
Average (all data)	59.9	16.7	15.3	0.224	0.926	0.317	5.86	0.304	1.22	1.56
Average representative Data	87.5	15.7	13.7	0.244	1.05	0.262	6.35	0.234	1.09	1.90

NOTES: 1. Data values enclosed in parentheses are assumed representative of the reservoir rock.
        2. Abbreviations have same meaning as described on Table I. In addition:

n = Slope on log k_{ro} vs log $(1 - S^*_w)$ plot = exponent in k'_{ro} equation
m = Slope on log k_{rw} vs log S^*_w plot = exponent in k'_{rw} equation
S'_{or} = Theoretical end-point S_{or} which is usually less than laboratory S_{or} value

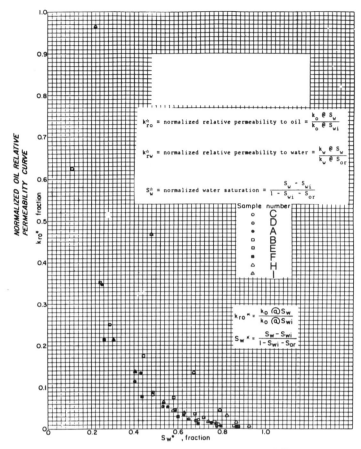

Figure 2–82. *Normalized oil relative permeability curve.*

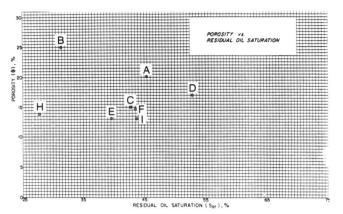

Figure 2–84. *Porosity vs. residual oil saturation.*

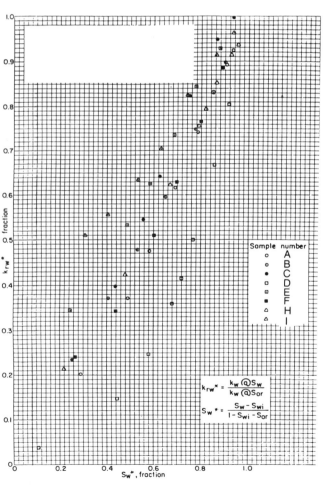

Figure 2–85. *Normalized water relative permeability curve.*

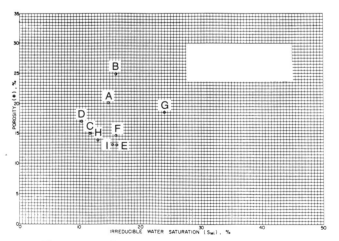

Figure 2–83. *Porosity vs. irreducible water saturation.*

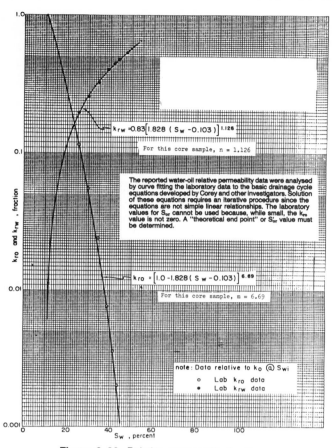

The reported water-oil relative permeability data were analysed by curve fitting the laboratory data to the basic drainage cycle equations developed by Corey and other investigators. Solution of these equations requires an iterative procedure since the equations are not simple linear relationships. The laboratory values for S_{or} cannot be used because, while small, the k_{ro} value is not zero. A "theoretical end point" or S'_{or} value must be determined.

$$k_{rw} = 0.83\left[1.828\left(S_w - 0.103\right)\right]^{1.126}$$

For this core sample, n = 1.126

$$k_{ro} = \left[1.0 - 1.828\left(S_w - 0.103\right)\right]^{6.69}$$

For this core sample, m = 6.69

note: Data relative to k_o @ S_{wi}
○ Lab k_{ro} data
● Lab k_{rw} data

Figure 2–86. *Relative permeability fractions.*

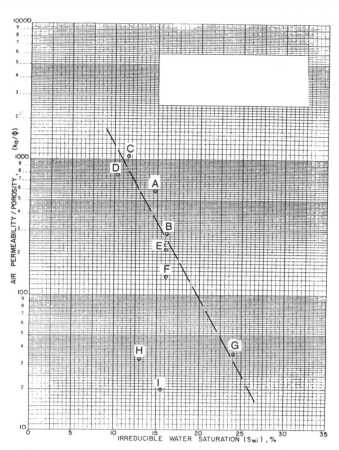

Figure 2–88. *Air permeability/porosity vs. irreducible water saturation.*

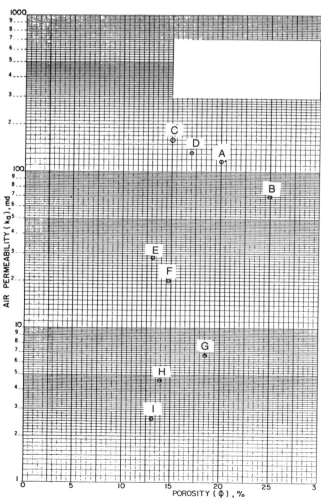

Figure 2–87. *Porosity vs. air permeability.*

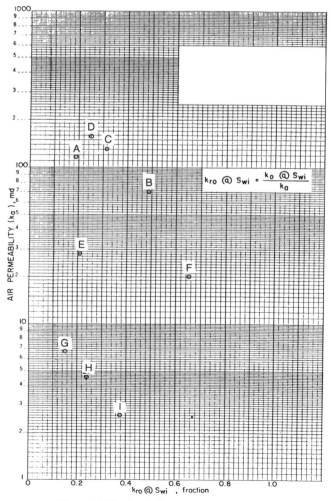

$$k_{ro} \text{ @ } S_{wi} = \frac{k_o \text{ @ } S_{wi}}{k_a}$$

Figure 2–89. k_{ro} @ S_{wi} *vs. air permeability.*

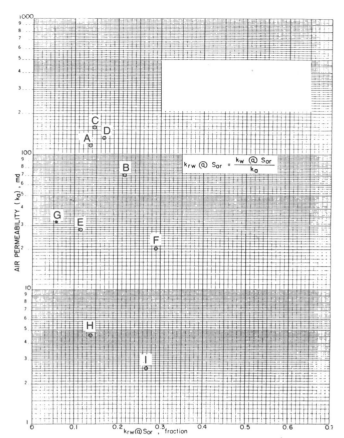

Figure 2–90. k_{rw} @ S_{or} vs. air permeability.

In the figure: k_{rw} @ S_{or} = $\dfrac{k_w\ @\ S_{or}}{k_o}$

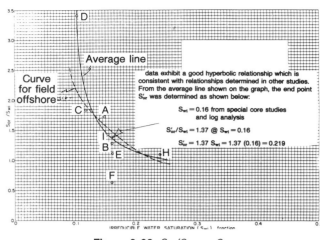

Figure 2–92. S_{or}/S_{wi} vs. S_{wi}.

data exhibit a good hyperbolic relationship which is consistent with relationships determined in other studies. From the average line shown on the graph, the end point S_{or} was determined as shown below:

$S_{wi} = 0.16$ from special core studies and log analysis

$S'_{or}/S_{wi} = 1.37$ @ $S_{wi} = 0.16$

$S'_{or} = 1.37\ S_{wi} = 1.37\ (0.16) = 0.219$

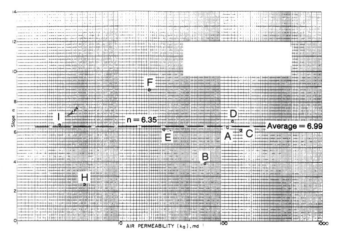

Figure 2–93. Slope n vs. air permeability.

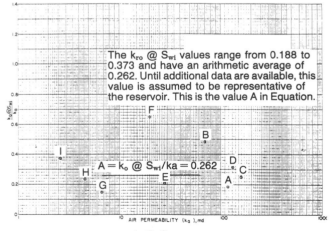

Figure 2–91. $\dfrac{k_o\ @\ S_{wi}}{k_\alpha}$ vs. air permeability.

The k_{ro} @ S_{wi} values range from 0.188 to 0.373 and have an arithmetic average of 0.262. Until additional data are available, this value is assumed to be representative of the reservoir. This is the value A in Equation.

$A = k_o\ @\ S_{wi}/ka = 0.262$

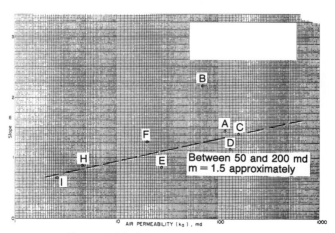

Figure 2–94. Slope m vs. air permeability.

Between 50 and 200 md m = 1.5 approximately

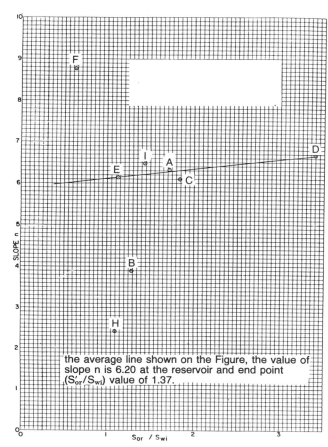

the average line shown on the Figure, the value of slope n is 6.20 at the reservoir and end point (S'_{or}/S_{wi}) value of 1.37.

Figure 2–95. S_{or}/S_{wi} *vs. slope n.*

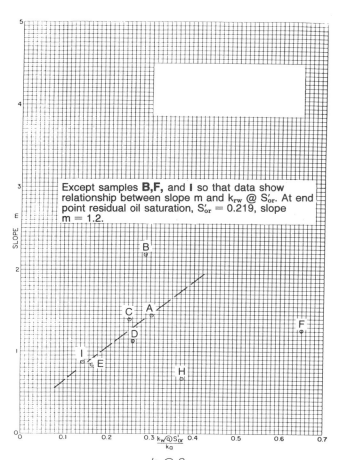

Except samples **B,F,** and **I** so that data show relationship between slope m and k_{rw} @ S'_{or}. At end point residual oil saturation, $S'_{or} = 0.219$, slope m = 1.2.

Figure 2–97. $\dfrac{k_w @ S_{or}}{k_a}$ *vs. slope m.*

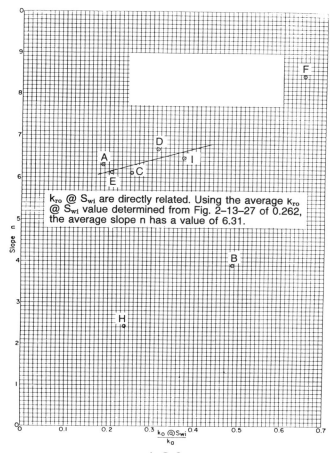

k_{ro} @ S_{wi} are directly related. Using the average k_{ro} @ S_{wi} value determined from Fig. 2–13–27 of 0.262, the average slope n has a value of 6.31.

Figure 2–96. $\dfrac{k_o @ S_{wi}}{k_a}$ *vs. slope n.*

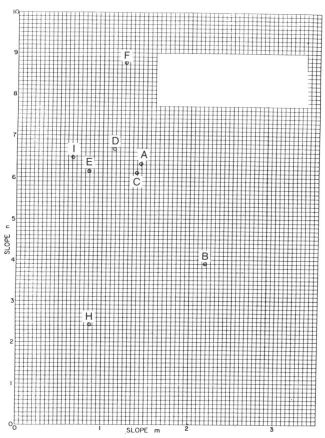

Figure 2–98. *Slope n vs. slope m.*

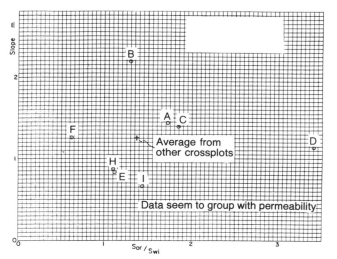

Figure 2–99. S_{or}/S_{wi} vs. slope m.

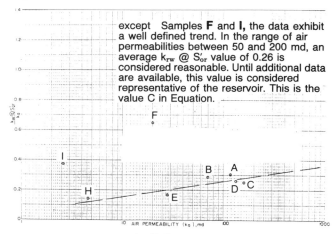

except Samples **F** and **I,** the data exhibit a well defined trend. In the range of air permeabilities between 50 and 200 md, an average k_{rw} @ S'_{or} value of 0.26 is considered reasonable. Until additional data are available, this value is considered representative of the reservoir. This is the value C in Equation.

Figure 2–100. $\dfrac{k_w \ @ \ S_{or}}{k_\alpha}$ vs. air permeability.

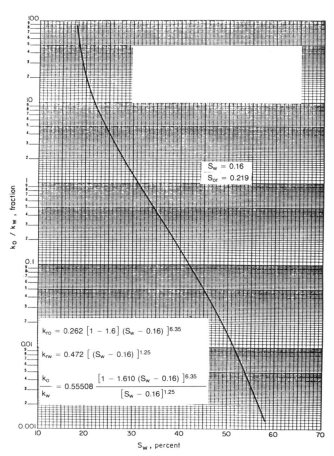

$$S_w = 0.16$$
$$S_{or} = 0.219$$

$$k_{ro} = 0.262 \left[1 - 1.6 \ (S_w - 0.16) \right]^{6.35}$$

$$k_{rw} = 0.472 \left[\ (S_w - 0.16) \ \right]^{1.25}$$

$$\frac{k_o}{k_w} = 0.55508 \ \frac{\left[1 - 1.610 \ (S_w - 0.16) \right]^{6.35}}{\left[S_w - 0.16 \right]^{1.25}}$$

Figure 2–101. *Field average relative permeability curve based on k_{ro} and k_{rw} equations shown in summary at the start of this section.*

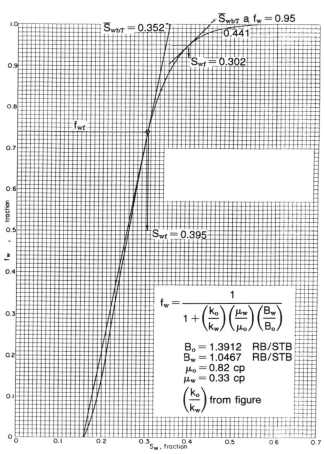

$\overline{S}_{wbT} = 0.352$

$\overline{S}_{wbT}$ a $f_w = 0.95$

0.441

$S_{wf} = 0.302$

f_{wf}

$S_{wf} = 0.395$

$$f_w = \frac{1}{1 + \left(\dfrac{k_o}{k_w}\right)\left(\dfrac{\mu_w}{\mu_o}\right)\left(\dfrac{B_w}{B_o}\right)}$$

$B_o = 1.3912$ RB/STB
$B_w = 1.0467$ RB/STB
$\mu_o = 0.82$ cp
$\mu_w = 0.33$ cp
$\left(\dfrac{k_o}{k_w}\right)$ from figure

Figure 2–102. *Fractional flow curve.*

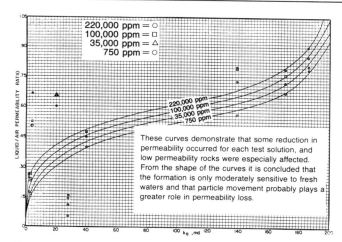

Figure 2–103. *Water sensitivity analysis results conventional core data.*

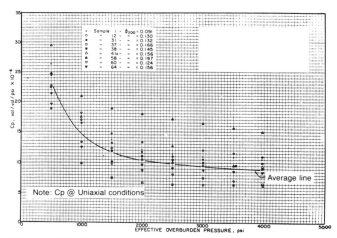

Figure 2–105. *Rock compressibility vs. effective overburden pressure special core analysis.*

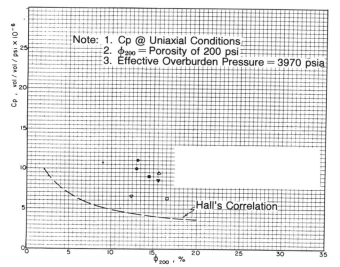

Figure 2–104. *Reported rock compressibility values; special core analysis.*

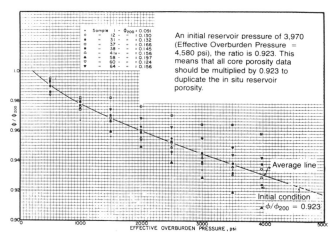

Figure 2–106. *Effective overburden pressure vs. ϕ/ϕ_{200} special core analysis.*

TABLE 4
Gas-Oil Relative Permeability Data

Sample number	Depth feet	k_a md	ϕ percent	S_{wi} percent	k_a/ϕ	$k_o @ S_{wi}$ md	$\dfrac{k_o @ S_{wi}}{k_a}$ fraction	S_{LR} percent	$\dfrac{k_g @ S_{LR}}{k_a}$ fraction	$\dfrac{k_o @ S_{LR}}{k_a}$ fraction	$k_g/k_o @$ ratio
A	10,100–01.0	106	15.5	19.1	683.8	88	0.830	51.9	0.531	0.0039	136
B	10,370–71.0	21	14.3	28.2	146.9	15	0.714	56.7	0.455	0.004	115
C	10,376–77.0	21	13.0	22.7	161.5	17	0.810	52.4	0.562	0.0038	147
D	10,380–81.0	34	15.7	24.6	216.6	28	0.824	55.8	0.533	0.004	123

TABLE 5
Summary of Waterflood Data

	Permeability, millidarcys								
To air k_a	To oil with connate water present k_o @ S_{si}	To water at residual oil saturation k_w @ S_{or}		Porosity, percent ϕ	Connate water, percent pore space S_{wi}	Oil in place, percent pore space S_o	Residual oil saturation, percent pore space S_{or}	Oil recovered, percent pore space $S_o - S_{or}$	$k_{a/c}$
		(1)	(2)						
A	67	42	43	14.6	18.9	81.1	20.2	60.9	486
B	23	18	18	13.9	18.4	81.6	13.5	68.1	237
C	23	11	11	14.5	30.1	69.9	11.1	58.8	186
D	12	6.8	6.9	12.7	27.5	72.5	18.4	54.1	125
E	11	5.9	6.2	12.4	26.6	73.4	15.9	57.5	112

Pay Thickness

The thickness of a formation usually is selected from the gamma ray and SP logs supplemented as necessary by other logs. The core analysis also can be helpful in making the determination.

The volumetric equation requires the use of a net thickness. The volume of hydrocarbon rock is desired. The thickness value chosen must be consistent with the porosity, in-situ water saturation, etc., used in the volumetric equation.

The material-balance equation depends upon volumetric concepts for products in place, but compressibility terms (water and rock expansion) must also involve water lying within and below the hydrocarbon volumes. An average value also is required for use in the radius of the investigation equation and in most flow and build-up analyses.

In tighter rocks, logs may not permit determination of the oil- or gas-bearing volumes since hydrocarbons and water are difficult to separate by means of logs. If the rock is extensively fractured so that effective communication is good over a large area and the hydrocarbon volume is relatively small in the sandwich of multiple shales and sands, an effective water displacement can occur due to vertical movement of water from shales into the hydrocarbon-bearing sands complemented by imbibition of the water pushed out of the shales into the sands (see Fig. 2–109).

Use of Figs. 2–110 and 111 can be illustrated by examples. Assume that a deviated hole is drilled into strata having a dip of 20° in the direction N30E, and that in the zone of interest, drift angle is 30° in the direction N30W.

If the measured thickness is 100 ft, what will be stratigraphic and vertical thickness?

Using Fig. 2–110, Angle A = 60°, the difference between N30E and N30W. Locate point 1 at angle A = 60° and angle B = θ = 20°.

From point 1, parallel to the latitude lines of the meridional net, count off a distance to the right equal to the drift angle, 30°, to locate point 2. This distance is measured by the meridians, the approximately north-south lines, which cross the latitude lines.

Angle α is the angle represented by the angle B scale at point 2. It is about 43°.

Using Fig. 2–111, stratigraphic thickness, ST, is found at point 3 on the curved lines scale where the vertical line for angle α, 43°, intersects the horizontal lines for measured thickness, 100 ft. ST is about 73 ft.

Vertical thickness, VT, is found at point 4 where the curved line representing ST, 73 ft, crosses the vertical line for the dip, 20°. It is about 78 ft.

Another example. Dip is 50° in the direction N80W. Deviation is 45° in the direction N60E. MT is 130 ft.

In this case, locate point 5 on Fig. 2–110 at A = 140° and B = θ = 50°.

Locate point 6 on the same latitude as point 5 and 45° to the right as explained in the previous example.

Angle α, represented by point 6, is about 29.5°.

On Fig. 2–111, ST is represented by point 7 and is about 113 ft. VT is represented by point 8 and is about 176 ft.

High angles. A special situation arises when dip and drift angle are very high and in the same general direction. It is not a common occurrence but does occasionally happen so its handling on the graph of Fig. 2–110 must be explained.

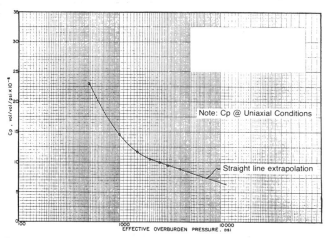

Figure 2–107. *Rock compressibility (extrapolated) vs. effective overburden pressure special core analysis.*

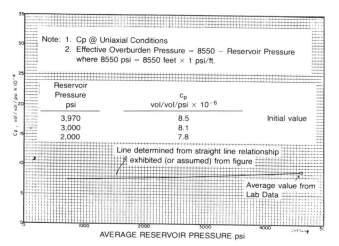

Figure 2–108. *Average rock compressibility vs. effective overburden pressure special core analysis.*

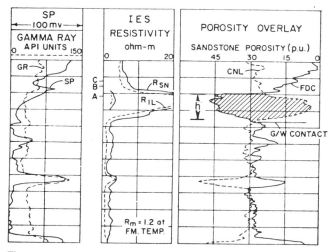

Figure 2–109. *CNL-FDC overlay showing gas zone in cleaner part of sand. As shown by the gamma ray, the upper part of the interval is shaly (courtesy Schlumberger).*

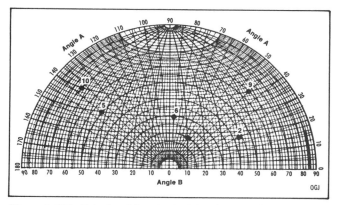

Figure 2–110. *Polar/meridional nets determine α angle (after Travis, OGJ, 2 July 1979).*

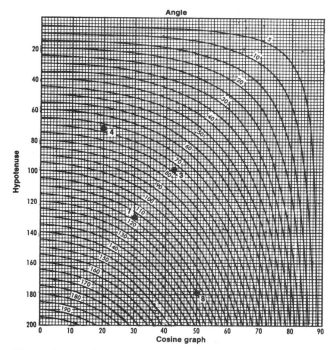

Figure 2–111. *Cosine graph showing thicknesses (after Travis, OGJ, 2 July 1979).*

For example, if dip were 65°, drift angle 60°, and angle A 45°, the first point would be plotted at 9. But note that in attempting to count 60° to the right along the same parallel of latitude, the perimeter of the graph is reached after counting 33½°.

In this event, counting is continued from the left perimeter of the graph for 26½° more along the same latitude.

The point sought would then be 10 and the angle indicated by the B scale would be about 70°.

The limit of the B scale is 90°, so when counting goes beyond the right perimeter, it means that angle α exceeds 90°. The angle indicated at point 10 is not angle α but its supplement. Angle α is 110°.

Note that point 10 is 20° from the left perimeter of the graph on the B scale which, added to the 90° on the right side, gives 110° for angle α.

The cosine of an angle is equal to the negative cosine of its supplement. Therefore, in these cases, the angle measured by the B scale, such as 70°, can be used to solve for ST and VT in the cosine graph of Fig. 2–111.

3 Oil and Gas Behavior

Flash and differential processes or a combination of the two may be used to separate the oil and gas. In the flash process, the volume of the fluids is increased by lowering the pressure without removing any part of the fluid. In the differential process, the pressure is reduced a small increment and the gas phase is removed after equilibrium has been attained. The pressure on the liquid phase is again reduced, the equilibrium gas is removed, and the process is repeated again and again. Fortunately, the total volume curve is not influenced by the liberation process until reservoir pressures have been reduced to 1,000–1,500 psi. The amount of gas liberated per unit volume of oil is greater for flash liberation than for differential liberation if a common bubble-point liquid is reduced to the same near atmospheric conditions using the two methods outlined above.

Fluids may be displaced from rocks by either miscible or immiscible processes as indicated by study of the following ternary diagram for oil and gas.

Although hydrocarbon mixtures are rarely approximated by three component mixtures, it is convenient to use the ternary diagram to obtain a picture of phase behavior. The three-component system modeled in the diagram consists of methane (C_1), the intermediates (C_2 through C_6), and the heavy components (C_7 and higher). Since the phase behavior of gases and liquids are a function of pressure, temperature, and composition, it is necessary that pressure and temperature be constant for any specific diagram. The critical pressures and temperature information indicate that methane is a gas at most field temperatures and pressures. The C_7^+ components would be liquid, and the C_2–C_6 or intermediates might be either gas or liquids, depending upon the field pressure and temperature. Oil is below bubble point in two-phase region.

Hydrocarbon liquids replace other hydrocarbon liquids by miscible displacement; but the displacement of oil by gas is immiscible and gas rapidly bypasses the oil. The displacement of a gas by a gas probably will be miscible and rather complete in the swept volume. Displacement of oil by water is usually immisicible.

Ternary Diagram

Bubble Point of Oils*

Bubble point pressure is defined as the pressure at which reservoir liquid is completely saturated at reservoir temperature. The bubble point or saturation pressure, thus, is the minimum pressure required to maintain a mixture of oil and gas in the liquid phase at a given temperature (see Figs. 3–1–5).

* This section has been published in "Continuous Tables" of *Petroleum Engineer*, November 1979.

131

Bubble point pressure may be determined in the laboratory PVT analysis of the reservoir fluids or combined surface sampled fluids since total volume curves, viscosity curves, etc., show a discontinuity at the bubble point. Accurate direct observation of the bubble point—start of free gas evolution—in the laboratory PVT cell is very difficult. Also, the laboratory-determined bubble point often is not completely reliable because it is often not possible to obtain a true subsurface sample (reservoir fluids are at or below the bubble point) and recombining of surface oil and gas samples in the laboratory is not easy. Gas separates slowly from the liquid near bubble point. The Y curve is useful, and manuals are available to guide sampling of fluids.

If a homogeneous reservoir contains a gas cap at initial conditions, fluids at the gas-oil contact should be near the bubble point. The initial pressure at the gas oil contact should approximate the bubble point. At lower depths in the oil column, the gas-oil ratio may be greater or less than the bubble point value depending on distribution and influence of shales, time sequence of oil and gas accumulations, etc.

Fluids that are undersaturated at reservoir conditions may be expected to reach bubble point pressure in the flow string as they slowly move to the surface. Flowing pressures carefully taken at various depths may be plotted against depth, and a slight discontinuity in the curve should be observed at the bubble point of the fluid if the well flow has been properly stabilized.

If a number of samples of reservoir fluid have been analyzed in the laboratory, agreement usually is not exact. The data may be plotted to obtain a curve of the saturation pressure versus gas-oil ratio. If the gas-oil ratio of the saturated fluid is known from field production history, the pressure corresponding to the field measured gas-oil ratio should approximate the bubble point pressure.

In a like manner, curves of the shrinkage factor at bubble point conditions, as obtained from a series of laboratory analysis of samples, may be plotted versus gas-oil ratio to obtain a value of the shrinkage factor (inverse of formation volume factor) which should associate with the field-measured value of gas oil ratio. If a curve of the shrinkage factor at bubble point conditions, as obtained from a series of laboratory analyses of reservoir samples, is plotted against the corresponding bubble point pressures, the pressure corresponding to the shrinkage factor may reflect the bubble point pressure.

The field performance curve of reservoir pressure versus cumulative oil produced usually shows a distinct change in slope at the bubble point. The point

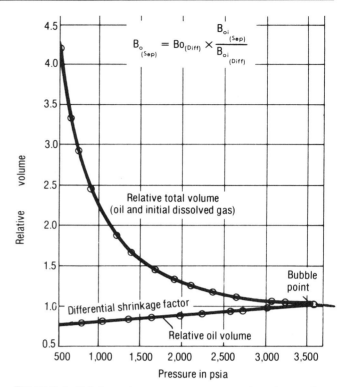

Figure 3–1. *Relative volumes versus pressure where base volume is that at bubble point (Petroleum Engineer). These curves and the basic data from which they are calculated all show a discontinuity at bubble point. The same holds for the viscosity curve.*

of intersection of the two curves approximates the saturation pressure. Total reservoir fluid withdrawal may be used instead of tank oil production. The method is useful only when initial reservoir pressure exceeds the bubble point pressure so that production both above and below bubble point is available.

Similar information may be obtained by dividing the cumulative production by the change in corresponding reservoir pressure, as determined by field measurements, and then plotting these data against cumulative production and/or time. The point at which the slope changes may be considered the bubble point. These methods use variations on a field-wide basis of the break in the total volume curve, which is evident in PVT analysis, when the bubble point is traversed.

The total volume factor and the Y-curve may be determined from laboratory data using accepted methods with the equation shown in Fig. 3–3.

In the absence of PVT data for the reservoir fluids, an analysis of fluid believed to be similar, possibly from an adjacent field, may be used. Some adjustment of the data may be required.

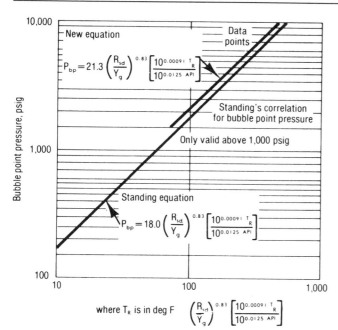

Figure 3–2. *Curves of the type suggested by Standing. The Standing curve shown is taken from Frick (SPE copyright). The new equation is for the field being studied.*

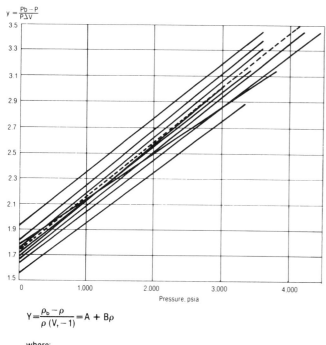

$$Y = \frac{\rho_b - \rho}{\rho(V_t - 1)} = A + B\rho$$

where:

Y = function defined by above equation,
ρ_b = bubble-point pressure, psia,
ρ = reservoir pressure, psia,
V_t = relative total volume or V/V_b where V_b is the bubble-point volume.

A plot on coordinate paper of Y versus p is often a straight line.

Figure 3–3. *The Y-curve. Illustrated are the range of values that can be obtained when oils at or below the bubble point are sampled in the field under controlled conditions.*

Using the equilibrium ratio and known constants, $k = y/x$, the composition of the liquid and the gas phases may be computed. If approximate methods are used to determine gas and liquid densities, total volume curves may be computed.

The log-log plot of ΔV versus Δp often is a straight line.

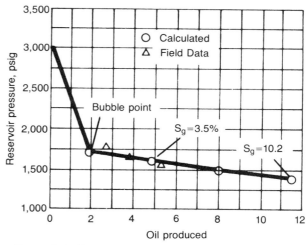

Figure 3–4. *Material balance calculations and performance.*

Required:
Bubble point pressure at 200 F of a liquid having a gas-ratio of 350 cu ft/bbl, a gas gravity of 075, and a tank-oil gravity of 30°API

Procedure:
Starting at the left side of the chart, proceed horizontally along the 350 cu ft/bbl line to a gas gravity of 075. From this point drop vertically to the 30° API line. Proceed horizontally from the tank-oil gravity scale to the 200 F line The required pressure is found to be 1930 psia.

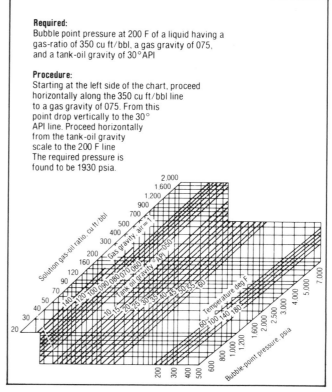

Figure 3–5. *Chart for calculating bubble-point pressure by Standing's correlation (© California Research Corp.)*

Adjustment of PVT Data for Bubble Point Pressure*

The data shown in Table 3–1 are the pertinent properties of a recombined surface sample from a field which was initially undersaturated. The sample was taken very early in the life of the field and the flowing bottom hole pressure during the sampling process was some 750 psi above the saturation pressure. A laboratory saturation pressure of 2,785 psia was obtained. Subsequent field data indicate that the saturation pressure was actually 2,500 psia, indicating the combined laboratory sample contained an excess of gas. The reference separator conditions for the field are 35 psig and 60 °F. The initial reservoir pressure was 3,240 psia and the reservoir temperature is 115°F.

Adjustment of Differential Vaporization Data

A. Multiply each differential gas liberation ratio by the unadjusted bubble point shrinkage, C_b, (Table 3–1, Col. IV & V and Table 3–2, Col. II & III) to convert the unadjusted differential gas liberation ratio to standard cubic feet per barrel of reservoir oil saturated at the unadjusted saturation pressure. Plot these ratios and the differential relative oil volumes, V_o, versus pressure (Fig. 3–6) and interpolate (or extrapolate) each curve to the new saturation pressure. The difference between the converted differential gas liberation ratio at the old saturation pressure and that at the new saturation pressure will be referred to as the corrective gas volume. The relative oil volume at the new saturation pressure as read from the relative oil volume versus pressure curve is the corrective relative oil volume.

B. Adjust the relative oil volume data (Table 3–1, Col. III) by dividing each of the unadjusted values by the corrective relative oil volume read from the relative oil volume curve obtained in Step I (Table 3–2, Col. IV & V).

C. Adjust the converted differential gas liberation ratios by subtracting (or adding) from each ratio the corrective gas volume obtained from the converted differential gas liberation ratio curve in Step I-A and by dividing each of these sums by the corrective relative oil volume. (See Table 3–2, Col. VI & VII) If the saturation pressure as determined from field data is lower than the laboratory saturation pressure, the sample contained an excess of gas and the corrective

gas volume is subtracted from each unadjusted differential gas liberation ratio. If the field saturation pressure is higher than the laboratory saturation pressure, the corrective gas volume is added to each unadjusted differential gas liberation ratio.

D. Plot the adjusted differential gas liberation ratios and relative oil volume values versus pressure to check the computations (Figure 3–7).

E. An adjustment in the gravity of the residual oil is not required.

Adjustment of Flash Vaporization Data

Based on differential vaporization data (preferred method).

1. Convert the unadjusted flash vaporization (Separator Tests) gas-oil ratios to ratios per barrel of saturated oil at the unadjusted saturation pressure by multiplying each ratio by the unadjusted flash vaporization shrinkage factor, C_b, at the indicated separator pressure. (Table 3–3, Col. II, III, IV)

2. Plot the converted gas-oil ratios obtained in Step II-A-1 and the unadjusted shrinkage factors versus separator pressure in order to obtain the unadjusted gas-oil ratio (expressed on a saturated oil basis) and the unadjusted shrinkage factor at the desired separator pressure (the average separator pressure prevailing during the productive history of the reservoir, (Figure 3–8).

3. From a plot of the adjusted differential vaporization data (Step I-D), determine a corrective gas volume and a corrective relative oil volume at the saturation pressure at which the flash data were obtained (Figure 3–7).

4. Multiply the converted unadjusted gas-oil ratio obtained in Step II-A-2 by the corrective relative oil volume determined in Step II-A-3 (Table 3–3, Col. V).

5. Subtract (if the PVT sample contains an excess of gas) from the gas-oil ratio determined in Step II-A-4 the corrective gas volume determined in Step II-A-3 to obtain the adjusted gas-oil ratio expressed relative to oil saturated at the new saturation pressure.

6. Adjust the shrinkage factor obtained in Step II-A-2 for the change in saturation pressure by multiplying it by the corrective relative volume obtained in Step II-A-3 (Table 3–3, Col. VI).

7. Convert the adjusted separator gas-oil ratio (expressed relative to oil saturated at the new saturation

---

* The material in this section is reproduced with the permission of Shell Oil Co.

TABLE 3–1
Properties of Recombined Oil Sample

I	II	III	IV	V
		Differential	Differential	
Pressure,	Flash relative	relative Oil	Gas Liberation	Reference
psia	total volume,	volume,	Ratio	shrinkage
p	V_t	V_o	D_F	C_b
2,785	1.000	1.000	0	0.681
2,740	1.004	—	—	
2,655	1.012	—	—	
2,510	1.027	—	—	
2,495	—	0.974	90	
2,395	1.041	—	—	
2,282	1.058	—	—	
2,210	—	0.958	161	
2,085	1.096	—	—	
1,930	1.133	—	—	
1,925	—	0.925	257	
1,710	1.208	—	—	
1,660	—	0.904	333	
1,510	1.299	—	—	
1,390	—	0.882	411	
1,385	1.375	—	—	
1,280	1.452	—	—	
1,110	—	0.859	491	
1,086	1.641	—	—	
961	1.831	—	—	
886	1.981	—	—	
828	—	0.835	572	
801	2.155	—	—	
741	2.326	—	—	
687	2.501	—	—	
642	2.675	—	—	
568	3.026	—	—	
550	—	0.810	652	
513	3.370	—	—	
468	3.703	—	—	
434	4.045	—	—	
310	—	0.784	732	

Separator Tests at 60° F.

Separator pressure, psig	Gas-oil ratio reference, D_b	Reference shrinkage, C_b	API gravity tank oil at 60°F.
0	1018	0.653	38.53
15	975	0.668	39.39
35	914	0.682	40.07
60	863	0.686	40.55
100	806	0.689	40.60

pressure) obtained in Step II-A-5 to a stock-tank basis by dividing it by the adjusted shrinkage factor obtained in Step II-A-6 (Table 3–3, Col. VII).

8. An adjustment in the gravity of the residual oil is not necessary.

Based on flash vaporization data (alternate method) (see Table 3–5 for illustration).

1. Convert the unadjusted flash vaporization gas-oil ratios (separator tests) to ratios per barrel of satu-

TABLE 3-2
Adjustment of Differential Vaporization Data

I	II III	IV V	VI VII
	Converted gas flashed ratio, D_F std cu ft/sat res bbl	Adjusted relative oil volume, V_o	Adjusted gas flashed ratio, D_F std cu ft/sat res bbl
Pressure, psia			
	@ 2785 psia		@ 2500 psia
2,495	90 × 0.681 = 61	0.974/0.976 = 0.998	(61–58)/0.976 = 3
2,210	161 × 0.681 = 110	0.958/0.976 = 0.982	(110–58)/0.976 = 53
1,925	257 × 0.681 = 175	0.925/0.976 = 0.948	(175–58)/0.976 = 120
1,660	333 × 0.681 = 227	0.904/0.976 = 0.926	(227–58)/0.976 = 173
1,390	411 × 0.681 = 280	0.882/0.976 = 0.904	(280–58)/0.976 = 227
1,110	491 × 0.681 = 334	0.859/0.976 = 0.880	(334–58)/0.976 = 283
828	572 × 0.681 = 390	0.835/0.976 = 0.856	(390–58)/0.976 = 340
550	652 × 0.681 = 444	0.810/0.976 = 0.830	(444–58)/0.976 = 395
310	732 × 0.681 = 498	0.784/0.976 = 0.803	(498–58)/0.976 = 451

rated oil at the unadjusted saturation pressure by multiplying each ratio by the corresponding unadjusted flash shrinkage factor at the indicated separator pressure.

2. Plot the converted separator gas-oil ratios obtained in Step II-B-1 versus separator pressure and extend the curve linearly from that ratio indicated at the highest separator pressure to zero cubic feet per barrel at the sample saturation pressure.

3. If necessary, extend the curve plotted in Step II-B-2 to its intersection with the new saturation pressure ordinate and read the corrective gas volume.

4. Plot the converted flash gas-oil ratios determined in Step II-B-1 versus separator pressure to obtain the unadjusted flash gas-oil ratio at the desired separator pressure (the separator pressure prevailing during the productive history of the reservoir). A

much shorter range of pressures is covered in this plot than in Step II-B-2 as the plot is not extended to the saturation pressure.

5. Plot the unadjusted shrinkage factors versus separator pressure to obtain the unadjusted shrinkage at the desired separator pressure.

6. Subtract the corrective gas volume obtained in Step II-B-3 from the unadjusted separator gas-oil ratio obtained in Step II-B-4 and divide this sum by the unadjusted shrinkage factor obtained in Step II-B-5 to determine the adjusted separator gas-oil ratio expressed relative to stock tank oil.

7. Determine the unadjusted change in volume per unit reservoir volume resulting from flash vaporizing the saturated oil at the highest separator pressure $(1 - C_b$ at the highest separator pressure).

8. From the "National Standard Petroleum Oil Ta-

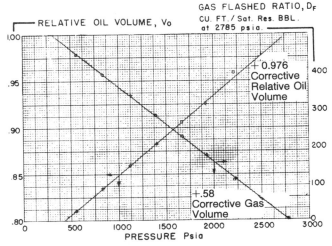

Figure 3-6. Differential relative oil volume vs. pressure.

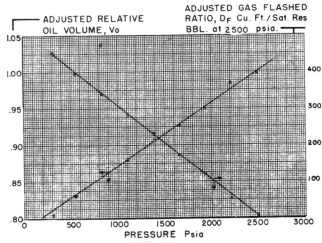

Figure 3-7. Adjusted differential gas liberation ratios and relative oil volume vs. pressure.

TABLE 3-3
Adjustment of Flash Vaporization Data

I Separator pressure, psig	II Gas-oil ratio	III Reference shrinkage factor, C_b	IV Converted gas-oil ratio cu ft/sat res bbl	V Adjusted gas-oil ratio cu ft/sat res bbl	VI Adjusted shrinkage, C_b	VII Adjusted gas-oil ratio, cu ft/stb
0	1018	0.653	$1018 \times 0.653 = 665$	$(665 \times 1.025) - 58 = 624$	$0.653 \times 1.025 = 0.669$	$624/0.669 = 933$
15	975	0.668	$975 \times 0.668 = 651$	$(651 \times 1.025) - 58 = 609$	$0.668 \times 1.025 = 0.685$	$609/0.685 = 889$
35	914	0.682	$914 \times 0.682 = 623$	$(623 \times 1.025) - 58 = 581$	$0.682 \times 1.025 = 0.699$	$581/0.699 = 831$
60	863	0.686	$863 \times 0.686 = 592$	$(592 \times 1.025) - 58 = 551$	$0.686 \times 1.025 = 0.703$	$551/0.703 = 784$
100	806	0.689	$806 \times 0.689 = 555$	$(555 \times 1.025) - 58 = 511$	$0.689 \times 1.025 = 0.706$	$511/0.706 = 724$

bles" (U.S. Department of Commerce Circular C 410, U.S. Government Printing Office, Washington, D.C.), determine the volume that one barrel of oil at reservoir temperature occupies at 60 °F.

9. Divide the unadjusted unit volume change at the highest separator pressure by the volume obtained in Step II-B-8.

10. Determine the change in unit volume caused by thermal shrinkage by subtracting the unadjusted unit volume change at the highest separator pressure from the value obtained in Step II-B-9.

11. Determine the unadjusted unit volume change due to gas liberation by subtracting the thermal shrinkage obtained in Step II-B-10 from the unadjusted unit volume change obtained in Step II-B-7.

12. Determine the shrinkage per cubic foot of gas liberated by dividing the unadjusted unit volume change due to gas liberation obtained in Step II-B-11 by the unadjusted gas-oil ratio at the highest separator pressure as obtained from Step II-B-1.

13. Determine the additional unit volume change due to liberation of the corrective gas volume by multiplying the value obtained in Step II-B-12 by the corrective gas volume as obtained in Step II-B-3.

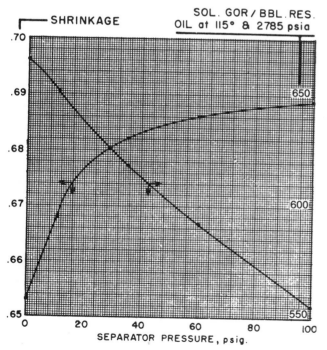

Figure 3-8. *Converted gas-oil ratios and unadjusted shrinkage factors vs. separator pressure.*

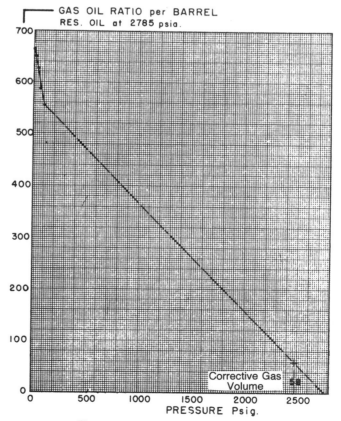

Figure 3-9. *Gas-oil ratio vs. pressure.*

14. Subtract the additional unit volume change obtained in Step II-B-13 from the unadjusted change in volume per unit volume (Step II-B-5) at the desired separator pressure and then divide this sum by one minus the additional unit volume change to determine the adjusted unit volume change.

15. Subtract the adjusted unit volume change determined in Step II-B-14 from one to obtain the adjusted shrinkage factor.

16. An adjustment in the gravity of the residual oil is not required.

Adjustment of Total Volume Data (see Table 3–4)

A. Using the unadjusted total volume data, calculate the values of the Y-function at each of the pressures using the equation:

$$Y = \frac{p_b - p}{p(V_t - 1)}$$

where p_b is the unadjusted saturation pressure.

B. Using the method of least squares, determine the best values of the constants, A and B, in the equation: $Y = A + B\,p$.

C. Calculate the adjusted values for V_t using the equation:

$$V_t = 1 + \frac{p_b - p}{p\,Y}$$

where p_b is the adjusted saturation pressure and $Y = A + B\,p$ so that

$$V_t = 1 + \frac{p_b - p}{A\,p + B\,p^2}.$$

TABLE 3–4
Adjustment of Total Volume Data

(A) Pressure, psia	Relative total volume, V_t	$P_b - p$ (2,785 − p)	$p(V_t - 1)$	Y
2,785	1.000	0		
2,740	1.004	45	10.960	4.10583
2,655	1.012	130	31.860	4.08035
2,510	1.027	275	69.025	3.98406
2,395	1.041	390	98.195	3.97169
2,282	1.058	503	132.356	3.80036
2,085	1.096	700	200.160	3.49720
1,930	1.133	855	256.690	3.33087
1,710	1.208	1075	355.680	3.02238
1,510	1.299	1275	451.490	2.82398
1,385	1.375	1400	519.375	2.69555
1,280	1.452	1505	578.560	2.60129
1,086	1.641	1699	696.126	2.44065
961	1.831	1824	798.591	2.28402
886	1.981	1899	869.166	2.18485
801	2.155	1984	925.155	2.14451
741	2.326	2044	982.566	2.08027
687	2.501	2098	1,031.187	2.03455
642	2.675	2143	1,075.350	1.99284
568	3.026	2217	1,150.768	1.92654
513	3.370	2272	1,215.810	1.86871
468	3.703	2317	1,265.004	1.83161
434	4.045	2351	1,321.530	1.77900

(B) $\Sigma\,Y = \Sigma\,B + \Sigma\,Ap$ $60.48111 = 22\,B + 30,268\,A$
     $\Sigma\,pY = \Sigma\,pB + \Sigma\,Ap^2$ $96,883.462 = 30,269\,B + 54,589,125\,A$

$$A = 0.001056139, \quad B = 1.296038$$

(C) $Y = 1.296038 + 0.001056139\,p$

$$V_t = 1 + \frac{p_b - p}{1.296038\,p + 0.001056139\,p^2}$$

TABLE 3-5
Based on Flash Vaporization Data (Alternate Method)

The following items, which are numbered identical
with those in the discussion, should
illustrate the method.

(1) Table 3–3.
(2) Fig. 3–9.
(3) Fig. 3–9.
(4) Not necessary in this case since 35 psig separator test data available.
(5) Same as (4).
(6) $(623 - 58)/0.682 = 828$.
(7) $(1 - 0.689) = 0.311$.
(8) 1 Bbl. at 150°F. = 0.9739 Bbl. at 60° F.
(9) $0.311/0.9729 = 0.3193$.
(10) $0.3193 - 0.3110 = 0.0083$.
(11) $0.3110 - 0.0083 = 0.3027$.
(12) $0.3027/555 = 0.0005454$.
(13) $(0.0005454)(58) = 0.0316$.

(14) $\dfrac{-0.0316 + (1.000 - 0.682)}{1 - 0.0316} = 0.2957$.

(15) $1.0000 - 0.2957 = 0.7043$.

Density of Oil, Gas, and Water

The gravity of an oil is generally expressed in the US in degrees API. The specific gravity of an oil at 60°F is related to the API gravity as follows: °API = (141.5/sp gr) − 131.5. or sp gr = (141.5/(131.5 + °API). Thus API gravities, ranging from 10 to 100, correspond to a range in specific gravity of 1.00 to 0.611 when expressed relative to water.

Also, °Be = 145 − (145/Sp. Gr.) when heavier than water.
°Be = (140/Sp. Gr.) − 130 when lighter than water.

Tables correcting measured API gravity in field at temperatures other than 60°F to 60°F are available from Frick (*Petroleum Production Handbook*). The density, specific gravity, or API gravity normally are determined by use of a hydrometer. Values may be measured in the field and in the laboratory.

The pressure gradient in the tubing string measured by a bottom-hole pressure gauge may be used to compute density values. As an example, if the pressure gradient is 30 psi per 100 feet of depth, the specific gravity is:

$(30/100) \times (453.6/(30.48 \times 6.452)) = 0.692$ gm/cc

where: 453.6 = grams per pound
30.48 = centimeters per foot
6.452 = square centimeters per square inch

For gas and water, suitable correlations are available (Table 3–5a).

The specific gravity of distilled water is one at 4°C. At 39.1°F, one pound per square inch is 2.3066 feet of water, so that 100 feet of water is equal to 43.3538 pounds per square inch. A cubic foot of water weighs 62.316 pounds at 68°F. Salt water has a somewhat higher density, depending upon what is dissolved in it, and values of about 48 pounds per 100 feet are commonly used.

The density of dry and wet gas can be read from Figs. 3–10 and 3–11.

Gas	ft³/lb	lb/ft³	Density
Air	13.14	0.07613	1.000
Methane	23.61	0.0424	0.556
Ethane	12.52	0.0799	1.049
Propane	8.47	0.1181	1.551
Butane	6.33	0.1580	2.076
Pentane	5.26	0.1901	2.497
CO			0.967
CO$_2$			1.53
N$_2$			1.25 at 0°
O$_2$			1.429

TABLE 3–5a
API Gravity Reduction to 60°F, ASTM—IP, 30–39° API.

Observed temper- ature, °F.	API gravity at observed temperature									
	30	31	32	33	34	35	36	37	38	39
	Corresponding API gravity at 60°F.									
50	30.7	31.7	32.7	33.7	34.7	35.7	36.7	37.8	38.8	39.8
51	30.6	31.6	32.6	33.6	34.6	35.7	36.7	37.7	38.7	39.7
52	30.5	31.5	32.6	33.6	34.6	35.6	36.6	37.6	38.6	39.6
53	30.5	31.5	32.5	33.5	34.5	35.5	36.5	37.5	38.5	39.5
54	30.4	31.4	32.4	33.4	34.4	35.4	36.4	37.5	38.5	39.5
55	30.3	31.3	32.3	33.4	34.4	35.4	36.4	37.4	38.4	39.4
56	30.3	31.3	32.3	33.3	34.3	35.3	36.3	37.3	38.3	39.3
57	30.2	31.2	32.2	33.2	34.2	35.2	36.2	37.2	38.2	39.2
58	30.1	31.1	32.1	33.1	34.1	35.1	36.1	37.1	32.2	39.2
59	30.1	31.1	32.1	33.1	34.1	35.1	36.1	37.1	38.1	39.1
60	30.0	31.0	32.0	33.0	34.0	35.0	36.0	37.0	38.0	39.0
61	29.9	30.9	31.9	32.9	33.9	34.9	35.9	36.9	37.9	38.9
62	29.9	30.9	31.9	32.9	33.9	34.9	35.9	36.9	37.8	38.8
63	29.8	30.8	31.8	32.8	33.8	34.8	35.8	36.8	37.8	38.8
64	29.7	30.7	31.7	32.7	33.7	34.7	35.7	36.7	37.7	38.7
65	29.7	30.7	31.7	32.7	33.6	34.6	35.6	36.6	37.6	38.6
66	29.6	30.6	31.6	32.6	33.6	34.6	35.6	36.6	37.5	38.5
67	29.5	30.5	31.5	32.5	33.5	34.5	35.5	36.5	37.5	38.5
68	29.5	30.5	31.5	32.4	33.4	34.4	35.4	36.4	37.4	38.4
69	29.4	30.4	31.4	32.4	33.4	34.4	35.3	36.3	37.3	38.3
70	29.3	30.3	31.3	32.3	33.3	34.3	35.3	36.3	37.2	38.2
71	29.3	30.3	31.3	32.2	33.2	34.2	35.2	36.2	37.2	38.2
72	29.2	30.2	31.2	32.2	33.2	34.1	35.1	36.1	37.1	38.1
73	29.1	30.1	31.1	32.1	33.1	34.1	35.1	36.0	37.0	38.0
74	29.1	30.1	31.0	32.0	33.0	34.0	35.0	36.0	37.9	37.9
75	29.0	30.0	31.0	32.0	32.9	33.9	34.9	35.9	36.9	37.9
76	28.9	29.9	30.9	31.9	32.9	33.9	34.8	35.8	36.8	37.8
77	28.9	29.9	30.8	31.8	32.8	33.8	34.8	35.7	36.7	37.7
78	28.8	29.8	30.8	31.8	32.7	33.7	34.7	35.7	36.7	37.6
79	28.8	29.7	30.7	31.7	32.7	33.7	34.6	35.6	36.6	37.6

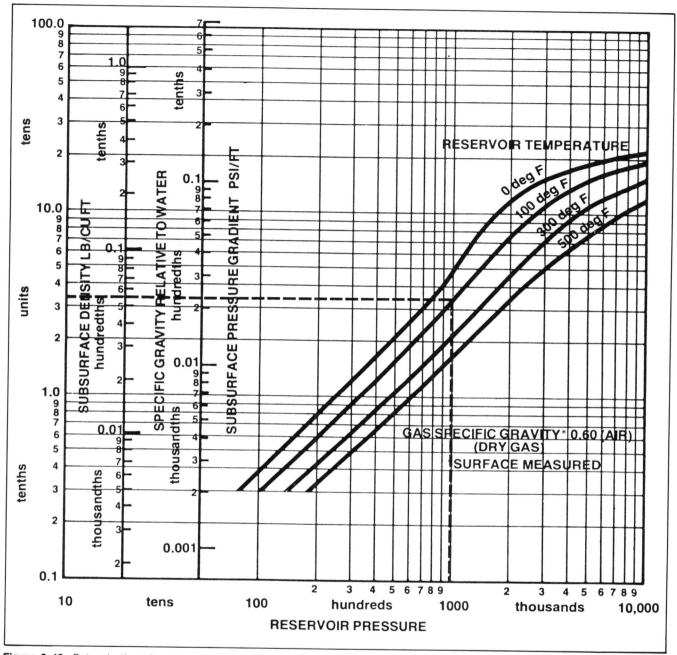

Figure 3–10. *Determination of subsurface density and specific gravity from surface-measured gas specific gravity for dry gas (courtesy Petroleum Engineer).*

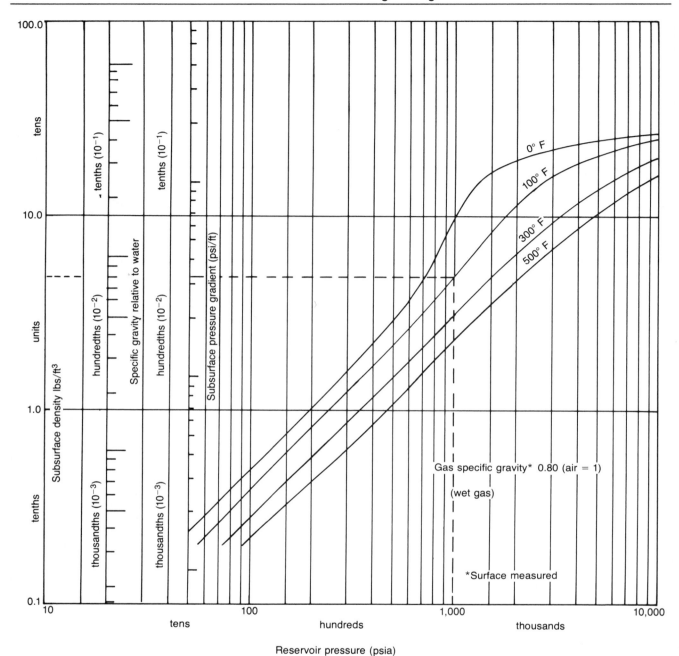

Figure 3–11. *Determination of subsurface density and specific gravity from surface-measured gas specific gravity for wet gas* (*courtesy Petroleum Engineer*). Density of wet gas.

Oil Formation Volume, Total Volume, and Formation Volume Factors for Condensate and Water

The oil formation volume factor is used to convert tank oil or oil measured at surface to subsurface conditions. The tank oil recovered at surface depends upon the separation methods employed, such as the number of separation stages, the operating temperature and pressure, production rates or overload of equipment, and general operations at separators and tanks. The proposed operating conditions are duplicated in the laboratory to obtain values for B_o at various stages during the operating life of a reservoir. Arps in Frick proposed a general correlation:

$$B_o = 1.05 + 0.0005\ R_s$$

The two-phase formation factor for oil and gas according to Arps can be calculated using the following equation:

$$B_t = B_o + 0.1781\ B_g\ (R_{sb} - R_s)$$

The type of curve obtained from laboratory analysis is shown on Fig. 3–12. Older correlations were presented as shrinkage factors or $1/B$.

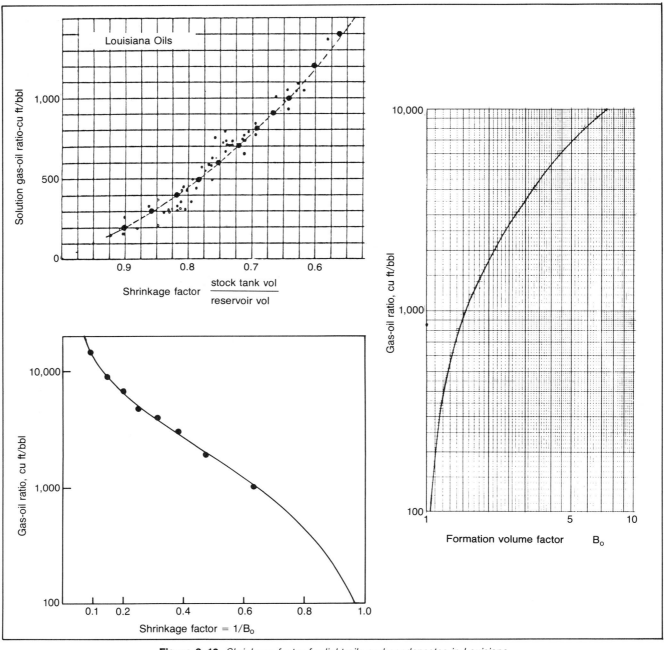

Figure 3–12. *Shrinkage factor for light oils and condensates in Louisiana.*

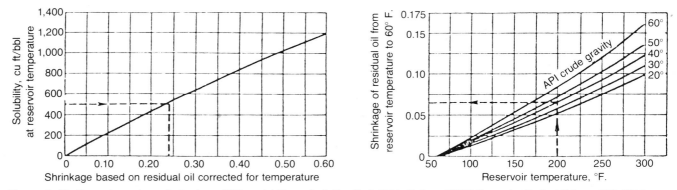

Figure 3–13. *Formation volume factor from GOR and API gravity (after Fertl, OGJ, 5 June 1978.) Example: To find B for a 40° API from a reservoir temperature of 200°F, and 2,000 psi. Solubility of 500 scf/stb. Reading lower chart for 500 cu ft in solution Sh 0.25; reading upper chart for drop from 2000°F to 60°F. gives Sh 0.068, therefore B = (1.065) (1.24) = 1.32.*

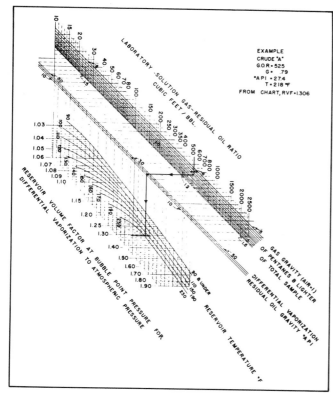

Figure 3–14. *Reservoir volume correlation (after Raza & Borden,* Petroleum Technology, *December 1950, © SPE-AIME).*

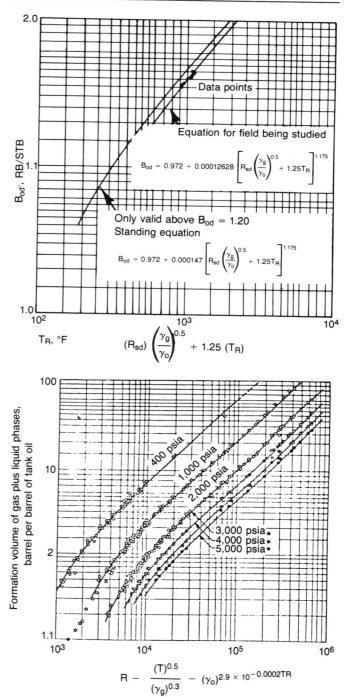

Figure 3–15. *Standing correlations for formation volume factors; base curves from Frick (© SPE-AIME, 1962).*

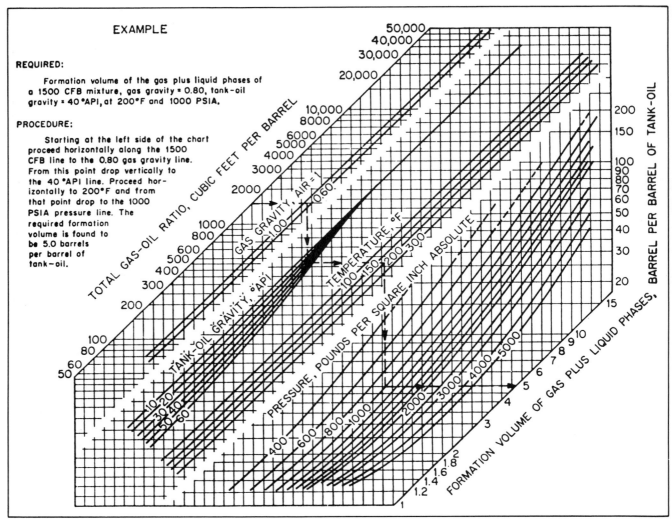

EXAMPLE

REQUIRED:

Formation volume of the gas plus liquid phases of
a 1500 CFB mixture, gas gravity = 0.80, tank-oil
gravity = 40°API, at 200°F and 1000 PSIA.

PROCEDURE:

Starting at the left side of the chart
proceed horizontally along the 1500
CFB line to the 0.80 gas gravity line.
From this point drop vertically to
the 40°API line. Proceed horizontally to 200°F and from
that point drop to the 1000
PSIA pressure line. The
required formation
volume is found to
be 5.0 barrels
per barrel of
tank-oil.

Figure 3–16. *Total formation volume by Standing (after Frick, SPE-AIME © 1962, using California Research Corp. © 1947).*

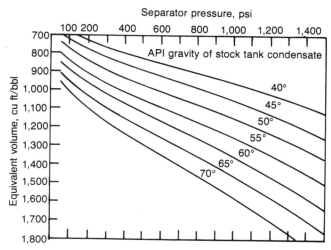

Figure 3–17. *Formation volume factors for condensate. Obtain equivalent gas volume for each barrel of stock-tank condensate from separator pressure and API gravity. Multiply monthly condensate production in barrels by equivalent gas volume. Add this total to metered dry-gas production for the month to obtain full well-stream volume (after Beal).*

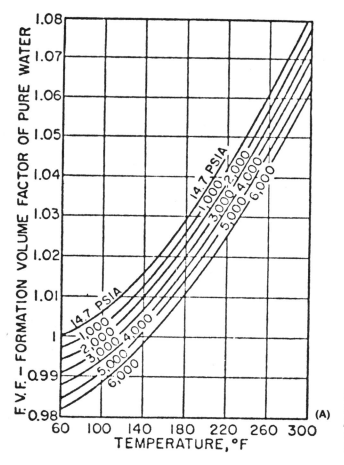

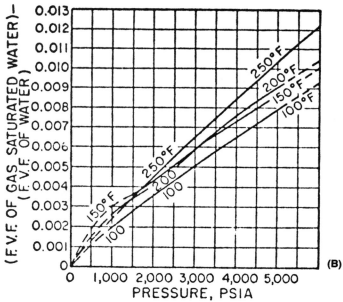

Figure 3–18. *Formation volume factors for water. (A) Formation volume factor of pure water. (B) Difference in FVF of gas-saturated pure water and pure water. For the FVF of the former, add this difference to the FVF in figure A (after Frick, SPE-AIME 2962; originally from* Thermodynamic Properties of Steam, *Wiley, 1936).*

Gas Formation Factor

The gas formation factor converts gas measured at surface to reservoir conditions. The conversion can be accomplished with reasonable accuracy by use of graphs in Frick as illustrated on Fig. 3–19. The gas gravity in many free-gas reservoirs often is in the range of 0.6, and values for gas released from solution often are in the range of 0.8. Gases are described by Boyle's and Charles' laws as $pv = nRT$

or
$$\frac{p_1 v_1}{Z_1 T_1} = \frac{p_2 v_2}{Z_2 T_2}$$

where pressure is absolute and temperature is in degrees Rankin ($= 460 + T$ in °F).

Natural gases require a correction factor defined as Z.

A value for Z can be obtained by using Fig. 3–20 or may be calculated using physical constants shown on Table 3–6 with the following procedure.

Component	Mol %	p_c	$p_c \times$ Mol %	T_c	$T_c \times$ Mol %
H_2S	2.10	1307	27.45	673	14.13
CO_2	2.50	1072	26.80	548	13.70
N_2	3.48	492	17.12	227	7.90
C_1	68.00	673	457.64	344	233.92
C_2	14.21	708	100.61	550	78.15
C_3	6.44	617	39.73	666	42.89
iC_4	0.50	530	2.65	733	3.67
C_4	1.34	551	7.38	766	10.26
iC_5	0.48	482	2.31	830	3.98
C_5	0.34	485	1.65	847	2.88
C_6	0.38	434	1.65	915	3.48
C_{7+}	0.23	397	0.91	973	2.24
TOTAL	100.00		685.90		417.20

Average $p_c = 685.90$
Average $T_c = 417.20$

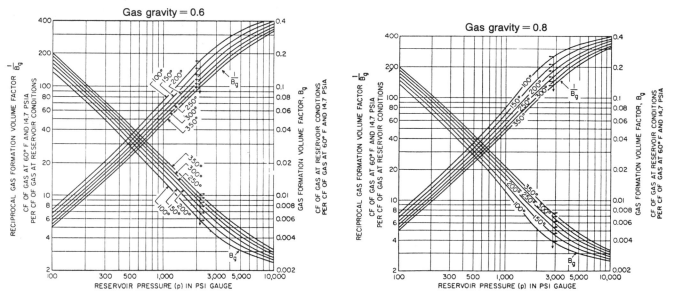

Figure 3–19. *Formation factor for gas (after Frick by Arps. © SPE-AIME).*

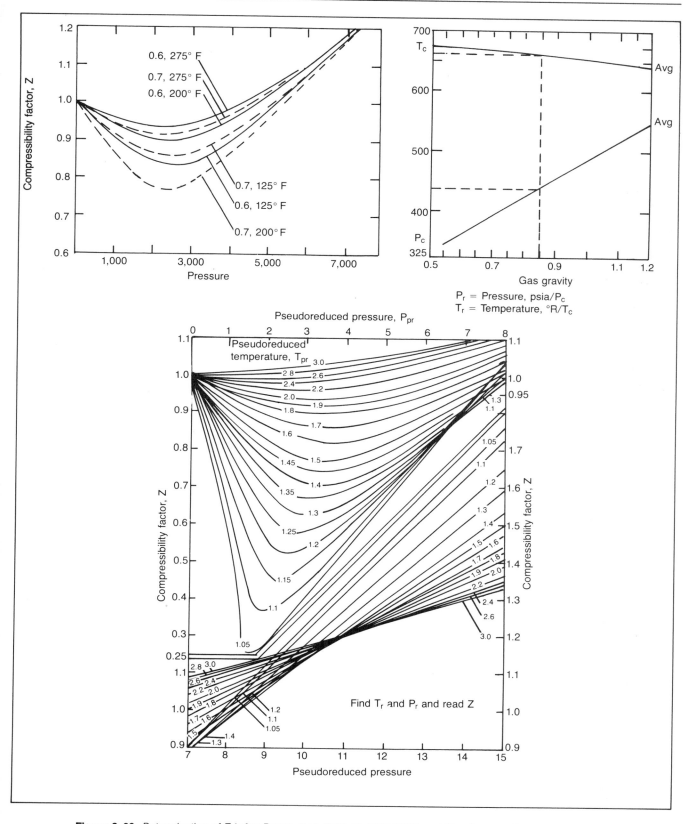

Figure 3–20. *Determination of Z (after Brown et al., OGJ, © SPE-AIME, and Standing and Katz, © SPE-AIME).*

TABLE 3–6
Physical Constants for Hydrocarbons

Symbol	C_1	C_2	C_3	$i\text{-}C_4$	$n\text{-}C_4$	$i\text{-}C_5$	$n\text{-}C_5$	C_6	C_7	C_{7+}
Molecular weight	16.04	30.07	44.10	58.12	58.12	72.15	72.15	86.17	100.20	M
1/(mol wt)	0.0623	0.0333	0.0227	0.0172	0.0172	0.0139	0.0139	0.0116	0.00998	1/M
Sp gr (water = 1)	0.300	0.378	0.509	0.5640	0.5840	0.6240	0.6310	0.6640	0.68800	S
1/(sp gr)	3.333	2.645	1.965	1.773	1.712	1.603	1.585	1.506	1.453	1/S
Mol wt/sp gr	53.46	79.54	86.67	103.0	99.50	115.7	114.4	129.8	145.6	M/S
Sp gr/mol wt	0.0187	0.0126	0.0116	0.00970	0.0100	0.00867	0.00877	0.00770	0.00686	S/M
Lb per gal liq	2.50	3.15	4.24	4.70	4.86	5.20	5.26	5.53	5.73	8.334 S
Lb per bbl liq	105	132	178	197	204	218	221	232	241	350 S
Lb per cu ft liq	18.70	23.56	31.73	35.16	36.41	38.90	39.34	41.39	42.89	62.34 S
Cu ft liq per lb	0.0535	0.0424	0.0315	0.0284	0.0275	0.0257	0.0254	0.0242	0.0233	0.01604/S
Cu ft gas per lb	23.61	12.62	8.60	6.52	6.52	5.27	5.27	4.40	3.78	379/M
Gas gravity (air = 1)	0.554	1.038	1.522	2.006	2.006	2.490	2.490	2.974	3.459	M/28.97
Lb per Mcf	42.33	79.35	116.4	153.4	153.4	190.4	190.4	227.4	264.4	2.639 M
Gal per Mcf	16.94	25.20	27.46	32.63	31.52	36.65	36.24	41.12	46.13	0.3168 M/S
Bbl per MMcf	403	600	654	777	751	873	863	979	1098	7.543 M/S
Cu ft gas per gal	59.03	39.78	36.32	30.62	31.57	27.37	27.69	24.31	21.66	3157 S/M
Mcf per bbl	2.48	1.67	1.54	1.29	1.33	1.15	1.16	1.02	0.910	132.6 S/M
Critical temperature, °R.	344	550	666	734	766	830	847	915	973	
Critical temperature, °F.	116.3	90.1	206.3	273.2	307.6	370.0	387.0	454.6	512.6	
Critical pres., psia	673.3	708.5	617.5	542.4	529.2	483.6	485.1	433.6	396.9	
Crit. vol-ft³/lb	0.0989	0.0789	0.0709	0.0685	0.0712	0.0685	0.0690	0.0685	0.0665	

One pound mole = 359 cubic feet of gas at 14.7 psi and 32°F.
= 379 cubic feet of gas at 14.7 psi and 60°F.

Compound	Critical pressure, psia	Critical temperature, °R.
Air	546	238
Nitrogen	492	227
Oxygen	731	278
Carbon dioxide	1072	548
Hydrogen sulphide	1307	673

Some gases react in an unusual manner, as shown in section on volatile oils.

An Equation for Z.

$$Z = 1 + (Z' - 1)(\sin 90 \, P_r/P_r')^N$$
$$N = [1.1 + 0.26 T_r + (1.04 - 1.42 T_r) \times P_r/P_r'] \, e^u/T_r$$
$$Z' = 0.3379 \ln(\ln T_r) + 1.091$$
$$P_r' = 21.46 Z - 11.9 Z'^2 - 5.9$$

where:

Z = gas compressibility
P_r = pseudoreduced pressure
T_r = pseudoreduced temperature
Z' = minimum compressibility for a constant T_r
P_r' = pseudoreduced pressure corresponding to Z'
$u = P_r/P_r'$

Good approximations for P_r and T_r for hydrocarbon gases with specific gravity less than 0.75 can be derived from the AGA "Standard Method." Results are:

$$P_r = P/(690 - 31 G)$$
$$T_r = T/(157.5 + 336.1 G)$$

where:

P = average flowing pressure, psia
T = flowing temperature, deg. R.
G = specific gravity (air = 1.0)

Equations are applicable when the temperature is less than 200°F and pressure is less than 2,000 psi.

Solubility of Gas in Oil and Water and Water in Gas

Gas solubility is the volume of gas liberated from a liquid at the separator measured at standard conditions of 14.7 psia and 60°F that is held in solution at reservoir pressure and temperature by that quantity of liquid which will occupy unit volume when produced. If tank conditions are not proper, additional gas may come out of solution at the tank and such gas must be added to that recovered at the separator. The need is for a conversion of the tank measured liquids to a designated subsurface condition for use in equations such as volumetric and material balance. The gas-oil ratio is an important parameter in determining such a conversion.

The best values can be obtained from PVT measurements made in the laboratory. Again, the analysis results are influenced by laboratory techniques, (liberation differential rather than flash), and separators (the number, type and operation anticipated in field separation). Samples are usually analyzed for several different conditions of separation so that the analyst can select values for different possible field operating conditions.

The initial gas-oil ratio (GOR) and later values found with production are measured routinely in the field. The GOR is constant when pressure is above the bubble point. It thereafter declines slightly, since relative permeability measurements indicate that free gas cannot move until a critical gas saturation is established in the field with a reduction of pressure below the bubble point. At later times during depletion by the solution-gas drive process, the GOR increases rapidly so that values of 15,000 to 20,000 are reached before it again declines because of a lack of gas as other driving forces become important.

The gas-oil ratio at the time of discovery can vary substantially. In many reservoirs, the oil is saturated at the existing pressure and temperature. In others, the oil is substantially undersaturated. When a free gas cap is present, oils at the gas-oil contact should be saturated. The oil below the contact may be at, above, or below this value, depending upon many factors including the existence of a common reservoir, type of migration, time of migration of gas and oils, and convection of fluids within the reservoir during geological time. Fields such as Elk Basin (Wyoming) and Rangely (Colorado) show separate accumulations with variations in GOR and oil gravity because of shale stringers and possibly different sources of oil. The La Paz Field (Venezuela) originally exhibited a constant GOR throughout a thick oil column. If gas is migrating underneath the oil at the time of first drilling, the GOR might show an increase with depth below the gas-oil contact. The Prudhoe Field, Alaska, apparently shows a decreasing GOR with depth.

Water can remove gas from oils, and tars and heavy oils are found at or slightly above the water-oil contacts in many fields. The tar zones may be at different subsurface elevations if the reservoir has been disturbed geologically since migration and accumulation.

A general correlation for gas in oils is shown as Fig. 3-21. An increase in temperature reduces the GOR and an increase in pressure increases the GOR if gas is available to keep the oil saturated as the depth of burial increases (Fig. 3-22). If reservoir pressure is decreased due to the removal of overburden, the oil may lose gas to a free gas cap.

The solubility of water in gas is illustrated by Fig. 3-23. Solubility of gas in water is illustrated by Fig. 3-24. Abnormally pressured waters contain gas that may be commercial at high gas prices.

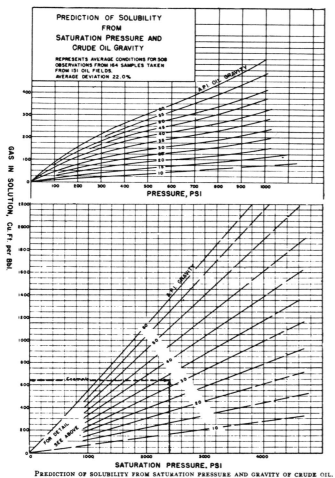

Figure 3–21. *Solubility of gas in oils (Beal, AIME* Trans. *165, 94, 1946).*

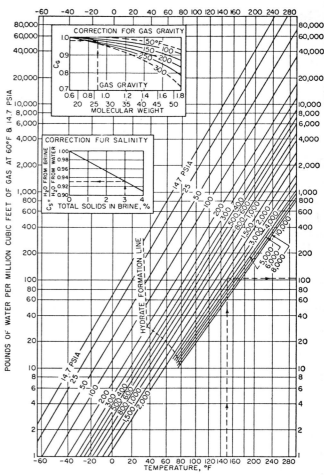

Figure 3–23. *Solubility of water in gas* (after Frick, © *SPE-AIME 1962, from McKetta and Wehe,* World Oil, *147, no. 7, 1958).*

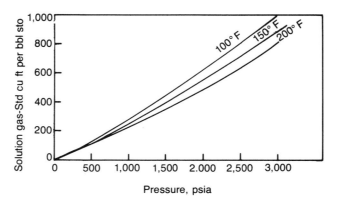

Figure 3–22. *Illustration of effect of temperature on gas in solution in oils.*

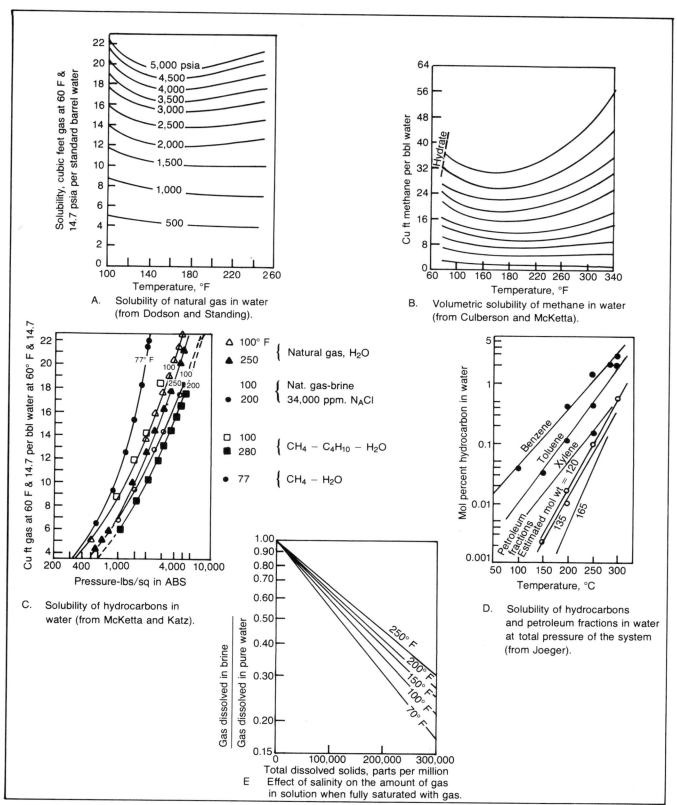

A. Solubility of natural gas in water (from Dodson and Standing).

B. Volumetric solubility of methane in water (from Culberson and McKetta).

C. Solubility of hydrocarbons in water (from McKetta and Katz).

△ 100° F
▲ 250 } Natural gas, H_2O

100
● 200 } Nat. gas-brine 34,000 ppm. N_ACl

□ 100
■ 280 } $CH_4 - C_4H_{10} - H_2O$

● 77 { $CH_4 - H_2O$

D. Solubility of hydrocarbons and petroleum fractions in water at total pressure of the system (from Joeger).

E. Effect of salinity on the amount of gas in solution when fully saturated with gas.

Figure 3–24. *Solubility of hydrocarbons in water (after Frick, SPE-AIME, 1962). To convert from mole-percent water to pounds water per 100 hydrocarbon, the reader may use the following relation:*

$$\frac{1\ lb\ water}{100\ lb\ hydrocarbon} = \frac{(mole\text{-}percent\ water)\,(18)\,(100)}{(100 - mole\text{-}percent\ water)\,(mole\ wt\ of\ hydrocarbon)}$$

Viscosity of Oils, Gas, Steam, and Water

Several types of fluids are recognized:

Newtonian—Unaffected by magnitude and kind of motion to which they are subjected.

Dilatant—Viscosity will increase as agitation is increased.

Plastic (Bingham)—Have a definite yield value which must be exceeded before flow starts. After flow starts, viscosity decreases with increase in agitation.

Pseudoplastic—Do not have a yield value, but do have decreasing viscosity with increase in agitation.

Thixotropic—Viscosity will normally decrease upon increased agitation, but this depends upon duration of agitation and viscosity of fluid and rate of motion before agitation.

Most oils and gases and water are Newtonian fluids. The viscosity of gas is much less than that of oil and water. Gas flows easier than the other two fluids, both in tubular configurations and in the rocks. Flow in rocks follows Darcy's law. When two fluids are flowing in the same general pore space, the following averaging method applies:

$$(k/\mu)_t = (k_o/\mu_o) + (k_g/\mu_g) + (k_w/\mu_w)$$
$$(k/\mu)_t = (162.6/mh)(B_o/q_o + B_g(q_{gt} - q_o R_s) + B_w/q_w)$$

where m is the slope of the flow or build-up curve.

The viscosity of oils, including the gas in solution, is measured during laboratory PVT testing. Viscosity decreases with pressure reduction when oils are above the bubble point and increases with a reduction in pressure when oils are below the bubble point Fig. 3–25. Fig. 3–26, which is a plot on unusual paper, shows that viscosity decreases substantially as temperature increases. One method for estimating viscosity of dead and live (gas in solution) oils is shown in Fig. 3–27.

The viscosity of gas depends upon the composition of the natural gas. When composition is known, papers by Carr et al., Trans. AIME, Vol. 201, 1954, should be consulted. When only gas gravity is known, data such as shown on Figs. 3–28 and 29 should be useful. The viscosity of steam is illustrated by Fig. 3–31 and is available more exactly from steam tables prepared by the utilities industry. The viscosity of water is shown on Fig. 3–30.

Viscosity is measured using several techniques, and definitions differ depending upon each type of US industry. Conversion techniques are shown in Figs. 3–32 and 33.

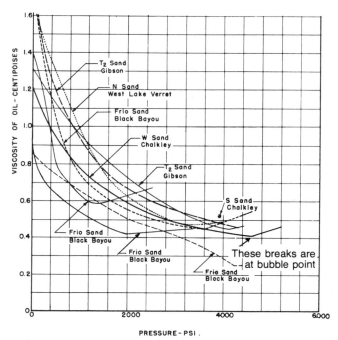

Figure 3–25. *Viscosity of oils in Louisiana determined by PVT analysis.*

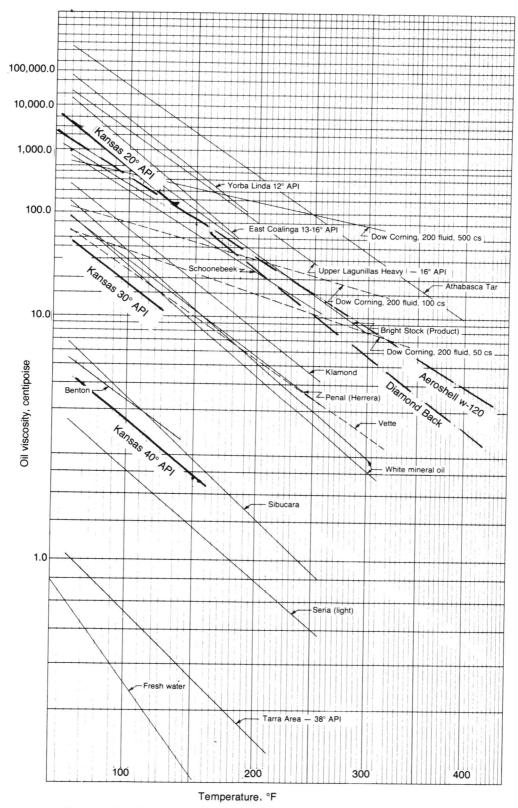

Figure 3–26. *Viscosity of heavy oils, gas-free oils (courtesy Shell Oil Company).*

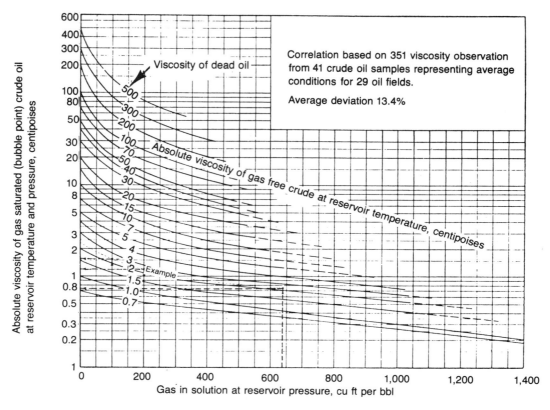

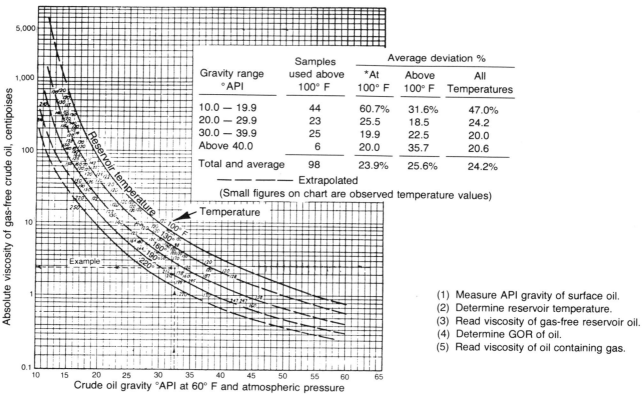

(1) Measure API gravity of surface oil.
(2) Determine reservoir temperature.
(3) Read viscosity of gas-free reservoir oil.
(4) Determine GOR of oil.
(5) Read viscosity of oil containing gas.

Gravity range °API	Samples used above 100° F	Average deviation %		
		*At 100° F	Above 100° F	All Temperatures
10.0 — 19.9	44	60.7%	31.6%	47.0%
20.0 — 29.9	23	25.5	18.5	24.2
30.0 — 39.9	25	19.9	22.5	20.0
Above 40.0	6	20.0	35.7	20.6
Total and average	98	23.9%	25.6%	24.2%

— — — — — Extrapolated
(Small figures on chart are observed temperature values)

Figure 3–27. *Viscosity of oil—Beal method for viscosity (© SPE-AIME).*

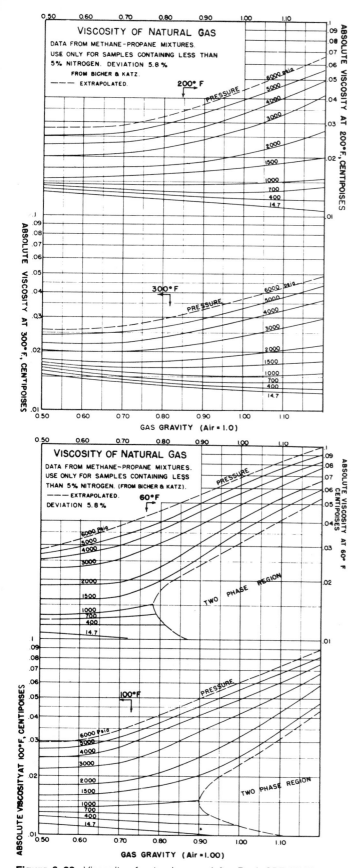

Figure 3–28. *Viscosity of natural gases (after Beal, SPE-AIME).*

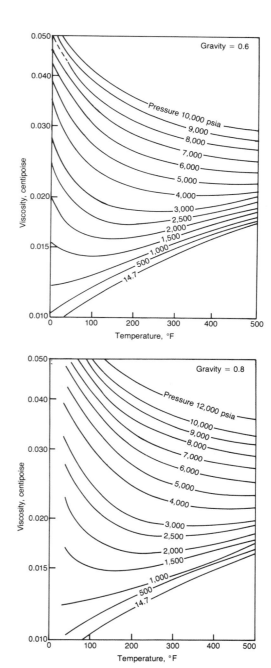

Figure 3–29. *Crossplot of viscosity data for gases (Bicher & Katz, Beal, © SPE-AIME).*

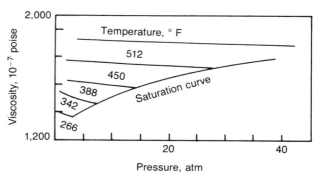

Figure 3–30. *Viscosity of steam.*

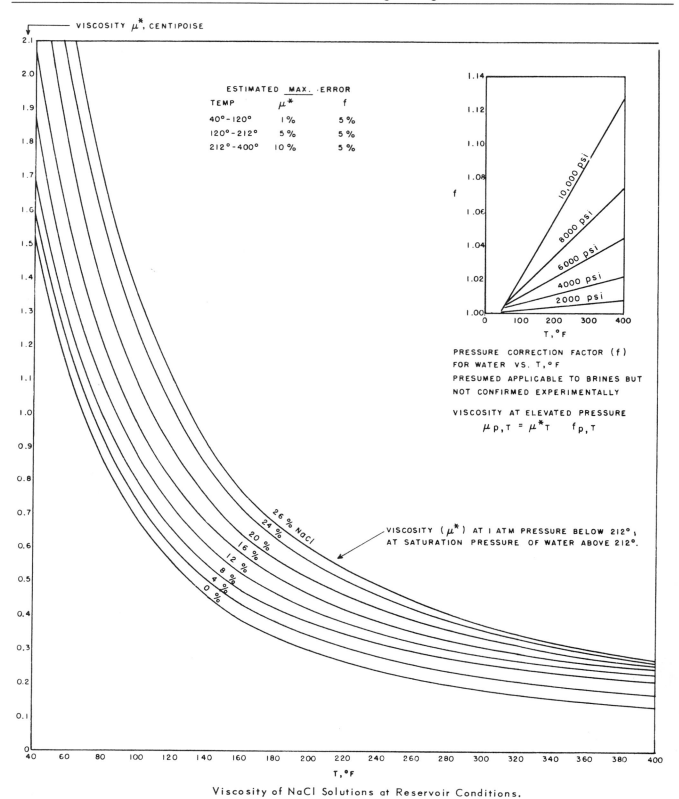

Viscosity of NaCl Solutions at Reservoir Conditions.

Figure 3–31. *Viscosity of water and saline solutions (courtesy Shell Oil Company and* Petroleum Engineer).

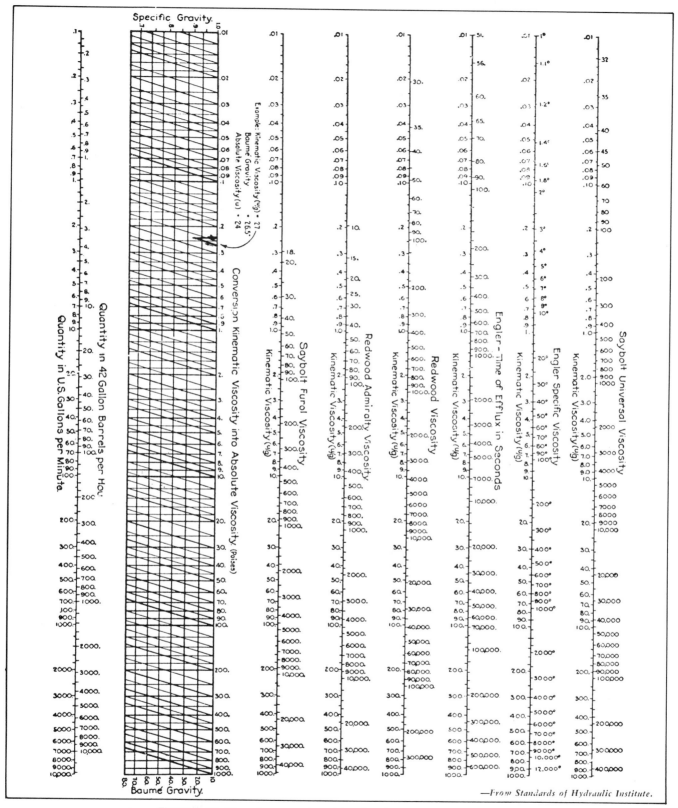

Figure 3–32. *Viscosity conversions (after Standards of Hydraulic Institute).*

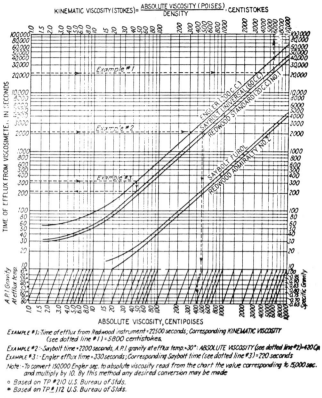

KINEMATIC VISCOSITY (STOKES) = $\dfrac{\text{ABSOLUTE VISCOSITY (POISES)}}{\text{DENSITY}}$, CENTISTOKES

EXAMPLE #1: Time of efflux from Redwood instrument = 22500 seconds; Corresponding KINEMATIC VISCOSITY (see dotted line #1) = 5800 centistokes.

EXAMPLE #2: Saybolt time = 7200 seconds, A.P.I. gravity at efflux temp = 30°: ABSOLUTE VISCOSITY (see dotted line #2) = 420 Cp.

EXAMPLE #3: Engler efflux time = 330 seconds; Corresponding Saybolt time (see dotted line #3) = 720 seconds.

Note: - To convert 150,000 Engler sec. to absolute viscosity read from the chart the value corresponding to 15,000 sec. and multiply by 10. By this method any desired conversion may be made.

o Based on TP #210 U.S. Bureau of Stds.

* Based on TP #112 U.S. Bureau of Stds.

Figure 3–33. *Viscosity conversions (courtesy Shell Oil Company).*

Compressibility of Oil, Gas, Water, and Rocks

For most situations, the total compressibility of the fluid and rock system is:

$$c_t = S_o c_o + S_g c_g + S_w c_w + c_f$$

The system is expressed in terms of pore volume, and compressibility often increases as porosity decreases.

Arps in Frick states that the typical values for oil is 0.000005 when gravity is low (contains little dissolved gas) to 25×10^{-6} for higher gravity oils. A figure of 0.000010 is a good average and may be used except in material balance equations when oil is above the bubble point. The compressibility of gases often is approximated by 1 divided by pressure, possibly adjusted for Z. Water has an average value of 0.000003 when gas in solution in water is very small. Rock compressibility is inversely related with porosity so that 0.000010 may be used when porosity is 2%, 0.0000048 for 10% porosity, and 0.0000034 for 25% porosity.

Greater accuracy often is possible from use of various graphs. For oils, SPE Monograph 1 includes the following example based on PVT analysis for oil below the bubble point. The following extract is taken from that source:

$$c_o = -\frac{1}{V} \times \frac{(V_1 - V_2)}{(p_1 - p_2)}$$

$$c_o = \frac{B_g}{B_o}\frac{dR_a}{dp} - \frac{1}{B_o}\frac{dB_o}{dp}$$

$$= \frac{12.9 \times 10^{-3}}{1.227}(0.0455)$$

$$- \frac{1}{1.227}(0.0001425)$$

$$= 0.0003622 \text{ psi}^{-1}$$

Values of (dR_s/dp) and (dB_o/dp) are obtained as the slopes of laboratory-determined curves of R_s and B_o

vs p; the slope is drawn at the estimated average pressure.

Since we do not have a gas analysis, we estimate $T_c = 464°R$ and $p_c = 655$ psia from the gas gravity of 0.93. Then $T_r = 720/464 = 1.55$, $p_r = 1,315/655 = 2.01$. $c_r = 0.56$. Then $c_g = c_r/p_c = 0.56/655 = 0.000854$ psia^{-1}.

We estimate $c_w = 3 \times 10^{-6}$ psi^{-1}, and $c_f = 4 \times 10^{-6}$ psi^{-1} at $\phi = 0.15$. Then $c_t = S_o c_o + S_g c_g + S_w c_w + c_f$, and for $S_o = 0.546$, $S_g = 0.204$ and $S_w = 0.25$, $c_t = 0.546 \ (0.0003622) + 0.204 \ (0.000854) + 0.25 \ (3 \times 10^{-6}) + 4 \times 10^{-6} = 0.000376$.

The characteristics of oils are rather variable, and no general relations have been published. Comparable oils may be used.

Gas compressibility varies with the inverse of pressure corrected by Z. The compressibility of gas may be read from Fig. 3–34. The compressibility for pure water and water saturated with gas is available from Figure 3–35. The compressibility of rocks is shown on Figures 3–36 through 39. The compressibility of fractures is large when porosity is limited to that of the fracture. Often the matrix contains water-filled porosity, and the values for fractures should be corrected accordingly. Material balance calculations, particularly when the reservoir oils are above bubble point, may be used to observe higher-than-usual compressibility if volumes are known with some degree of accuracy. The compressibility equation is based on saturation of unit bulk volume. Often, intermediate beds of shale and adjacent areas, particularly downdip, make contributions to pressure behavior. Equations such as material balance, particularly when above bubble point, must be adjusted accordingly. In abnormally pressured reservoirs, the contribution from shale can be substantial.

Well-consolidated deep rocks including shales are relatively rigid, and they react slowly to changes in reservoir pressure. As illustrated on Figs. 3–36 and 37, unconsolidated or friable sands and shales exhibit a high degree of compressibility, which is effective during the production from oil, gas, and water reservoirs. Literature of the ground-water industry should be reviewed when these conditions are encountered. Substantial subsidence of the surface of the earth has been observed, which is caused by withdrawal of fluids from reservoirs at substantial depth, near Houston and in California. The associated economic risks can be substantial in populated areas.

Compressibility of a Fracture*

For flow through a narrow duct, an expression for effective fracture clearance develops directly.

$$w^3 = \frac{12 \ Q \mu L}{D \ \Delta p}$$

Using as a model a cylindrical rock sample of diameter D and containing a single, planar fracture along its axis, w is defined as an idealized fracture clearance or thickness. It denotes the clearance that would exist in a smooth, planar fracture of constant rectangular cross-section Dw and length L, which would conduct fluid of viscosity μ at a flow rate Q under a pressure differential Δp, the same as the core containing the true fracture and depending only on the fracture for fluid conductivity. The actual clearance of the fracture in a real core would, of course, be somewhat different from the calculated ideal w because of surface roughness.

By considering this fractured rock as a porous medium, although flow only occurs through the fracture,

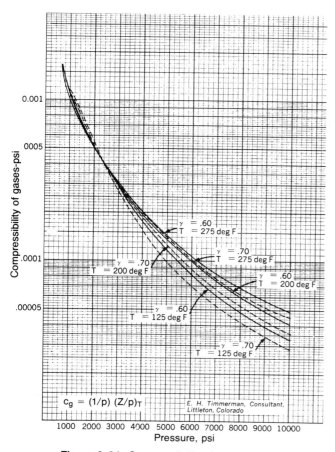

Figure 3–34. *Compressibility of natural gas.*

* This section is taken from Jones, *JPT*, January 1975, © SPE-AIME.

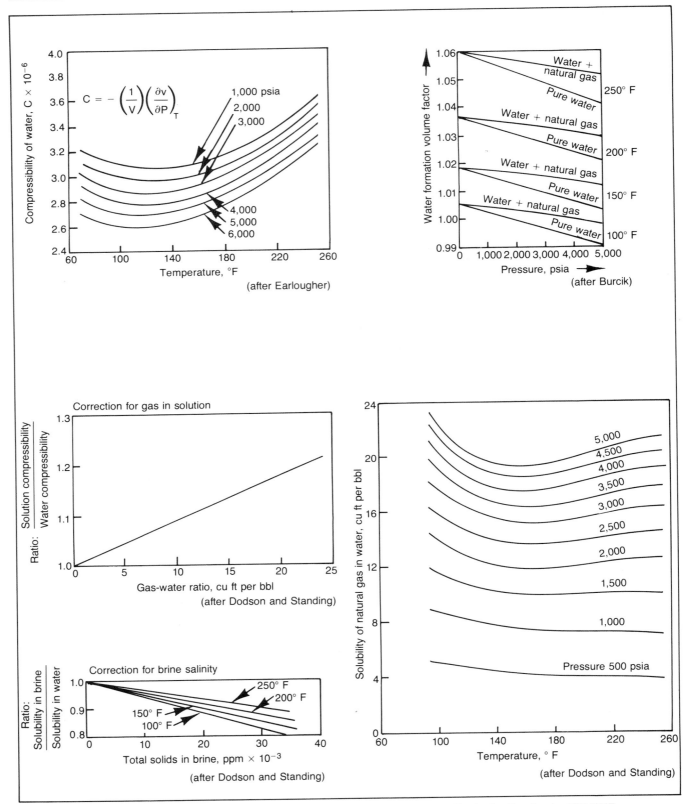

Figure 3–35. *Compressibility of water (after Earlougher from Frick,* Petroleum Production Handbook, *SPE-AIME).*

$c_w = (1/B_w)\,(\partial B_w/\partial p) + (B_g/B_w)\,(\partial R_{sw}/\partial p)$

$c_w = (1/B_w)\,(\partial B_w/\partial p)$ *without gas*

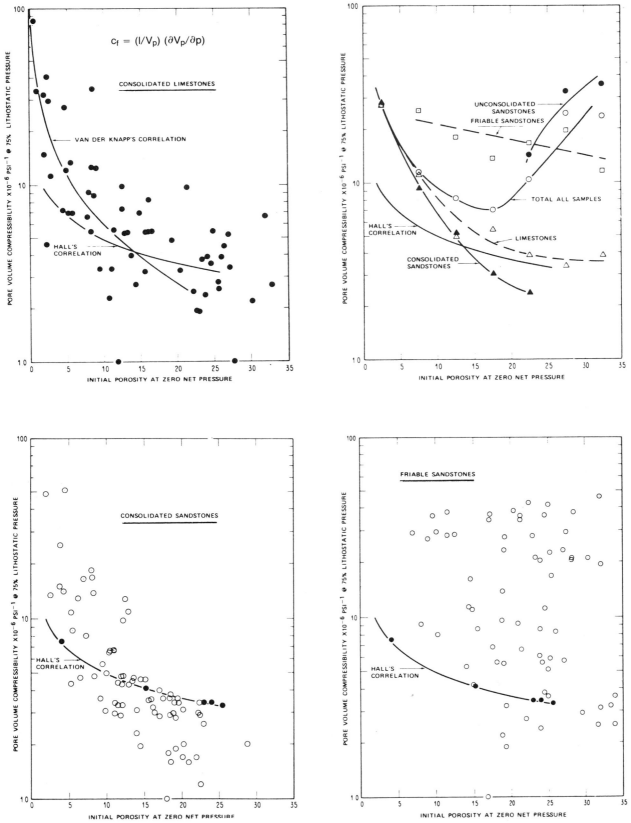

Figure 3–36. *Rock compressibility (after Newman, JPT, February 1973, © SPE-AIME).*

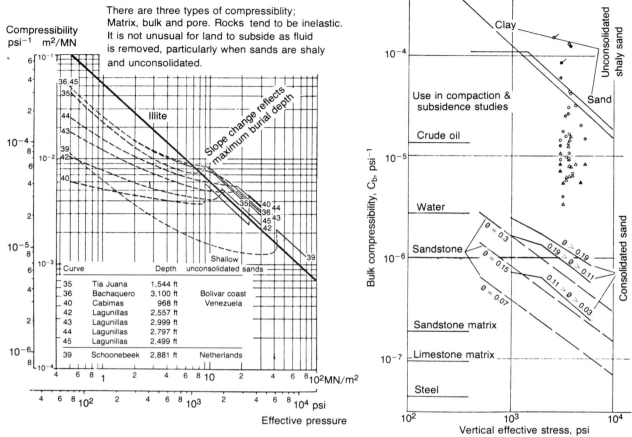

Figure 3–37. *Compressibility of rocks (courtesy Shell Oil Co.)*

the flow can also be expressed by the Darcy equation

$$k = \frac{Q \mu L}{A \Delta p}$$

The factor A in this equation represents the cross-sectional area of the cylindrical rock sample whose flow capacity depends solely upon the fracture. An expression relating fracture clearance and Darcy permeability of the rock can be derived after noting that the right-hand members of the two previous equations differ only by constant factors, so that

$$w^3 = \frac{12A \times 10^{-8}}{D} k \quad \text{and} \quad w \propto k^{1/3}$$

$$c_f = \frac{1}{\left(\dfrac{\phi}{\phi_i}\right)} \cdot \frac{d\left(\dfrac{\phi}{\phi_i}\right)}{dp_k} = \frac{0.4343}{p_k \log(p_k/p_h)}$$

From this equation, fracture compressibility in a 10,000-ft dense carbonate reservoir is estimated at 96×10^{-6} psi^{-1} at initial pressure that decreases to about 72 $\times 10^{-6}$ psi^{-1} at depletion. Reservoir rock

pore volume compressibility usually ranges approximately from 2×10^{-6} psi^{-1} to 15×10^{-6} psi^{-1}.

Land Subsidence in Oil and Water Fields*

	Subsidence (ft)	Through year
Oil field		
Wilmington, Calif. (see Gilluly and Grant, 1949)	29	1966
California water fields		
Santa Clara Valley (San Jose)	13	1967
San Joaquin Valley		
Los Banos—Kettleman hills area	26	1966
Tulare—Wasco area	12	1962
Arvin—Maricopa area	8	1965

* This section is taken from Wilmington, *JPT*, February 1972. (Lohman, *Ground Water Hydraulics*, USGS, 1970, and Symposium on Transient Ground Water Hydraulics, Colorado State University, Fort Collins, July 1963.)

Lohman (1961) showed that for elastic confined aquifers for which C may be assumed to equal 1,

$$\frac{b}{E_s} = S/\gamma - \theta\, b\beta$$

and that Hooke's Law (strain is proportional to stress, within the elastic limit) may be expressed:

$$\Delta b = \frac{b}{E_s}\Delta p$$

where Δb = change in b, in ft
Δp = change (generally decline) in artesian pressure, lb in.$^{-2}$

Combining equations:

$$\Delta b = \Delta p(S/\gamma - \theta\, b\beta)$$

This equation gives the amount of subsidence, Δb, for an elastic confined aquifer of known S, θ, and β, for a given decline in artesian pressure, Δp. For example, assume $S = 2 \times 10^{-4}$, $\theta = 0.3$, $b = 100$ ft, $\Delta p = 100$ lb in.$^{-2}$, and note that it is convenient to use $1/\gamma$, which equals 2.31 ft lb^{-1} in.^{2}. Then,

$$\Delta b = 10^2 \text{ lb in.}^{-2} (2 \times 10^{-4} \times 2.31 \text{ ft lb}^{-1} \text{ in.}^2 - 0.3 \times 10^2 \text{ ft} \times 3.3 \times 10^{-6} \text{ in.}^2 \text{ lb}^{-1})$$
$$= 0.04 \text{ ft}$$

Similarly, for $b = 1,000$ ft, $\Delta p = 1,000$ lb in.$^{-2}$, $S = 10^{-3}$, $\theta = 0.3$, $\Delta b = 1.3$ ft.

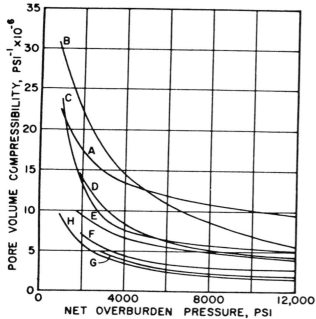

Figure 3–38. Rock compressibility vs. pressure (after Fatt, Trans. AIME, © SPE-AIME). A: Basal Tuscaloosa sandstone, Mississippi, porosity—13 percent. B: Sandstone from wildcat, Santa Rosa County, Fla., porosity—15 percent. C: Sandstone from Ventura Basin field, Calif., porosity—10 percent. D: Sandstone from West Montalvo area field, Calif., porosity—12 percent. E: Weber sandstone, Rangely field, Colo., porosity—12 percent. F: Sandstone from Nevada Wildcat, porosity—13 percent. G: Strawn sandstone, Sherman field, Tex., porosity—13 percent. H: Bradford sandstone, Pennsylvania, porosity—15 percent.

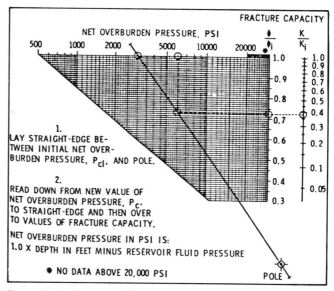

Figure 3–39. Compressibility of a fracture (after Jones, JPT, © SPE-AIME).

The m(p) Curve

Gas reservoirs were being developed in rocks having a permeability less than 10 md during the late 1950s on 640-acre spacing. The static pressures and build-up pressures measured in the field deviated from results observed when permeability was higher. In tight pays, the values for Z and viscosity both vary with distance from the well bore since these terms are pressure dependent. The pressure drop in tight pays can be substantial near the well bore, as discussed in other sections of this text. Russell, Goodrich, and Bruskotter of Shell Oil published a method for averaging pressures that solved the problem (*JPT*, January 1966). Ramey et al. published an m(p) method, summarized in the section in *JPT*, May 1966, and in a number of later issues. Ramey presented a method for calculating the m(p) function and a graph based on critical values. Using these techniques, Timmerman prepared information that determines the m(p) curve for various gas gravities and temperatures as presented in Figs. 3–43 and 44.

Many studies have shown that the p^2 methods are applicable when all pressures are less than about 1,000 psia. The pressure may be used directly when all pressures are above 3,000 psi. The m(p) curves are applicable for all pressure values since changes in Z and μ are adjusted by the m(p) function, which relates with pressure directly.

The m(p) Function of Ramey et al.*

For many gases, $z\mu$ approaches a constant at pressures below 1,000 psi and varies directly with pressure above 3,000 psi, as first indicated by Wattenbarger et al. and illustrated in Fig. 3–40. At higher pressures, $z\mu$ is roughly proportional to p or $p/z\mu$ is constant and well test plots in terms of p are valid regardless of the pressure gradient:

$$m(p) = \frac{2p_i}{z_i\mu_i} \int_o^p dp' = \frac{2p_i}{z_i\mu_i} p$$

At lower pressures, $z\mu$ is almost constant and the ideal gas method (plotting p^2) applies below 1,000 psi:

$$m(p) = \frac{2}{z_i\mu_i} \int_p^p p' \ dp' = \frac{1}{z_i\mu_i} p^2$$

The formula for determining flow capacity below 1,000 psi is:

$$kh = \frac{5.792 \times 10^4 \ q_{sc}p_{sc}T}{T_{sc} \ (\text{slope of } p^2 \text{ plot})} \ (z\mu)_{\text{wellbore}}$$

* This section is taken from Ramey et al., *JPT*, May 1966.

Flow capacity above 3,000 psi can be determined from:

$$kh = \frac{5.792 \times 10^4 \ q_{sc}p_{sc}T}{T_{sc} \ (\text{slope of } p \text{ plot})} \left(\frac{z\mu}{2p}\right)_{\text{wellbore}}$$

In these equations, the slopes of the p and p^2 plots and the fluid properties are taken at a single point or series of points on the drawdown or buildup curve. Such plots should not be used in many cases to determine skin, and the m(p) curve should be used to determine skin effect.

These equations may be compared with the normal equations using the average values of Russell et al. and the m(p) function of Ramey et al.:

$$kh = \frac{162.6 \ q\mu B}{m}$$

where m is the slope of the Russell build-up curve. The 162.6 is for q, in bbls. The amount of 28,958 should be used when q is in cubic feet.

$$kh = 1,637 \frac{qT}{-b}$$

where b is the slope of the Ramey build-up curve and q is in *Mcf*.

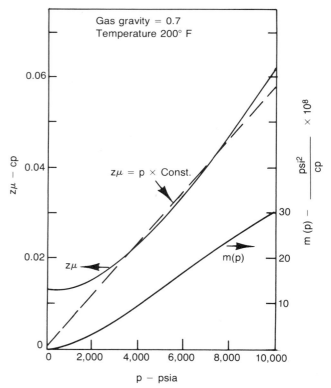

Figure 3–40. *Plot of $Z\mu$ vs. pressure (after Ramey et al.).*

1. Reduced pressure and temperature method of Ramey (Fig. 3–41).

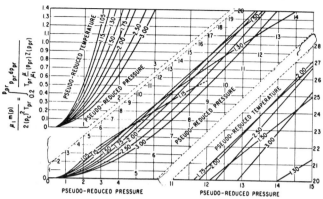

Figure 3–41. *Reduced pressure and temperature method of Ramey (after Ramey et al., JPT, May 1966).*

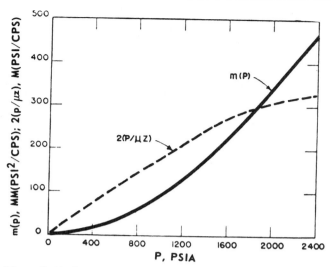

Figure 3–42. *Calculation of m(p) (after Ramey et al., JPT, May 1966).*

2. Calculation of m(p) when gas characteristics are known (Fig. 3–42).

Gas Properties

p, psia	z	Viscosity, cp
400	0.95	0.0117
800	0.90	0.0125
1,200	0.86	0.0132
1,600	0.81	0.0146
2,000	0.80	0.0163
2,400	0.81	0.0180

Find $m(p)$ vs. p for the gas properties tabulated above, using the equation defining $m(p)$. The quantity $2(p/\mu z)$ can be calculated and plotted vs. pressure, as shown below. Integration can be performed in a tabular calculation by reading midpoint values of $2(p/\mu z)$ from the graph and multiplying by Δp. The computed $m(p)$, psi^2/cp, is also shown on the figure. This curve can be used for future tests with this well or other wells producing the same gas at the same formation temperature. Often, only gas gravity is available. In this case, $m(p)$ can be found without integration from a graph (Figs. 3–43 and 44).

p, psia	z	μ cp	$2(p/\mu z)$ psi/cp	Mean $2(p/\mu z)$	Δp, psi	$2(p/\mu z)$ $(x/\Delta p)$	$m(p)$ psi^2/cp
400	0.95	0.0117	71,975	35,988	400	14.4×10^6	14.4×10^6
800	0.90	0.0125	142,222	107,099	400	42.9×10^6	57.3×10^6
1,200	0.86	0.0132	211,416	176,819	400	70.7×10^6	128.0×10^6
1,600	0.81	0.0146	270,590	241,003	400	96.5×10^6	224.5×10^6
2,000	0.80	0.0163	306,748	288,669	400	115.5×10^6	340.0×10^6
2,400	0.81	0.0180	329,218	319,000	400	127.6×10^6	467.6×10^6

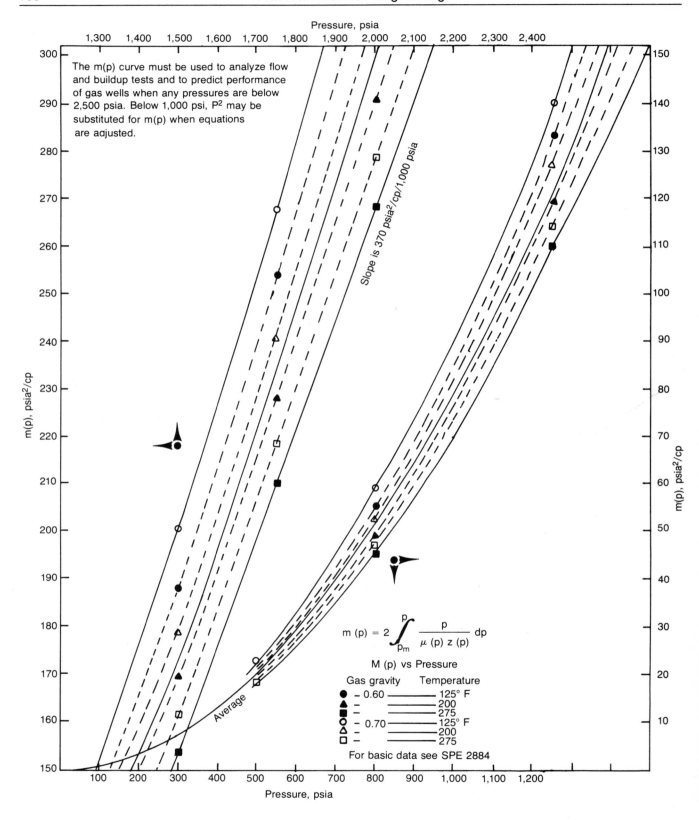

Figure 3–43. *m(p) curve (courtesy* Petroleum Engineer).

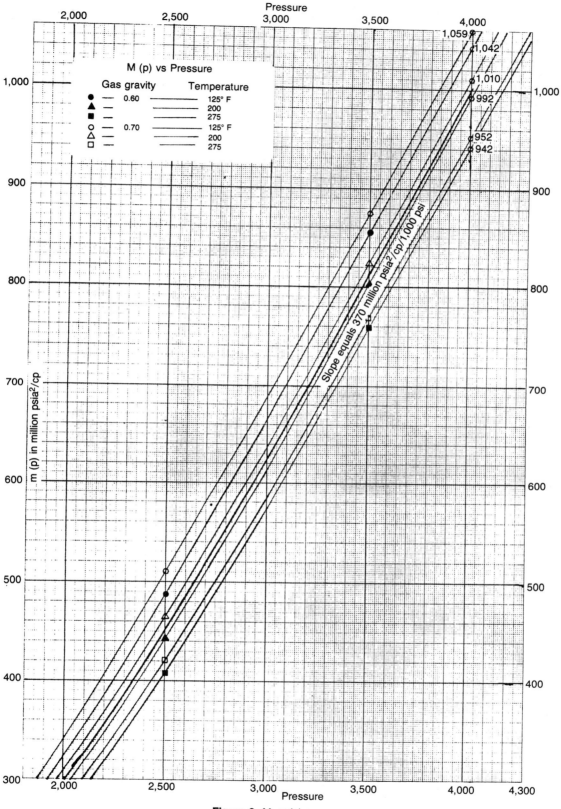

Figure 3–44. m(p) curve.

PVT Adjustments Related to Separator Conditions

The laboratory PVT data often include analysis at various or different separator conditions so that data consistent with separator operating conditions to be used in the field can be selected. Vazquez and Beggs in SPE Paper 6719, "Correlations for Fluid Physical Prediction," reviewed this problem. The gas specific gravity relates directly with the separator pressure and temperature. Oil volatility represented as a function of API also is important. The applicable concepts and constants obtained from regression analysis follow for determining gas gravity:

$$\left(\frac{\gamma_g(114.7)}{\gamma_g(p)}\right) = c_1 + c_2 \log\left(\frac{p}{114.7}\right)$$

where $\gamma_g(114.7) =$ gas specific gravity that would result from a separator pressure of 100 psig (114.7 psia) and separator temperature T, °F

$\gamma_g(p) =$ gas specific gravity measured at the given separator pressure, p

$$\left(\frac{\gamma_g 114.7}{\gamma_g p}\right) = 1 + 0.5912 \,(API)\,(T) \log\left(\frac{p}{114.7}\right) 10^{-4} \,\gamma_g(114.7)$$

$$= \left(\frac{\gamma_g(114.7)}{\gamma_g(p)}\right)(\gamma_g(p))$$

API = stock-tank API oil-gravity
$T =$ separator temperature, °F

The solution gas-oil ratio can be obtained in similar manner:

$$R_s = R_{sp} - \left[(R_1)_{st}\right]\frac{B_{ofb}}{B_{odb}}$$

where: $R_s =$ gas in solution per stock-tank barrel of oil, scf/STB

$R_{sp} =$ total gas liberated at the separator per stock-tank barrel of oil by flashing bubble-point oil, scf/STB

$(R_1)_{st} =$ the standard cubic feet of gas liberated by a differential process per barrel of residual oil at standard conditions, scf/STB

$B_{ofb} =$ bubble-point oil required to yield 1 bbl of stock-tank oil when flashed through the separator to stock-tank conditions, bbl/STB

$B_{odb} =$ bubble-point oil required to yield 1 bbl of stock-tank oil when differentially liberated to stock-tank conditions, bbl/STB

$$\frac{R_s}{\gamma_g} = c_1 \,(p)^{c_2}\, 10^{c_3 \left(\frac{API}{T}\right)}$$

so that $\log\left(\frac{R_s}{\gamma_g}\right) = \log(c_1) + c_2 \log(p) + c_3 \left(\frac{API}{T}\right)$.

In like manner, the formation volume factor for below bubble point is:

$$B_o = \frac{B_{ofb}}{B_{odb}} B_{od}$$

where: $B_{od} =$ oil at reservoir pressure p required to yield 1 bbl of stock-tank oil when differentially liberated to stock-tank conditions, bbl/STB

$B_o =$ oil at reservoir pressure p required to yield 1 bbl of stock-tank oil when flashed through the separator, bbl/STB; the term often referred to as simply the flash formation volume factor

$$B_o = c_1 + c_2(R_s) + c_3(T-60)\left(\frac{API}{\gamma_g}\right) + c_4(R_s)(T-60)\left(\frac{API}{\gamma_g}\right)$$

The formation volume factor above bubble point is:

$$c = \frac{\ln\left(\frac{B_{ob}}{B_o}\right)}{(P_b - p)}$$

A similar correlation for viscosity is determined.

$$\mu_o = \mu_{ob}\left(\frac{p}{P_b}\right)^m$$

where: $\mu_o =$ viscosity (cp) of undersaturated oil
$\mu_{ob} =$ viscosity of bubble-point oil
$p =$ reservoir pressure above bubble point
$P_b =$ bubble-point pressure of the crude oil

$$\log(m) = c_1 + c_2\,(p) + c_3 \log(p)$$

These correlations should be used with caution since the range of usefulness of the regression equations is somewhat unknown. The method may be used with another data base for the regression analysis, and the general technique and equations are valuable in obtaining equations for data in computer operation.

Volatile Oils and Condensates

Oils and potential oils include kerogen, which is a hydrocarbon such as oil shale that can be converted to oil with high temperature, tars, which have a very high viscosity, heavy oil, normal oils with API gravity of say 20 to 40 degrees, volatile oils that normally have gravity higher than 35° API, condensates, and dry gases. The dry gases are composed largely of methane so that no liquids are recovered at normal separator conditions.

Condensates usually are a mixture of lighter hydrocarbons with small heptanes plus fractions. Liquid recovery with field separators depends strongly upon operating pressure and temperature. These gases also may exhibit retrograde condensation characteristics, as illustrated by Fig. 3–45. The heavier hydrocarbons simply condense in the pore space with reduction in pressure (Fig. 3–46). Such condensation represents a loss in potential recovery and can be complicated by inability of flow at economical rates at pressures as high as 5,000 psi when condensate content is high, say 300 bbl/MMcfg. This problem is related to relative permeability and viscosity. Liquids can condense near the well bore in quantities that prevent gas production at economical rates when permeability is low, say less than 25 md, and pay thickness is not extremely large.

Determining whether a hydrocarbon mixture is in liquid or gas state at reservoir conditions can be diffi-

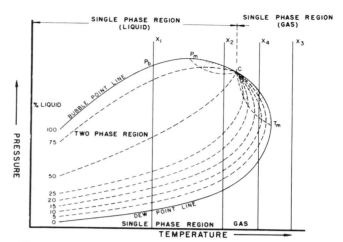

Figure 3–45. *Retrograde condensation (courtesy Shell Oil Co.)*

cult. Light oils at reservoir conditions exhibit characteristics somewhat different from condensate, as illustrated by curves shown on Fig. 3–47. The portion of heptanes plus fraction is substantial compared to that of a condensate reservoir. The separator operating conditions continue to be critical. The formation volume factor is high—often above 1.5 at discovery conditions—compared to that of normal oils. Recovery by gas injection can be high.

Field/reservoir	Av. depth, feet	Lithology	Recovery efficiency (%) Primary	With injection Method	With injection RE	Remarks
Blackjack Creek/Smackover	15,700	Dolomite	19[x]	Waterflood	39*	Av. perm. 112 md. 100 MMstb OOIP. 10.3% H_2S. Peripheral waterflood. 320 acres/well.
Jay-LEC/Smackover	15,400	Dolomite	17[x]	Waterflood	48*	Av. perm. 35 md. 728 MMstb OOIP. 9% H_2S. Staggered line drive waterflood. 160 acres/well.
Pickton/Rodessa	7,900	Oolitic limestone	19[x]	Gas injection**	61†	Av. perm. 250 md. Inj. @ 500 psi below P_b. Crestal gas injection. 40 acres/well.
Raleigh/Hosston	12,600	Sandstone	32[x]	Gas injection**	66[xxx]	Av. perm. 41 md. Inj. @ 2,000 psi above P_b. Crestal gas injection. 40 acres/well.
Shoats Creek/Cockfield	8,950	Sand	<20[x]	Gas injection**	41	Av. perm. 35 md. Inj. @ 1,000 psi above P_b. Peripheral gas injection. 140 acres/well.
North Louisiana/Smackover	10,000	Oolitic limestone	22[xx]	....	....	Av. perm. 174 md. RE extrap. from 700 psi. Cum. GOR = 8.79 Mcf/stb. 160 acres/well.
Unnamed/A	8,200	Sand	25[xx]	....	....	Av. perm. 750 md. RE extrap. to 500 psi from 1544 psi BHP. Cum. GOR = 8.73 Mcf/stb. 308 acres/well.

[x] Estimated. Reservoir produced with indicated injection method.
[xx] Produced by depletion. RE of STOOIP.
[xxx] RE of propanes-plus.

* Estimated. Floods recently initiated.
** See Table 4 for statistical summary of these projects.
† RE of [STOOIP + NGL]. RE for STOOIP is 52%.

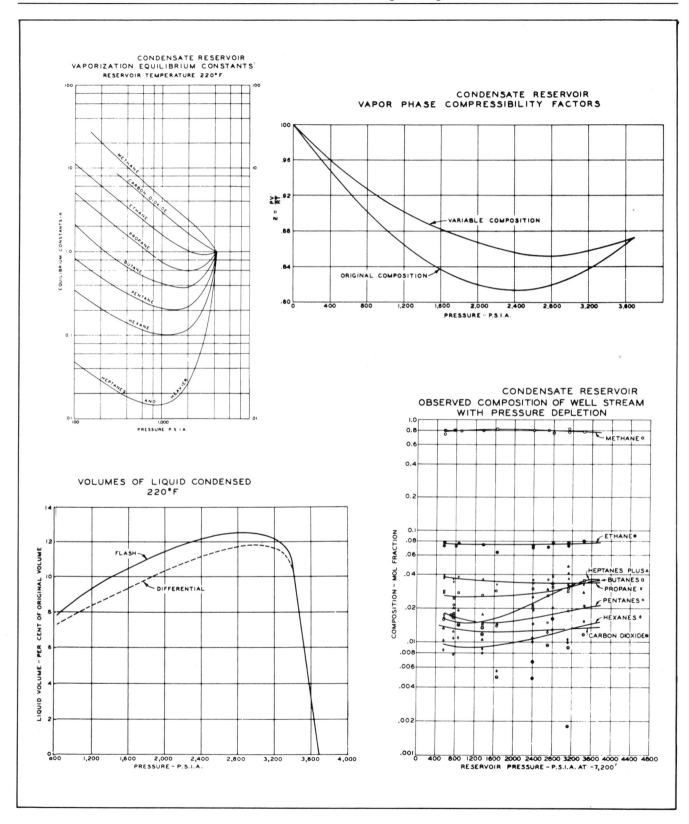

Figure 3–46. *Behavior of condensate reservoir (after Allen & Roe,* Trans. AIME, 1950). *also see Standing et al., AIME JPT (May 1944) and Roland et al., OGJ (27 March 1941) for calculations using equilibrium constants.*

Characteristics of Natural Gas

Natural gases often condense as the pressure is reduced and thereafter again vaporize as the pressure is reduced further. The process is known as retrograde condensation and is unusual except for hydrocarbon gases. Retrograde condensation reduces gas volumes slightly, and in tight formations may severely reduce the effective permeability near the well bore so that abandonment pressures in rock with low permeability and porosity may be in the order of many thousand psi. Effective permeability in some formations may again improve as the condensate vaporizes.

Fig. 3–45 should explain the phenomenon of retrograde condensation if characteristics of line X_4 are observed.

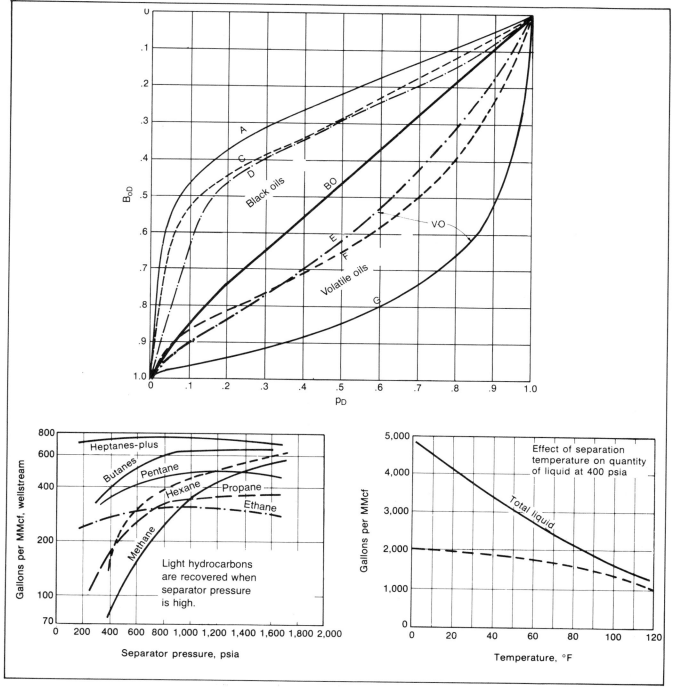

Figure 3–47. *Behavior of volatile oils (Cronquist, World Oil, April 1979).*

Analyzing PVT Data and Determining Computer Data Equations

Glaso in *JPT*, May 1980, analyzed North Sea oils using methods proposed by Standing. The basic laboratory data shown in Table 3–7 have been converted to graphs from which equations can be computed by

TABLE 3–7
Surface Properties and Experimentally Determined Bubble-Point Pressure and FVF

Sample	Gas/oil ratio, R (scf/STB)	Average surface gas gravity, γ_g (air = 1)	Stock-tank oil gravity γAPI	Stock-tank oil gravity γ_o	Corrected stock-tank oil gravity $\gamma API, cor$	Corrected stock-tank oil gravity $\gamma_o, corr$	Reservoir temperature, T (°F)	Experimental bubble-point pressure, p_b (psig)	Experimental bubble-point FVF, B_{ob} (bbl/STB)
1	1,409	0.799	42.5	0.813			260	4,494	1.854
2	756	0.759	35.1	0.849			230	3,501	1.442
3[a]	1,924	0.935	38.0	0.835			245	4,497	2.210
4	950	0.980	31.7	0.867			230	3,501	1.589
5	1,623	0.761	39.0	0.829			250	5,405	1.918
6	909	0.732	38.0	0.835			180	3,796	1.434
7	1,280	0.756	38.6	0.832			180	4,735	1.664
8	1,052	0.767	37.4	0.838			193	4,011	1.577
9	1,039	0.946	39.2	0.829			210	3,158	1.596
10	2,060	0.753	38.5	0.832			254	6,641	2.186
11	1,450	0.793	42.9	0.811			245	4,498	1.846
12	181	1.024	37.6	0.837			100	500	1.146
13	90	1.269	37.6	0.838			80	150	1.092
14	338	0.998	42.9	0.811			155	1,000	1.204
15	169	1.265	42.5	0.813			80	250	1.087
16	200	1.054	38.2	0.834			100	500	1.105
17	860	0.758	37.6	0.837			192	3,683	
18	1,361	0.721	37.3	0.838			270	5,545	1.784
19	256	1.049	37.4	0.837			225	1,169	1.218
20	1,417	0.755	37.5	0.837			180	5,232	
21	1,328	0.750	38.2	0.834			180	4,810	
22	648	0.695	33.2	0.859			200	3,546	
23	2,036	0.760	36.2	0.844			280	7,127	2.110
24[a]	1,950	0.894	32.5	0.863			270	5,545	1.784
25[a]	2,216	0.909	33.6	0.857			250	5,405	2.160
26[a]	2,637	0.889	34.8	0.851			254	6,641	2.588
ME	770	0.729	44.8	0.803	36.6	0.843	210	3,247	1.450
US	889	0.849	43.3	0.809	40.9	0.821	147	2,420	1.474
US	459	0.650	22.3	0.920	23.7	0.912	211	3,814	1.230
ZK-U	267	1.173	34.8	0.851	30.6	0.873	210	1,126	1.217
ZK-4	853	1.158	41.7	0.817	36.7	0.841	230	2,850	1.515
ZK-4	1,258	1.033	40.8	0.821	35.9	0.845	230	3,748	1.755
ZK-5	1,124	1.023	41.5	0.818	38.5	0.832	230	3,493	1.680
FA-1[b]	840	1.248	27.9	0.888	34.9	0.850	240	1,361	1.593
PB-3[c]	891	0.975	28.1	0.885	26.1	0.878	200	4,215	1.380
PB-4[c]	887	0.968	28.4	0.883	26.1	0.899	200	4,210	1.410
EB-4[d]	1,718	1.000	42.4	0.814	41.0	0.820	235	4,005	1.996
EB-1[d]	1,850	1.053	40.3	0.824	38.9	0.830	223	4,186	2.126
UA-1	228	1.276	32.6	0.862	30.9	0.871	200	931	1.162
Texas	780	0.868	40.4	0.823	39.7	0.827	220	2,620	1.483
Texas	620	0.973	29.9	0.881	29.1	0.881	220	2,695	1.382
GL-1[e]	1,344	0.985	48.1	0.788	45.2	0.801	248	4,620	1.850
GL-2[e]	1,452	1.034	47.7	0.790	44.7	0.803	249	4,432	1.901
California	688	0.710	36.6	0.842	37.8	0.836	242	3,395	1.406
Mississippi	326	0.863	37.2	0.839	39.1	0.829	253	1,525	1.326

Note: Numbered samples are North Sea oil.
[a] Volatile oil.
[b] 12 mol% CO_2 and 55 mol% H_2S in total surface gases.
[c] 18 mol% CO_2 in total surface gases.
[d] 10 mol% H_2S in total surface gases.
[e] 26 mol% N_2 in total surface gases.

regression analysis to express the PVT characteristics of the oils. Data shown are from flash separation data.*

Equation for Saturation Pressure (Fig. 3–48)†

A sample yielded these data: (1) saturation pressure at each temperature, (2) volume at saturation pressure, (3) compressibility of undersaturated oil (above saturation pressure), and (4) the Y factor below saturation pressure used to describe the volumetric behavior of both gas and liquid phases.

Each sample then was flash-separated in two stages, with separation conditions held constant as follows.

First Stage. p = 400 psig (2758 kPa), and T = 125°F (51.7°C).

Second Stage. p = 0 psig (6.895 Pa), and T = 125°F (51.7°C).

Oil-field units are used where p_b is the bubble-point (saturation) pressure in psia, R is the producing gas-oil ratio in scf/STB, $\overline{\gamma}_g$ is the average specific gravity of the total surface gases, T is the reservoir temperature in °F, and °API is the stock-tank oil gravity. The term p_b^* is the correlating number to calculate bubble-point pressure.

Formation Volume Factor (FVF) at Bubble Point (Fig. 3–49)

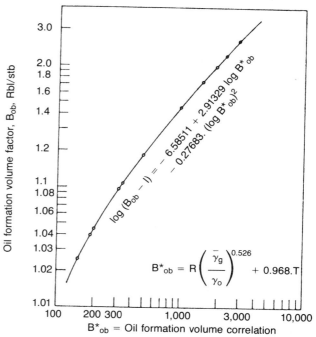

Figure 3–49. *Prediction of oil FVF at saturation pressure from gas-oil ratio, total surface-gas gravity, stock-tank oil gravity, and reservoir temperature (after Glaso, JPT, May 1980).*

Total Volume Factor, B_t

$$B_t^* = R\frac{(T)^{0.5}}{(\overline{\gamma}_g)^{0.3}} \cdot \gamma_o^{2.9 \cdot 10^{-0.00027 \cdot R}} \cdot p^{-1.1089}$$

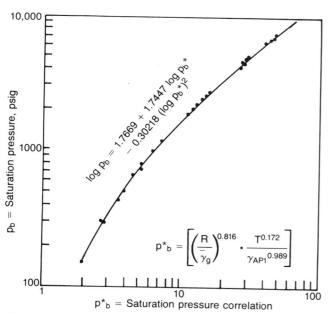

Figure 3–48. *Prediction of saturation pressure from gas-oil ratio, total surface-gas gravity, stock-tank oil gravity, and reservoir temperature (after Glaso, JPT, May 1980).*

* Meehan presented a computer program for PVT correlations in the *Oil and Gas Journal*, Oct. 27, 1980.

† The following section is taken from Glaso, *JPT*, May 1980, © SPE-AIME.

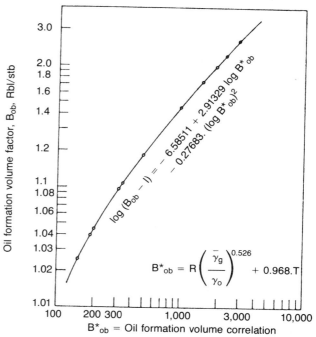

Figure 3–50. *Prediction of two-phase FVF from gas-oil ratio, total surface-gas gravity, stock-tank oil gravity, reservoir temperature, and pressure (after Glaso).*

Correcting Flash API Gravity for Paraffinicity

Figure 3–51 shows the relation between dead oil viscosity, API gravity, and temperature. The curve is specifically for North Sea oils (i.e., for oils with a K_{uop} factor of 11.9) in the temperature range of 50 to 300°F. The equation that gives a best-fit

$$\mu_{oD} = c(\log \gamma_{API})^d$$

where $c = 3.141(10^{10})(T)^{-3.444}$

$d = 10.313(\log T) - 36.447.$

Rearrange to solve directly for dead oil gravity,

$$\gamma^*_{API,corr} = \text{antilog}\ [(a\mu_{oD})^b]$$

where $a = 3.184(10^{-11})(T)^{3.444}$

$b = [10.213(\log T) - 36.447]^{-1}.$

One can correct the API gravity of stock-tank oil from a reservoir with paraffinicity not characterized with a K_{uop} factor of 11.9, extending the use of the developed PVT correlations to all oil types.

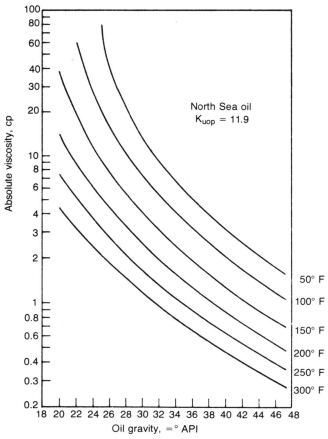

Figure 3–51. *Viscosity of gas-free crude oils at atmospheric pressure vs. API gravity and temperature (after Glaso).*

The method for correcting the stock-tank oil gravity requires (1) differential separation viscosity and (2) residual oil gravity. Using the following relation, one can calculate the corrected flash stock-tank oil gravity, $\gamma_{API,corr}$, to be used:

$$\gamma_{API,corr} = \frac{\gamma^*_{API,corr}}{\gamma^*_{API}} \cdot \gamma_{API}$$

where γ^*_{API} = gravity of residual oil from differential separation, $\gamma^*_{API,corr}$ = corrected gravity of residual oil from differential separation, using differential dead oil viscosity measured at reservoir temperature and atmospheric pressure, and γ_{API} = gravity of stock-tank oil from flash separation.

If the K_{uop} factor is assumed to be constant within a specific region, the determined ratio $(\gamma^*_{API,corr}/\gamma^*_{API})$ also is a constant and can be used to correct other samples in that region to a K_{uop} factor of 11.9 (North Sea) for use in the present correlations.

Correcting Saturation Pressure for Nonhydrocarbons

The effect of nonhydrocarbons on saturation pressure can range from minimal to extreme, depending on the type of nonhydrocarbon, the quantity with which it is found in the reservoir oil, temperature, and stock-tank oil gravity.

Specifically, this study analyzed the effects of CO_2, N_2, and H_2S on saturation pressure. The following relations resulted.

$$p_{b_{est,CO_2}} = \frac{p_{b_{CO_2}}}{p_{b_h}} \times p_{b_{est}}$$

$$p_{b_{est,N_2}} = \frac{p_{b_{N_2}}}{p_{b_h}} \times p_{b_{est}}$$

$$p_{b_{est,H_2S}} = \frac{p_{b_{H_2S}}}{p_{b_h}} \times p_{b_{est}}$$

$P_{b_{est}}$ represents the estimated saturation pressure calculated previously. This value is corrected for content of CO_2, N_2, and/or H_2S in surface gases by correction factors $p_{b_{CO_2}}/p_{b_h}$, $p_{b_{N_2}}/p_{b_h}$, and/or $p_{b_{H_2S}}/p_{b_h}$, respectively, developed from laboratory tests.

It was found that not only N_2 mole percent in surface gases affected saturation pressure but also reservoir temperature and API gravity of the stock-tank oil.

The effect of H_2S on saturation pressure is shown in Fig. 3–54. The only parameter to alter the effect of H_2S on saturation pressure was API gravity of the stock-tank oil.

Best-fit equations are given here

Nitrogen (Fig. 3–52)

$$\frac{p_{bN_2}}{p_{bh}} = 1.0 + [(-2.65 \times 10^{-4}\,\gamma_{API} + 5.5$$

$$\times 10^{-3})\,T + (0.0931\,\gamma_{API} - 0.8295)]$$
$$\cdot y_{N_2} + [(1.954 \times 10^{-11}\,\gamma_{API}^{4.699})\,T$$
$$+ (0.027\,\gamma_{API} - 2.366)](y_{N_2})^2,$$

where y_{N_2} is mole fraction N_2 in total surface gases.

Carbon Dioxide (Fig. 3–53)

$$\frac{p_{bCO_2}}{p_{bh}} = 1.0 - 693.8 y_{CO_2} \cdot T^{-1.553},$$

where y_{CO_2} is mole fraction CO_2 in total surface gases.

Hydrogen Sulfide (Fig. 3–54)

$$\frac{p_{bH_2S}}{p_{bh}} = 1.0 - (0.9035 + 0.0015\,\gamma_{API}) y_{H_2S}$$

$$+ 0.019(45 - \gamma_{API})(y_{H_2S})^2,$$

where γ_{H_2S} is mole fraction H_2S in total surface gases.

Volatile Oils

It was found that the temperature effect introduced gave an estimate bubble-point pressure too high for

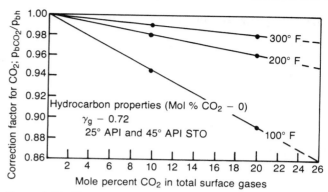

Figure 3–53. *Effect of CO₂ on saturation pressure for two API-gravity oils at various temperatures (after Glaso).*

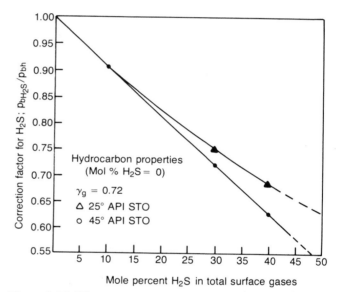

Figure 3–54. *Effect of H₂S on saturation pressure for two API-gravity oils (temperature has no effect) (after Glaso).*

these samples. This effect is expected due to the flattening of the pressure-temperature phase diagram near the critical point.

By a slight change in the temperature exponent, it is possible to estimate the bubble-point pressure for volatile oil within a deviation similar to that for black oils. The alternative equation to calculate the bubble-point pressure correlation number for volatile oils is

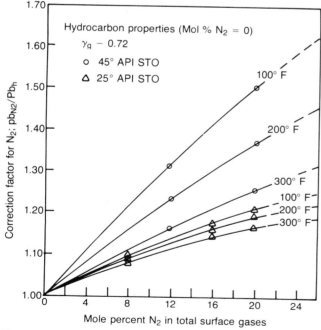

Figure 3–52. *Effect of N₂ on saturation pressure for two API-gravity oils at various temperatures (after Glaso).*

$$p_b^* = \left(\frac{R}{\gamma_g}\right)^{0.816} \times \frac{T^{0.130}}{\gamma_{API}^{0.989}}$$

Glaso also presented the prior data as the mono-graphs in Figs. 3–55 and 56.

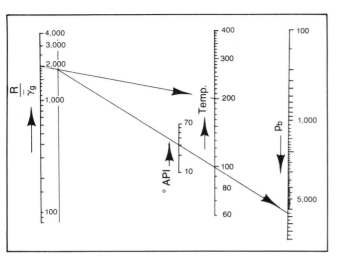

Figure 3–55. *Bubble-point pressure correlation for North Sea hydrocarbon systems—flash separation. Estimate bubble-point pressure at 200°F of a reservoir fluid with GOR 1,750, 0.875 average gas gravity, and tank gravity of 30°API (after Glaso).*

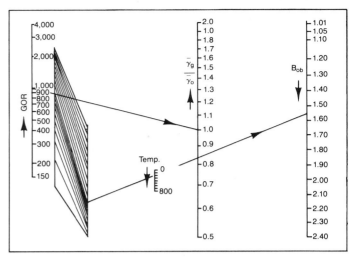

Figure 3–56. *FVF of bubble-point liquid correlation for North Sea hydrocarbon systems—flash separation (after Glaso).*

An analysis by Wayne Beeks, Keplinger & Assoc., Inc., Denver, using routine analysis by Core Laboratories is presented below:

TABLE 3–8
Separator Tests of Reservoir Fluid Sample

Separator pressure, psi gauge	Separator temperature, °F.	Separator gas/oil ratio[1]	Stock tank gas/oil ratio[1]	Stock tank gravity, °API @ 60°F.	Shrinkage factor, V_R/V_{SAT}[2]	Formation volume factor, V_{SAT}/V_R[3]	Specific gravity of flashed gas
60	90	509	12	26.9	0.7987	1.252	0.638
300	90	451	58	27.1	0.8052	1.242	0.590
750	90	373	134	27.2	0.8050	1.242	0.588
1000	90	340	168	27.1	0.8049	1.242	0.584

(1) Separator and stock tank GOR in cubic feet of gas @ 60°F. and 14.7 psi absolute per barrel of stock tank oil @ 60°F.

(2) Shrinkage Factor: V_R/V_{SAT} is barrels of stock tank oil @ 60°F. per barrel of saturated oil @ 3891 psi gauge and 187°F.

(3) Formation Volume Factor: V_{SAT}/V_R is barrels of saturated oil @ 3891 psi gauge and 187°F. per barrel of stock tank oil @ 60°F.

TABLE 3-9
Solution, PVT Tabular Solution

Pressure, psig	B_t, rb/stb	B_o, rb/stb	R_s, scf/stb	μ_o, cp
6,000	1.22106	1.22106	509	1.523
5,600	1.22511	1.22511	509	1.494
5,200	1.22914	1.22914	509	1.465
4,800	1.23314	1.23314	509	1.426
4,400	1.23710	1.23710	509	1.379
4,000	1.24098	1.24098	509	1.333
3,891	1.24200	1.24200	509	1.320
3,800	1.24742	1.23781	496	1.323
3,700	1.25383	1.23314	484	1.333
3,600	1.26077	1.22844	469	1.357
3,500	1.26828	1.22372	458	1.360
3,400	1.27644	1.21899	443	1.375
3,200	1.29492	1.20951	418	1.424
3,000	1.31687	1.20000	390	1.480
2,800	1.34311	1.19046	364	1.543
2,600	1.37474	1.18091	336	1.606
2,400	1.41324	1.17134	310	1.676
2,200	1.46066	1.16176	283	1.766
2,000	1.51986	1.15217	267	1.844
1,800	1.59502	1.14257	231	1.940
1,600	1.69242	1.13296	204	2.042
1,400	1.82194	1.12334	178	2.165
1,200	2.00011	1.11371	149	2.301
1,000	2.25654	1.10407	122	2.468
800	2.65014	1.09443	93	2.636
600	3.31678	1.08478	63	2.838
400	4.65498	1.07512	37	3.102
200	8.54949	1.06546	15	3.485
0	114.0*	1.05500	0	4.340

* Value at 0 psig calculated from B_t equation
$[B_t = B_o + B_g (R_{si} - R_s)]$

where: $B_g = 0.00504 \left(\dfrac{647}{14.7}\right)$

log 1 = zero and Δv is read at Δp = one.

$Y = (p_b - p)\, B_{oi}/p(B_t - B_{oi}) = (p_b - p)/p(B_t - 1)$
$= A + Bp$

p_b = pressure at bubble point

p = pressure

B_t = two-phase formation volume factor $B_t = B_o + 0.1781 B_g(R_{sb} - R_s)$ (approx.)

B_o = oil formation volume factor $B_o = 1.05 + 0.0005 R_s$ (approx.)

R_s = gas solubility factor

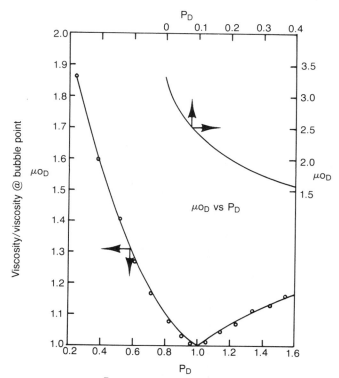

Figure 3–57. *Analysis after Beeks.*

TABLE 3–10
Reservoir Fluid Sample Tabular Data

Pressure psi gauge	Pressure-volume relation @ 187° F., relative volume of oil and gas, V/V_{SAT}	Viscosity of oil @ 187° F., centipoises	Differential liberation @ 187° F.		
			Gas/oil ratio liberated per barrel of residual oil	Gas/oil ratio in solution per barrel of residual oil	Relative oil volume, V/V_R
6020		1.53			
6000	0.9836				1.230
5610		1.49			
5500	0.9872				1.235
5200		1.45			
5000	0.9910				1.240
4810		1.41			
4500	0.9949				1.245
4430		1.37			
4400	0.9957				1.246
4300	0.9965				1.247
4200	0.9974				1.248
4100	0.9982	1.33			1.249
4000	0.9990				1.250
3891	1.0000	1.32	0	518	1.251
3881	1.0005				
3859	1.0015				
3817	1.0034				
3749	1.0068				
3700		1.32			
3660	1.0118				
3650			30	488	1.241
3500		1.35			
3490	1.0217				
3360			69	449	1.226
3213	1.0416				
3200		1.42			
3057			110	408	1.211
2855	1.0748				
2800		1.54			
2661			161	357	1.193
2473	1.1250				
2400		1.68			
2262			213	305	1.174
2104	1.1978				
2000		1.85			
1858			267	251	1.154
1710	1.3197				
1500		2.11			
1437			322	196	1.134
1305	1.5328				
1045			372	146	1.116
1000		2.46			
953	1.8810				
715	2.3224				
704			415	103	1.100
550	2.8762				
500		2.96			
395			456	62	1.086
385	3.8768				
200			483	35	1.075
0		4.34	518	0	1.055
				@ 60°F. =	1.000

Gravity of residual oil = 26.9° API @ 60°F.

V = Volume at given pressure.
$V_{SAT.}$ = Volume at saturation pressure and the specified temperature.
V_R = Residual oil volume at 14.7 PSI absolute and 60°F.

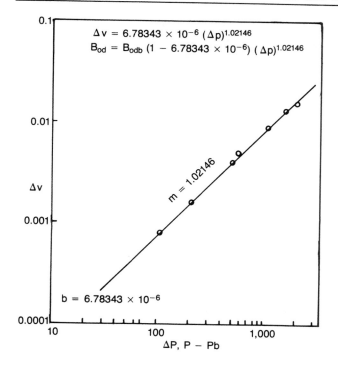

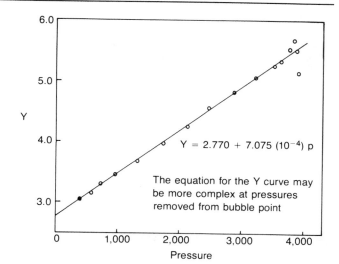

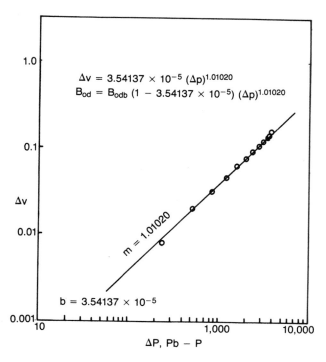

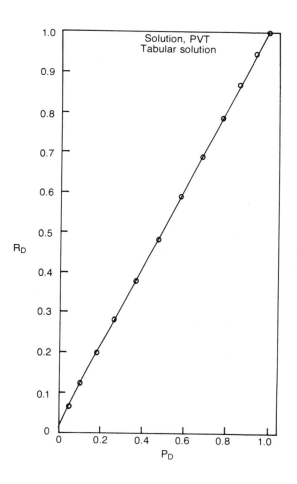

4   Determination of Various Engineering Parameters

Subsurface Temperature

A maximum recording thermometer or a continuous recording thermometer may be used to measure subsurface temperatures. These instruments are often run at the time of bottom-hole pressure measurements so that temperatures are available in most active fields. A curve of temperature versus depth can be constructed, expressed either as temperature gradients or temperature values (Fig. 4–1). The temperature normally increases with depth at a rate of 1.2 to 2.2°F per 100 feet of depth, starting with a surface temperature near 74°F in temperate climes such as Texas. In Alaska, permafrost at about 32°F exists to a depth of say 1,500 feet below surface.

Temperature gradients are not uniform. As an example, temperatures at about 2,000 feet in Yellowstone National Park and other geothermal areas are comparable to the temperatures at 25,000 feet in some parts of West Texas and Oklahoma. Earth temperatures are influenced by water migration from surface and subsurface sources, formation type, moving faults, closeness to the earth's mantle, radioactivity of minerals, barriers to normal fluid migration, and other causes. Flowing and injected fluids including drilling muds usually lower temperatures for a period of time, depending upon the length of the period of injection. Log analysis and other observations should use temperatures applicable at the time of measurement in the area affected by the tools. Some rocks and fluids are better insulating materials than others, and a change in temperature gradient with depth in a well is not unusual. Some of the effects of operations on temperature are shown in Figs. 4–2 and 4–3.

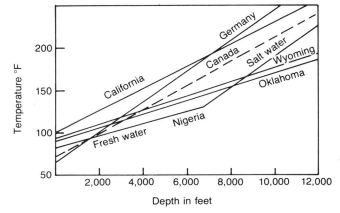

Figure 4–1. *A method of presenting static pressure.*

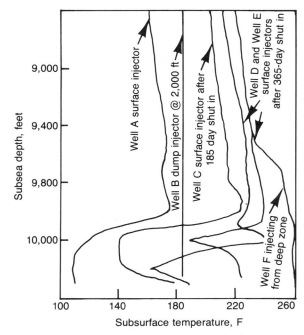

Figure 4–3. *Temperatures in injecting and shut-in wells subjected to injection from surface and subsurface source.*

182

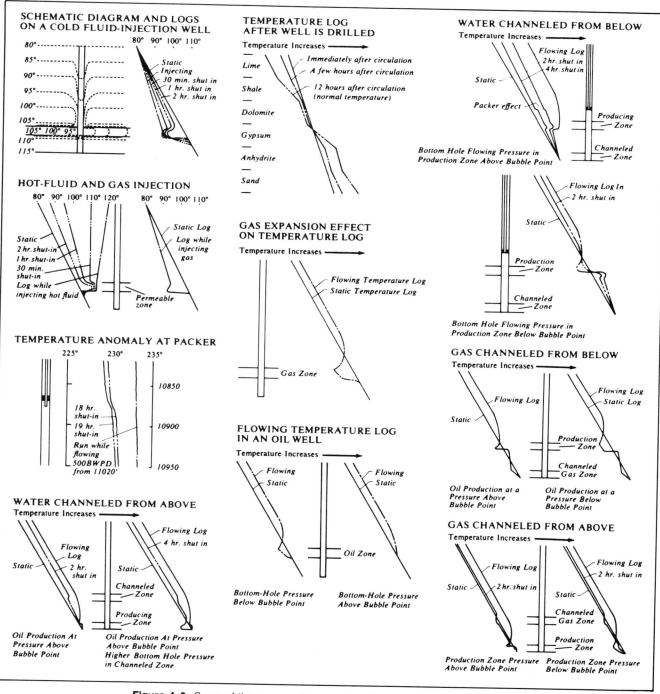

Figure 4–2. *Some of the causes of temperature abnormalities (after Layne Wells).*

Shut-in or Static Pressure

A recording pressure gauge—usually an Amerada—is lowered into a well to the desired depth to obtain the pressure. In reasonably tight pays (permeability less than about 25 md), the time that the well must be closed in to reach a pressure representative of its drainage area may be greater than desired to be consistent with mechanical problems, economic losses, and proration conditions. Build-up techniques can reduce the shut-in time if reservoir conditions are uniform but may not completely solve the problem of attaining a reliable static or shut-in pressure in many of the tighter reservoirs.

Recording gauges are available with accuracy greater than about four significant figures. The gauges may record more digits, but accuracy should be carefully checked before using the additional values. Moon effects can be observed, and changes in barometric pressure can be seen if the well is open at the surface. Also, transients from changes in operations of other

When the surface elevation is above sea level, reservoir pressures at shallow depth take the form shown as Fig. 4–7. Plotting the data versus subsea depth often results in a straight line, which starts at a pressure greater than zero at subsea depth. In some areas the gradient from surface is equal to that of water, but along the US Rockies the gradient may be 0.32 psi/foot or less at depths some distance below the surface. Deeper zones often reflect a normal pressure gradient, resulting from water head that implies poor local connection between zones in a vertical direction. Possibly a regional shale member separates the deeper from shallower porous formations.

In shut-in gas wells or in oil wells where the casing is known to be filled with gas to the base of the tubing (usually oil wells where the gas oil ratio is normal to high when well has been producing for some time without production packer), the bottom-hole pressure may be approximated by the formula:

$$\text{Subsurface pressure} = \begin{matrix}\text{Surface} \\ \text{tubing} \\ \text{or} \\ \text{casing} \\ \text{pressure}\end{matrix} + \begin{matrix}\text{Depth to} \\ \text{base of} \\ \text{gas column}\end{matrix} \times \begin{matrix}\text{Average} \\ \text{gradient} \\ \text{of the} \\ \text{gas column}\end{matrix} + \begin{matrix}\text{Depth of} \\ \text{liquid} \\ \text{column}\end{matrix} \times \begin{matrix}\text{Average} \\ \text{gradient} \\ \text{of liquid}\end{matrix}$$

wells adjacent to the well being measured are at times observed, even with the Amerada gauge. All gauges must adjust to the temperature at depth of measurement, and the time lapse can be in hours rather than in seconds. Design of pressure measurements requires care.

Reservoir pressures are often related to the hydrostatic pressure of the fluid column from the surface to the reservoir, particularly when sand in a sand-shale sequence is greater than 75% and faults or other avenues are available for movement of fluids that are forced out of rocks during normal compression as burial depth increases. Corrections are necessary when using this gradient at elevations above sea level, as illustrated in Figs. 4–4 and 5. Water gradient is about 0.45 per foot. When shale predominates in a section, escape of fluids is often impossible and the initial reservoir pressure is greater than indicated by a water gradient. Porosity also is often greater than normal for the depth, and abnormally high porosity is reflected in porosity logs and possibly in drilling time. A typical curve is shown for California conditions as Fig. 4–6.

The subsurface pressure measured by this equation must be identical with the pressure at comparable depth obtained at identical time from the flow and/or build-up test. An echometer or similar device may be used to confirm the depth of gas-oil contact in a well. The casing need not be gas filled to the depth desired for pressure measurement, but the accuracy usually is improved since oil density when gas is present can be difficult to estimate. Presence of water is an additional complication.

Pressure build-up tests including DST often are used for determination of bottom-hole pressures. Such curves should be extrapolated to the time to reach the drainage boundary calculated from the radius of drainage equation when the reservoir is circular or square. The time to reach the drainage boundary is inserted into $t + \Delta t / \Delta t$ scales, and the average pressure for the drainage area of the well is read directly rather than using the methods for correcting p^*, which are necessary for more unusual shapes (Fig. 4–8).

A static pressure map is used to obtain field-wide average reservoir pressures and to obtain the pressure of untested locations. Averaging methods include:

$$\text{Well average pressure} = \frac{\sum\limits_{0}^{n} p_i}{n}$$

$$\text{Areal average pressure} = \frac{\sum\limits_{0}^{n} p_i A_i}{\sum\limits_{0}^{n} A_i}$$

$$\text{Volumetric average pressure} = \frac{\sum\limits_{0}^{n} p_i A_i h_i}{\sum\limits_{0}^{n} A_i h_i}$$

A computer program for predicting bottom-hole pressures in gas well was presented by Arnindin, van Poollen, and Farshad in *Petroleum Engineer,* November 1980.

The gravity of the gas changes with gas composition, pressure, and temperature. Various curves, such as presented on Figure 4–9, can be used to obtain the pressure head created by the gas column. The density of oils can be measured by obtaining pressures at various depths when bottom-hole pressures are measured. Oil and gas mixtures—gas in solution and in free state within the oil column—can result in a variable pressure gradient that is difficult to estimate.

The casing often is filled with gas, and use of casing pressures can give results even in flowing and pumping wells.

Proper interpretation of the data is essential. The hydraulic gradient existing naturally in many formations is often related to the weight of a column of water at the depth chosen for measurement. The density of gas and oil is less than that of water. The hydraulic gradient should be used only at the gas or oil-water interface. The pressure at the top of the oil or gas column will be somewhat abnormal on the high side since the pressure at the water contact is reduced by the gas or oil density multiplied by the column height. The hydraulic pressure depends upon the density of water. The normal gradients per foot of depth are 0.435 psi for fresh water, 0.46–0.52 for salt water, 0.25–0.40 for oils, and 0.075–0.20 for gas. Oil gravity is important together with gas in solution and in free state. The density of gas is almost proportional to the pressure. The overburden gradient—rock plus fluids—often is near one psi per foot. Large withdrawals from aquifers such as the East Texas Woodbine formation have lowered the regional reservoir pressure. Also, some regions have abnormal fluid pressures, particularly when shale content in vertical section is high. Other regions such as the foothills of the U.S. Rockies have pressures of only 0.3–0.4 psi per foot at the shallower depths.

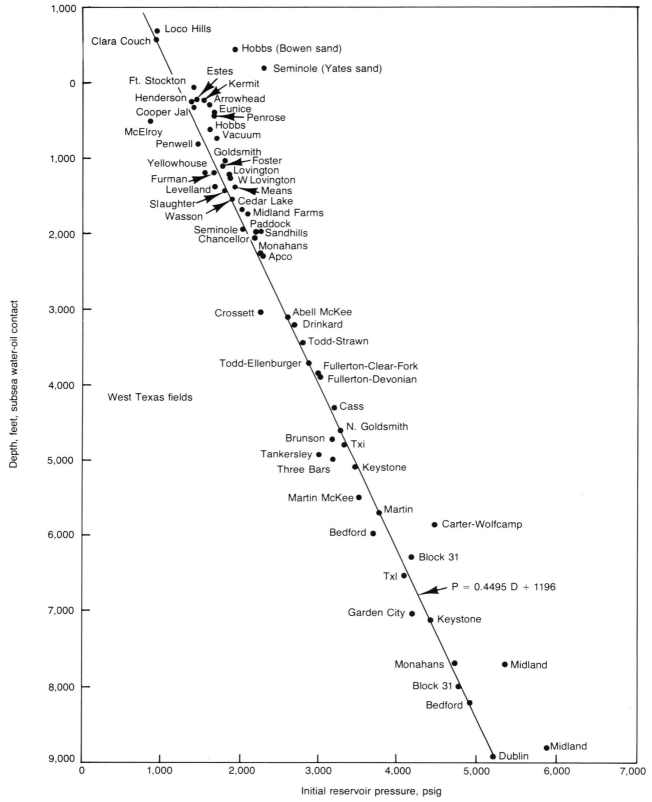

Figure 4–4. *Initial static pressure for West Texas fields (after van Everdingen et al., courtesy Shell Oil Co.)*

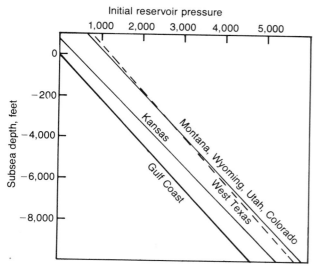

Figure 4–5. *Initial static pressure based on averages available in the early 1950s vs. subsea depth. The data is often proportional to water density times subsea depth after adjusting by a constant.*

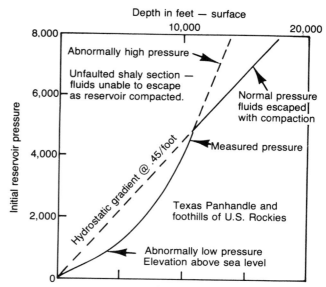

Figure 4–7. *Initial static pressure for the foothills of the U.S. Rockies versus depth measured from surface.*

A rapid method for obtaining average reservoir pressure in a well's drainage area:

1. Determine the time to first reach the end of true transient flow
   $t = \phi\mu c_t r^2_e/0.00264k = 339$ hours
2. Plot normal build-up curve using any reasonable time for t
3. Read average pressure at time of 339 hours

Gas Well

Time obtained from past production/last production   50,800 hours
Time required to investigate 640 acres   998 hours
Time required to reach end of transient flow   399 hours

Curve	Flow Time hours	Slope psi/cycle	kh md ft	p* psia	p̄, psig Horner	Miller et al.	Timmerman
A	50,800	17	581	2,895	2,859	2,860	2,859
B	998	17	581	2,869			2,859
C	399	17	581	2,861			2,859

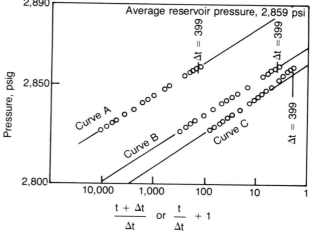

Figure 4–8. *Method for obtaining static pressure from build-up test.*

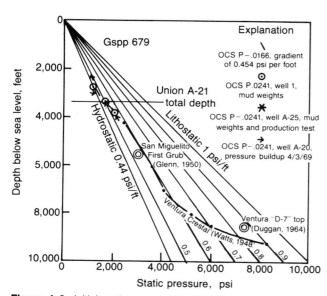

Figure 4–6. *Initial static pressure in California (after USGP 679).*

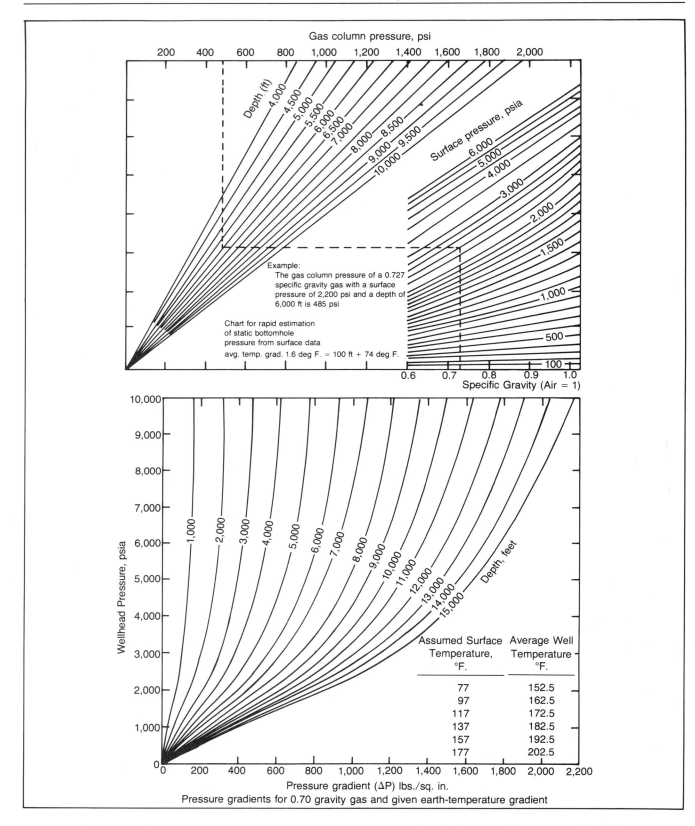

Figure 4–9. *Two approaches to determining static pressure in gas wells (after Raza and Katz, Trans. AIME, 1945).*

Flowing Bottom-Hole Pressures

Many of the characteristics of flowing wells are reviewed in *Production Log Interpretation* by Schlumberger. Some of these principles are summarized in the section following for flow in both the well bore and in the reservoir. These principles control or determine the flowing bottom-hole pressure at the sand face and at other depths within the well and at distances from the well in the reservoir.

The bottom-hole pressure may be measured with a suitable pressure-sensitive instrument at any desired depth with the well flowing at the desired rate. High rates of flow cause complications since the instrument can be disturbed by the flow rate so that wirelines are broken and lowering the instrument while the well is flowing may not be possible. The well may be shut in for a short time to lower the instrument into the well, and special cradles may be placed into the tubing below the fluid intake point so that pressures may be measured at the bottom of the tubing string even when flow rate is substantial. If the well is shut in, care must be taken for restabilization of the well before using measured pressure values. Wireline instruments can be read at the surface.

The beam pump uses rods that prevent lowering the pressure instrument through the tubing string. The instruments may be lowered through the annulus, but risks of wireline break and bomb hangup are often substantial. Pressure sensors can be permanently mounted at desired depths, and the pressure information can be transmitted to the surface either through a conducting cable or a pressure tube strapped to the tubing string. Such connectors are difficult to run into the well, and breakage in pumping wells is rather frequent. Cost is also high.

The surface pressure on either tubing or casing may be converted to subsurface conditions using the equation previously presented under static pressures when the density of the fluids within the column can be estimated. Fluid characteristics within the tubing can be rather variable, but the casing, if not sealed by a packer, often is filled with gas in both flowing and pumping wells to the base of the tubing or to the depth of pump intake. For the pumping well, gas may be flowing in the annulus and some friction loss may be present. The depth to the fluid level can be determined inexpensively by using an echometer or similar device—possibly a float. Curves may be used to obtain the density of the gas with reasonable accuracy, but determining liquid density is more difficult. Service companies have developed gradiometers, flowmeters, and radio activity devices for interpretation of flow behavior in wells. The flow may be rather variable and unstable in various sections of the flow string of deep wells, and the manuals of the service companies should be consulted.

The pressure in the well bore at the sand face is balanced between the pressure controlled by the inflow into the well bore, based on reservoir conditions, and the pressure head friction surface pressure in the well bore. When reservoir conditions are adequately known, the reservoir controlled rate of flow-pressure relation can be calculated using equations that are discussed in detail in Chapter 5.

Concepts Relating Pressure and Drainage Area–Distance from Well Bore

A skin usually causes a pressure drop near the well bore, where

$$\Delta p_{skin} = s(q\mu/2\pi kh)$$

(after Hurst, Petroleum Engineer, October 1953).

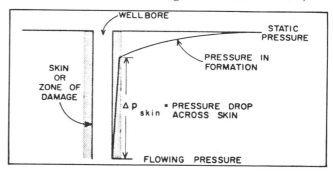

A fracture causes a negative skin since the initial flow has a linear rather than a radial configuration.

$$r_{inv} = \sqrt{0.00105kt/\phi\mu c_t}$$

Pressure varies as a natural logarithm of the distance from the well.

Distance, ft	Log	Natural Logarithm
1	0	0.0
10	1	2.3026
100	2	4.6052
1,000	3	6.9078
10,000	4	9.210

The area drained by wells also is proportional to the flow rates of the wells. At the drainage boundary, dp/dx adjusts to zero. The area varies as πr^2; the circumference of a circle varies as $2\pi r$.

Drainage area		½ side of square, ft	Radius of circle, ft	Ratio
Acres	ft²			
5	217,800	233	263	0.707
10	435,600	330	372	0.707
20	871,200	467	526	0.707
40	1,742,400	660	745	0.707
80	3,483,800	933	1,054	0.707
160	6,969,600	1,320	1,490	0.707
320	13,939,200	1,867	2,106	0.707
640	27,878,000	2,640	2,978	0.707
1,280	55,756,800	3,733	4,212	

Information using these equations may be plotted to obtain the graph in Fig. 4–10 for a skin-free well.

Integration of the following curve indicates that the average reservoir pressure of a flowing well may be read (for a circular flow configuration) at a distance approaching 0.6 of the drainage distance. Also, the average reservoir pressure is approximately 0.94 of the drainage boundary pressure when the well spacing is equal to field conditions.

In the curve, assume the pressure drop is weighted by an area influenced by drop. The following data is used:

Plotting these data and reading the graph, the following conclusions can be made:

1. Average reservoir pressure may be read from a pressure map at 0.59 times the distance to the drainage boundary.

2. The average reservoir pressure is the boundary pressure minus $(1 - 0.94)$ times the total pressure change to the boundary.

When pay is very poor, the p_T function is preferred over the ln and Ei functions.

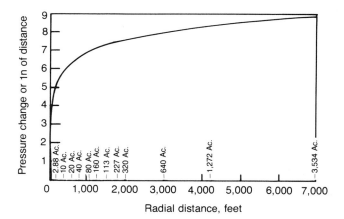

Acreage drained		Location of average pressure		Feet to avg. press.	Acreage to avg. press.	Pressure change at avg. press.
Feet	Acres	Feet	Acres			Pressure change to drainage boundary
372	10	200	2.88	0.50	0.29	0.87
2106	320	1500	113	0.57	0.35	0.94
2978	640	1775	227	0.59	0.35	0.94
7000	3534	4200	1272	0.60	0.36	0.94

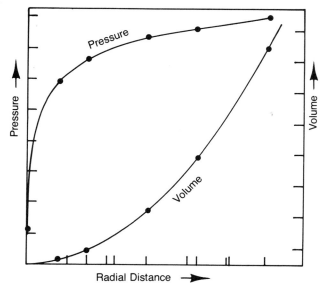

Figure 4–10. Graph for a skin-free well.

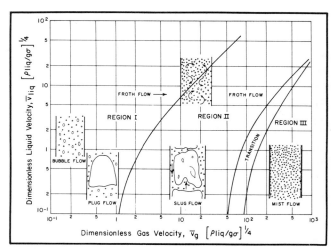

Figure 4–11. Types of flow possible in the well bore (after Schlumberger, "Production Log Interpretation," JPT, October 1961).

The pressure gradients in many flowing wells have been measured and the data plots often take the form shown in Fig. 4–12. Computer programs are available. Articles such as "Pressure Traverse Found for Two-phase Vertical Flow" (Borgia and Gottardi, *OGJ*, September 15, 1980) and "Computer Program helps Analyze Unsteady Flowing Wells" (Meldahl et al., *OGJ*, September 8, 1980) give additional information.

Radial Flow at Constant Rate

The value t_{De} must be less than 0.1 and flow configuration must be radial with small afterflow effect: t_{Drw} = 25+. Radius of investigation must exceed the total fracture length; $p_i = p_e = p_b$ = pressure at drainage boundary at start of test.

Oil:

$$p_{wf} = P_i - \frac{70.6q\mu B}{kh}\left[-Ei\left(-\frac{\phi\mu cr_w^2}{0.00105kt}\right) + 2S\right]$$

$$p_{wf} = P_i - \frac{162.6q\mu B}{kh}\left[\log\frac{kt}{\phi\mu cr_w^2} - 3.23 + 0.87s\right]$$

$$kh = \frac{162.6q\mu B}{m} \text{ when } p \text{ is plotted vs log of } t.$$

Gas:

$$m(p_b) - m(p_{wf}) = 1.637\, q_{sc}T/kh\left[\log\frac{kt}{\phi\mu c_t r_w^2} - 3.23 + 0.87s\right.$$
$$\left. + 0.87Dq_{sc}\right]$$

$$p_b^2 - p_{wf}^2 = (1.637\, q_{sc}\mu_{\bar{p}}z_{\bar{p}}T/kh)\left[\log\frac{kt}{\phi\mu c_t r_w^2} - 3.23\right.$$
$$\left. + 0.87s + 0.87Dq_{sc}\right]$$

$$p_b - p_{wf} = (818q_{sc}\mu_{\bar{p}}z_{\bar{p}}T/\tilde{p}kh)\left[\log\frac{kt}{\phi\mu c_t r_w^2} - 3.23\right.$$
$$\left. + 0.87s + 0.87Dq_{sc}\right]$$

where $\tilde{p} = (\bar{p} + p_w)/2$; in bracket, μ and c_t are at $\bar{p}$.

For Flow

$$s = 1.15\left[\frac{p_b - p_{1hr}}{m} - \log\left(\frac{k}{\phi\mu cr_w^2}\right) + 3.23\right]$$

For Buildup

$$\text{skin} = s = 1.151\left[\frac{p_{1hr} - p_{wf}}{m} - \log\left(\frac{k}{\phi\mu cr_w^2}\right) + 3.23\right]$$

Radial Flow at Constant Sand-Face Pressure

$$1/q_{sc} = 162.6(B\mu/kh)(p_b - p_{wf})\,[\log kt/\phi\mu c_t r_w^2$$
$$- 3.23 + 0.87s]$$

$$m = 162.6(B\mu/kh)(p_b - p_{wf}) \text{ when } 1/q$$
$$\text{is plotted vs. log of } t$$

$$s = 1.151[1/q_{sc}@1hr)/m - \log(k/\phi\mu c_t r_w^2) + 3.23]$$

The equations for gas are similar to those for oil, when constants and terms are adjusted as in the above.

Semisteady State Radial Flow

t_{De} must exceed 0.3 p_i = initial pressure

Oil

$$p_i - p_{wf} = \frac{qt}{\pi\phi chr_e^2} + \frac{q\mu}{2\pi kh}\left[\ln\frac{r_e}{r_w} - \frac{3}{4} + s\right]$$

$$\bar{p} - p_{wf} = 141.2(q_{sc}B\mu/kh)\left[\ln\frac{r_e}{r_w} - 0.75 + s\right]$$

The 0.75 is for a circle with no flow across the outer boundary.

Gas

$$m_{(\bar{p})} - m_{(p_{wf})} = (1{,}423\, q_{sc}T/kh)\left[\ln\frac{r_e}{r_w} - 0.75 + s + Dq_{sc}\right]$$

$$\bar{p}^2 - p_{wf}^2 = (1{,}423\, q_{sc}\mu_{\bar{p}}z_{\bar{p}}T/kh)\left[\ln\frac{r_e}{r_w} - 0.75 + s + Dq_{sc}\right]$$

$$\bar{p} - p_{wf} = (711\, q_{sc}\mu_{\bar{p}}z_{\bar{p}}T/\tilde{p}kh)\left[\ln\frac{r_e}{r_w} - 0.75 + s + Dq_{sc}\right]$$

Other Equations

$$r_{inv} = \sqrt{\frac{0.000264\,kt}{0.25\phi\mu c_t}} = \sqrt{0.00105\frac{kt}{\phi\mu c_t}}, \text{ ft}$$

$$\left(\frac{k}{\mu}\right)_t = \left(\frac{k_o}{\mu_o} + \frac{k_g}{\mu_g} + \frac{k_w}{\mu_w}\right)$$

$$c_t = c_o S_o + c_w S_w + c_g S_g + c_f$$

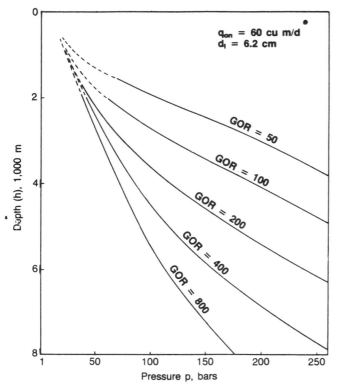

Figure 4–12. *Pressure traverses in flowing oil wells (after Borgia and Gottardi,* OGJ, *15 September 1980).*

For build-up, plot pressure versus:

Linear $\sqrt{t+\Delta t}-\sqrt{\Delta t}$ and $\sqrt{\Delta t}$

Spherical $1/\sqrt{\Delta t}-1/\sqrt{t+\Delta t}$ and $1/\sqrt{\Delta t}$

Radial $\log[(t/\Delta t)+1]$ and log of Δt

Curve match $\log \Delta p$ vs $\log t$ and $\log[(t/\Delta t)+1]$

Coordinate p versus t

When t is very large relative to Δt, Δt may be used instead of $(t+\Delta t)/\Delta t$ (often possible when Δt is less than t).

$$\Delta p_{skin} = m(0.87s)$$

When skin is moderately negative (skin −4+).

$$r_w' = r_w e^{-s} = \tfrac{1}{2} \times x_{fr} + \tfrac{1}{4} \text{ (total fracture length)}$$

$$\text{Flow efficiency} = \frac{p^* - p_{wf} - \Delta p_{skin}}{p^* - p_{wf}}$$

$$\ln \frac{r_e}{r_w} = \ln \frac{r_r}{r_w} + s$$

"Time of afterflow" in hours is:

$$t_{af} + 2 \times 10^3 \mu V_w c_w /hk$$

where V_w is the volume (bbl) and c is the compressibility (psi^{-1}) of fluid in the wellbore. If there is a large volume of gas in the wellbore, it will usually be satisfactory to take $c_w = c_g$ rather than c_t.

Linear Flow

Oil $$p_t = p_b = p_e$$

$$p_b - p_{wf} = 16.26(q_{sc}B\mu/A)\sqrt{t/k\phi\mu c_t}$$
$$m = 16.26(q_{sc}B\mu/A)\sqrt{1/k\phi\mu c_t}$$

During linear flow, slope of log-log plot approximates ½; slope $\sqrt{t}$ is straight line.

Gas

$$m_{(p)} - m_{(p_{wf})} = 163.7(q_{sc}T/A)\sqrt{t/k\phi\mu c_t}$$
$$m = 163.7(q_{sc}T/A)\sqrt{1/k\phi\mu c_t}$$

$$M = \frac{16.3q_B}{2X_f h}\left(\frac{\mu}{K\phi C_e}\right)^{1/2}$$

and

$$X_f = \frac{8.15_{qB}}{M_1 h}\left(\frac{\mu}{K\phi C_e}\right)^{1/2}$$

Its intercept at zero shut-in time is equivalent to ΔP_{skin}. A positive value for ΔP_{skin} indicates a damaged condition around the wellbore and a negative value for ΔP_{skin} indicates an improved condition around the wellbore. The slope of the straight line can be used either to calculate the permeability of the formation or the surface area of the matrix exposed to the fracture depending on which information is known from other analysis.

Spherical Flow

$$k^{3/2} = 2,453(q_{sc}B\mu \sqrt{\phi\mu c_t})/m$$
    when pressure is plotted versus $1/\sqrt{\text{hours}}$ for oil
$$k^{3/2} = 24,700(q_{sc}T \sqrt{\phi\mu c_t})/m$$
    when $m_{(p)}$ is plotted versus $1/\sqrt{\text{hours}}$ for gas

Pseudospherical flow may occur between end linear and start radial flow.

$$s = 34.7r_{ew} \sqrt{\frac{\phi\mu c}{k}}\left[\frac{P_{\Delta t} - p_{wf}}{m} + \frac{1}{\sqrt{\Delta t}}\right] - 1.0$$

$$r_{ew} = \frac{b}{2 \ln\left(\frac{b}{r_w}\right)}$$

b = thickness of the zone open to flow (for hemispherical flow, replace b by $2b$)

r_w = actual wellbore radius

Radial Flow in Dimensionless Terms

Oil

$p_{t_D} = kh(p_e - p_{wf})/141.3 \ q_{sc}B\mu$ so that $\log p_{t_D}$
$= \log (p_e - p_{wf}) + \text{constants}$
$t_D = 0.000264 \ kt/\phi\mu c_t r_w^2$ so that $\log t_D = \log t + \text{constants}$

Gas

$$p_{t_D} = kh[m_{(P_c)} - m_{(p_{wt})}]/1,422 \ q_{sc}T$$

Unsteady Flow in Wells*

The gas within the oil tends toward PVT equilibrium when a well is shut in. The gas segregates so that a free gas column is present above a relatively dead oil in the flow string. When the well is opened to flow, the pressure at the surface quickly drops to reflect the weight of the rather heavy gas-poor oil and then builds as the fluid in the column is replenished by the oil from the formation. Flow instability is caused by the segregation of free gas from the liquid in the rising column and segregation of free gas from liquid at restrictions such as the tubing intake. A packer shutting off the casing annulus may be beneficial. A third type of heading may be caused when two or more zones having differing permeability are produced simultaneously.

Formation heading. The equilibrium situation in an oil well producing from one reservoir can be established simply by graphical means (Fig. 4–13).

Gilbert represented the productivity index (PI) of a well with tubing intake or bottom hole pressures plotted against liquid production rates. The tubing outlet or well head pressure (curve B) may be obtained from curve A by subtracting the pressure loss in the vertical string for each of several rates of flow.

This can be done if a comprehensive set of vertical gradient curves is available. Superimposing performance curve C for a 10/64-in. bean, it can be concluded that the well should flow at about 100 b/d with wellhead pressure of about 840 psi.

Points 1 and 2 are possible equilibrium points. More detail is shown for the same well equipped with a 13/64-in. bean (Fig. 4–14).

Intersection 1 provides a stable flowing condition. If a tendency develops to increase the flow rate, the bean imposes more pressure resistance than the well can sustain. If a tendency develops to reduce the flow rate, the well develops a higher tubing pressure than the bean requires.

* This section is taken from "Computer Program Helps Analyze Unsteady Flowing Wells" by A. W. Grupping et al., *Oil and Gas Journal,* September 8, 1980.

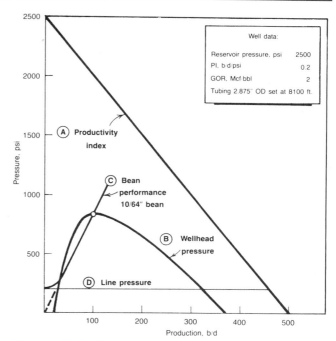

Figure 4–13. *Equilibrium conditions, flowing oil well (after Grupping et al., OGJ).*

In each case a change of flowing rate generates a pressure differential that will return the well to its equilibrium producing condition.

Intersection 2 is an unstable equilibrium point. Any deviation from the indicated flow rate brings into action a pressure differential that will increase the deviation and cause the well either to flow faster or to die, depending on the direction of the initial deviation.

Consider the equilibrium situation in a well that is capable of producing simultaneously from two reservoirs with different reservoir pressures and GORs

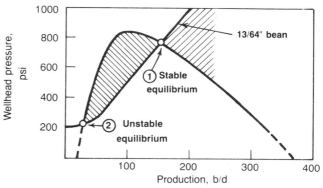

Figure 4–14. *Stable and unstable equilibrium (after Grupping et al., OGJ).*

(Fig. 4–15). The combined inflow performance of the two reservoirs is represented by curve PI_{1+2}.

At bottom hole pressure (BHPs) higher than the reservoir pressure of reservoir 2, interflow will take place from reservoir 1 into reservoir 2. If the injectivity index (II) of reservoir 2 is equal to its (PI), the combined inflow performance curve may be continued as a straight line into the region below 1,200 b/d.

At rates lower than 1,200 b/d only reservoir 1 will produce, the GOR being 300 cu ft/bbl (Fig. 4–15). At higher rates, both reservoirs will produce simultaneously.

Because their GORs are different, the GOR of the combined production varies with the production rate. This can be calculated with the formula:

$$GOR_{comb} = [(GOR\ 1\ (Q1) + GOR\ 2\ (Q2)]/[(Q1 + Q2)]$$

where $Q1$ and $Q2$ are the oil production rates of the reservoirs 1 and 2.

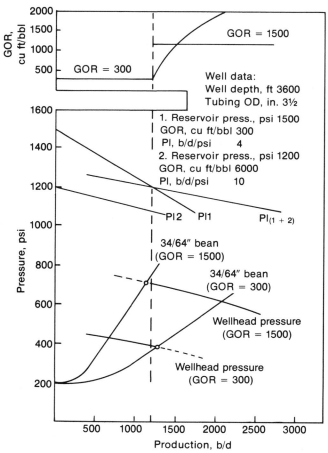

Figure 4–15. *Pseudoequilibrium, two reservoirs (after Grupping et al., OGJ).*

To investigate the equilibrium situation, it is assumed that the GOR of the combined production does not increase gradually at rates higher than 1,200 b/d but shows a sudden stepwise increase at that rate to a constant value of 1,500 cu ft/bbl.

With this simplification, wellhead pressure curves for GORs of 300 and 1,500 may be drawn in the pressure-production diagram (Fig. 3).

For steady production with a GOR of 300 cu ft/bbl, the corresponding wellhead pressure curve must be intersected with a bean performance curve (for oil with a GOR of 300).

For steady production with a GOR of 1,500, the corresponding wellhead pressure curve must be intersected with a bean performance curve (for oil with a GOR of 1,500). It can be shown that there are bean sizes for which neither of these two intersections exist.

For example, two bean performance curves of a 34/64-in. bean are shown. The two are for GORs of 300 and 1,500.

The low GOR curve intersects the low GOR wellhead pressure curve in the high GOR area. The high GOR curve intersects the high GOR wellhead pressure curve in the low GOR area.

It is clear that no stable equilibrium conditions exist for a 34/64-in. bean. How does the well react to this situation?

Well reaction. A bean size is selected for which the low GOR performance curve intersects the low GOR wellhead pressure curve in the low GOR area (Fig. 4–16). A 30/64-in. bean fulfills this requirement. The well then produces at conditions represented by point A.

When the bean size is increased to 34/64-in., the production rate and wellhead pressure instanta-

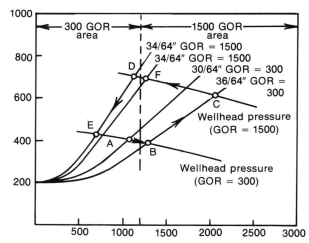

Figure 4–16. *Formation heading cycle (after Grupping et al., OGJ).*

neously move to conditions represented by point B.

This is because the bean passes low GOR oil, while the string is still filled with low GOR oil. Since point B is located in the high GOR area, the well begins to fill up from the bottom with high GOR oil. As this happens, conditions move from point B to point C.

When the well is completely filled with high GOR oil and this oil begins to flow through the bean, conditions suddenly move from point C to point D. The well then fills with low GOR oil, the wellhead bean still passes high GOR oil and conditions and move from D to E.

When the low GOR oil reaches the surface bean the well returns to the initial situation at B. It appears that conditions will never stabilize but forever follow path B-C-D-E.

The remedy for this unstable situation is to bean the well up further until the bean performance curve for high GOR oil intersects the wellhead pressure curve for high GOR oil in the high GOR area.

A 36/64-in. bean makes this intersection at F. With this size bean conditions will move along the path A-B' B' to F and stabilize there.

The diagram represents conditions in the well at any time (Fig. 4–16). Actually, there is no sudden jump to a GOR of 1,500. The GOR of the mixture flowing up the well increases or decreases gradually as the combined production rate increases or decreases.

Near point F, the GOR of the oil in the well is not constant but still highly variable. In practice it is uncertain whether or not the well would stabilize at a 34/64-in. bean.

Interflow between reservoirs often plays a significant role in formation heading (Fig. 4–17). Shown is a pressure-production diagram for a two-reservoir system with different reservoir pressures and PIs.

If the injectivity index II2 of the reservoir with the lower reservoir pressure is equal to its productivity index PI2, the combined productivity index PI_{1+2} may be continued as a straight line towards zero production rate. It intersects the pressure axis at 1,700 psi, which is the pseudo-reservoir pressure.

At this pressure the injectivity index II2 intersects the index PI1 of the higher-pressure reservoir. The corresponding flow rate of 140 b/d is the rate of interflow between the two reservoirs when the well is shut in.

If the II of the reservoir with the lower pressure is not equal to its PI (different viscosities or GORs of produced and injected oil), the pseudo-reservoir

pressure and rate of interflow assume different values (Fig. 4–17).

Methods Used to Improve Accuracy of Pressure Measurements*

Walker's Method. In Walker's method, the casinghead pressure is recorded. The liquid level in the annulus is determined acoustically while the well is pumping at a steady rate.

* This section is taken from "Laboratory Work Improves Calculations" by Podio, Tarillion, and Roberts, OGJ, August 25, 1980.

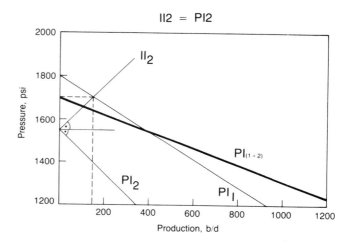

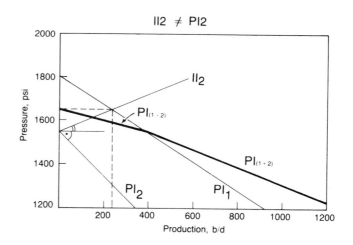

Figure 4–17. *Interflow in a two-reservoir system (after Grupping et al., OGJ).*

The back pressure on the casing is then changed. The pump stroke and speed are not changed. The well is permitted to restabilize at the same liquid production rate and the annulus fluid level is again determined. (Fig. 4–18).

If the specific gravity is assumed to be constant and the production rate unchanged, the bottomhole flowing pressure must be the same in the two tests.

The two unknowns, specific gravity of the annulus fluid and the flowing BHP, can be calculated:

$$Pbh = Pc + Pfg + H (SG) (0.433)$$

where:

> *Pbh* is the flowing BHP, psia
> *Pc* is the casinghead pressure, psia
> *Pfg* is the pressure drop due to the flowing gas column, psia
> *H* is the height of the gas/liquid mixture level, ft
> *SG* is the average specific gravity of the gas/liquid mixture

The main drawback associated with Walker's method is the time required to reestablish the initial liquid production rate after the casinghead pressure has been changed, if indeed it can be reestablished.

Gas Blow-around Method. The casinghead pressure is increased until the gas in the annulus begins to blow around the foot of the tubing. This condition will be marked by the sudden collapse of the area on the dynamometer card.

The back pressure on the casing is then reduced so that the annulus fluid level is only slightly above the pump intake level (Fig. 4–19). The well is allowed to pump steadily for a few hours to reestablish an equilibrium condition. The procedure is repeated.

This establishes the casinghead pressure when the annulus is filled with gas. The pump intake pressure is then equal to the casinghead pressure plus the pressure due to the flowing gas column.

This method obtains the pump intake pressure which may or may not be the BHP, depending on whether or not the pump is set at the same level as the producing formation. The method can be time consuming, especially when dealing with wells having low casinghead gas production.

Gradient Correction Method. The pressure gradient in the gas/liquid mixture is calculated from the density of the liquid and the amount of gas present as determined from the measurement of the flow rate of produced casinghead gas.

For a constant standard rate of casinghead gas, the actual volume of gas in the mixture in the annulus is a function of pressure. The gas volume decreases as the pressure increases.

Correspondingly, the gradient of the mixture increases as pressure increases. It is therefore not accurate to assume a uniform gradient as a function of depth.

For small gas rates and considerable depth the mixture gradient will approximate the liquid gradient. At high gas rates or for shallow depth the mixture gradient will be significantly smaller than the liquid gradient.

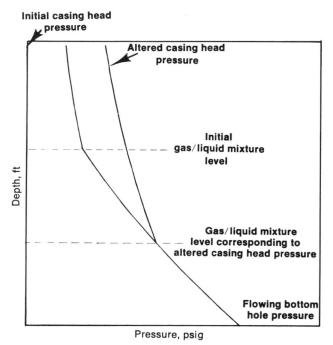

Figure 4–18. *Walker's method (after Podio, Tarillion, and Roberts, OGJ).*

The best method to account for this effect is the calculation of the annular pressure vs. depth traverse in a stepwise manner, starting where the pressure is known. This point generally corresponds to the liquid level, where the pressure $Pf1 = P1$ has been calculated from the casinghead pressure by adding the hydrostatic pressure due to the gas column (Fig. 4–20).

A small pressure increment ΔP (order of 25 psi) is considered and an average pressure for this interval is calculated by $Pav = P1 + \Delta P/2$. This average pressure is used to obtain the mixture gradient (Gm) from

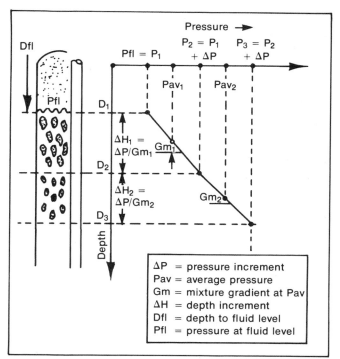

Figure 4–20. *Gradient correction method (after Podio, Tarillion, and Roberts, OGJ).*

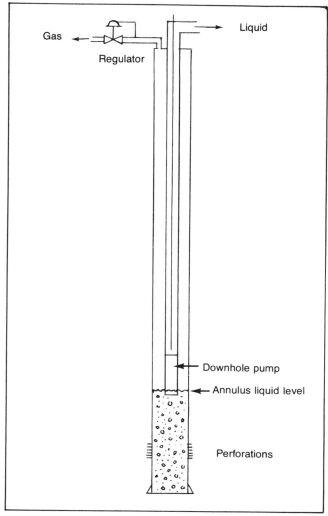

Figure 4–19. *Gas blow-around method (after Podio, Tarillion, and Roberts, OGJ).*

one of the correlations that expresses the mixture gradient correction factor (GCF) as a function of pressure and gas flow rate.

The change in elevation corresponding to the pressure step ΔP is calculated by $\Delta H = \Delta P/Gm$. This distance is added to the depth of the previous pressure point.

The procedure is repeated until the cumulative depth equals or exceeds the depth to the perforations.

The assumption is made that for small pressure steps the gradient of the mixture does not change significantly and can be represented by an average value calculated at the mid point of the pressure step.

Means have to be available to calculate this gradient from the known quantities. This is done by relying on empirical correlations. Two correlations have been published for calculating the pressure gradient in static annular gas/liquid columns.

The S curve gives a liquid gradient correction factor (ratio of the gas/liquid mixture density to the liquid density) determined as a function of the quantity $Q/AP^{0.4}$, where:

$Q =$ flow rate of gas bubbling through the liquid, *Mscfd*
$A =$ cross-sectional area of the flow conduit, sq in.
$P =$ pressure at the point where the GCF is to be determined, psia

It can be concluded that liquid viscosity has an effect on the GCF. The effect appears to be small in the range from 1 to 50 *cp*. It could probably be neglected for viscosities less than 10 *cp*.

The experimental data can be made to fit the modified "S" curve by defining a corrected superficial gas velocity:

$$Vsg' = Vsg/\mu^{0.06} \text{ where } 1 \le \mu \le 50 \; cp$$

and using this new parameter in determining the correction factor from the modified "S" curve.

Example: In calculating BHP in pumped wells producing casing head gas, the modified "S" curve is used to determine the gas/oil mixture gradient. Well data are:

Casing—4.5-in. OD, 15.10 lb/ft, 3.826-in. ID
Tubing—2.375-in. OD
Depth to pump and perforations—3,750 ft
Producing casing head pressure—55 psig
Producing annulus fluid level—3,000 ft
Casing head gas rate—30 Mcfd, 0.8 specific gravity
Annular liquid—oil, 0.85 specific gravity
Annular temperature—80°F.

The pressure at the fluid level can be calculated by one of several relations that give pressure increase due to static gas columns. One such relation suitable for hand calculation is:

$$Pfl = Pc \, (1 + Dfl/40,000)$$

where: Pfl = gas pressure at the fluid level, psia
Pc = casing head pressure, psia
Dfl = depth to fluid level, ft

For computer solution, the Rzasa-Katz method is recommended because it includes the effect of temperature, gas gravity, and Z factor.

Using the data furnished, $Pfl = (55 + 14.7) \, (1 + 3000/40000) = 74.9$ psia.

The pressure traverse calculation starts at the initial pressure of $Pfl = 74.9$ psia. A pressure step $\Delta P = 25$ psi is used. The average pressure for the first interval is:

$$Pav = (74.9 + 74.9 + 25)/2 = 87.4 \text{ psia}$$

The correlating parameter for the GCF is calculated at this pressure. With the modified "S" curve the correlating parameter is the superficial gas velocity:

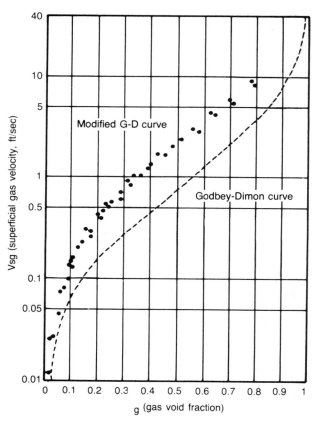

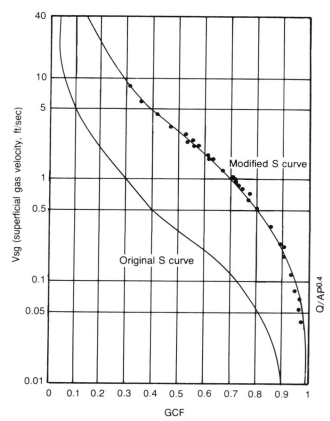

Figure 4–21. *Modified curves (after Podio, Tarillion, and Roberts, OGJ).*

$Vsg = Qg/A$ actual cu ft/sec/sq ft
$A = \pi/4 \ (3.826^2 - 2.375^2)(1/144) = 0.0491$ sq ft

Converting standard gas rate to actual rate: $(30000/86400)(14.7/8.74)(460 + 80)/(460 + 60) = 0.0626$ cu ft/sec.

$$Vsg = 0.0626/0.0491 = 1.275 \text{ ft/sec.}$$

Using this value in Fig. 4–21 gives the GCF = 0.625.
The mixture gradient is calculated by the equation $Gm = Gl(\text{GCF})$ where:

Gl = liquid gradient
$= (0.433)(0.85) = 0.368$ psi/ft
$Gm = (0.368)(0.625) = 0.230$ psi/ft

The distance corresponding to the 25 psi pressure step is: $\Delta H = 25$ psi/0.230 psi/ft $= 108.7$ ft.

The continuation of the stepwise procedure is shown in Table 4–1. The BHP at 3,750 ft is calculated to be 273 psia.

For comparison the corresponding value for the Godbey-Dimon correlation would be 214 psia.

A method for calculating pressures in flowing gas wells is given as Fig. 4–22.

TABLE 4–1
Pressure Traverse Steps

Depth, ft	Pressure, psia	Pressure step	Average Pressure, psia	Superficial gas velocity, ft/sec	Gradient correction factor	Mixture gradient, psi/ft	Depth step, ft
3,000	74.9						
		25	87.4	1.275	0.625	0.230	108.7
3,109	99.9						
		25	112.4	0.991	0.660	0.243	102.9
3,212	124.9						
		25	137.4	0.811	0.695	0.256	97.7
3,310	149.9						
		25	162.4	0.686	0.725	0.267	93.7
3,404	174.9						
		25	187.4	0.595	0.745	0.274	91.2
3,495	199.9						
		25	212.4	0.525	0.765	0.282	88.8
3,584	224.9						
		25	237.4	0.469	0.780	0.287	87.1
3,671	249.9						
		25	262.4	0.425	0.795	0.293	85.5
3,757	274.9						
		25	287.4	0.388	0.805	0.296	84.4
3,841	299.9						

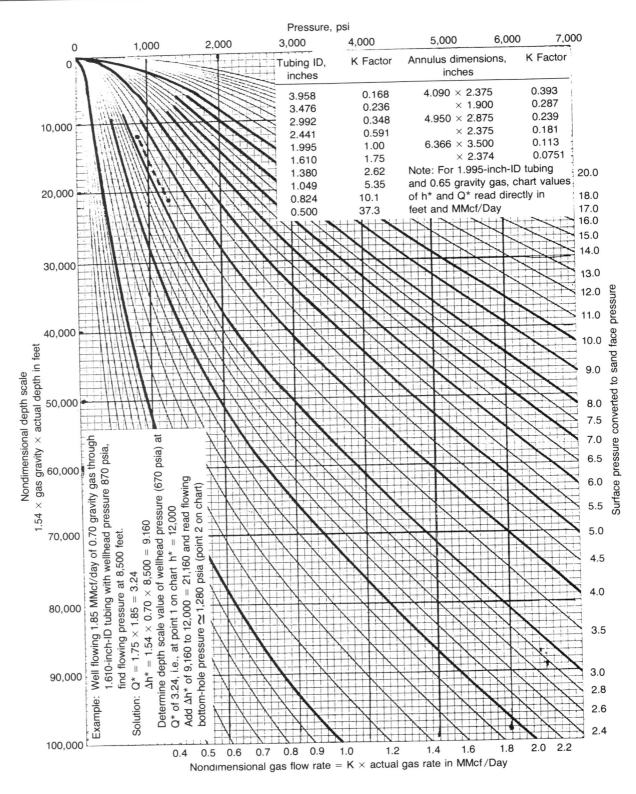

The chart contains the following labeled text and data:

Pressure, psi (top axis): 0, 1,000, 2,000, 3,000, 4,000, 5,000, 6,000, 7,000

Tubing ID, inches	K Factor	Annulus dimensions, inches	K Factor
3.958	0.168	4.090 × 2.375	0.393
3.476	0.236	× 1.900	0.287
2.992	0.348	4.950 × 2.875	0.239
2.441	0.591	× 2.375	0.181
1.995	1.00	6.366 × 3.500	0.113
1.610	1.75	× 2.374	0.0751
1.380	2.62		
1.049	5.35		
0.824	10.1		
0.500	37.3		

Note: For 1.995-inch-ID tubing and 0.65 gravity gas, chart values of h* and Q* read directly in feet and MMcf/Day

Left vertical axis: **Nondimensional depth scale** — 1.54 × gas gravity × actual depth in feet. Values: 0, 10,000, 20,000, 30,000, 40,000, 50,000, 60,000, 70,000, 80,000, 90,000, 100,000

Right vertical axis: **Surface pressure converted to sand face pressure**. Values: 20.0, 18.0, 17.0, 16.0, 15.0, 14.0, 13.0, 12.0, 11.0, 10.0, 9.0, 8.0, 7.5, 7.0, 6.5, 6.0, 5.5, 5.0, 4.5, 4.0, 3.5, 3.0, 2.8, 2.6, 2.4

Bottom axis: **Nondimensional gas flow rate = K × actual gas rate in MMcf/Day**. Values: 0.4, 0.5, 0.6, 0.7, 0.8, 0.9, 1.0, 1.2, 1.4, 1.6, 1.8, 2.0, 2.2

Example: Well flowing 1.85 MMcf/day of 0.70 gravity gas through 1.610-inch-ID tubing with wellhead pressure 870 psia, find flowing pressure at 8,500 feet.

Solution: Q* = 1.75 × 1.85 = 3.24
Δh* = 1.54 × 0.70 × 8,500 = 9.160

Determine depth scale value of wellhead pressure (670 psia) at Q* of 3.24, i.e., at point 1 on chart h* = 12,000
Add Δh* of 9,160 to 12,000 = 21,160 and read flowing bottom-hole pressure ≅ 1,280 psia (point 2 on chart)

Figure 4–22. *Converting flowing gas well pressures to subsurface condition (courtesy Shell Oil Co.)*

Geopressured Reservoirs

Shales and sands compact and lose porosity and in situ water as the weight of overburden increases with depth of burial. Water is within the pore space, and some is chemically/physically combined with the shale molecules. Simple change of porosity forces water from the pore space, but the water attached to the shale molecules is more difficult to remove. Shale dehydration occurs during diagenesis of clayey sediments at temperatures of 225–285°F at rather high pressures. The water then released from shales is in the range of 10–15% of the compacted bulk volume. In-situ oil, gas, and water are forced out of the shale and sands during periods of loss in porosity and compaction. Changes in oil character also occur at the elevated temperatures and pressures.

The water can migrate to the surface only when flow channels are present. When sand in a sand-shale sequence exceeds 50% and sand is displaced opposite sand at the faults, few geopressured reservoirs are encountered. With greater amounts of shale, the flow channels are not available and geopressured reservoirs are relatively common. Also, the sand stringers are often thin and the chances of small reservoirs are increased.

Abnormally high pressures occur when the water is unable to escape. Porosity does not decrease in a normal manner, and the resistivity reflects a higher water saturation, both of which are illustrated by Figure 4–23. The higher-than-normal pressure must be offset by a heavier drilling mud but the mud must not be capable of fracturing the rocks. Normal compaction with burial decreases the porosity so that the salt water seen by the logs is less. In geopressured

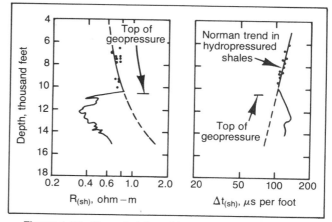

Figure 4–23. *Characteristics of logs in geopressured zones.*

shales and sands, porosity is retained and resistivity is lower. The sonic and other porosity logs reflect the change in porosity trend.

Figure 4–24 presents a method used to determine breakdown pressure. The pressure at which a rock fractures is required to prevent severe loss of drilling mud and loss of injected fluids in EOR projects. The surface pressure reflects a severe loss or fluid injection increases substantially at same pressure when rocks rupture.

Jim Bugbee in the early 1940s concluded correctly that abnormal pressures were likely when the sand in a sand-shale sequence became less than 75% and when shale members were very thick. Today, the list of possible criteria include:

Data source	Characteristic	Time data available	Parameter
Analogy	Sand-shale ratio	Prior to drilling	Fluid escape
Geophysics	Velocity, gravity, etc.	Prior to drilling	Porosity, compaction
Drilling	Rate, torque, drag	During drilling	Hardness, compaction
Mud	Gas, mud weight, kicks, temperature, chemical variations, mud loss	During drilling	Pressure, water, chemicals, attempts to blowout, mud loss, fracturing,
Cuttings	Shale-sand ratio, bulk density, porosity, shape and size, compaction	During drilling	sand-shale ratio, porosity, compaction
Logging	Resistivity and porosity, bulk density, composition changes in water with depth	After drilling	Sand-shale ratio, porosity, water salinity, resistivity
Production	Pressure, permeability, water salinity, reservoir size	After completion	Formation characteristics, fluids

The geopressure condition in Louisiana has been studied in considerable detail with conclusions illustrated in Figs. 4–25 through 31. These data suggest that geopressured reservoirs are often small in size and hydrocarbons are absent when normal R_{sh}/observed resistivity of shale exceeds 3.5. High tempera-

tures often are associated with gas reservoirs, and aquifers filled with gas-saturated water are found when both pressures and temperatures are high.

Field tests indicate that the adjustment during production occurs at a slow rate and often does not influence well and field performance. The compressibility is increased, but permeability is relatively unaffected during the productive life of most geopressured reservoir.

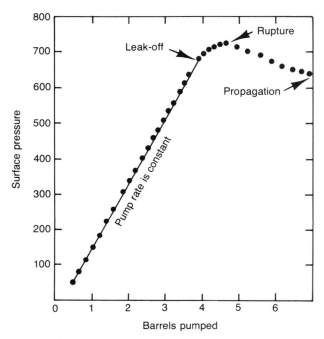

Figure 4–24. *Test for rock rupture or fracture.*

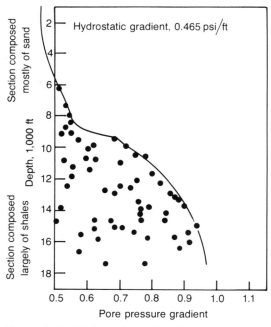

Figure 4–25. *Hydrostatic gradient increases for geopressured reservoirs (courtesy JPT).*

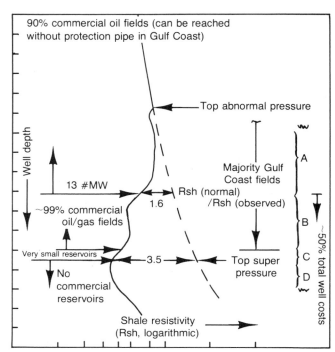

Figure 4–26. *Shale resistivity profile. Over 90% of Gulf Coast reservoirs do not show geopressures (courtesy JPT).*

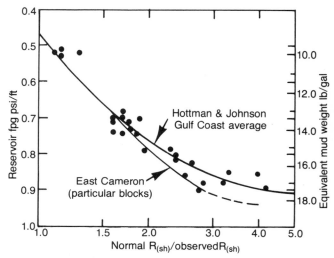

Figure 4–27. *Relationship between shale resistivity parameter Rn(sh)/Rab(sh) and reservoir fluid pressure gradient (courtesy JPT).*

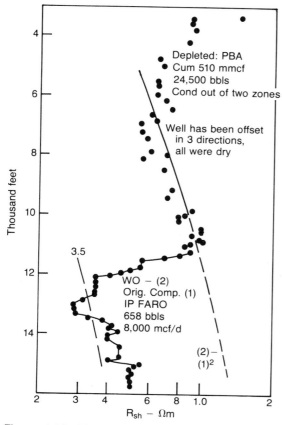

Figure 4–28. *Limited-size reservoirs, Gulf Coast area* (*courtesy* JPT).

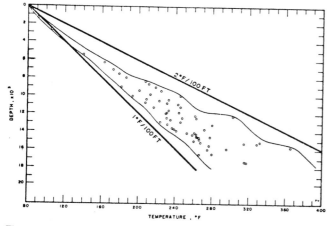

Figure 4–29. *Temperature vs. depth for 60 geopressured reservoirs* (*courtesy* JPT).

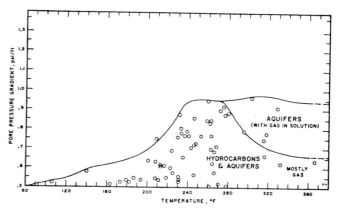

Figure 4–30. *Reservoir content relations* (*courtesy* JPT).

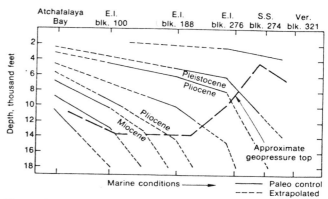

Figure 4–31. *Geopressures vs. marine shale. Geopressures are nearer the surface in more recent deposits* (*courtesy* JPT).

These figures are taken from the *Journal of Petroleum Technology*, August 1971, © SPE-AIME. Also see the AAPG publication, *Problems in Petroleum Migration*.

Geological Concepts

Engineers should be able to talk with geologists, so some understanding of the methods and terms used is required for efficient communication. Also, the engineer must prepare geological studies when the aid of regional and local structural geologists is not available. Geology, too, is a large field of expertise, and texts available from the AAPG should be consulted as necessary. The following brief summary of geological concepts directly applicable to understanding flow in reservoirs is basic.

Most reservoirs are not uniform over the area covered by the oil or gas accumulation. The variation in pay development is reflected in well inflow capacity and well performance with time. Proper use of logs with geological concepts can determine the nature of the porous medium and its distribution within the area of interest. The quality and quantity of the reservoir rock, its permeability and porosity, are necessary to determine net pay, permeability, and porosity. The accuracy with which the geologist, engineer, and petrophysicist describe and understand the geometry, discontinuities, and the vertical and horizontal distribution of net pore volume and its porosity and permeability distribution determines the success of operational plans during both primary and secondary recovery operations. All data sources must be combined to obtain an accurate description of the reservoir before performance predictions can be successful.

The generalized stratigraphy of each area has been published in a form similar to that illustrated for part of New Mexico in Fig. 4–32. These presentations are based on the fossils and fauna found by paleontologists in cores and cuttings. When combined with logs, these data permit the selection of general zones of interest—probable oil and gas productive zones—and a general understanding of the probable characteristics of each interval based on analogy.

Oil and gas accumulations are found on structures and other reservoir traps. Understanding the nature of these traps, including probable displacement mechanism, is basic to reservoir engineering and development drilling. The types of traps have been classified by Dr. Wilhelm, as shown in Fig. 4–33.

The computer has been used to enhance seismos information so that cross sections, such as shown in Fig. 4–34, can be constructed. These sections give a picture of regional geology, including probable traps, and reduce risk in exploratory drilling and define probable reservoir boundaries. Unfortunately, the process is an improvement of old techniques and not

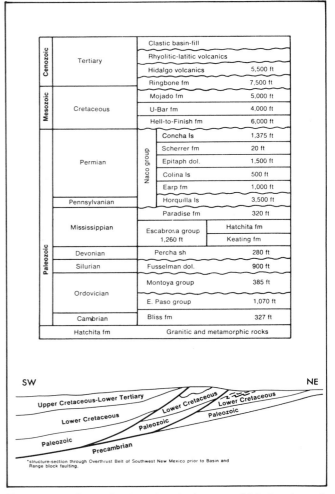

Figure 4–32. *Generalized stratigraphy* (*courtesy* Oil & Gas Journal, *21 April 1980*).

a new tool, so that few new reservoirs have been found in older developed areas.

Calcareous reservoirs have been classified in Fig. 4–34A and Table 4–1A, and a similar classification of sandstone reservoirs is presented in Table 4–2. These classifications offer a tool for selection of probable reservoirs and a general understanding of the nature and distribution-location of porous zones.

Logs and geological concepts can be used to select probable distribution of porous intervals. Typical illustrations are presented as Fig. 4–35. The typical curve shapes are summarized in Fig. 4–36 as an aid in selecting probable sand distribution from shape of logs.

A typical distribution of calcareous rocks is shown in Fig. 4–37, using information for the Means Field

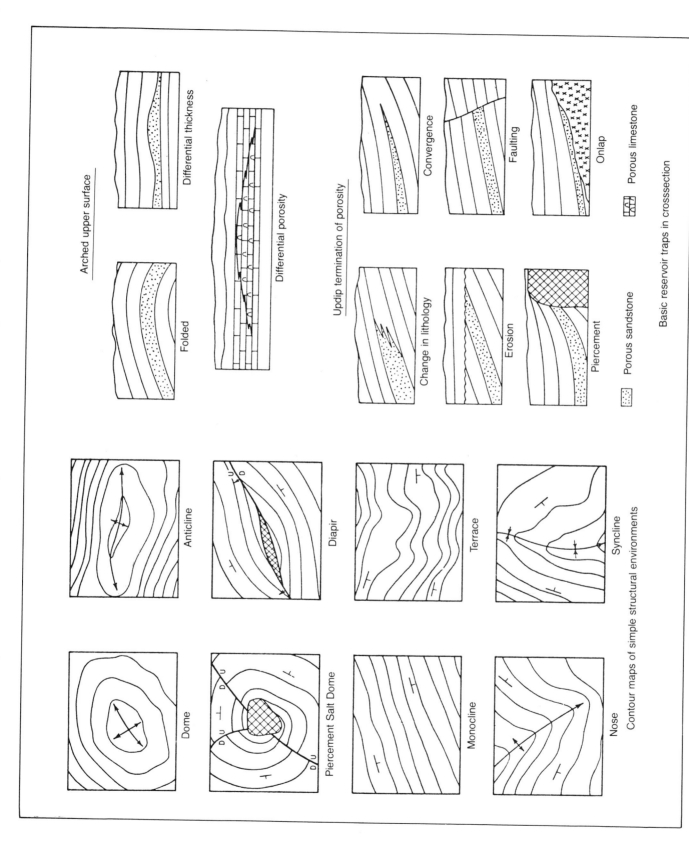

Figure 4-33. *Maps and cross sections of typical traps, prepared by Dr. Wilhelm of Shell Oil Co.*

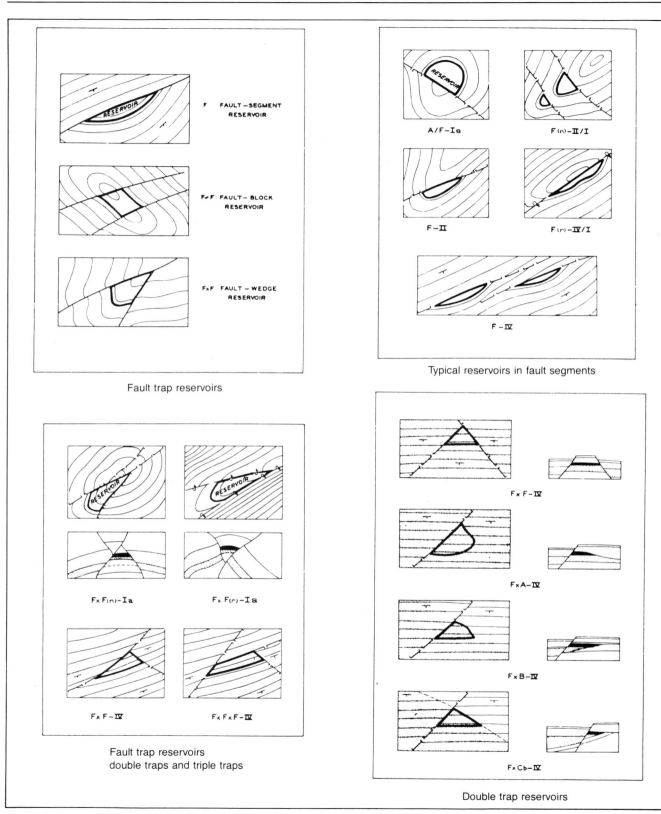

Fault trap reservoirs

Typical reservoirs in fault segments

Fault trap reservoirs
double traps and triple traps

Double trap reservoirs

Figure 4–33 *cont'd*

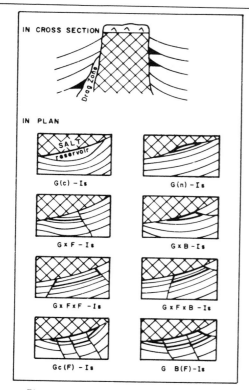

Piercement trap reservoirs on shallow
piercement type salt domes

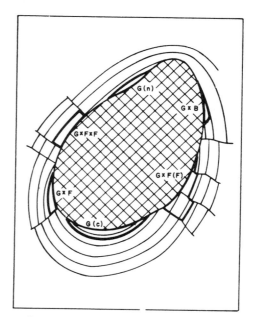

Structural arrangement of piercement traps

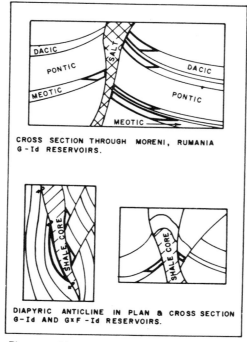

Piercement trap reservoirs in diapyric anticlines

Figure 4–33 *cont'd*

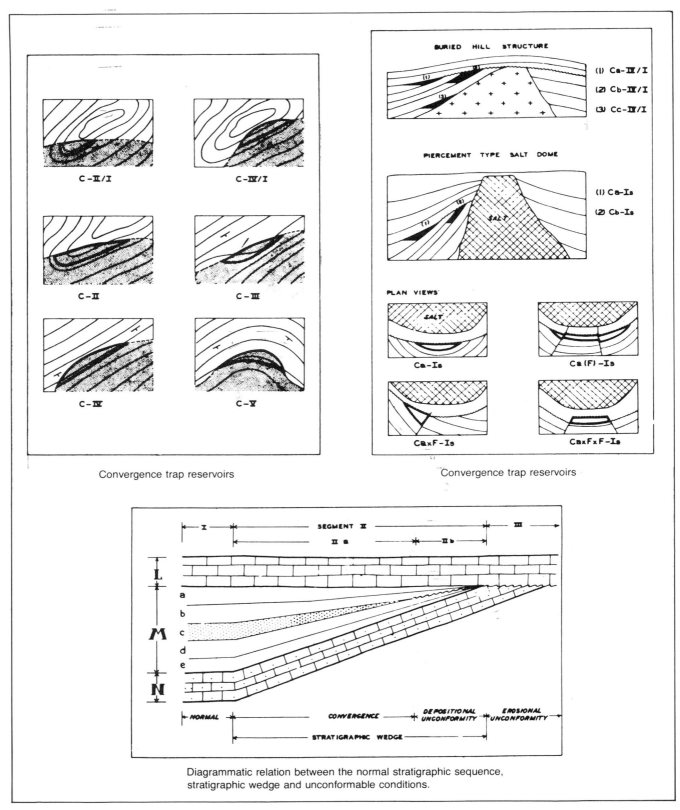

Convergence trap reservoirs

Convergence trap reservoirs

Diagrammatic relation between the normal stratigraphic sequence,
stratigraphic wedge and unconformable conditions.

Figure 4–33 *cont'd*

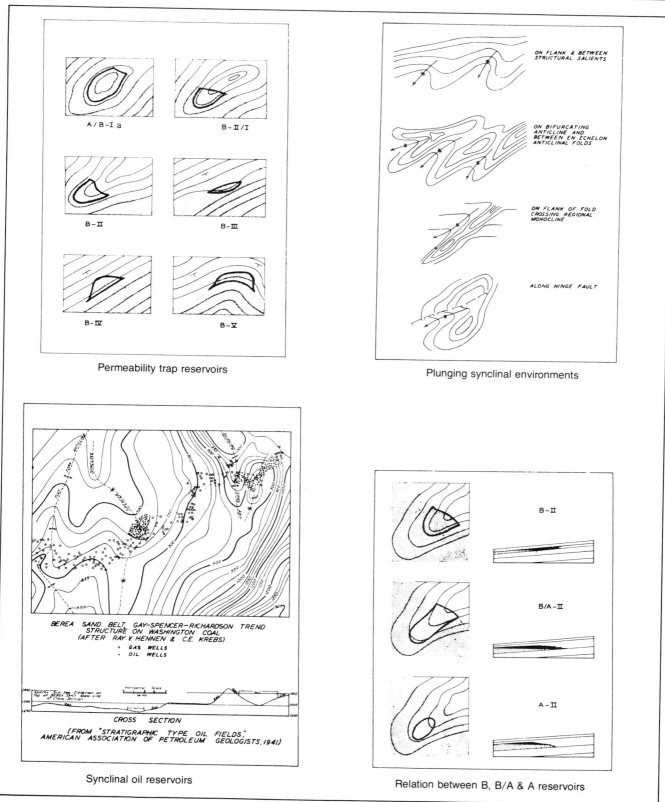

Permeability trap reservoirs

Plunging synclinal environments

ON FLANK & BETWEEN STRUCTURAL SALIENTS

ON BIFURCATING ANTICLINE AND BETWEEN EN ECHELON ANTICLINAL FOLDS

ON FLANK OF FOLD CROSSING REGIONAL MONOCLINE

ALONG HINGE FAULT

BEREA SAND BELT GAY-SPENCER-RICHARDSON TREND STRUCTURE ON WASHINGTON COAL (AFTER RAY V HENNEN & C.E. KREBS)
○ GAS WELLS
• OIL WELLS

CROSS SECTION
(FROM "STRATIGRAPHIC TYPE OIL FIELDS," AMERICAN ASSOCIATION OF PETROLEUM GEOLOGISTS, 1941)

Synclinal oil reservoirs

Relation between B, B/A & A reservoirs

B-II

B/A-II

A-II

Figure 4–33 *cont'd*

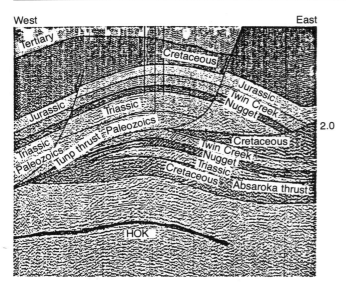

Figure 4–34. *Typical cross section prepared from enhanced seismos survey and geology.*

in West Texas. Similar data for sands in a marsh-marine environment are illustrated on Figure 4–38. The *Oil and Gas Journal* issue of October 31, 1977, contains a study of a lower delta plain, which combines many of the ideas used to study a sand-shale sequence of beds.

Shales separate possible sand reservoirs, and a shale-calcareous sequence in vertical direction is frequently indicated by logs. If the sands are thin, say less than 10 feet thick, intermixing of sand and shale is probable and a sandy shale or shaley sand is deposited. Mixing also occurs at the top and the base of sands.

The size of a continuous shale deposit, also a sand deposit, depends upon its thickness and the environment under which it was deposited. Marine-deposited shales are more extensive than deltaic and fluvial-river shales. As examples, a marine-deposited shale one foot thick may continuously cover an area of one to three miles. Thicker shales may cover areas as large as 60 or more square miles. The shales are barriers to the vertical flow of fluids during both the natural accumulation of hydrocarbons into reservoirs and during the exploitation of the oil and gas by man. The surface area of the shale lenses is very large, and fluid movement even with very low permeability may be substantial unless restricted by capillary and similar forces. Fortunately, these forces are present and oil and gas can be trapped in reservoirs lying beneath thick shale stringers. The membrane effect permits gas to migrate through some shales.

The river and delta tend to be migratory so that sands and shales are eroded. Possibly only 15% of the original deposits remain after burial by subsequent deposits. The size of the shale and sand deposits is controlled by the size of channels that comprise the drainage network. The distribution of the meandering belts depend upon the flow rates and width of the area subject to flow of water in the rivers and deltas. Often, the meander-belt widths over which sandstone and minor shale are distributed are 15 to 20 times channel width. A width of 1,000 to 3,000 feet may be associated with a channel sand thickness of 25 to 50 feet. In deltas, the sand width to sand thickness ratio is 40 to 60 in the upper delta plain and 15 to 150 in the lower delta plain where stream flow rates are less. Sands and calcareous zones often contain clays and shales intermixed with the sands. In many depositional systems, shales occurring in sand deposits are limited areally, and shale stringers may not be present in adjacent wells. The thickness of such deposits may be only one to two feet, the

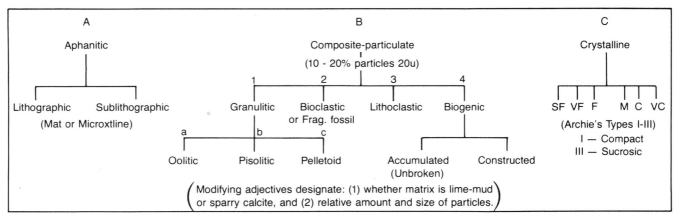

Figure 4–34A. *Classification of carbonate rocks (after Archie, courtesy Shell Oil Co.)*

TABLE 4–1A
Generalized Classification and Symbols (after Archie)

Class		Crystal or grain size, mm	Usual appearance luster	Approximate matrix porosity, % Not visible (12X–18X) A	Visible porosity (% of cutting surface) size of pore, mm <0.1 B	0.1–2.0 C	>2.0 D	Approximate total porosity A + B	A + C
I	L*	0.4+	Resinous	2	e.g. 10%	e.g. 15%	***	12%	17%
	M	0.2	to						
Compact	F	0.1							
	VF	0.05	Vitreous**	5	e.g. 10%	e.g. 15%		15%	20%
II Chalky	VF	0.05–	Chalky**	15	e.g. 10%	e.g. 15%		25%	30%
	F	0.1	Finely Sucrose	10	e.g. 10%	e.g. 15%		20%	25%
III Sucrose	M	0.2							
	L	0.4	Coarsely Sucrose	5	e.g. 10%	e.g. 15%		15%	20%

* L = large (coarse); M = medium; F = fine; VF = very fine.
** Where cuttings are between vitreous and chalky in appearance, designate as I-II or II-I. Samples are considered in the VF group when the grain or crystal size is difficult to distinguish (12X–18X). Put in F group if grains are easily distinguished.
*** When pores are greater than about 2.0 mm and therefore the pores occur at the edge of the cuttings (for example, subcavernous pores) amount of such porosity is indicated by the % of cuttings in an interval showing evidence of large pores.
Symbols:

$III\ F\text{-}B_{10}$ = Finely sucrose (therefore, matrix porosity about 10%), visible porosity about 10%, total porosity about 20%.
(II-I)VF-A = Chalky to vitreous, very fine texture (therefore, matrix porosity about 8%), no visible porosity, total porosity about 8%.

widths range from 25 to 100 feet, and the lengths may be 100 to 1000 feet. Careful interpretation of log and core information in relation to the above geological knowledge is essential since fluids tend to flow through channels of highest permeability, when connected, and the displacement of oil by liquids may be poor when preferential flow channels are present. The gas displacement process when gas is in solution in oil is less affected. The shales deposited under a marine environment are usually rather continuous, and shale alone may be deposited in deep water where water movement is slow or nonexistent. The composite sand connected in river channels may extend for ten or more miles.

A study of 17 braided stream channels of the Prudhoe Bay type suggests that the shale lense width approximates 40 times the channel deposit thickness and that length is about three times the width. The height-width-length ratio for sand in Uinta Basin of Utah, based on surface observations, is about 1–22–250. This ratio results in following values for Utah:

Thickness	Width	Length
1	22	250
5	110	1,250
10	220	2,500
25	550	6,250
100	2,200	25,000

Engineers should study these relations so that a better understanding of the importance of shale stringers relative to flow and recovery is attained.

A basic understanding of the sources of oil and gas and their migration into traps is necessary in exploration and in determining possibility of natural water displacement during production. Today, everyone agrees that oil and gas are a stored energy resource created years ago by the sun from remains of algae, plants, and animals. *World Oil*, March 1979, contains a discussion that traces the changes in the deposits of algae, plants, and animals by aerobic oxidation,

TABLE 4–2
Genetic Classification of Sand Deposits (after Shell)

Category A	Category B	Category C			Category D
Continental Deposits	Fluvial Deposits	Alluvial Plain or Cone Aggradation	Meander Belt Complex		Point Bar Channel Fill Sand Splay on Flood Plain
		Valley Fill	Braided Stream Complex		Channel Fill Channel Lateral Accretion Deposit Sand Splay on Flood Plain
		Terrace Complex			Terrace Deposit
		Coalescent Alluvial Fan Complex			Alluvial Fan
	Aeolian Deposits	Unstabilized Dune Aggradational Complex (desert conditions—little vegetation)			Longitudinal Dune Ridge Transverse Dune (Scalloped Sand Wave) Barchan Dune Hummocky Dune
		Stabilized Dune Aggradational Complex (coastal conditions—appreciable vegetation)			Transverse Dune (Lobate Sand Wave) Parabolic Dune Hummocky Dune
	Lacustrine Deposits	contains sand deposit types in Coastal and Marine categories			
	Glacial Deposits	not investigated			
Coastal Deposits	Deltaic Deposits	Alluvial-Deltaic Plain or Cone Aggradation	Meander Belt Complex		Point Bar Channel Fill Sand Splay on Flood Plain
			Braided Stream Complex		Channel Fill Channel Lateral Accretion Deposit Sand Splay on Flood Plain
		Distributary Channel and Delta-Marine Fringe Complex	Distributary Channel Deposits		Point Bar (Limited) Channel Fill Channel Lateral Accretion Deposit Subsiding Channel Bottom Deposit
			Delta-Marine Fringe	Inner Fringe	Distributary Mouth Bar Interdistributary Accumulation of Sand Lenses Beach Accretion Deposit Barrier Bar Tidal Channel Fill Tidal Channel Lateral Accretion Deposit Tidal Channel Point Bar Tidal Bar
				Outer Fringe	Shallow-Water Accumulation of Sand Lenses Shallow-Water Bar Tidal Bar Tidal Channel Fill
		Coalescent Delta Fan Complex			Delta Fan
		Coalescent Spill-Over Delta Complex			Spill-Over Delta ("Gilbert" Delta)
	Non-Deltaic Shoreline Deposits	Exposed Shoreline Complex			Barrier Bar Nearshore Bar
				Cape Complex	Barrier Bar Beach Accretion Deposit Protruding Shoal Deposit
					Beach Accretion Deposit Beach Sand Deposit on Transgressive Unconformity Sheet Sand Deposit on Transgressive Unconformity
		Protected Shoreline Deposits	Lagoonal Deposit		Lagoonal Delta (Low Energy) Tidal Delta Tidal Channel Fill Tidal Channel Lateral Accretion Deposit Tidal Channel Point Bar Tidal Bar Beach Accretion Deposit
			Estuarine Deposit	Drowned Valley Fill	Estuarine Flood Delta Tidal Delta Tidal Channel Fill Tidal Bar Beach Accretion Deposit
				Abandoned Distributary Channel Fill	
Marine Deposits	Normal Surface and Bottom Current Deposits	Shallow-Water Offshore Deposits	Shallow-Water Slope or Flat Deposits		Sheet Sand, Sand Lens Tidal Bar Tidal Channel Fill Sheet Sand Deposit on Transgressive Unconformity
			Shallow-Water Bank Deposits		Tidal Bar Tidal Channel Fill Winnowed Sand Sheet
	Density Current Deposits	Deep-Water Offshore Deposits (Includes Some Sand Reworked by Bottom Currents)			Axial Turbidite Slope Turbidite Canyon Turbidite Submarine Fan

NOTE: This classification is concerned with the three-dimensional distribution of sands and sandstones, and with the mechanisms and processes responsible for the accumulation of sand. The main mechanism is the depositional current system or depositional regimen. This system, consisting of interrelated and interacting currents with their characteristic velocities, directions and durations, gives rise to sand deposits with characteristic textures, sedimentary structures and vertical sequences.

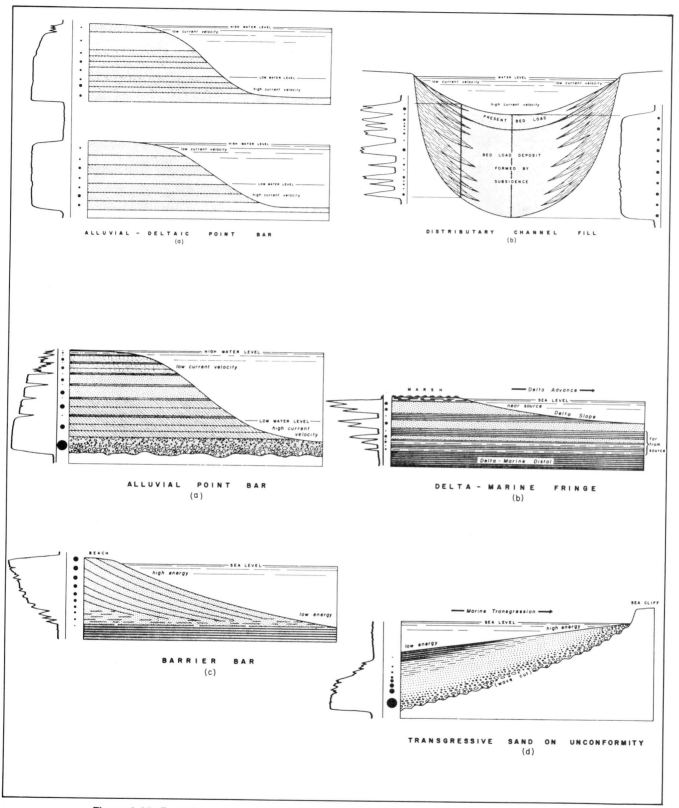

Figure 4–35. *Examples of the use of logs to define characteristics of sands (courtesy Shell Oil Co.)*

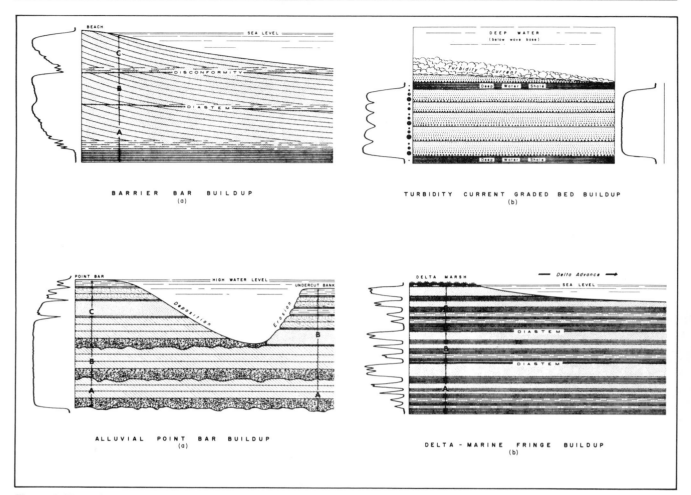

Figure 4–35 *cont'd*

sulfate reduction bacteria, fermentation, and thermal maturation at high pressure. The quantity of raw material must be sufficient to permit migration into available traps, and the entire process is dependent upon a rate of burial that prevents destruction of the raw material before oil and gas can form by fermentation and thermal maturation. Whittaker and Dyson, *Oil and Gas Journal*, September 22, 1980, classify the types of kerogen (an oil shale intermediate such as found in Colorado and other parts of the world), as illustrated in Fig. 4–39. Methane can form at rather shallow depths as a result of fermentation, and methane gas in land fills containing garbage is a problem.

At shallower depths of burial, microbial activity is important in the catalytic diagenesis of organic material to kerogen. Such activity declines with burial depth. Surface active clays such as montmorillonite are capable of filtering fatty acids and polar organic material from moving water displaced from sands and shales as burial depth increases. Long chained hydrocarbons are thus produced and trapped. At a later time, temperature (80°C to 200°C) and pressure increase with additional burial so that the change of montmorillonite to illite releases high-calcium water in an amount equal to about 15% of bulk volume. The hydrocarbons in the shale source rocks are carried to traps by the moving water. The hydrocarbons such as kerogens are reduced to oil and gas by the pressure-temperature conditions in the presence of catalysts. Source rocks or shales must contain high amounts of potential kerogen, have conditions suitable for conversion of algae, plant, and animal remains to hydrocarbons, move water compressed from the rocks during natural burial, and have traps if hydrocarbons are to be present when formation are penetrated by the drill.

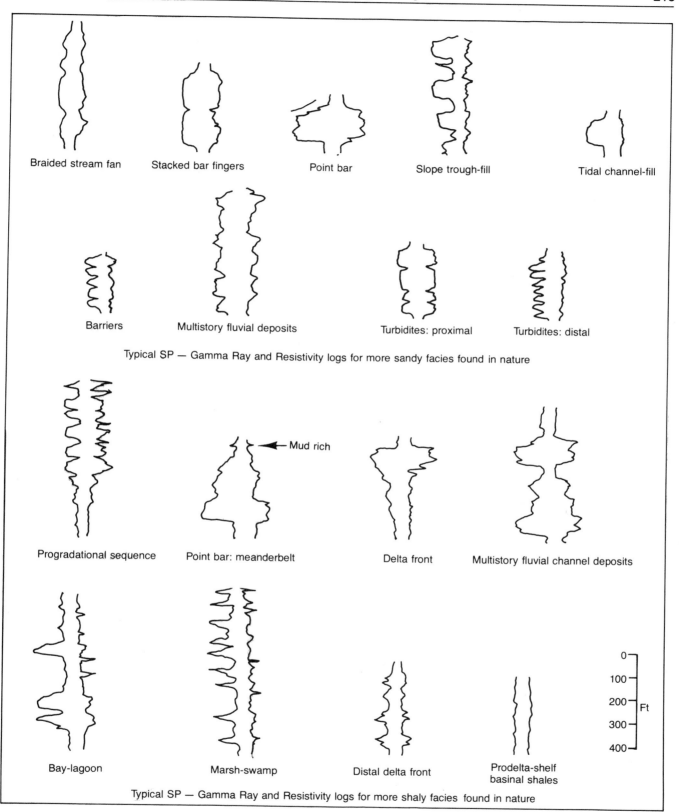

Figure 4-36. *Summary of log shapes related with pay facies.*

TABLE 4–3
General Shale Continuity in Subenvironment (after George et al.)

Principal Depositional Environment	Continuous	Discontinuous	Some Considerations in Shale Barrier Description
Dunes (Aeolian)		Interdune playas and wadi (stream) deposits	Shales rare
Alluvial fan	Lower fan "sheets"	Upper fan channels	Channel spacing and gradients; debris flow deposits
Braided stream		Channel braids and lateral secondary channels	Channel depth (based on fundamental bed thicknesses), valley width
Meandering stream	Flood basin (remnant in meandering belt complexes)	Top part of point bars and abandoned channels	Channel depth (based on fundamental bed thicknesses), valley width
Upper deltaic plain	Flood basin (remnants; as above)	Upper point bar, abandoned channels, channel braids	Channel depth, coastal plain vs valley confinement
Lower deltaic plain	Interdistributary areas	Distributary channels	Channel depth, spacing of distributaries
Delta fringe	Areas marginal to river mouth bars	River mouth bars	Size of feeder river, wave, tidal, and strength of currents along the shore
Beaches-barrier island	Lagoons, shoreface	Tidal inlet channel, shoreface	Shales uncommon in lower, rare in upper parts
Tidal flat	High flat	Low flat and channels	Tide-range
Submarine fan	Lower fan "sheet"	Upper fan	Feeder canyon size, particle-size range

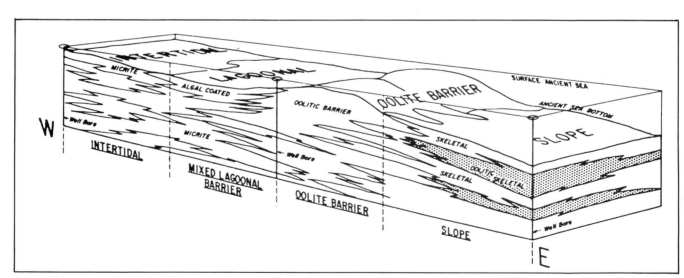

Figure 4–37. *Typical calcareous deposit (after George et al., JPT, November 1978). "Deposition here was in an intertidal-lagoon-bank sequence. The best porosity generally is in the oolitic facies, which was deposited in shallow water and formed an offshore bank protecting the lagoon from waves. A skeletal facies was deposited in front of the oolite bank and also has good porosity, mainly secondary porosity formed by leaching of the skeletal material. The lagoonal facies has lesser porosity and is characterized by numerous thin porosity zones interbedded with shales and carbonate muds. Shelfward from the lagoonal facies is the intertidal facies, composed of anhydrite and micritic dolomite with little or no porosity."*

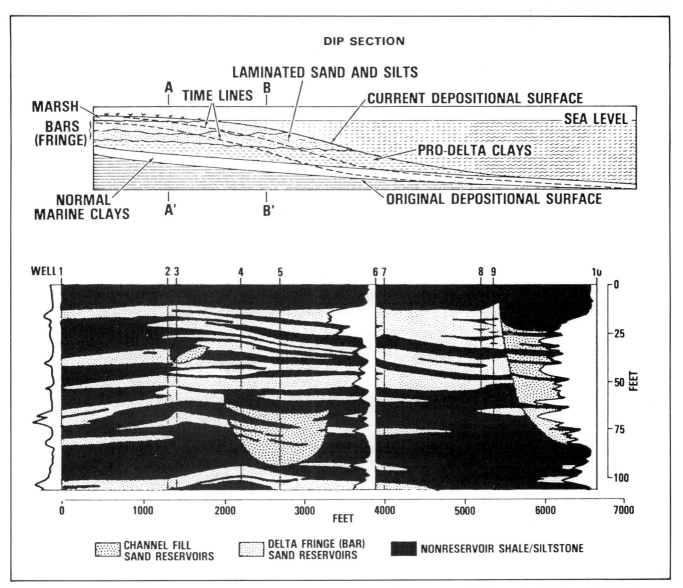

Figure 4–38. *Sand-shale sequence in marsh-marine environment (after Sneider et al.,* JPT, *November 1978,* © *SPE-AIME).*

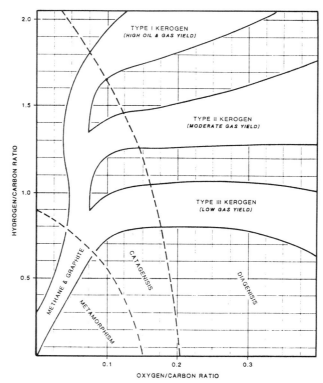

*Showing the thermal progression of kerogens

Figure 4–39. *Types of kerogen. Kerogen is often called oil shale and can be converted to oil and gas by nature and man (after A. H. Whittaker and P. Dyson,* Oil and Gas Journal, *September 22, 1980).*

Determination of Areas and Volumes

The applicable formula for many geometric shapes are included in Chapter 1. When these equations are not applicable, approximate values for area and volumes may be used. These methods are based on the construction of a contour map, such as illustrated by Fig. 4–40. Either the gross or net pay thickness is determined from logs (SP, gamma ray, etc.) at sufficient points so that average thickness can be determined for each contour. At times an isopach map is constructed showing net pay, gross pay, or possibly oil-filled porosity, $h\phi(1 - S_w)$. Such maps may be planimetered, or other suitable approaches may be used to obtain area and/or volumes. One such method involves the use of average length times average width of each contour. If suitable contours are planimetered using maps based on the top and base of the sand,

the sand volume can be indicated on a plot (Fig. 4–41). The pay thickness often is presented as a cross section.

Determination of individual parameters such as area, volume and $\phi(1 - S_w)$ often is preferred to use of only $\phi(1 - S_w)h$, since interpretation of data is improved. The map may be sliced in either the vertical or horizontal direction. Both approaches often are required to properly interpret water and gas displacement information.

Methods for Determining Area and Volume

The area may be obtained with a planimeter, or the following method may be used for data shown on Fig. 4–40:

Contour interval	Length	Width	Acres between contour (top of pay)
Above 3,700	490	150	1.7
3,700–3,750	1,230	240	6.8
3,750–3,800	1,950	245	11.0
3,800–3,850	2,570	225	13.3
			Total 32.8

Content	Contour interval	Approx. area	Avg. thick.	Gross ac. ft.
Oil Sand	3,850–3,800	12.5	50	625
	3,800–3,750	9.0	50	450
	Total Oil	21.5	50	1,075
Gas Sand	3,750–3,690	4.0	60	240

The volume can be derived using two methods.
Method 1: Map sliced horizontally

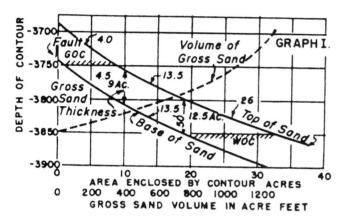

Area Calculations Using Graph

Contour	Area	
3,850–3,800	[(32.5 − 18.5) + (19.0 − 8.5)]/2	= 12.5
3,800–3,750	[(19.0 − 8.5) + (9.5 − 1.0)]/2	= 9.0
3,750 plus	(8.5 − 0.5)/2	= 4.0

Method 2: Map sliced vertically

Content	Contour interval	Approx. area	Avg. thick.	Gross ac. ft.
	3,850–3,800	32.8 − 19.5 = 13.3	$\dfrac{0+52}{2}$	365
Oil Sand	3,800–3,750	19.5 − 8.5 = 11.0	$\dfrac{52+50}{2}$	560
	3,750–3,700	8.5 − 1.7 = 6.8	$\dfrac{50+50}{2\times 2}$	170
	Total Oil			1,095
Gas Sand	3,750–3,700	8.5 − 1.7 = 6.8	$\dfrac{50+50}{2\times 2}$	170
	Above 3,900	1.7 − 0 = 1.7	$\dfrac{50+50}{2}$	85
	Total Gas			255

This assumes that the productive area is divided into vertical segments at the point of contour interval. Basic data are taken from Figs. 4–40 and 41.

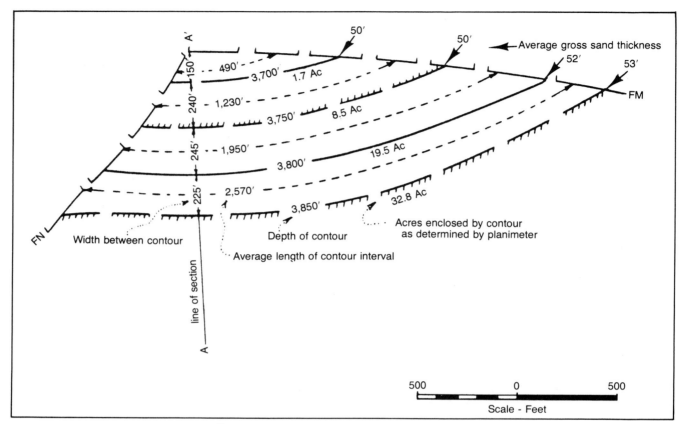

Figure 4-40. *Basic map contoured on top of H-sand.*

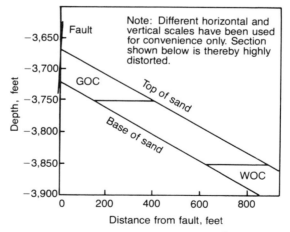

Figure 4-41. *Cross section of A-Á.*

Tools For Comparing Log and Core Information

Each engineer and geologist has his own techniques. Maps are used to indicate structure, gross and net pay, porosity and permeability distribution, in-situ water content, and many other parameters. Again, we should mention that the lazy person who does not look at information available from various data sources and compare accuracy and usefulness is likely to make poor decisions. Cross sections are useful in presenting structure, faults, unconformities, and a picture of the reservoir.

The petrophysical data sheet, such as used by Archie in the early 1940s, continues to be most valuable for comparing logs with core information. A typical example is included as Fig. 4–42. The method is used to compare log information, both basic logs and calculated data, available from the respective logs with each other and with the information available from cores. Porous and permeable intervals, oil and gas contacts, and contacts with water can be selected using this technique. The graphs can be used to compare any

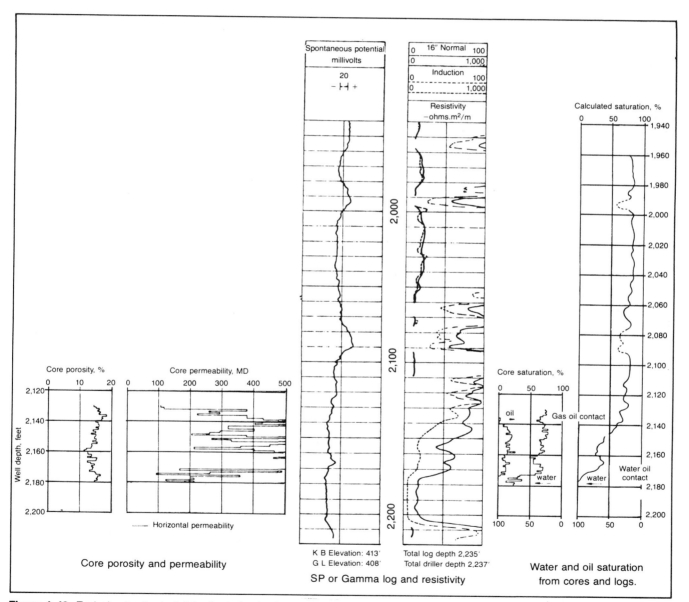

Figure 4–42. *Typical petrophysical data sheet showing log and core data. Core and log information should be compared before a detailed study of either set of information.*

information from two or more data sources when depth is a variable. Often, original logs are simply used, but these must be permanently combined when a permanent record or presentation to others is desired. Cross sections are useful.

Stick charts based on some useable datum can be used to correlate data and pick common tops of pays and other types of information. A typical example used to pick formation zones is shown as Fig. 4–43.

The block or fence diagram, such as illustrated by Fig. 4–44, is used when detailed correlation is required. Such detail is justified when making plans for waterflood and EOR projects where capital expenditures are large. The success of such projects often depends upon an awareness of zones of higher permeability and detailed location of shales and the depositional environment. For the case shown on Figure 4–44, the channel will flood differently than the surrounding, more marginal, pays. Models using plastic are constructed to present complicated conditions.

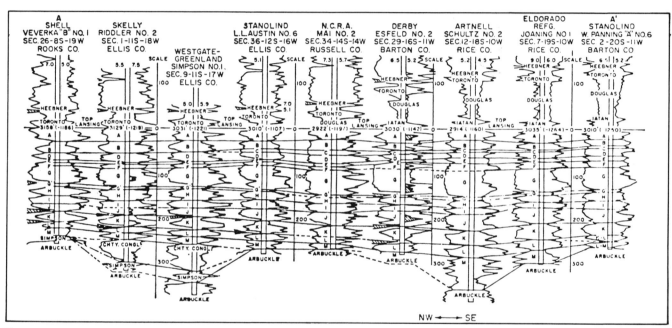

Figure 4–43. *Stick diagrams; penetration correlation chart by zoning. Gamma-ray-neutron cross section, Lower Pennsylvania system, Central Kansas (from Frick,* Petroleum Production Handbook, *copyright SPE-AIME, 1962).*

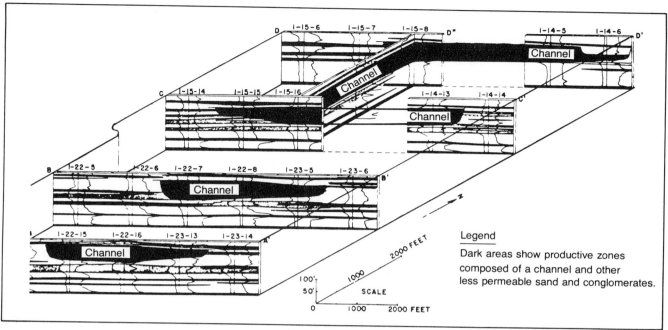

Figure 4-44. *Block or fence diagram (courtesy Shell Oil).*

Performance Curves for Reservoirs

General knowledge regarding the shapes of production performance curves is required for use in determining the type of natural energy or drive available. Also, only a very small part of the reservoir is sampled by logs, cores, and other techniques. Sound values for the terms appearing in the applicable equations, which relate to the types of drives available, may not be found during a study of the available data. A match between calculated results and actual field performance during the past life of the reservoir is essential before a forecast of future performance can be made. The computer is often used to obtain the best possible match, but it is not essential. Matching curves without careful analysis of values for each of the terms appearing in the equations is absurd and can result in useless forecasts.

The pressure and gas-oil ratio versus cumulative oil produced for a gas displacement drive (solution gas drive) are shown on Fig. 4-45. The initial pressure drops rapidly when the oil is not saturated with gas at initial conditions. Thereafter, the pressure plots as a rather straight line for a period of time before movement of free gas becomes large. The pressure declines rapidly after free gas bypasses the in-place oil. Careful measurement of GOR for the time shortly after the pressure is below the bubble point will show a slight decline in GOR because the gas first released from the oil does not move. A critical gas saturation is necessary before gas can move. Gas bypassing increases according to the relative permeability ratios and viscosity ratios until gas released from the oil is produced. The final GOR often is rather low.

The oil and water produced versus time for another solution-gas drive reservoir is shown in Fig. 4-46. Oil produced increases during the time when new wells are drilled and, when not prorated, usually declines after drilling is completed. The decline during later life often approximates a straight line when oil rate is plotted on a semilog scale and time is plotted on a coordinate scale. The semilog plot is often used to forecast the future recovery when solution-gas drive mechanism is operative.

Figure 4-46 also shows probable performance of a reservoir that has been substantially depleted by solution-gas drive mechanism and thereafter is subjected to a successful waterflood operation. During a waterflood operation, oil production begins to increase shortly after the start of water injection. The oil rate peaks rather quickly and then goes into a decline period, which often is a straight line when

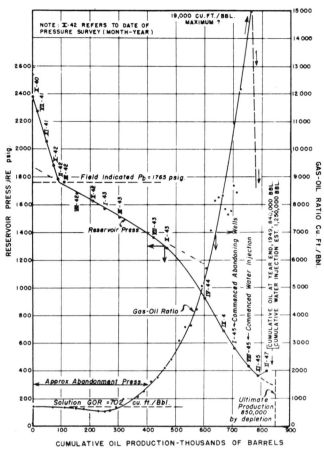

Figure 4–45. *Pressure and GOR for a solution-gas drive reservoir (courtesy Shell Oil Co.)*

the oil rate on the semilog scale is plotted versus cumulative oil recovered on coordinate scale. A plot of percent oil in total liquid versus cumulative oil using coordinate paper is useful for water-drive reservoirs.

Typical production performance curves for reservoirs operating under a partial natural water drive mechanism are plotted on Figure 4–47. The pressure declines rapidly initially but often becomes constant with time—possibly with rate—when the rate is relatively slow. The pressure versus time plot shows little or no decline, and pressure becomes rate dependent. Water was injected to increase the production rate without pressure decline in both of the reservoirs. In one, the pressure increased, and in the other it was maintained at a constant value. Maintaining the pressure retains gas in solution and usually increases recovery, as discussed in later sections of this book.

Production data is available from Petroleum Information, Denver, and from the G.E. Worldwide Mark 111 Network. A telephone call can obtain information of various types from these sources.

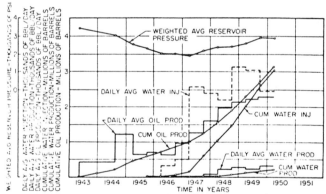

Pressure and production statistics for the Bacon Lime reservoir.

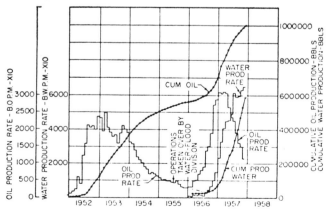

Figure 4–46. *Characteristics of a solution-gas reservoir with time (after Riley, JPT, from Frick, SPE-AIME).*

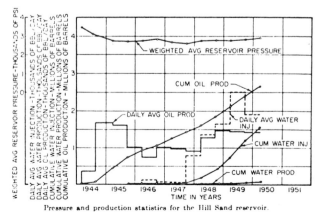

Pressure and production statistics for the Hill Sand reservoir.

Figure 4–47. *Performance of partial water-drive reservoirs supplemented by water injection (after Trube and DeWitt, Trans. AIME, Vol. 189, 1950).*

Water Cut in Oil Produced

Interpretation of the results of drill-stem tests and initial flow tests requires an understanding of the fractional flow theory. The water in the pay is nearly 100% near the water-oil contact, and the water content gradually decreases based on capillary and gravity forces until a height at which irreducible water remains in the pore space. Water is producible with oil in the transition heights. The fractional flow equations should normally be applicable if coning is not probable. Water cut is defined as:

$$\text{Water cut} = \frac{F_{w/o}}{1 + F_{w/o}}$$

where: $F_{w/o} = A k_{rw}/k_{ro}$
A = viscosity ratio, μ_w/μ_o
k_{rw} and k_{ro} = relative permeabilities to water and oil

The equation is used in the following form:

$$\frac{\text{Water}}{\text{oil} + \text{water}} = f_w$$

$$= \frac{1}{1 + (k_o/k_w)(\mu_w/\mu_o)} \text{ (at reservoir conditions).}$$

The relative permeability relations differ for various rocks and the viscosities of the oils varies substantially. General relations should be used cautiously, and the results shown as Fig. 4–48 should be used as an example of the type of curves that can be constructed for use by engineers during analysis of field test results. Accuracy is improved when curves are calculated using the best data available for existing conditions. For the charts shown, relative permeability is calculated from the equations:

$$k_{rw} = \left(\frac{S_w - (S_w)_{irr}}{1 - (S_w)_{irr}} \right)^3$$

and

$$k_{ro} = \left(\frac{0.9 - S_w}{0.9 - (S_w)_{irr}} \right)^2$$

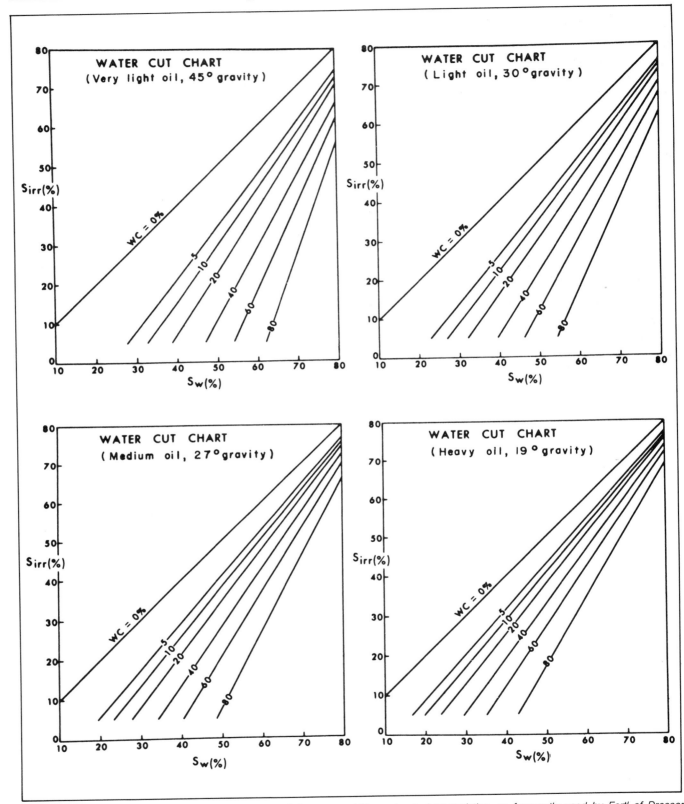

Figure 4–48. *Typical water-cut curves for assumed relative permeability and pvt characteristics, as frequently used by Fertl of Dresser Atlas, Schlumberger, and others. This example is taken from Schlumberger manuals.*

5 Performance of Flowing and Shut-in Wells

The flow history of an unfractured well may be divided into three periods for analysis, as shown in Fig. 5–1. The transient or early flow period is usually used to analyze flow. The late transient period is more complicated, and the semisteady state flow period is used in reservoir limit tests. Fig. 5–1 depicts a radial flow configuration of circular shape. Radial flow is preceded by a period of linear flow when wells contain fractures. If the pay interval is partly penetrated or perforated, a spherical flow-dominated period should be expected between the linear and radial flow times. Also, the first flow unloads the well while accepting a contribution from the reservoir. Thus, a group of curves must be constructed to analyze well tests properly. Flow tests may better represent well performance than build-up tests since particle movement, turbulence, and capillary constrictions are then included.

The buildup characteristics of a well may be idealized as shown in Fig. 5–2. For the figure shown, Miller, Dyes and Hutchinson scales are used. The Horner plot is equally useful and is often preferred. Some of the better references are listed below. A complete list is available in General Index, Society of Petroleum Engineers, published in books covering years 1953–66 and 1967–74.

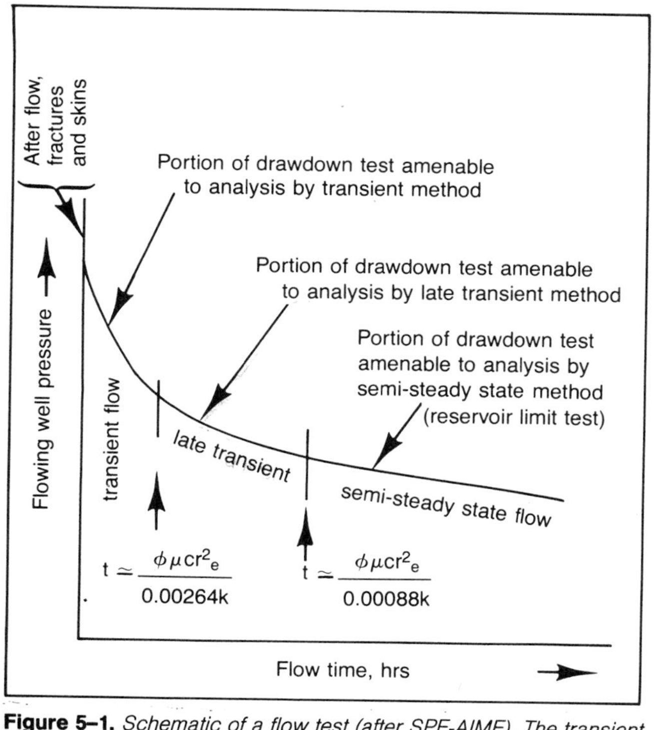

Figure 5–1. *Schematic of a flow test (after SPE-AIME). The transient flow period will be a straight line if pressure is plotted versus the log of time.*

Al-Hussainy, Ramey, and Crawford. "The Flow of Real Gases Through Porous Media." *JPT*, May 1966.

Bixel and van Poollen. "Pressure Drawdown and Buildup in Presence of a Radial Discontinuities." *SPEJ*, September 1967.

Carslaw and Jaeger. *Conduction of Heat in Solids.* Oxford Press.

Chatas. "Unsteady Spherical Flow in Petroleum Reservoirs." *SPEJ*, June 1966.

Culham. "Pressure Buildup Equations for Spherical Flow Regime Problems." *SPEJ*, December 1974.

Earlougher. "Advances in Well Test Analysis." SPE Monograph, Vol. 5, 1977.

Earlougher and Kersch. "Short-Time Transient Test Data." *JPT*, July 1974.

Frick et al. *Petroleum Production Handbook.* This is a basic book by Society of Petroleum Engineers in 1962.

Havlena, D., and Odeh, A. S. "The Material Balance as a Straight Line." *JPT*, August 1963 and July 1964.

Horner, "Pressure Buildup in Wells." E. J. Brill, Leiden, 1951. Third World Petroleum Congress Proceedings.

Matthews, C. S., and D. G. Russell. "Pressure Buildup and Flow Tests in Wells." SPE Monograph 1. The references included in each section are complete, 1967.

Miller, Dyes, and Hutchinson. "Estimation of Permeability and Reservoir Pressure from Bottom Hole Pressure Buildup Characteristics." *Trans.* AIME, 189, 1950.

Millhelm and Cichowic. "Testing and Analyzing Low Permeability Fractured Gas Wells." *JPT*, February 1968.

Moran and Finklea. "Theoretical Analysis of Pressure Phenomena Associated with the Wireline Formation Tester." *JPT*, August 1962.

Muskat, M. *Flow of Homogeneous Fluids.* McGraw-Hill Book Co.

Muskat, M. *Physical Principles of Oil Production.* McGraw-Hill Book Co., New York, 1949, reprinted by SPE.

Prats, M. "Method for Determining the Net Vertical Permeability near a Well from In-situ Measurements." *JPT*, May 1970.

Prats and Levine. "Effect of a Vertical Fracture on Reservoir Behavior-Results in Oil and Gas Flow." *JPT*, October 1963.

Ramey et al. "Unsteady State Pressure Distributions Around a Well with a Single Infinite Conductivity Vertical Fracture." *SPEJ*, August 1974.

Ramey et al. "Well Test Analysis For a Well in a Constant Pressure Square." *SPEJ*, April 1974.

Russell, Goodrich, Perry, and Bruskotter. "Methods for Predicting Gas Well Performance." *JPT*, January 1966.

Truitt and Russell, "Transient Pressure Behavior in Vertically Fractured Reservoirs," *JPT*, October 1964.

van Everdingen. "The Skin Effect and its Influence." *Trans.* AIME 198, 1953.

van Everdingen. A. F., and W. Hurst. "Application of Laplace Transformations to Flow Problems." *Journal of Petroleum Engineering,* December 1948.

van Everdingen and Meyer. "Analysis of Build-up Curves Obtained after Well Treatment." *JPT*, April 1971.

van Everdingen, Timmerman, and McMahon. "Application of Material Balance Equation to a Partial Water Drive Reservoir." *Pet. Trans.*, Vol. 198, 1953.

Wattenbarger and Ramey. "Gas Well Testing with Turbulence, Damage, and Well Storage." *JPT*, August 1968.

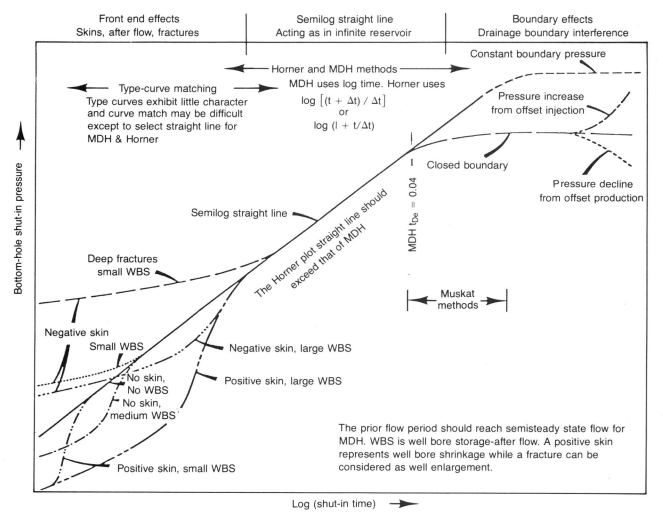

Figure 5–2. *Schematic of build-up test (after SPE-AIME). The prior flow period should reach semisteady state flow for MDH. WBS is well-bore storage afterflow. A positive skin represents well bore shrinkage while a fracture can be well enlargement.*

Evolution of Well Testing

Oil and gas-well testing techniques changed as the basic theory of flow of fluids was developed. Also, well-test procedures were revised as the properties of the reservoirs encountered changed from highly porous and very permeable to low permeability—fractured and unfractured.

Gas Wells

It is impossible in this short text to recognize the contribution made by thousands of authors in many thousands of papers. However, three major advances should be understood to appreciate the advances in theory necessary to include tighter pays developed at later time.

The first analysis was based on the emperical method applicable to very porous and permeable reservoirs developed by Rawlins and Schellhardt, "Back Pressure Data on Natural Gas Wells and Their Application to Production Practices," Monograph 7, U.S.B.M. This method today is known as the four-point (sometimes as the one-point) method. The square of the average reservoir pressure minus the square of the flowing sand-face pressure is plotted versus the respective flow rates on log-log paper. The maximum rate is read at the pressure equal to the average reservoir pressure after a straight line is drawn through test points for four semistabilized flow rates. Theory developed at a later date has proven the technique to be reliable for very good reservoirs where short tests investigate large areas.

The deficiency of the four-point test was recognized as field development and production included tighter pays, say in the range of 10 to 50 md. The isochronal test that includes the time factor of transient flow was developed by Cullender in "The Isochronal Performance Method of Determining Flow Characteristics of Gas Wells," *Trans.* AIME (1966) 204. In this method, the pressure and flow rates are measured, as in the four-point method, at specified times and a series of log-log curves are plotted so that the curve related with semisteady state production can be determined.

Development of even tighter gas wells was common during the late 1950s and fracturing with large amounts of sand was routine. Pressure difference across the drainage area often was great. By 1966, a group of engineers working with Russell, Shell Oil, and an independent group working with Dr. Ramey published articles using basic flow equations applicable to all gas wells, regardless of the permeability and fractures used by the operators. The state of the art was summarized in 1967 in "Pressure Buildup and Flow Tests in Wells" by Matthews and Russell, SPE Monograph 1, Henry L. Doherty Series. The state of the art was again reviewed in 1977 in "Advances in Well Test Analysis" by Earlougher in SPE Monograph 5.

Oil Wells

Similar changes were made in the analysis of oil wells with time. Test procedures and analysis must include a term known as relative permeability to allow for flow of two or more fluids at the same time. A method known as the productivity index was devised at an early time, and there is a reluctance on the part of field operators to change to the use of the basic flow equations. During the early life of the oil business, PI was simply defined as barrels per pound drop in pressure with measurements made at surface. Later, the concept was defined as

$$J_o = PI = (stb/d)/(\bar{p} - p_{wf}) = q_o/(\bar{p} - p_{wf})$$

Later development of theory proved this concept to be applicable when pay was permeable and a strong water drive maintained pressure, as shown by the semisteady state flow equation:

$$(\bar{p} - p_{wf})/q_{sc} = B\mu/2\pi kh[\ln(r_e/r_w) - 3/4]$$

When relative permeability and PVT factors are changing, the equation takes the form:

$$q_o = (2\pi kh/\ln(r_e/r_w)\int_{p_w}^{p_e} (k_{or}/\mu_o B_o)\,dp$$

where average values are used. The subject is discussed in a later section.

Basic Dimensionless Pressure vs. Time Relations

Van Everdingen-Hurst first proposed these curves in 1949. Since that time, other authors have used the computer to calculate similar curves for other boundary conditions. The following graph shows that many of these computer calculations are mere repetitions of the earlier work. As expected, required changes in the time scale resulted in comparable curves within accuracy of assumptions used in the computer calculations. The results of these conversions and methods used are shown in Fig. 5–3 for the radial flow case. Comparable data for the linear flow case are shown in Fig. 5–4.

These curves are used in both flow and build-up analysis. Ramey, Earlougher, and others have used these curves in a curve match technique. Unfortunately, the dimensionless curves show little character in the range of application so that curve matches are difficult to establish. The dimensionless pressure is not matched by either the E_i nor log function until the dimensionless time approaches 50^+, as shown by comparing Curve 3 with Curves 4 and 5. Fractures are beneficial to flow, as indicated on the graph.

The shape of the build-up curves depends upon the flow time used in the Horner calculations when flow time is short. The importance of this factor can be established by a review of "Flow Time for Use with Build-up Testing." When permeability is high, the drainage boundary and interference conditions can be reached in a short flow or build-up time.

Theoretical p_D versus t_D plot*

Basic curves for use in curve matching. Square and radial configurations (Fig. 5–3).

Curves 1: P_{tD} vs t_D for a vertically fractured well-analytical solution for square by Ramey. Pressure at $0.8\ x_f$ which approximates a

propped hydraulic fracture of infinite conductivity. True infinite conductivity fracture. *SPEJ*, Aug. 1975.

Curves 2: P_{tD} vs t_D for a vertically fractured well—analytical solution for square by Ramey. Pressures for a natural, unpropped fracture. *SPEJ*, Aug. 1975.

Curves 3: P_{tD} vs t_D for a radial system as proposed by van Everdingen-Hurst in AIME Trans. 1949. Also, van Everdingen-Roberts in *JPT*, April 1971.

Curve 4: $\frac{1}{2}|E_i(-\frac{1}{4}\ t_D|$ as calculated by Ramey. When t_D is greater than 80, the p_{tD}, the E_i and log functions should result in identical values.

Curve 5: $\frac{1}{2}\ (\ln\ t_D + .809)$ as computed by van Everdingen et al. Timmerman has proposed that the straight line portions of Horner and other semilog plots be extrapolated and then plotted on this curve scale to aid in the curve match selection by matching log and P_D curves.

Note: These curves are consolidated into a single set of curves as suggested by Timmerman. The Russell-Truit curves and tables presented in *JPT*, Oct. 1964, should be adjusted by $r'_w = \frac{1}{2}x_{fr}$ and the observation that the p_{tD} and t_D scales must reflect circular rather than a square drainage pattern. $A_{re} = 0.785\ A_{xe}$ and p_D revision 221.8/141.2.

The Ramey et al curves referenced in *JPT*, Oct. 1976, page 1265, should be adjusted by $r'_w = \frac{1}{2}x_{fr}$ and the t_D scales adjusted for circular versus square drainage. *SPEJ*, Aug. 1975.

The van Everdingen-Hurst curves then fall on these adjusted curves within accuracy of calculation procedures and field data without any adjustment, and they may always be used.

* This section taken from *Petroleum Engineer.*

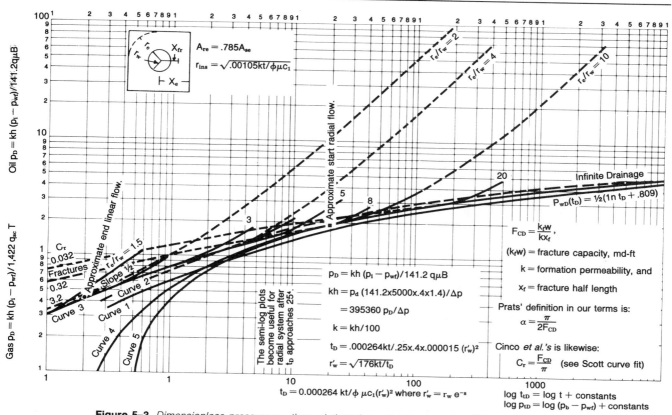

Figure 5–3. *Dimensionless pressure vs. time relations for radial flow (courtesy Petroleum Engineer).*

Q_T = Cumulative fluid flow in reduced units

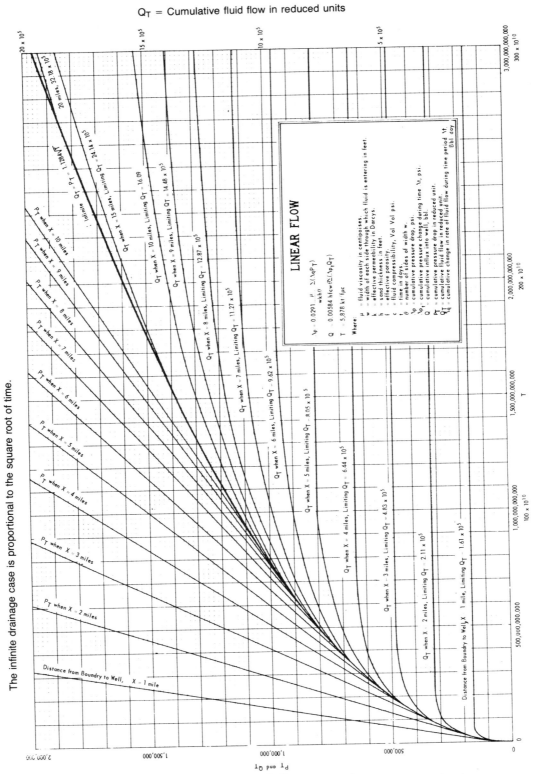

Figure 5-4. Dimensionless pressure versus time relations for linear flow (courtesy Shell Oil Co.) Homogeneous reservoir, slow change in fluid compressibility, and no skin.

Flow and Build-up Test Equations

Early petroleum engineers such as Muskat studied the laws of nature and applied concepts known in hydrodynamics, heat conduction, and electricity to flow of fluids through rock using Darcy's law to derive the following equations expressed as partial derivatives:

a. For linear flow:

$$\frac{\partial^2 p}{\partial x^2} + \frac{\partial^2 p}{\partial y^2} = (\phi\mu c/k)\,\partial p/\partial t$$

b. For radial flow:

$$\frac{1}{r}\frac{\partial}{\partial r}\left(r\frac{\partial p}{\partial r}\right) + \frac{1}{r^2}\frac{\partial^2 p}{\partial \theta^2} = (\phi\mu c/k)\,\partial p/\partial t$$

c. For spherical flow the reader is referred to Muskat's text entitled, *Flow of Homogeneous Fluids.*

Many of the techniques available for applying the equations to the reservoirs using proper boundary conditions and transformations are summarized by Carslaw and Jaeger in *Conduction of Heat in Solids,* 2nd Edition, Oxford Press. Relative permeability was devised for two-phase flow.

Frequent Assumptions

Flow is into a well, which is open throughout the entire pay interval.

The pay is a uniform, homogeneous-anisotropic porous medium with uniform thickness.

The porosity and permeability are constant throughout area and independent of pressure.

The reservoir fluids have a small and constant compressibility.

The fluids have constant viscosity and other characteristics throughout test and area.

The pressure gradients are relatively small and gravity terms are negligible.

The actual assumptions made depend upon the derivation proposed by each author of papers. In some cases, equations must be revised or input must be adjusted to obtain reliable results. The boundary assumptions frequently used in analysis of flow and build-up tests include flow into a centrally located wellbore at either constant sand-face pressure or alternatively at constant flow rate at the sand face. When neither pressure nor rate is constant, $\Delta p/q$ and/or superposition can often salvage the use of test data. The external or drainage boundary conditions also must be defined. The well may be in an infinite reservoir having uniform properties and containing uniform fluid content. Alternately, the well boundary may be limited by faults, lack of sand, and interference from other wells. If flow rate in adjacent wells is changing, the drainage boundary of the test well may adjust during the test to reflect its relative flow rate. During fluid injection, the boundary pressure may be constant. The analyst must be alert and use the proper equations or make adjustments, such as discussed in later sections of this chapter, when designing and analyzing well tests.

When the well or field radius is large, the p_{tD} versus t_D relations must be used instead of point source solutions. Also, when pressure drops are large, *PVT* functions are often not constant throughout the area influenced by the test and $m(p)$ methods must be used for gas wells. The reliability of any study depends upon sound thinking, use of proper techniques and equations, and hard work. Often there are no easy short cuts.

TABLE 5–1
Conversion of Units Used in Test Measurements to Units Used in Equations

Symbol	Darcy units	Multiply by	Practical U.S. field units
p = pressure	atmospheres	14.7	psi
c = compressibility	vol/vol/atm	1/14.7	vol/vol/psi
ϕ = porosity	fraction	none	fraction
h = pay thickness	cm	1/30.48	feet
k = permeability	darcy	1000	md
	0.9869×10^{-8} cm^2		
μ = viscosity	cp	none	cp
q = flow rate	cc/sec	1/1.8401	Bbl/day or B/D
Q = cum. prod.	cc	1/158987	Bbl.
l = length	cm	1/30.48	feet
one barrel is equal to 5.6146 cu. ft. and 42 gallons			
t = time	sec.	1/3600	hours

Scaling or Dimensionless Groups

Scaling groups behave as a single variable in the equations, and calculation time can be reduced when scaling groups are used. Also, scaling groups should be used when designing and interpreting laboratory experiments, comparing laboratory data with field data, comparing data from two reservoirs, and when plotting graphs. Some of the scaling groups used in flow studies follow:

Constant Terminal Rate Case

Linear Flow: Conversion: $14.7 \times 1.8401/(1/1000) \times 30.48 \times 30.48 = 29.114$

cgs units: $\Delta p = p_i - p_w = (q\mu/khw)\, p_D$
$\qquad = A_1\, B\, \Sigma(\Delta q\, p_D)$

Practical units: $A_1 = 29.114\, \mu/khw$ (per sand face)

Radial Flow: Conversion: $14.7 \times 1.8401/(2 \times 3.1416 \times (1/1000) \times 30.48 = 141.24$

Group	cgs units	Practical field units	Conversion
Time group, t_D			
linear flow	$kt/\phi\mu c$	$0.2449\, kt/\phi\mu c$	$(1/1000)3600(1/14.7)$
radial flow	$kt/\phi\mu cr^2$	$0.000264\, kt/\phi\mu cr^2$	$\dfrac{3600}{1000 \times 14.7 \times 30.48 \times 30.48}$
spherical flow	$kt/\phi\mu c\,(r - r_w)^2$	$0.000264\, kt/\phi\mu c\,(r - r_w)^2$	
Area group, A			
linear	wl		
radial	$0.7854d^2 = 3.1416r^2$		
spherical	$3.1416d^2$		
Volume group, V	add an h to area group		
Distance group, r_D			
radial	r/r_w, often use r_e/r_w		
Transmissibility group, J	kh/μ	$0.304\, kh/\mu$	$(1/1000)\, 30.48$
Storage group, S	ϕch	$2.073\, \phi ch$	$(1/14.7)\, 30.48$
Flow group, q_D			
linear flow	$q\mu/khw$	$1.9805\, q\mu/khw$	$\dfrac{1.840}{(1/1000) \times 30.48 \times 30.48}$
radial flow	$q\mu/2\pi kh$	$9.607\, q\mu/kh$	$\dfrac{1.840}{2 \times 3.1416 \times (1/1000) \times 30.48}$

q is at sand-face conditions, and B often is included to convert to surface measurements.

Cumulative production, Q_D	
Linear flow	$q/hw\phi c$
Radial flow	$Q/2\pi\phi chr_e^2$ where r_e is radius to boundary
Skin, $\overline{S}$	$\dfrac{kh\Delta p_{skin}}{141.4\, q\mu B}$
Storage, $\overline{C}$	$C/2\pi h\phi cr_w^2$

Gas conversion group, B_g

$$B_g = \frac{V_{reservoir}}{V_{surface}} = \frac{p_{sc}\, T\, z_{\bar p}}{\bar p\, z_{sc}\, T_{sc}} = \frac{14.7\, T\, z_{\bar p}}{\bar p \times 1 \times 520} = 0.02827\, T z_{\bar p}/\bar p \quad \text{(Use when } p_e - p_w \text{ is small)}$$

When $p_e - p_w$ is large:

$\qquad B_g = (p_{sc}\, T\, z_{\bar p})/(\bar p\, T_{sc}\, z_{sc})$
$\qquad \bar p = $ average reservoir pressure
$\qquad \bar p = (\bar p + p_w)/2$ if $2p/z\mu$ is a straight line for $\bar p + p_w$
$\qquad$ Otherwise, use $m(p)$ function of Ramey

Compressibility group

$\qquad c_t = S_o c_o + S_g c_g + S_w c_w + c_f$
$\qquad$ When gas, $c_t = S_g/p = S_g c_g$

Pressure group, p_D

cgs units: $\Delta p = p_i - p_w = (q\mu/2\pi kh)p_D = \theta A_2 \Sigma(\Delta q p_D)$

practical units: $A_2 = 141.4 \ B\mu/kh$ (for full circle)

Constant Terminal Pressure Case

Linear Flow: Conversion: $14.7 \times 30.48 \times 30.48/(14.7 \times 158987) = .005843$

cgs units: $\Delta q = Q = \phi chw \Sigma(\Delta p Q_D) = \theta B_1 \Sigma(\Delta p Q_D)$

practical units: $B_1 = 0.005843 \ \phi chw$ (per sand face)

Radial Flow: Conversion: $2 \times 3.1416 \times 30.48 \times 30.48 \times 30.48 \times 14.7/14.7 \times 158987 = 1.1191$

cgs units: $\Delta q = Q = 2\pi\phi cr_b^2 h\Sigma(\Delta p Q_D) = \theta B_2 \Sigma(\Delta p Q_D)$

practical units: $B_2 = 0.1119 \ \phi cr_b^2 h$ (for full circle)

where r_b is radius to boundary of reservoir
and θ is fraction of circle,
and linear flow is one face only
p_D and Q_D are available from tables

Flow Equations for Incompressible Flow–No Depletion Occurs in Reservoir

Linear flow:
cgs units:

$$p_e - p_w = (q\mu/k)(L/A)$$

Practical units:

Liquid: $p_e - p_w = 887.45 \ (q\mu/k)(L/A)$

Conversion: $14.7 \times 1.84 \times 1000 \times 30.48/30.48 \times 30.48 = 887.45$

A hydraulic fracture has two faces so that area, $A = 2 \times h \times L_f$

q is measured at $(p_e - p_w)/2$

Radial flow:
cgs units:

$$p_e - p_w = (q\mu/2\pi kh) \ln (r_e/r_w)$$

Practical units:

Liquids $p_e - p_w = 141.24(q\mu/kh)\ln (r_e/r_w)$
$= 325.22(q\mu/kh)\log (r_e/r_w)$

Conversion: $14.7 \times 1.84 \times 1000/2 \times 3.14 \times 30.48 = 141.24$
$141.24 \times 2.303 = 325.2$
$141.24/2 = 70.62$

Spherical flow:
cgs units:

$p_e - p_w = (q\mu/4\pi k)(1/r_w - 1/r_e) = (q\mu/4\pi kr_w)$ when r_e is large

Practical units for Liquids:

$$p_e - p_w = 70.62(q\mu/k)(1/r_w - 1/r_e)$$

Conversion: See radial flow.

Incompressible flow may be approached during water flooding operations.

Flow Equations for Slightly Compressible Fluids Flow–Reservoir is Being Depleted

Transient linear flow
cgs units:

$$p_e - p_w = (2q\mu/kA)\sqrt{kt/\pi\phi\mu c_t}$$

The boundary pressure is often expressed as $p_e = p_i = p_b$

Conversions: $14.7 \times 1.84 \times 1000 \times 2/30.48 \times 30.48 = 58.22$
$3600/1000 \times 3.1416 \times 14.7 = 0.0795$
and
$\sqrt{0.0795} = 0.2792$ and $0.2792 \times 58.22 = 16.26$

Practical units for liquids:

$p_e - p_w = (16.26 \ q_{sc} \ B\mu/kA)\sqrt{(kt/\phi\mu c_t)}$
$= (16.26q_{sc}B\mu/A)\sqrt{(t/k\phi\mu c_t)}$
the slope of plot of pressure versus $\sqrt{t}$ is
$m = (16.26q_{sc}B\mu/A)\sqrt{1/k\phi\mu c_t}$.

This equation contains two unknowns, k and A. k is usually determined before fracturing well.

Practical units for gas:

Conversion: $58.22 \times 1000/5.614 = 10370$ and
$10370 \times .2792 = 2896$

$p_e - p_w = 2896 \ (q_{sc}B_g\mu/kA)\sqrt{kt/\phi\mu c_t}$
$p_e^2 - p_w^2 = 5792(q_{sc}B_g\mu/kA)\sqrt{kt/\phi\mu c_t}$
$p_e^2 - p_w^2 = 5792(q_{sc}\mu/kA)(p_{sc} T \ Z_{\bar p}/T_{sc}Z_{sc})\sqrt{kt/\phi\mu c_t}$
$p_e^2 - p_w^2 = 163.7(q_{sc} T \ \mu_{\bar p}Z_{\bar p}/kA)\sqrt{(kt/\phi\mu c_t}$
$m(p_e) - m(p_w) = (163.7 \ q_{sc}T/kA)\sqrt{kt/\phi\mu c_t}$
when $p_{sc} = 14.7$ and $T_{sc} = 520 \ R$
$\bar p(p_e - p_w) = [(p_e + p_w)/2](p_e - p_w) = (p_e^2 - p_w^2)/2$
The slope of $m(p)$ versus $\sqrt{t}$ plot is
$m = (163.7 \ q_{sc} \ T/kA)\sqrt{k/\phi\mu c_t} = (163.7q_{sc}T/A)\sqrt{1/k\phi\mu c_t}$

μ and Z usually are at an average pressure

The $m(p)$ function contains Z and μ and may be read from curves based on calculations using gas gravity and temperature relations for gases. Early flow from a well is complicated by wellbore conditions, afterflow as wellbore unloads, and other factors. A fractured well often has a negative skin, which is handled within the A term. The fracture has two faces, and A must so consider the data. The linear flow period gradually disappears as the flow streamlines become more spherical and radial with flow time.

Transient spherical flow

cgs units: Various equations appear in literature adjusted to specific application.

$$p_{(r,t)} = p_e - Aq\mu B/4\pi kr \, erfc \, [(B_1 r^2/4kt/\phi\mu c)^{1/2}], \, t > 0$$
$$p_e - p_{r,t} = (p_e - p_w)(r_w/r) \, erfc\sqrt{(\phi\mu c/k)(r - r_w)^2/4t}$$
$$p_e - p_w = (q\mu/4\pi kr_w)[1 - e^{kt/\phi\mu cr_w^2} erfc\sqrt{kt/\phi\mu cr_w^2}]$$

Practical units: Spherical flow may occur between linear and radial flow times.

$$p_e - p_w = (2453 Bq\mu/k^{3/2})\sqrt{\phi\mu c_t} \, [1/\sqrt{t}] \text{ for liquids when}$$
t is in hours.
$k^{3/2} = 2453 \, (q_{sc}B\mu\sqrt{\phi\mu c_t})/m$ when pressure is plotted versus $1/\sqrt{hours}$
$$s = 34.7 \, r_{ew}\sqrt{\phi\mu c/k}[(p_{\Delta t} - p_{wf})/m + 1/\sqrt{\Delta t}] - 1.0$$

Conversions: $A = 14.7 \times 1.84 \times 1,000 \times 30.48/$
$30.48 \times 30.48 = 887.22$
$B_1 = 30.48 \times 30.48 \times 3600/14.7 =$
$227,548$ and $\sqrt{227,548} = 477.02$
$(887.22 \times 477.02)/(4 \times 3.1416^{3/2}) =$
$19,000$ when t is in minutes and
$19000/\sqrt{60} = 2453$ in hrs.
$477.02/\sqrt{3.1416} \times \sqrt{60} = 34.7$

$k^{3/2} = 24,700(q_{sc}T\sqrt{\phi\mu c_t})/m$ for gas when $m(p)$ is plotted versus $1/\sqrt{hours}$.

Conversions: $2,453 \times 1,000 \times 14.7/5.614 \times 520 = 12352$ when p is used; $24,700$ when p^2 and $m(p)$ are used.

In the above equation for skin,

$$r_{ew} = b/2 \ln(b/r_w)$$

where b is thickness of zone open to flow. For hemispherical flow or partial penetration replace b by $2b$. r_w is actual well bore radius.

$$k_{av} = k_h \, k_h \, k_v$$

Transient radial flow

Use data only before $t_{De} = 0.1$ Also, avoid data for afterflow, linear, etc.
In cgs units:

$$p_e - p_{wf} = (q\mu/4\pi kh)[\ln(kt/\phi\mu cr_w^2) + 0.809 + 2s + 2D_{qsc})]$$

where s is the skin effect and D_{qsc} represents possible turbulent flow. The total compressibility, c_t or $c = S_o c_o + S_g c_g + S_w o_w + c_f$ where c_f is for formation.

Practical Units:
Liquids.
Conversions: $14.7 \times 1.84 \times 1000/4 \times 3.1416 \times$
$30.48 = 70.62$;
$70.62 \times 2.303 = 162$.
$3600/1000 \times 14.7 \times 30.48$
$\times 30.48 = 0.000264$;

$\ln .000264 = -8.23$
$-8.23 + 0.809 = -7.43$
$-7.43 \times 0.4343 = -3.23$
$\log x = 0.4343 \ln x$
$2x \, 0.4343 = 0.87$
$1/.4343 = 2.303$

At constant rate flow at sand face.

$$p_{wf} = p_e - (70.6 \, q\mu B/kh)[-Ei(-\phi\mu cr_w^2/0.00105 \, kt) + 2s]$$
$$= p_e - (70.6 \, q\mu B/kh)[\ln (0.000264 kt/\phi\mu c_t r_w^2)$$
$$+ 0.809 + 2s]$$
$$= p_e - (162.6 \, q\mu B/kh)[\log (kt/\phi\mu c_t r_w^2) - 3.23$$
$$+ 0.87 \, s]$$
$m = 162.61 \, q\mu B/kh = $ slope of the pressure versus log t plot
$kh = 162.6 \, q\mu B/m$ and $s = 1.151[(p_e - p_{1hr})/m$
$$- \log(k/\phi\mu c_t r_w^2) + 3.23]$$

At constant pressure at the pay face.

$$1/q_{sc} = 162.6 \, B\mu/kh(p_e - p_{wf})[\log (kt/\phi\mu c_t r_w^2)$$
$$- 3.23 + 0.87s + 0.87D_{qsc}]$$
$m = 162.6\mu B/kh(p_e - p_{wf})$ when $1/q_{sc}$ is plotted vs. log t.

When flow is slightly variable, $1/q$ may be replaced by $(p_e - p_{wf})/q$ and equations are adjusted accordingly.

$$s = 1.15[(1/q_{sc \, at \, 1 \, hr})/m - \log(k/\phi\mu c_t r_w^2) + 3.23]$$

Gas.
Conversions: $162.6 \times 1000 \times 14.7/5.614 \times 520$
$= 818.5$
$\bar{p}(p_e - p_{wf}) = (p_e^2 - p_{wf}^2)/2$
$818.5 \times 2 = 1637$

At constant rate flow at sand face

$$p_e - p_{wf} = (818 \, q_{sc}\mu_{\bar{p}}Z_{\bar{p}}T/\bar{p}kh)[\log (kt/\phi\mu c_t r_w^2)$$
$$- 3.23 + 87s + 87D_{qsc}]$$

Define $\tilde{p} = (\bar{p} - p_{wf})/2$

$$[p_e - p_{wf} = 818.6 \, q_{sc} \, \mu_{\tilde{p}} \, Z_{\tilde{p}} \, T/\tilde{p} \, kh]$$
$$[p_e^2 - p_{wf}^2 = 1637 \, q_{sc} \, \mu_{\bar{p}} \, Z_{\bar{p}} \, T/kh]$$
$$[m(p_e) - m(p_{wf}) = 1637 \, q_{sc} \, T/kh]$$

$$\left[\begin{array}{c} \log (kt/\phi\mu c_t r_w^2) - 3.23 + 0.87s + 0.87D_{qsc} \\ \text{The average or boundary value} \\ \text{should be used for terms in bracket.} \end{array} \right]$$

Determine slope, *kh*, and skin as for oil using adjusted equations. Gas at constant sand-face pressure:

$$1/q_{sc} = [1637(T/kh)/(m(p_e) - m(p_{wf}))]$$
$$[\log (kt/\phi\mu c_t r_w^2) - 3.23 + 0.87s + 0.87D_{qsc}]$$

Determine slope, *kh*, and skin as for oil using

adjusted equations as required. Other equations often found in literature.

$$p_e - p_w = 25152(q_{sc}\mu/kh)(p_{sc}TZ/T_{sc}\,\bar{p})$$
$$[\tfrac{1}{2}\ln(0.000264kt/\phi\mu c_t r_w^2) + 0.4045 + s + D_{q_{sc}}]$$

$$p_e - p_w = 12576\,q_{sc}\,B_g\mu/kh$$
$$[\ln(0.000264kt/\phi\mu c_t r_w^2) + 0.809 + 2s + 2D_{q_{sc}}]$$

$$p_e - p_w = 28958\,q_{sc}\,B_g\,\mu/kh$$
$$[\log(kt/\phi\mu c_t r_w^2) - 3.23 + 0.87s + 0.87\,D_{q_{sc}}]$$

$$p_e^2 - p_w^2 = 25152(q_{sc}\mu/kh)(p_{sc}TZ/T_{sc})$$
$$[\ln(0.000264kt/\phi\mu c_t r_w^2) + 0.809 + 2s + 2D_{q_{sc}}]$$

$$p_e^2 - p_w^2 = 57920(q_{sc}\mu/kh)(p_{sc}TZ/T_{sc})[\log(kt/\phi\mu c_t r_w^2)$$
$$- 3.23 + 0.87s + 0.87D_{q_{sc}}] = m(p_e) - m(p_w)$$

$$m(p_e) - m(p_w) = 1637(q_{sc}T/hk)[\log(0.000264kt/\phi\mu c_t r_w^2)$$
$$+ 0.3513 + 0.87s + 0.87D_{q_{sc}}]$$

Late Transient Flow Period

Use when t_{De} is between 0.1 and 0.3.
Define in cgs units $\hat{p} = \bar{p} - q\mu/2\pi kh$ $(\ln(r_e/r_w) - 3/4 + s)$ so that $\hat{p}$ is p_{wf} during semisteady flow, then in practical units $\log(p_{wf} - \hat{p}) = \log(118.6q_{sc}\mu B/kh) - 0.00168kt/\phi\mu c_t r_e^2$. The plot shown as Fig. 5–5 can be made. In practical units:

$$kh = 118.6\,q_{sc}\,\mu B/b$$

where b is intercept as shown on Fig. 5–5. The contributory pore volume is

$$V_p = 0.1115\,q_{sc}B/bc_t \text{ slope of curve}$$
$$s = 0.84\,(\bar{p} - \hat{p})/b - \ln(r_e/r_w) + 3/4$$
$$p(\text{skin}) = bs/0.84$$

The technique for obtaining the most likely pressure is used in many situations, such as fall-off analysis of injectors.

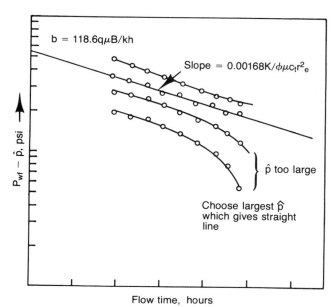

b = 118.6qµB/kh

Slope = 0.00168K/φµc$_t$r^2$_e$

$\hat{p}$ too large

Choose largest $\hat{p}$ which gives straight line

$P_{wf} - \hat{p}$, psi

Flow time, hours

Figure 5–5. *Schematic late transient drawdown analysis plot (Monograph 1, © SPE-AIME, 1967).*

Semisteady State Flow

Use after t_{De} is greater than 0.3.

$$p_i - p_{wf} = q_{sc}t/\pi\phi c_t hr_e^2 + q\mu/2\pi kh$$
$$[\ln(r_e/r_w) - 3/4 + s] \text{ in cgs units.}$$

In this equation kh, r_e, and s are all often unknowns. Two of these must be determined using independent techniques before the equation can be used to determine the last. A plot of p_{wf} versus t should be a straight line with slope

$$B_L = q/\pi\phi c_t hr_e^2 \text{ in cgs units}$$

The contributory pore space in reservoir barrels is in practical units.

$$V_p = 0.0418\,q_{sc}B/B_L\,c_t$$

where B_L is slope of plot when pressure in well is in psi and t is in hours when p is plotted versus t.

Often, reservoir drainage is not circular and boundaries may not be complete. The equations that describe semisteady-state flow are:

Linear flow in practical units for liquids flow.

$$p_e - p_{wf} = 443\,q_{sc}\mu BL/kA \text{ where } A \text{ is for two faces for most field conditions}$$

Radial flow for liquids during semisteady-state flow.

$$p_{avg} - p_{wf} = (141.2\,q_{sc}\,\mu B/kh)[\ln(r_e/r_w) - 0.75 + s + D_{q_{sc}}]$$

The constant 0.75 is for circular flow. Conversion is

$$14.7 \times 1.8401 \times 1,000/2 \times 3.1416 \times 30.48$$
$$= 141.24 \text{ and } 1/141.24 = 0.00708$$

Radial flow for gas during semisteady-state flow.

$$[\bar{p} - p_{wf} = (711\,q_{sc}\,\mu_{\bar{p}}Z_{\bar{p}}T/kh\,\bar{p})]$$
$$[\bar{p}^2 - p_{wf}^2 = (1,422\,q_{sc}\,\mu_{\bar{p}}Z_{\bar{p}}T/kh)]$$
$$[m(\bar{p}) - m(p_{wf}) = (1,422\,q_{sc}T/kh)]$$
$$(\bar{p} + p_{wf})/2 = \bar{p}$$
$$[\ln(r_e/r_w) - 0.75 + s + D_{q_{sc}}]$$

The average reservoir pressure, $\bar{p}$, often is a fair approximation of the pressure at drainage boundary so that $\bar{p} = 0.90 - 0.95\,p_e$.

Build-up Equations for Slightly Compressible Fluids—Depletion Occurs in Reservoir

Transient linear build-up in practical units for liquids:

$$p_e - p_{wf} = (16.26\,q_{sc}\mu B/kA)(\sqrt{k/\phi\mu c_t})(\sqrt{t + \Delta t} - \sqrt{\Delta t})$$

The slope of plot of pressure versus $\sqrt{\Delta t}$ or $(\sqrt{t + \Delta t} - \sqrt{\Delta t})$ should be a straight line so that

$$m = (16.26\,q_{sc}\,\mu B/kA)(\sqrt{k/\phi\mu c_t})$$

Linear flow occurs during early part of build-up test and often is complicated by wellbore effects. A and k are unknowns, and k should be determined before fracturing the well. Transient linear build-up in practical units for gas:

$$m(p_e) - m(p_{wf}) = (163.72 \ q_{sc} \ T/kA)(\sqrt{k/\phi\mu c_t})(\sqrt{t + \Delta t} - \sqrt{\Delta t})$$
$$m = (163.72 q_{sc} T/kA)(\sqrt{k/\phi\mu c_t})$$

Transient spherical build-up for liquids in practical units:

$$p_e - p_{wf} = (2453 \ q_{sc} \ \mu B/k^{1.5})(\sqrt{\phi\mu c_t})(1/\sqrt{\Delta t} - 1/\sqrt{t + \Delta t})$$

The plot of p_{wf} or p_{ws} versus $(1/\sqrt{\Delta t} - 1/\sqrt{t + \Delta t})$ will give a straight line with slope m.

$$m = (2,453 \ q_{sc} \ \mu B/k^{3/2})(\sqrt{\phi\mu c_t})$$

For hemispherical flow or partial penetration of well, the constant becomes 4,906.

$$\text{skin} = s = 34.7 \ r_{ew}\sqrt{(\phi\mu c_t/k)}[(p_{\Delta t} - p_{wf})/m + 1/\sqrt{\Delta t}] - 1$$
$$r_{ew} = b/(2 \ln(b/r_w))$$
b = thickness of zone open to flow (for hemispherical, $b = 2b$)

Transient spherical flow for gas in practical units:

$$k^{3/2} = k^{1.5} = 24700 \ q_{sc} \ T(\sqrt{\phi\mu c_t}/m)$$

Transient radial build-up for liquids in practical units:

$$p_e - p_{wf} = (162.6 \ q_{sc} \ \mu B/kh) \log [(t + \Delta t)/\Delta t]$$

The slope of the proper straight line when pressure is plotted versus $\log \Delta t$ or $\log (1 + t/\Delta t)$

$$m = 162.6 \ q_{sc} \ \mu B/kh \text{ and } kh = 162.6 \ q_{sc} \ \mu B/m$$
Total skin = $s = 1.151[(p_{1 \text{ hr}} - p_{wf})/m - \log(k/\phi\mu c r_w^2 + 3.23] = (k/k_s - 1) \ln(r_s/r_w)$
Also, $r_w' = r_w \ e^{-s}$ where r_w' is effective well radius of fractured well.
$$p_{\text{skin}} = 0.87 \ sm$$

Flow efficiency has been defined as $1/[\text{damage ratio}] = J_{\text{actual}} - J_{\text{ideal}}$

$$J_{\text{actual}} = q/(p^* - p_{wf}) \text{ and } J_{\text{ideal}} = q/(p^* - p_{wf} - \Delta p_s)$$

Transient radial build-up for gases:

$$m(p_e) - m(p_{wf}) = 1637(qT/kh) \log [(t + \Delta t)/\Delta t]$$
$$\text{slope} = m = 1637 \ qT/kh \text{ and } kh = 1637 \ qT/m$$

Previous sections should be consulted when other pressure forms are used.

The Time Equation

$$t_D = 0.000264 \ kt/\phi\mu c r^2 \quad \text{in practical units.}$$

True transient flow ends when $t_D = 0.1$ and semi-steady-state flow begins when $t_D = 0.3$. The radius of investigation is different, depending upon the author, but $t_D = 0.25$ is generally accepted. Here, r is the radius of the drainage boundary.

Radius of Investigation Equation

$$r_{\text{inv}} = \sqrt{0.00105 \ kt/\phi\mu c_t} \quad \text{which is derived from } t_D = 0.25$$

Analysis of Drill Stem Tests

Flow during drill stem tests and buildup time often is short so that special analysis techniques often can be used. Johnston offers the following equations.

$$\text{damage} = EDR = (p_e^2 - p_{ws}^2)/m(\log t + 2.65)$$

where $m = (p_e^2 - p_{ws}^2)/\log$ cycle = slope of buildup curve

$$\text{Transmissibility} = kh/\mu Z = 1637 \ T \ q_{sc}/m$$

Equations are for gas well

$$\text{Radius of investigation} = kt/(40 \ \phi(1 - S_w)\mu c_g)$$

where t is time in days.

Damage Ratio

DR = 1/flow efficiency
$= (kh/\mu)_{BU}/(kh/\mu)_{PI}$
$= 0.183 \ (p_{ws} - p_{wf})/m$ for use with drill-stem tests
$= (p_{ws}^2 - p_{wf}^2)/m(\log t + 2.65)$

Average When Two or More Fluids are Flowing at Same Time

$(k/\mu)_t = k_o/\mu_o + k_g/\mu_g + k_w/\mu_w$
$= 162.6/mh \ (B_o q_o + B_g(q_{gt} - q_o R_s) + B_w q_w)$

where measured rates are at surface.

There are many variations of these equations, and the reader is referred to individual papers and SPE Monographs 1 and 5 for references. SPE Monograph contains detailed examples of the application of build-up methods to field data. Many examples also are given in later parts of this text.

The Productivity Index or PI

$$PI = (b/d)/(\bar{p} - p_{wf}) = J_o$$

Many variations of this concept have been used by different operators during the past 50 years, and it is obvious that the concept or ratio is a small part of the transient and semisteady-state equations.

Time and Radius of Investigation Equations

The time equation, radius of investigation, distance to faults and boundaries is generally accepted as:

$$t_D = 0.000264 \; kt/\phi \mu c_t r^2$$

This equation relates distance with time when other parameters are known. The equation may be used with the well radius, r_w, as well as with the drainage boundary radius, r_e. SPE Monograph 1 suggests that transient flow ends when t_{De} has a value of 0.1 or $t = \phi \mu c_t r_e^2/0.00264 \; k$, and that semisteady-state flow begins when $t_{De} = 0.3$ or $t = \phi \mu c_t r_e^2/0.00088 \; k$. Other authors have used somewhat different values for t_{De}. The generally accepted equation for determination of radius of investigation was proposed by van Poollen as follows:

$$r_{inv} = \sqrt{0.000264 \; kt/0.25 \; \phi \mu c_t}$$
$$= \sqrt{0.00105 \; kt/\phi \mu c_t} \quad \text{in practical units}$$

Other authors have used values for the constant as shown below:

Comparison of Published Radius of Investigation Equations When $HKVP = 1.0$

Source	Equation	Radius comparison	Area comparison
Tek et al.	$\sqrt{0.00485 \dfrac{kt}{\phi \mu c}}$	2.15	4.62
Jones	$\sqrt{0.00422 \dfrac{kt}{\phi \mu c}}$	2.00	4.00
Hurst	$\sqrt{0.00184 \dfrac{kt}{\phi \mu c}}$	1.32	1.74
van Poollen	$\sqrt{0.00105 \dfrac{kt}{\phi \mu c}}$	1.00	1.00
Hutchinson et al.	$\sqrt{0.000593 \dfrac{kt}{\phi \mu c}}$	0.75	0.56

These equations are for a circular uniform reservoir. An example using the van Poollen equation follows:

If the test, either flow or build-up, is 72 hours and kh = 172 md/cp with $\phi c_t = 0.00000232$ psi^{-1},

$$radius \; of \; drainage, \; r_d = 0.029 \sqrt{\frac{(172)(72)}{0.00000232}}$$

or 2,100 feet. This estimate is valid if no boundaries are within the 2,100 ft and no other wells, which may cause interference, are operating within about 4,200 feet.

As shown in Fig. 5–6, the plot of radius versus time on log-log paper is a straight line.

These equations are applicable for determination of stabilization time—time to reach semisteady-state flow, time to end of true transient flow, time for pressure transients to reach boundaries, etc.

Earlougher and Kazemi in *JPT*, January 1980, conclude that both flow and build-up portions of a test must exceed the time to reach boundaries if the boundary is to be observed in the buildup test. A lengthy buildup test following a short flow period may not observe the fault. The radius of investigation during flow must be at least four times the distance to a fault (note that the equations differ). These authors use the following equations for radius of investigation:

$$r_{inv} = 1.784 \sqrt{kt/\phi \mu c_t} \quad \text{in standard international units}$$
$$= \sqrt{0.000839 \; kt/\phi \mu c_t} \quad \text{in practical units (0.00088 by Russell)}$$

The distance to a linear fault is calculated from:

$$L = 7.493 \sqrt{k \; t_x/\phi \mu c_t} \quad \text{where } t_x \text{ is time to intersection of curves.}$$
$$= \sqrt{0.000148 \; kt_x/\phi \mu c_t} \quad \text{in practical units (Same as SPE Monograph 1)}$$

For a linear fault in circular drainage area in uniform pay, the second slope is exactly twice that of the first slope. Fig. 5–7 shows computer-generated curves when a fault is 92.37 meters or 303.4 feet from a well with calculated results as follows:

Curve	Flow time		Time to curve break		Calculated L		r_{inv} by flow	
	SI	Hours	SI	Hours	Meters	Feet	Meters	Feet
1	10.8	3					130.6	428.6
2	21.6	6	45.00	12.5	112.0	367.5	184.7	606.1
3	86.4	24	30.31	8.42	91.9	301.6	369.5	1212.2

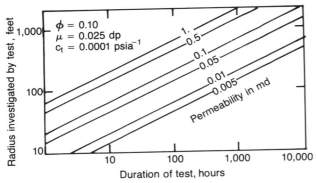

Figure 5–6. *Radius of investigation based on van Poollen equation.*

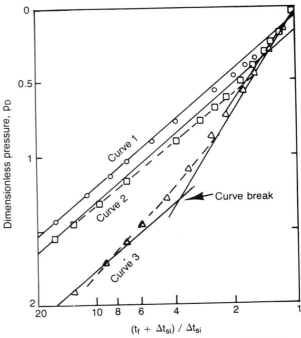

Figure 5–7. *Horner fault plot (after Earlougher et al., JPT, January 1980).*

Each author would probably draw slightly different curves, but it is obvious that flow time must be sufficient if sound build-up analysis is to be attained. In the field, conditions are not ideal and a slope change of exactly two may not be observed when flow time is sufficient. Also, the slope changes may accurately reflect only the nearest boundary, which may be fault, pay change, etc.

Radius of Investigation Equation

Plots of change in pressure difference between surface and sand-face pressures versus time are useful in determining length of afterflow periods. The radius of investigation equation is particularly useful. Linear flow should end, and the transition between linear and radial flow should start after the distance investigated is a short distance from the fracture face. Also, work of Prats et al., suggests that flow should have a relatively radial configuration when radius investigated exceeds twice the half fracture length.

Interpretation of well flow and build-up tests is an art rather than an exact science. Computer techniques are very valuable in preparation of data, but all results should be carefully interpreted using all diagnostic tools available. The methods used will vary somewhat with each test. The quality of the interpretation is heavily dependent upon the experience, integrity, intellect, willingness to analyze detail, and ability of the analyst to study and apply the geological, logging, and reservoir concepts involved to obtain the best possible fit of all available information. Recognition and use of type curves is also essential.

Determination of Straight Lines for Equations (for use in flow and build-up analysis methods)

Many factors influence the pressure being measured in the test well, and the simple application of Horner and Miller, Dyes, and Hutchinson analysis methods can at times be insufficient. Flow in the test well may not be at constant rate nor alternately at constant pressure. The superposition technique and alternates such as use of $\Delta p/q$ methods may correct this problem. Pressure transients created by operations at adjacent wells may be measured during the test. Documented problems such as wellbore effects, afterflow, well damage, and positive skins caused by few perforations that are equivalent to a smaller wellbore; negative skins caused by wellbore enlargement and fractures that may result in a linear flow configuration; partial pay penetration, limited interval perforated, slanted or directional wellbores; and changes in permeability, porosity, pay thickness and fluids must be considered when relating proper equations with the slopes read from straight lines indicated on various plots of the data. At late parts of the test, well interference (the drainage area of wells adjusts to their relative rates of flow); pay pinchouts, shale stringers, changes in permeability, changes in fluids, and normal irregularities in pay and fluids with distance and in vertical direction influence pressure measurements for both flow and build-up tests. Flow test data are preferred since capillary restraints that may be effective, particle movement, and turbulence may be reflected in the flow data. Flow between layers during build-up can present problems.

Interpretation of pressure changes of less than a few pounds per square inch require special consideration. Changes in atmospheric pressure when the well is open at the surface, the gravity of the moon, and small pressure changes caused by irregularities in flow rate and pay and fluid characteristics can be measured. Instrument characteristics and stabilization must be considered.

Proper test design is very important if the correct slopes are to be measured during flow and build-up tests, but the major problem often is in selection of the straight-line portion of the curves and application in the proper equations. The start of the straight line for radial flow is discussed in SPE Monographs 1 and 5. Wellbore aftereffects become relatively unimportant when:

$$t = 0.00002 \, \mu \, V_w c / kh \quad \text{for a skin-free well}$$

2. The linear flow plots should change from the initial straight line.
3. The slope of the plot of Δp versus Δt on log-log paper should be less than one half.

The time in hours when turbulence may end in a fractured well is given by the equation:

$$t = 40,000 \, \phi \mu c L_f / k$$

where L_f is ½ fracture length on both sides of well

It also has been observed that early flow in a fractured well gradually changes from a linear-elliptical to a rather radial flow configuration with time so that an effective well radius, r'_w, may be used in the equations. Review of mathematical and model studies indicates that the radius of drainage obtained from the radius of an investigation equation using T_{De} of 0.25 should equal two times the fracture radius, $r'_w = r_w e^{-s}$, before reliable radial flow information is obtainable. The use of r'_w replaces the skin term, or s is omitted when r'_w is used in flow equations.

At least four plots of the pressure history are desired when the radial plot, Horner and MDH, suggest an unusual character. The log-log or curve match data requires a reliable starting relation between time and pressure. The spherical plot is relatively independent of flow time used. Pay thickness is absent in the spherical flow equation.

The respective flow and buildup time scales are shown below:

Plot type or flow configuration	Pressure	Time scales to be used in plots of pressure versus time			Where:
		Flow test	Buildup test		
			Horner	MDH	
Linear flow	p	$\sqrt{t}$	$\sqrt{(t+\Delta t)} - \sqrt{\Delta t}$	$\sqrt{\Delta t}$	t = flow time
Spherical flow	p	$1/\sqrt{t}$	$1/\sqrt{\Delta t} - 1/\sqrt{t+\Delta t}$	$1/\sqrt{\Delta t}$	Δt = shut-in time
Radial flow	p	$\log t$	$\log (1 + t/\Delta t)$	$\log \Delta t$	
Curve match (log-log plot)	$\log \Delta p$	$\log t$	$\log (1 + t/\Delta t)$	$\log \Delta t$	

In this equation, t is the afterflow time in hours, V_w is the wellbore volume in barrels, and c is the compressibility of the wellbore fluids in psi^{-1}. If two or more fluids are flowing, the total mobility $(k/\mu)_t$ should be used where k is in md and h is in feet.

Other procedures used to determine after flow include:

1. The plot of BHP–THP or BHP–CHP versus time should become rather flat.

The linear flow configuration dominates during the early flow period when fractures are present. Fractures in lenses can result in difficult to analyze shifts in curves. The flow gradually shifts to a spherical configuration, particularly when the interval is partly penetrated or when only part of the pay is perforated. The characteristics of spherical flow are noted on most plots, even when spherical flow should not be expected and the radial flow equations should not be used until a time after the severe curvature of the spherical plot. The curve match or log-log plot shows a slope of one during the afterflow period, and a slope of one-half suggests linear flow. The severe curvature of the spherical flow plot usually occurs after the log-log plot has a slope of less than one-half. The part of the radial flow plot used in the equations should be a straight line. Accurate starting relations are required for the log-log plots.

At late times, the radial flow plot may be distorted by boundaries. Transient flow on the Horner build-up plot probably ends before t_{De} has a value of 0.1 or $\Delta t = \phi \mu c_t r_e^2 / 0.00264\ k$. The end point for the MDH plot is less, around $t_{De} = 0.04$. The end point for the flow tests is less than those for the Horner build-up test, probably about that of the MDH build-up plot.

Very tight and very permeable reservoirs also require special analysis. In very tight pays, the radius

investigated by long-time tests may be only a short distance—say 100 feet—from the wellbore. If the pay in addition is variable in quality in either horizontal or vertical direction, both flow and build-up data can be distorted. There may also be interference between the flow transients and the build-up transients created by the test well. Careful study of the data at times can create confidence in such test results, even when wells have been fractured through multizones and lenses. When the permeability is very large and the test results are obtainable only because flow rates are very high, time to reach boundaries can be very short. Also, the slopes of the Horner plots increase as the flow time used becomes less when values for flow time are less than about 500 hours. The well quickly reaches semisteady-state conditions—often in less than 15 hours—so that the use of flow times in excess of 500 hours do not realistically represent field condi-

Legend:
Time obtained from past production/last production   13,630 hours
Time obtained from investigating drainage of 160 acres   146 hours
Time required to reach end of transient flow   59 hours

Curve	Flow hours	Slope, psi/cycle	kh md ft	p* psig	p̄, psig			k md
					Horner,	Miller et al.,	Timmerman	
A	13,630	70	528	4,585	4,419	4,417–4,461	4,420	7.65
B	146	77	480	4,457			4,420	6.95
C	59	84	440	4,435			4,420	6.38

Data for Curve A was obtained from Appendix B, Example 1 and Appendix C, SPE Monograph 1, Henry Doherty Series. Other curves use identical data except for production

time, t. Use of psig rather than psia does not affect curves, etc. The slope increases as flow time is decreased.

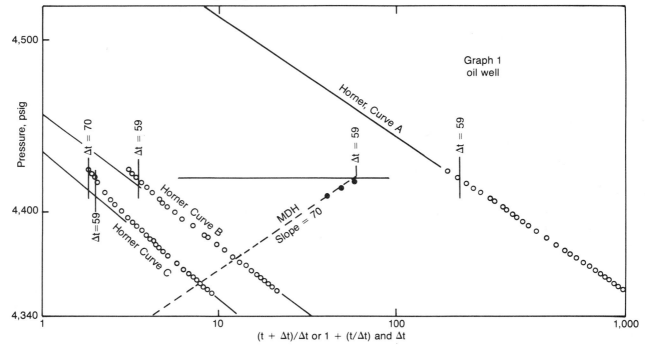

Figure 5–8. *Horner and MDH plots.*

tions. A suggested technique involves the plot of MDH and the plot of Horner using various flow times less than 500 hours. The Horner slopes are read for each assumed flow time and permeability and the times to reach the end of transient flow are calculated. The curve that relates with the time to the end of transient flow would appear realistically to fit the field data. Under these conditions, the Horner curves at later times of the test curve toward the final pressure, and the Horner slope used in *kh* calculations should not include such curvature. The MDH calculations usually seem to give a slope less than that of the suggested useable Horner plot. The average pressure can be read from the Horner plots at the time required to reach the end of transient flow, as shown in Fig. 5–8.

Often the average reservoir pressure and skin are known with some reliability. The Horner plots when flow time is based on time to reach the end of the transient flow boundary should approximate the average reservoir pressure when the Horner time scale has a value between 1.0 and 2.0, depending upon the actual drainage radius. At high drainage value, a time scale value of one is indicated. Likewise, the calculated skin should be realistic when the chosen slopes are reasonable approximations of true values. The permeability should be determined independently when using the linear equations.

Sound analysis, experience, and hard work often can result in the selection of the most probable values for permeability and skin. In some cases, a limited range of values must be accepted.

Design of Flow and Build-up Tests

Tests must be executed safely and without damage to the environment. Employees must be protected; fires, escape of hydrocarbons to the air, and contact of salt water and oil with the ground are not permitted. The produced fluids must be measured at the surface. Flow at high rates can cause problems due to temperature changes in equipment and freezing of chokes. Produced sand can cut out and damage equipment. Time and flow rate must be accurately measured at the surface, and subsurface pressure measurements must be accurately related with surface time. As much data as possible should be collected during the test, and all data relative to the test and well performance should be made available to the analyst. The surface arrangement of facilities used to test a high pressure gas well is shown below:

Test design should include a set of calculations similar to those made during analysis of results, using likely assumed values so that flow rates and time are chosen properly. Such somewhat lengthy and detailed calculations may be substituted by analogy if prior tests have been successfully made and analyzed for the well and reservoir conditions. One simply must be certain that test results can be adequately analyzed before spending the time and money, including risks, in the field.

Cooperation between the drilling, production and engineering groups is essential.

Gas flow rates must be measured. Tables have been published by manufacturers of flow equipment for various size flow orifices. These flow rates must be corrected when gas gravity and temperature differ from conditions specified for each set of tables.

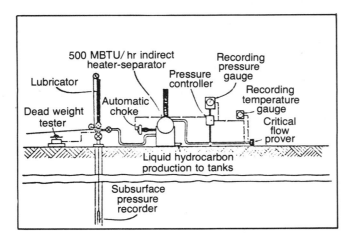

TABLE 5–2
Flowing Temperature Factors*

$$Factor = \sqrt{\frac{520}{460° + T}}$$

Temp. °F	Factor	Temp. °F	Factor	Temp. °F	Factor	Temp. °F	Factor
20	1.0408	65	0.9952	110	0.9551	155	0.9195
21	1.0398	66	0.9943	111	0.9543	156	0.9187
22	1.0387	67	0.9933	112	0.9534	157	0.9180
23	1.0376	68	0.9924	113	0.9526	158	0.9173
24	1.0365	69	0.9915	114	0.9518	159	0.9165
25	1.0355	70	0.9905	115	0.9510	160	0.9158
26	1.0344	71	0.9896	116	0.9501	161	0.9150
27	1.0333	72	0.9887	117	0.9493	162	0.9143
28	1.0323	73	0.9877	118	0.9485	163	0.9135
29	1.0312	74	0.9868	119	0.9477	164	0.9128
30	1.0302	75	0.9859	120	0.9469	165	0.9121
31	1.0291	76	0.9850	121	0.9460	166	0.9112
32	1.0281	77	0.9840	122	0.9452	167	0.9106
33	1.0270	78	0.9831	123	0.9444	168	0.9099
34	1.0260	79	0.9822	124	0.9436	169	0.9092
35	1.0249	80	0.9813	125	0.9428	170	0.9085
36	1.0239	81	0.9804	126	0.9420	171	0.9077
37	1.0229	82	0.9795	127	0.9412	172	0.9069
38	1.0219	83	0.9786	128	0.9404	173	0.9063
39	1.0208	84	0.9777	129	0.9396	174	0.9055
40	1.0198	85	0.9768	130	0.9388	175	0.9048
41	1.0188	86	0.9759	131	0.9380	176	0.9042
42	1.0178	87	0.9750	132	0.9372	177	0.9035
43	1.0168	88	0.9741	133	0.9364	178	0.9028
44	1.0157	89	0.9732	134	0.9356	179	0.9020
45	1.0147	90	0.9723	135	0.9348	180	0.9014
46	1.0137	91	0.9715	136	0.9341	181	0.9007
47	1.0127	92	0.9706	137	0.9333	182	0.9000
48	1.0117	93	0.9697	138	0.9325	183	0.8992
49	1.0107	94	0.9688	139	0.9317	184	0.8985
50	1.0098	95	0.9680	140	0.9309	185	0.8979
51	1.0088	96	0.9671	141	0.9301	186	0.8972
52	1.0078	97	0.9662	142	0.9293	187	0.8965
53	1.0068	98	0.9653	143	0.9284	188	0.8958
54	1.0058	99	0.9645	144	0.9279	189	0.8951
55	1.0048	100	0.9636	145	0.9271	190	0.8944
56	1.0039	101	0.9627	146	0.9263	191	0.8937
57	1.0029	102	0.9618	147	0.9255	192	0.8931
58	1.0019	103	0.9810	148	0.9247	193	0.8923
59	1.0010	104	0.9602	149	0.9240	194	0.8916
60	1.0000	105	0.9592	150	0.9233	195	0.8910
61	0.9990	106	0.9585	151	0.9225	196	0.8903
62	0.9981	107	0.9576	152	0.9217	197	0.8896
63	0.9971	108	0.9568	153	0.9210	198	0.8889
64	0.9962	109	0.9559	154	0.9202	199	0.8882
						200	0.8876

* The following tables based on data from the American Meter Co.

TABLE 5-3
Specific Gravity Factors

$$Factor = \sqrt{\frac{0.60}{Sp.\ Gr.}}$$

Sp. Gr.	Factor	Sp. Gr.	Factor
0.500	1.0954	0.725	0.9097
0.505	1.0900	0.730	0.9066
0.510	1.0847	0.735	0.9035
0.515	1.0794	0.740	0.9005
0.520	1.0742	0.745	0.8974
0.525	1.0690	0.750	0.8944
0.530	1.0640	0.755	0.8914
0.535	1.0590	0.760	0.8885
0.540	1.0541	0.765	0.8856
0.545	1.0492	0.770	0.8827
0.550	1.0445	0.775	0.8793
0.555	1.0398	0.780	0.8771
0.560	1.0351	0.785	0.8743
0.565	1.0304	0.790	0.8715
0.570	1.0260	0.795	0.8687
0.575	1.0215	0.800	0.8660
0.580	1.0171	0.805	0.8635
0.585	1.0127	0.810	0.8607
0.590	1.0084	0.815	0.8580
0.595	1.0041	0.820	0.8554
0.600	1.0000	0.825	0.8528
0.605	0.9958	0.830	0.8502
0.610	0.9918	0.835	0.8476
0.615	0.9877	0.840	0.8452
0.620	0.9837	0.860	0.8353
0.625	0.9798	0.880	0.8257
0.630	0.9759	0.900	0.8165
0.635	0.9721	0.920	0.8076
0.640	0.9682	0.940	0.7989
0.645	0.9645	0.960	0.7906
0.650	0.9608	0.980	0.7825
0.655	0.9571	1.000	0.7746
0.660	0.9535	1.020	0.7669
0.665	0.9498	1.040	0.7595
0.670	0.9463	1.060	0.7523
0.675	0.9427	1.080	0.7453
0.680	0.9393	1.100	0.7385
0.685	0.9359	1.120	0.7319
0.690	0.9325	1.140	0.7255
0.695	0.9292	1.160	0.7192
0.700	0.9258	1.180	0.7131
0.705	0.9225	1.200	0.7071
0.710	0.9193	1.220	0.7013
0.715	0.9161	1.240	0.6956
0.720	0.9129	1.260	0.6901

TABLE 5-4
Four-Inch Well Tester Flow Rate, Mcfd Based on: 14.65 psia, 60°F, 0.60 SP. GR.

Pressure, inches of mercury	Orifice diameter, inches				
	1	1¼	1½	2	2½
1.0	89	140	204	377	638
1.1	94	147	214	396	669
1.2	98	154	224	414	700
1.3	102	160	233	430	728
1.4	106	166	242	447	756
1.5	109	172	250	463	783
1.6	113	178	259	479	810
1.7	117	183	267	493	835
1.8	120	189	275	508	859
1.9	124	194	282	523	884
2.0	127	199	290	536	907
2.2	133	209	304	563	952
2.4	139	219	318	588	995
2.6	145	228	331	613	1040
2.8	151	237	344	637	1080
3.0	156	245	357	660	1120
3.2	161	253	369	682	1150
3.4	167	262	380	704	1190
3.6	171	269	392	725	1230
3.8	176	277	403	746	1260
4.0	181	285	414	766	1300
4.5	193	303	440	814	1380
5.0	203	320	465	860	1450
5.5	214	336	489	904	1530
6.0	224	352	512	947	1600
6.5	234	367	534	988	1670
7.0	243	382	555	1030	1740
8	261	410	597	1100	1870
9	278	437	636	1180	1990
10	295	463	673	1250	2110
11	310	488	709	1310	2220
12	326	512	745	1380	2330
13	341	535	778	1440	2440
14	355	558	811	1500	2540
15	369	580	843	1560	2640
16	383	602	875	1620	2740
17	396	623	906	1680	2840
18	410	644	936	1730	2930
19	423	664	966	1790	3020
20	436	684	995	1840	3120
21	448	704	1020	1900	3210
22	461	724	1050	1950	3300
23	473	743	1080	2000	3380
24	485	762	1110	2050	3470
25	497	781	1140	2100	3560
26	509	800	1160	2150	3640
27	521	819	1190	2200	3730
28	533	837	1220	2250	3810
29	544	855	1240	2300	3890
30	556	873	1270	2350	3980

TABLE 5-4 *(continued)*
Four-Inch Well Tester Flow Rate, Mcfd
Based on 14.65 psia, 60°F, 0.60 SP. GR.

Pressure, inches of mercury	Orifice diameter, inches				
	1	1¼	1½	2	2½
31	567	891	1300	2400	4060
32	578	909	1320	2450	4140
33	590	926	1350	2490	4220
34	601	944	1370	2540	4300
35	612	961	1400	2590	4380
36	623	979	1420	2630	4450
37	634	996	1450	2680	4530
38	645	1010	1470	2730	4610
39	656	1030	1500	2770	4690
40	666	1050	1520	2820	4770
41	678	1068	1551	2873	4855
42	689	1085	1576	2921	4935
43	701	1103	1602	2969	5017
44	713	1121	1628	3018	5099
45	724	1139	1655	3066	5181
46	736	1157	1681	3115	5263
47	746	1174	1705	3160	5339
48	758	1192	1732	3209	5422
49	769	1210	1758	3258	5505
50	780	1227	1783	3304	5582
51	792	1246	1809	3353	5666
52	803	1263	1834	3399	5743
53	814	1280	1859	3445	5820
54	824	1297	1884	3491	5898
55	835	1314	1908	3537	5975
56	846	1331	1933	3583	6054
57	857	1348	1958	3629	6132
58	868	1365	1983	3675	6209
59	879	1382	2008	3722	6288
60	889	1398	2031	3764	6359
61	900	1415	2056	3810	6437
62	911	1433	2081	3857	6516
63	922	1450	2106	3903	6595
64	932	1466	2129	3946	6666
65	943	1483	2154	3992	6745

TABLE 5-4 *(continued)*
Four-Inch Well Tester Flow Rate, Mcfd
Based on 14.65 psia, 60°F, 0.60 SP. GR.

Pressure, inches of mercury	Orifice diameter, inches				
	1	1¼	1½	2	2½
66	953	1499	2177	4035	6817
67	964	1516	2202	4082	6896
68	974	1532	2225	4124	6967
69	985	1549	2250	4171	7047
70	995	1565	2273	4213	7118
71	1005	1581	2296	4255	7189
72	1016	1598	2322	4303	7270
73	1026	1614	2345	4345	7341
74	1037	1632	2370	4393	7422
75	1047	1647	2393	4435	7493
76	1059	1665	2419	4482	7573
77	1069	1681	2441	4525	7645
78	1080	1699	2467	4573	7726
79	1090	1714	2490	4615	7798
80	1101	1732	2516	4663	7879
81	1111	1748	2539	4706	7951
82	1121	1764	2562	4748	8022
83	1133	1782	2588	4796	8104
84	1143	1797	2611	4839	8175
85	1153	1813	2634	4481	8247
86	1163	1829	2657	4924	8319
87	1174	1847	2683	4972	8401
88	1184	1863	2706	5015	8473
89	1194	1879	2729	5057	8544
90	1205	1895	2752	5100	8617
91	1216	1913	2778	5149	8699
92	1226	1928	2801	5191	8770
93	1236	1944	2824	5234	8843
94	1247	1962	2851	5283	8925
95	1258	1978	2874	5326	8998
96	1268	1994	2897	5368	9070
97	1279	2012	2923	5417	9153
98	1290	2028	2946	5460	9225
99	1300	2044	2969	5503	9297
100	1311	2062	2996	5552	9380

TABLE 5-5
Two-Inch Orifice Well Tester Flow Rate, Mcfd (Based on: 14.65 psia, 60°F, 0.60 SP. GR.)

Pressure, inches of mercury	Orifice size, inches							
	⅛	¼	⅜	½	¾	1	1¼	1½
1.0	1.91	6.23	13.2	23.1	52	96	162	258
1.1	2.01	6.53	13.8	24.2	55	100	170	271
1.2	2.11	6.83	14.5	25.2	58	105	178	283
1.3	2.19	7.14	15.1	26.3	60	109	185	295
1.4	2.26	7.37	15.6	27.4	62	113	192	306

TABLE 5–5 *(continued)*
Two-Inch Orifice Well Tester Flow Rate, Mcfd (Based on: 14.65 psia, 60°F, 0.60 SP. GR.)

Pressure, inches of mercury	Orifice size, inches							
	1/8	1/4	3/8	1/2	3/4	1	1 1/4	1 1/2
1.5	2.36	7.67	16.3	28.4	64	117	199	317
1.6	2.42	7.89	16.7	29.5	67	121	206	328
1.7	2.50	8.11	17.2	30.5	69	125	212	338
1.8	2.59	8.41	17.9	31.6	71	129	219	348
1.9	2.65	8.63	18.3	32.6	73	132	225	358
2.0	2.73	8.86	18.8	33.7	75	136	231	367
2.2	2.86	9.31	19.8	34.7	78	143	242	386
2.4	2.98	9.69	20.5	35.8	82	149	253	403
2.6	3.12	10.1	21.5	37.9	85	155	264	420
2.8	3.23	10.5	22.3	38.9	89	161	274	446
3.0	3.35	10.9	23.1	40.0	92	167	284	451
3.2	3.47	11.3	23.9	43.1	95	173	293	467
3.4	3.58	11.6	24.7	44.2	98	179	303	482
3.6	3.68	11.9	25.3	45.2	101	184	312	497
3.8	3.79	12.3	26.2	46.3	104	189	321	510
4.0	3.89	12.6	26.8	47.4	107	194	330	524
4.5	4.14	13.4	28.5	50.5	113	206	350	558
5.0	4.37	14.2	30.1	52.6	120	218	370	589
5.5	4.60	14.9	31.7	56.8	126	229	389	619
6.0	4.80	15.6	33.2	59.0	132	240	407	648
6.5	5.01	16.3	34.6	61.0	137	251	425	676
7.0	5.22	17.0	36.0	63.5	143	261	442	704
8	5.59	18.3	38.6	68.1	154	280	475	756
9	5.95	19.3	41.2	72.8	164	298	506	806
10	6.33	20.5	43.8	77.4	173	316	536	854
11	6.65	21.6	45.9	80.9	183	333	565	899
12	7.00	22.7	48.3	85.5	192	349	593	944
13	7.29	23.6	50.4	89.0	200	365	620	986
14	7.62	24.8	52.7	93.6	209	381	646	1025
15	7.93	25.8	54.7	97.0	217	396	672	1068
16	8.22	26.8	56.8	100.0	225	411	691	1110
17	8.49	27.6	58.6	104	233	425	721	1148
18	8.80	28.7	60.8	107	241	439	746	1185
19	9.08	29.5	62.7	110	249	453	769	1223
20	9.36	30.5	64.6	115	256	467	793	1262
21.	9.60	31.2	66.4	118	264	481	816	1296
22.	9.89	32.1	68.4	120	271	494	838	1333
23.	10.1	32.9	70.0	123	278	507	861	1370
24.	10.4	33.9	71.9	127	285	520	883	1404
25.	10.6	34.6	73.6	131	292	533	905	1440
26.	10.9	35.5	75.4	133	300	546	927	1474
27.	11.2	36.4	77.2	136	306	559	948	1508
28.	11.5	37.2	78.9	140	313	571	969	1540
29.	11.7	37.9	80.5	142	320	583	990	1576
30.	11.9	38.7	82.2	145	327	596	1011	1608
31.	12.2	39.6	84.0	150	334	608	1032	1640
32.	12.4	40.5	86.0	152	340	620	1052	1673
33.	12.6	41.2	87.4	155	347	632	1073	1709
34.	12.9	42.0	89.0	157	353	644	1093	1740
35.	13.2	42.7	90.6	160	360	656	1113	1771

TABLE 5-5 *(continued)*
Two-Inch Orifice Well Tester Flow Rate, Mcfd (Based on: 14.65 psia, 60°F, 0.60 SP. GR.)

Pressure, inches of mercury	Orifice size, inches							
	⅛	¼	⅜	½	¾	1	1¼	1½
36.	13.4	43.5	92.2	164	366	668	1133	1802
37.	13.6	44.2	93.9	167	373	679	1153	1835
38.	13.8	45.0	95.5	169	379	691	1173	1865
39.	14.1	45.7	97.0	172	386	703	1193	1896
40.	14.3	46.4	98.7	175	392	714	1212	1928
41.	14.7	47.5	101	178	400	729	1240	1957
42.	14.9	48.2	102	180	406	740	1250	1990
43.	15.1	48.9	104	184	413	752	1270	2019
44.	15.3	49.6	105	187	418	763	1300	2052
45.	15.5	50.3	107	189	426	775	1320	2080
46.	15.7	51.2	108	191	431	787	1337	2112
47.	15.9	51.9	109	193	437	798	1348	2145
48.	16.3	52.6	112	198	442	809	1369	2172
49.	16.5	53.3	113	200	450	819	1390	2203
50.	16.7	54.1	115	202	455	830	1411	2230
51.	16.9	54.7	116	205	461	841	1422	2260
52.	17.1	55.4	118	207	466	853	1441	2293
53.	17.3	56.3	119	211	473	865	1463	2320
54.	17.6	57.0	121	213	480	877	1485	2350
55.	17.8	57.7	122	216	487	887	1506	2380
56.	18.0	58.3	124	219	492	897	1527	2410
57.	18.2	59.0	125	222	497	907	1538	2440
58.	18.4	59.7	126	224	504	918	1559	2470
59.	18.6	60.4	129	226	509	929	1578	2500
60.	18.8	61.1	130	229	515	940	1599	2530
61	19.0	61.8	132	231	521	950	1610	
62	19.2	62.5	133	235	526	962	1632	
63	19.4	63.1	134	237	533	971	1653	
64	19.6	63.9	136	240	538	985	1664	
65	19.9	64.5	137	242	545	993	1686	
66	20.1	65.3	139	245	552	1000	1706	
67	20.3	66.0	140	247	557	1010	1727	
68	20.5	66.7	141	250	563	1020	1738	
69	20.7	67.4	143	253	568	1040	1759	
70	20.9	68.0	144	255	573	1050	1780	
71	21.1	68.8	146	259	579	1055	1790	
72	21.3	69.4	148	261	585	1062	1812	
73	21.6	70.1	149	263	591	1072	1832	
74	21.8	70.7	151	266	595	1083	1843	
75	22.0	71.6	152	269	603	1094	1864	
76	22.2	72.2	154	272	609	1104	1885	
77	22.4	72.8	155	274	615	1115	1907	
78	22.6	73.4	156	276	619	1126	1917	
79	22.8	74.1	158	278	624	1136	1939	
80	23.0	75.0	159	280	632	1147	1959	
81	23.3	75.5	160	284	636	1157	1970	
82	23.4	76.0	162	286	642	1168	1991	
83	23.7	77.0	163	289	650	1179	2011	
84	23.9	77.5	165	291	654	1188	2025	
85	24.0	78.1	167	294	658	1198	2045	

TABLE 5–5 *(continued)*
Two-Inch Orifice Well Tester Flow Rate, Mcfd (Based on: 14.65 psia, 60°F, 0.60 SP. GR.)

Pressure, inches of mercury	Orifice size, inches							
	⅛	¼	⅜	½	¾	1	1¼	1½
86	24.3	79.0	168	297	665	1210	2065	
87	24.4	79.6	169	299	670	1220	2073	
88	24.6	80.2	171	302	676	1230	2093	
89	24.9	81.0	172	305	683	1241	2116	
90	25.1	81.4	173	307	687	1251	2127	
91	25.3	82.3	175	309	693	1262	2148	
92	25.5	82.9	176	312	700	1273	2170	
93	25.7	83.5	177	313	704	1284	2180	
94	25.9	84.3	179	316	710	1294	2200	
95	26.1	84.9	180	318	715	1305	2222	
96	26.3	85.7	183	321	722	1314	2232	
97	26.5	86.3	184	324	727	1325	2255	
98	26.8	87.0	186	326	734	1335	2275	
99	27.0	87.5	187	329	739	1346	2286	
100	27.2	88.3	188	331	744	1357	2306	

Density of mercury is 13.546; Water is one. Fresh water head equals .435 psi/foot.

TABLE 5–6
Three-Inch Orifice Well Tester Flow Rate, Mcfd (Based on: 14.65 psia, 60°F, 0.60 SP. GR.)

Pressure, inches of mercury	Orifice diameter, inches				
	¾	1	1¼	1½	2
1.0	50	90	143	211	419
1.1	53	94	150	222	440
1.2	55	99	157	232	460
1.3	57	103	163	241	479
1.4	59	106	169	251	497
1.5	62	110	175	259	515
1.6	64	114	181	268	533
1.7	66	118	187	277	549
1.8	68	121	192	285	565
1.9	70	124	198	293	581
2.0	71	128	203	301	597
2.2	75	134	213	315	626
2.4	78	140	223	330	655
2.6	82	146	232	344	682
2.8	85	152	241	357	709
3.0	88	157	250	370	734
3.2	91	162	258	382	759
3.4	94	168	266	395	783
3.6	96	173	274	406	807
3.8	99	178	282	418	830
4.0	102	182	290	429	852
4.5	108	194	308	458	906
5.0	114	205	325	482	957
5.5	120	215	342	507	1010
6.0	126	226	358	531	1050

TABLE 5–6
Three-Inch Orifice Well Tester Flow Rate, Mcfd (Based on: 14.65 psia, 60°F, 0.60 SP. GR.)

Pressure, inches of mercury	Orifice diameter, inches				
	¾	1	1¼	1½	2
6.5	131	235	374	554	1100
7.0	137	245	389	576	1140
8	147	263	418	619	1230
9	157	280	445	660	1310
10	166	297	471	698	1390
11	175	313	497	736	1460
12	183	328	521	772	1530
13	192	343	545	807	1600
14	200	358	568	841	1670
15	208	372	590	875	1740
16	215	386	613	908	1800
17	223	399	634	940	1870
18	230	413	656	971	1930
19	238	426	676	1010	1990
20	245	439	697	1030	2050
21	252	452	717	1060	2110
22	259	464	737	1100	2170
23	266	476	757	1120	2230
24	273	489	776	1150	2280
25	280	501	796	1180	2340
26	286	513	815	1210	2400
27	293	525	834	1230	2450
28	300	537	852	1260	2510
29	306	548	871	1290	2560
30	313	560	889	1320	2610

TABLE 5–6 *(continued)*
Three-Inch Orifice Well Tester Flow Rate, Mcfd (Based on: 14.65 psia, 60°F, 0.60 SP. GR.)

Pressure, inches of mercury	Orifice diameter, inches				
	¾	1	1¼	1½	2
31	319	571	907	1340	2670
32	325	583	925	1370	2720
33	332	594	943	1400	2770
34	338	605	961	1420	2830
35	344	616	979	1450	2880
36	350	627	996	1480	2930
37	356	638	1010	1500	2980
38	363	650	1030	1530	3030
39	369	660	1050	1550	3080
40	375	671	1070	1580	3130
41	383	686	1090	1615	3188
42	389	698	1108	1641	3240
43	396	709	1126	1668	3294
44	403	721	1145	1696	3348
45	409	733	1163	1723	3402
46	416	744	1181	1750	3456
47	422	755	1199	1776	3506
48	428	767	1217	1803	3560
49	435	778	1236	1830	3614
50	441	789	1253	1856	3665
51	447	801	1272	1884	3721
52	453	812	1289	1910	3771
53	460	823	1307	1936	3822
54	466	834	1324	1962	3873
55	472	845	1341	1987	3924
56	478	856	1359	2013	3975
57	484	867	1376	2039	4026
58	490	878	1394	2065	4077
59	497	889	1412	2091	4129
60	502	899	1427	2115	4176
61	508	910	1445	2141	4227
62	515	921	1463	2167	4279
63	521	933	1480	2193	4330
64	526	943	1497	2217	4378
65	532	954	1514	2243	4429

TABLE 5–6 *(continued)*
Three-Inch Orifice Well Tester Flow Rate, Mcfd (Based on: 14.65 psia, 60°F, 0.60 SP. GR.)

Pressure, inches of mercury	Orifice diameter, inches				
	¾	1	1¼	1½	2
66	538	964	1530	2267	4476
67	544	975	1548	2293	4529
68	550	985	1564	2317	4575
69	556	996	1582	2343	4627
70	562	1007	1598	2367	4674
71	568	1017	1614	2391	4721
72	574	1028	1632	2418	4774
73	580	1038	1648	2441	4821
74	586	1049	1666	2468	4874
75	592	1059	1682	2492	4920
76	598	1071	1700	2518	4973
77	604	1081	1716	2542	5020
78	610	1092	1734	2569	5073
79	616	1103	1750	2593	5121
80	622	1114	1769	2620	5174
81	628	1124	1785	2644	5221
82	634	1134	1801	2667	5268
83	640	1146	1819	2695	5321
84	646	1156	1835	2719	5369
85	651	1166	1851	2743	5416
86	657	1176	1867	2766	5463
87	663	1188	1886	2794	5517
88	669	1198	1902	2818	5564
89	675	1208	1918	2842	5611
90	680	1218	1934	2866	5659
91	687	1230	1953	2893	5712
92	693	1240	1969	2917	5760
93	699	1251	1985	2941	5807
94	705	1262	2004	2968	5861
95	711	1272	2020	2992	5909
96	716	1283	2036	3016	5960
97	723	1294	2055	3044	6010
98	728	1304	2070	3068	6058
99	734	1315	2087	3092	6105
100	741	1326	2106	3120	6160

* Actual Pipe Diameter, 3.068 inches.

Flow Time for Build-up Testing*

The relations between the respective time scales used in flow and build-up test analysis are shown below for flow times of 15, 50, and 500 hours.

Radial Flow Configuration									
Shutin Time, Δt, hrs	0.25	1.25	2.50	5.0	10.00	18.00	24.00	5 to 18	10 to 24
$\Delta (\log \Delta t)$	0	0.699	1.000	1.301	1.602	1.857	1.982	0.556	0.380
$\Delta (\log (15/\Delta t + 1))$	0	0.671	.940	1.183	1.387	1.523	1.573	0.340	0.186
$\Delta (\log (50/\Delta t + 1))$	0	0.690	.981	1.262	1.525	1.726	1.814	0.464	0.289
$\Delta (\log (500/\Delta t + 1))$	0	0.698	.998	1.297	1.593	1.842	1.962	0.545	0.369

Linear Flow Configuration							
$\Delta (\sqrt{\Delta t})$	0	0.618	1.081	1.736	2.662	3.743	4.399
$\Delta (\sqrt{15 + \Delta t} - \sqrt{\Delta t})$	0	0.500	0.81	1.180	1.57	1.91	2.07
$\Delta (\sqrt{50 + \Delta t} - \sqrt{\Delta t})$	0	0.550	0.92	1.410	2.01	2.59	2.89
$\Delta (\sqrt{500 + \Delta t} - \sqrt{\Delta t})$	0	0.600	1.03	1.630	2.45	3.35	3.88

Spherical Flow Configuration							
$\Delta (1/\sqrt{\Delta t})$	0	1.106	1.368	1.533	1.684	1.764	1.796
$\Delta (1/\sqrt{\Delta t} - 1/\sqrt{15 + \Delta t})$	0	1.110	1.35	1.51	1.62	1.67	1.70
$\Delta (1/\sqrt{\Delta t} - 1/\sqrt{50 + \Delta t})$	0	1.105	1.365	1.547	1.672	1.744	1.771
$\Delta (1/\sqrt{\Delta t} - 1/\sqrt{500 + \Delta t})$	0	1.106	1.368	1.553	1.683	1.763	1.795

Remarks: The proper flow time should approximate the time required to reach the drainage boundary using the radius of investigation equation $t = \phi \mu c_t r_e^2/0.00088k$, which is applicable when flow is circular or square.

The value Δt may replace $(t + \Delta t)/\Delta t$ during a limited early period when flow time has high value, say 500 hours.

The slope of the radial flow Horner plot becomes independent of flow time after t exceeds 500 hours. $\sqrt{\Delta t}$ may replace $\sqrt{t + \Delta t} + \sqrt{\Delta t}$ only at very very early shut-in time.

$1/\sqrt{\Delta t}$ may replace $1/\sqrt{\Delta t} - 1/\sqrt{t + \Delta t}$ during the first 24 hours of shut-in time when flow time exceeds 15 hours.

The various time scales need not be calculated if the measured pressures are placed in following table for the respective times indicated.

The correct flow time is needed for linear, spherical and radial plots. Various techniques were used to determine flow time when Shell, during the late 1940s, began using flow and build-up tests to calculate permeability and skin.

In 1951, Horner suggested that, in absence of reliable flow time starting with stabilized conditions, the flow time be approximated by dividing cumulative production by a constant production rate measured during about one week prior to shutin for build-up test. This was a reasonable method since most wells

* This section is taken from "Continuous Tables," *Petroleum Engineer.*

in the U.S. in those days were being flowed at low rate. The build-up was seldom being used to analyze good reservoirs having permeability above 500 md but instead was used to determine problems in poor wells having permeability in the 10 to 100 md range.

Time to reach steady state drainage boundary is long when permeability is small and all values calculated by the Horner method usually were based on flow time in excess of 500 hours. For these conditions, the Horner method is a reasonable approximation. Unfortunately, the Horner method becomes unreliable when permeability is large and time required to reach drainage boundary is small.

Miller, Dyes, and Hutchinson in 1950 showed that for the case of no influx at the boundary, true transient circular flow ends before $t_D = 0.000264kt/\phi \mu c_t r_e^2 = 0.1$. Also, semisteady state flow starts when $t_D = 0.2$ to 0.3. Other authors have selected somewhat different values in this general range.

A well producing at semisteady state is draining an area defined by the relative production rates of the wells located within the affected area. Use of a flow time in excess of the time required to reach the drainage boundary is not sound. Kazemi, at least in part, recognized this fact in 1974. The correct flow time for use in Horner buildup plot is the time to reach the drainage boundary, and this time can be approximated by use of equations $t = \phi \mu c \, r_e^2/0.001056k$ to $t = \phi \mu c r_e^2/0.00088k$ which are based on $t_D = 0.25$ and 0.3, respectively. The MDH and Horner techniques both require data limited to transient build-up period. Approximate values for the desired flow time can be read for a typical oil reservoir from Fig. 5–9. When permeability is above 500 md

Well _____ Flow Rate _____
Test Date _____ P_{wf} _____ , Datum _____

Time scales for flow and build-up.

$\Delta(m(p)/q$	q	p_i-p_{wf}	m(p)	Pressure	Δt	$\sqrt{\Delta t}$	$1/\sqrt{\Delta t}$	Radial $(t+\Delta t)/\Delta t$			Linear $\sqrt{t+\Delta t}-\sqrt{\Delta t}$			Spherical $1/\sqrt{\Delta t}-1/\sqrt{t+\Delta t}$		
								t=15	t=50	t=500	t=15	t=50	t=500	t=15	t=50	t=500
					0.25	0.500	2.000	61.00	201.00	2001.00	3.41	6.59	21.87	1.74	1.859	1.955
					0.50	0.707	1.414	31.00	101.00	1001.00	3.23	6.40	21.67	1.16	1.273	1.369
					0.75	0.866	1.154	21.00	67.67	667.00	3.10	6.26	21.51	0.90	1.014	1.109
					1.00	1.000	1.000	16.00	51.00	501.00	3.00	6.14	21.38	0.75	0.860	0.955
					1.25	1.118	0.894	13.00	41.00	401.00	2.91	6.04	21.27	0.63	0.724	0.859
					1.50	1.224	0.816	11.00	34.33	334.00	2.84	5.95	21.17	0.57	0.677	0.771
					1.75	1.323	0.756	9.50	29.57	286.00	2.77	5.87	21.08	0.52	0.617	0.711
					2.00	1.414	0.707	8.50	26.00	251.00	2.71	5.80	20.99	0.47	0.568	0.662
					2.50	1.581	0.632	7.00	21.00	201.00	2.60	5.67	20.84	0.39	0.494	0.587
					3.00	1.732	0.577	6.000	17.67	168.00	2.51	5.55	20.70	0.34	0.440	0.532
					3.50	1.871	0.534	5.29	15.28	143.00	2.43	5.44	20.57	0.30	0.397	0.489
					4.00	2.000	0.500	4.75	13.50	126.00	2.36	5.35	20.45	0.27	0.364	0.455
					4.50	2.121	0.471	4.33	12.11	112.00	2.30	5.26	20.34	0.24	0.336	0.426
					5.00	2.236	0.447	4.00	11.00	101.00	2.23	5.18	20.24	0.23	0.312	0.402
					6.00	2.449	0.408	3.50	9.33	84.33	2.13	5.03	20.05	0.19	0.274	0.363
					7.00	2.646	0.378	3.14	8.14	72.43	2.04	4.90	19.87	0.17	0.246	0.334
					8.00	2.828	0.354	2.88	7.25	63.50	1.97	4.79	19.71	0.14	0.223	0.310
					9.00	3.000	0.333	2.67	6.56	56.55	1.90	4.68	19.56	0.13	0.203	0.289
					10.00	3.162	0.316	2.50	6.00	51.00	1.84	4.58	19.42	0.12	0.187	0.272
					12.00	3.464	0.289	2.25	5.17	42.67	1.74	4.41	19.16	0.10	0.162	0.245
					14.00	3.742	0.267	2.07	4.57	36.71	1.64	4.26	18.93	0.08	0.142	0.233
					16.00	4.000	0.250	1.94	4.13	32.25	1.57	4.12	18.72	0.07	0.127	0.206
					18.00	4.243	0.236	1.83	3.78	28.78	1.50	4.00	18.52	0.07	0.115	0.192
					20.00	4.472	0.224	1.75	3.50	26.00	1.45	3.90	18.33	0.05	0.104	0.180
					22.00	4.690	0.213	1.68	3.27	23.73	1.39	3.79	18.16	0.05	0.095	0.169
					24.00	4.899	0.204	1.63	3.08	21.83	1.34	3.70	17.99	0.04	0.088	0.160
					30.00	5.477	0.183	1.50	2.67	17.67	1.23	3.47	17.55		0.071	0.140
					40.00	6.324	0.158	1.38	2.25	13.50	1.09	3.16	16.91		0.053	0.115
					50.00	7.071	0.141	1.30	2.00	11.00	0.99	2.93	16.38		0.041	0.098
					100.00	10.00	0.100	1.15	1.50	6.00	0.72	2.25	14.50		0.018	0.059
					250.00	15.81	0.063	1.06	1.20	3.00	0.47	1.51	11.58		0.005	0.026
					500.00	22.36	0.045	1.03	1.10	2.00	0.33	1.09	9.62		0.002	0.013
					1000.00	31.62	0.032	1.02	1.05	1.50	0.24	0.78	7.11		0.001	0.006

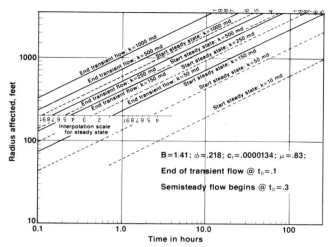

Figure 5–9. *Drainage area vs. time for assumed conditions* (*courtesy* Petroleum Engineer).

and spacing is 160 acres, the flow time required to reach the drainage boundary may approximate 10 hours. The time may approach 500 hours when k is 10 md.

Theoretical consideration indicates that log Δt may be substituted for log $((t/\Delta t) + 1)$ when flow time is large relative to shut-in time. This criteria often is not satisfied, however, when the proper flow time associated with the drainage boundary is required.

Data indicate that use of $\sqrt{\Delta t}$ for linear flow is not permissible unless the flow time to reach drainage boundary exceeds 500 hours. Use of $1/\sqrt{\Delta t}$ as the spherical flow plot is a reliable approximation for almost all flow times for build-up analysis purposes. Use of Δt in the radial flow plots is reliable during a short range of build-up time if correct flow time exceeds 500 hours. At lesser flow time, the MDH plot will suggest slopes that are below the correct value. The radial plot using Horner scales is independent of flow time after flow time exceeds 500 hours.

Selection of pressures compatible with time values permits the use of the time scales directly. Also, $p* = \bar{p}$, as was recognized by Kazemi, if t is correct.

References

Mathews and Russell, Pressure Buildup and Flow Tests in Wells. Monograph Volume 1, Society of Petroleum Engineers.

Kazemi, Determining Average Reservoir Pressure from Pressure Buildup Tests, SPE Transactions, Vol 255, 1974.

Miller, C. C., Dyes, A. B. and Hutchinson, C. A., Jr.: "The Estimation of Permeability and Reservoir Pressure from Bottom Hole Pressure Buildup Characteristics," Trans., AIME (1950) Vol. 189, 91–104.

Horner, D. R.: "Pressure Build-Up in Wells," Proc., Third World Pet. Cong. (1951) Sec. II, 503–523.

Shapes of Build-up Curves

Long-term build-up curves tend to bend downward when reservoirs are bounded, well interference is present, phase separation occurs, and mobility of fluids increases in the area influenced by the test. A decrease in porosity and a decrease in permeability (boundary when porosity approaches zero) also cause curves to bend downward as time increases. Typical examples are shown in Fig. 5–10.

Faults, partial boundaries, stratified layers without cross flow, lateral decrease in mobility, increases in porosity and permeability, lenses, irregular well locations or drainage areas, unconnected zones with widely differing pressure, and use of improper flow times will cause build-up curves to bend upward with time. Examples of build-up curves are illustrated in Fig. 5–11.

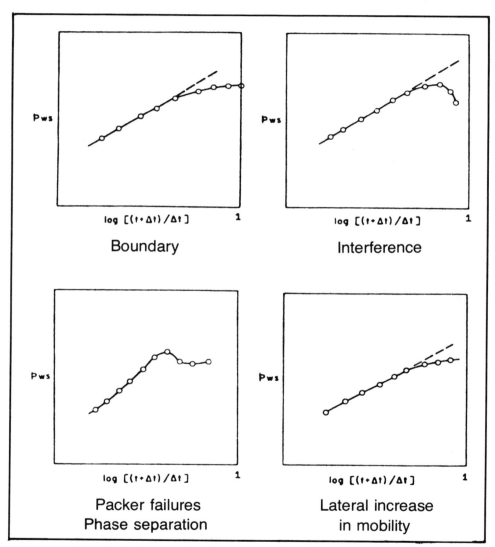

Figure 5–10. *Downtrending Horner plots (from SPE Monograph 1, 1968).*

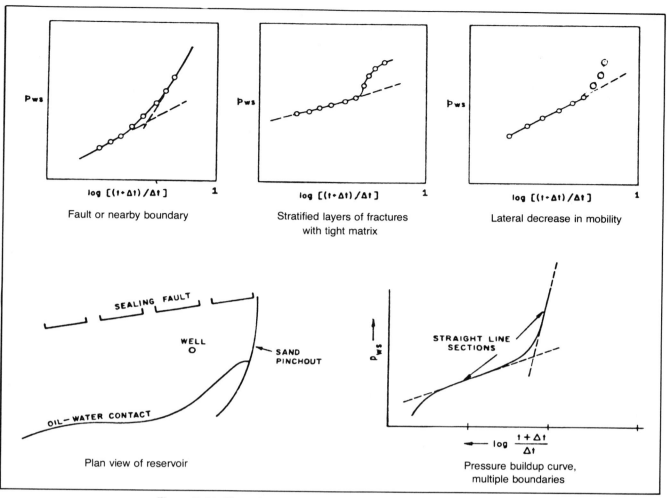

Figure 5–11. *Uptrending Horner curves (from SPE Monograph 1, 1967).*

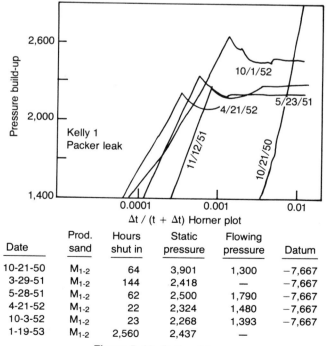

Date	Prod. sand	Hours shut in	Static pressure	Flowing pressure	Datum
10-21-50	M$_{1-2}$	64	3,901	1,300	−7,667
3-29-51	M$_{1-2}$	144	2,418	—	−7,667
5-28-51	M$_{1-2}$	62	2,500	1,790	−7,667
4-21-52	M$_{1-2}$	22	2,324	1,480	−7,667
10-3-52	M$_{1-2}$	23	2,268	1,393	−7,667
1-19-53	M$_{1-2}$	2,560	2,437	—	

Figure 5–12. *Packer failure.*

Parameters in Equations

The following figures should be studied to determine relative importance of each parameter appearing in the equations.

Varying porosity and connate water by 10% has little effect on the pressure (Fig. 5–13). A lower initial pressure relates with less gas in place and the pressure is lower but shape of curve is changed little. Temperature and gravity have little effect (Fig. 5–14).

A lower permeability is related with a lower cumulative recovery and rate of production at any given time (Fig. 5–15). A positive skin lowers the ability of the well to produce (Fig. 5–16).

Drainage area does not appear in the true transient flow period but does become important in late transient and semi-steady state flow (Fig. 5–17). Permeability appears twice in the transient flow equation and a lower permeability drops the pressure more quickly both because of the log term and because of kh, which is most important (Fig. 5–18).

Russell, Goodrich, Perry, and Bruskotter in *JPT* (January 1966) propose a method for averaging pressures when reservoirs are tight so that pressure drop in a reservoir is severe. Fig. 5–19 suggests that this method is useful. Ramey et al. in *JPT* (August 1968) agree with this conclusion when the pressure function

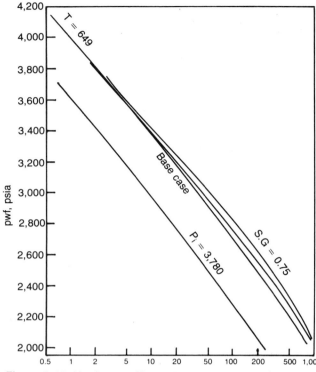

Figure 5–14. *Varying specific gravity, temperature, and initial pressure by 10% (courtesy Shell Oil Co.)*

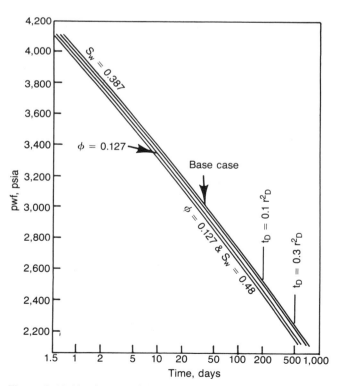

Figure 5–13. *Varying porosity and water saturation by 10% (courtesy Shell Oil Co.)*

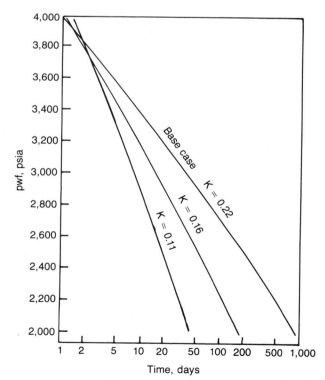

Figure 5–15. *Relation of lower permeability (courtesy Shell Oil Co.)*

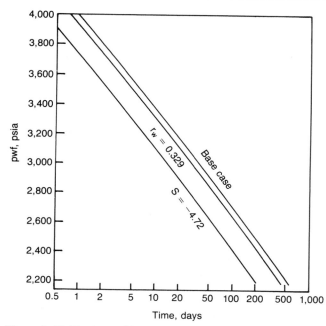

Figure 5–16. *Varying well-bore radius and skin factor by 10% (courtesy Shell Oil Co.)*

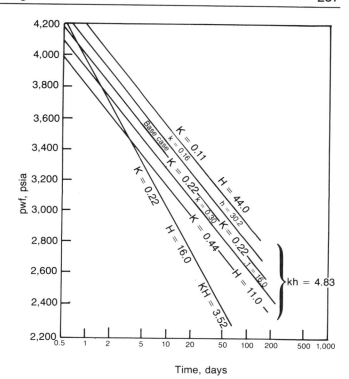

Figure 5–18. *Varying permeability and net pay to depict effect on slope (courtesy Shell Oil Co.)*

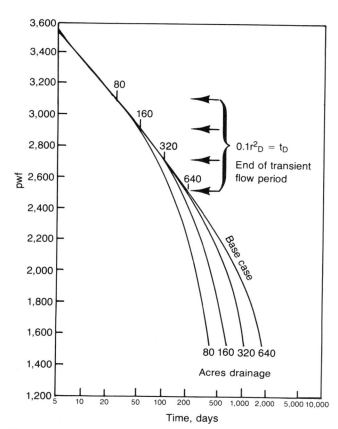

Figure 5–17. *Effect of varied drainage area (courtesy Shell Oil Co.)*

$m(p)$, does not change direction from uptrending to downtrending. Apparently, the $m(p)$ function is somewhat more reliable and should be used when available.

Wattenbarger and Ramey in *JPT* (August 1968) investigated the influence of various parameters using a dimensionless function $m_D(1, t_D)$ which is the equivalent to the p_D function for liquids.

$$m_D(1, t_D) = \tfrac{1}{2} \ln t_D + 0.4045$$

The slope of the semilog plots, MDH and Horner, is $2.303/2 = 1.151$ as in the liquid case. When the pressure drop within the formation is large, the $m(p)$ function or real gas considerations are important, and these curves use this concept. A constant rate at sand face is used throughout the study.

Fig. 5–20 shows that gas when properly expressed in the $m(p)$ function behaves as the liquid case for both injection and production. In Fig. 5–21, skins displace the curves as in the liquid case. The early slope reflects the damaged zone and the later slope represents chiefly the undamaged zone:

$$s = [(k - kd)/k_d]\ln(r_d/r_w)$$

Some correction for viscosity-compressibility may be required when well-bore storage is considered (Fig. 5–22):

$$m_D(1, t_D) = (1/\overline{C})t_D$$

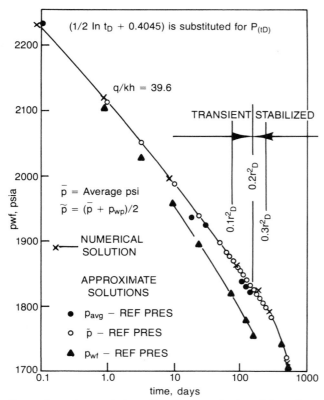

— Comparison of numerical results with approximate solutions for varied reference pressures during transient flow.

Figure 5–19. *Numerical results and approximate solutions for varied pressures during transient flow (courtesy Shell Oil Co.).*

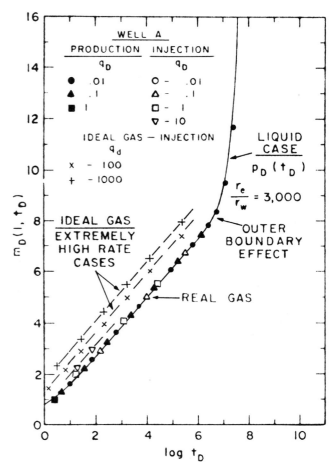

Figure 5–20. *Gas behaves as liquid for injection and production (after Wattenbarger & Ramey, JPT, August 1968, © SPE-AIME).*

Turbulence stabilizes after a rather short flow period or before $t_D = 2,000$ (Fig. 5–23). The data also show that the normal plot does not give the proper slope when turbulence is present in gas wells. The viscosity in the turbulent area differs from the usual value and causes the slope to increase with an increase in flow rate and turbulence in the zone near the well.

$$m_D(1, \ t_D) = (\tfrac{1}{2} \ln t_D + 0.4045) + \text{skin}$$
$$+ \ D_i(\mu_i/\mu_{1 \ \text{atm}})q_{sc}(1 - r_w/r_d)$$

where D_i represents D calculated at initial conditions and r_d is the drainage boundary. Apparently the turbulence stabilizes before $t_D = 2,000$, and turbulence depends on flow rate per unit thickness and is rather independent of k.

Build-up

Ramey et al. also presented Figs. 5–24 and 25 for build-up conditions. When the flow rate for a gas well

is relatively low, the early build-up data all have a slope of 1.151 and are the same as for the liquid case. These curves are for increasing flow times and differ because of the state of depletion of the reservoir. When the flow rate is relatively high, the slope is slightly on the high side, but kh values should be reasonably good. The curves in Fig. 5–26 for formation damage are similar to those for the liquid case. The $m(p)$ data can be used in MDH and Horner techniques. Curves for well-bore storage (Fig. 5–27) also are similar to those for the liquid case. For build-up, turbulence gradually dies out, as shown in Fig. 5–28. Turbulence can be analyzed during a test at two flow rates and may be represented as a skin caused by flow. The 0.75 usually appears in semisteady state equations for a closed circular reservoir since the radius of drainage, r_d, becomes fixed at 0.472 r_e. For earlier transient times, $r_d = 1.5 r_s \sqrt{t_D}$.

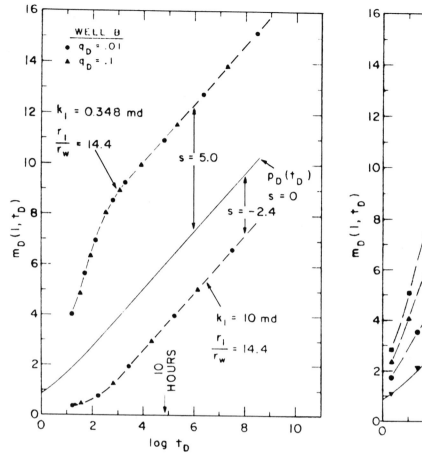

Figure 5–21. *Damaged and undamaged zones (after Wattenbarger and Ramey, JPT, August 1968).*

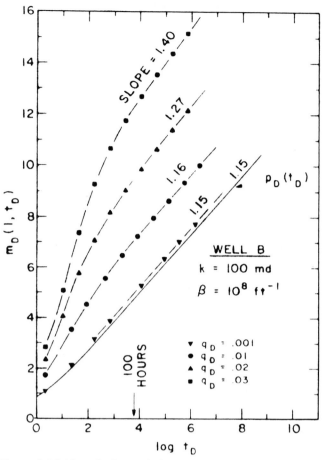

Figure 5–23 *Viscosity in a turbulent area (after Wattenbarger and Ramey, JPT, August 1968).*

All figures copyrighted by SPE-AIME.

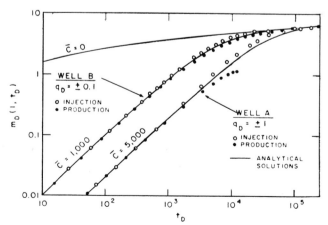

Figure 5–22. *Correction for viscosity/compressibility (after Wattenbarger and Ramey, JPT, August, 1968).*

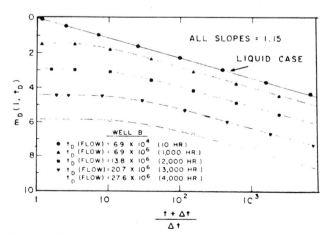

Figure 5–24. *Build-up conditions (after Ramey et al., JPT, August 1968).*

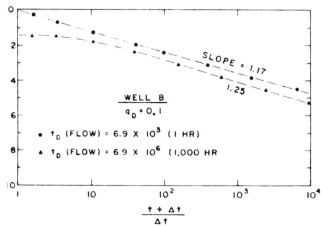

Figure 5–25. *Build-up conditions* (*after Ramey et al., JPT, August 1968*).

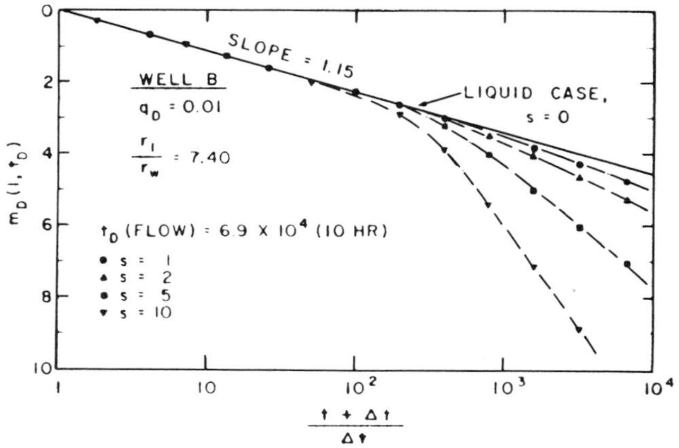

Figure 5–26. *Curve for formation damage* (*after Wattenbarger and Ramey, JPT, August 1968*).

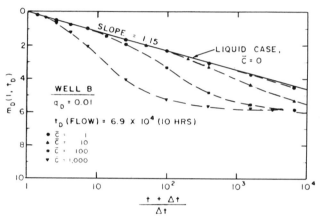

Figure 5–27. *Curve for well-bore storage* (*after Wattenbarger and Ramey, JPT, August 1968*).

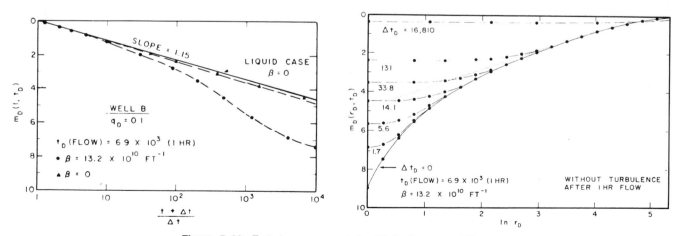

Figure 5–28. *Turbulence curves* (*after Wattenbarger and Ramey, JPT, August 1968*).

False Straight Lines

Plots for linear, spherical, radial and log-log conditions have been arbitrarily constructed using information that results in a straight line when pressure is plotted versus log time. These curves are the only true straight lines. Study of the remaining graphs shows many other apparent straight lines, which result from scales used and plotting errors. Also, at early time the slope of the log-log plot approaches ½. The effect of the early time relation between pressure and time is illustrated. The initial time and pressure must be reliable if the log-log plot is to be meaningful.

Van Everdingen and Hurst in 1949 showed that the dimensionless pressure $p_{tD} = (kh/141.2qB\mu)/(p_i - p_{wf})$ when plotted versus the log $t_D = 0.000264\ kt/\phi\mu c_t r_w^2$ results in a straight line when t_D has a value greater than 25–50 and a single well is producing

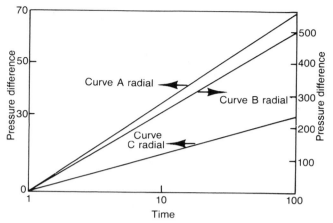

Figure 5–29. *Semilog plot. All curves are straight lines.*

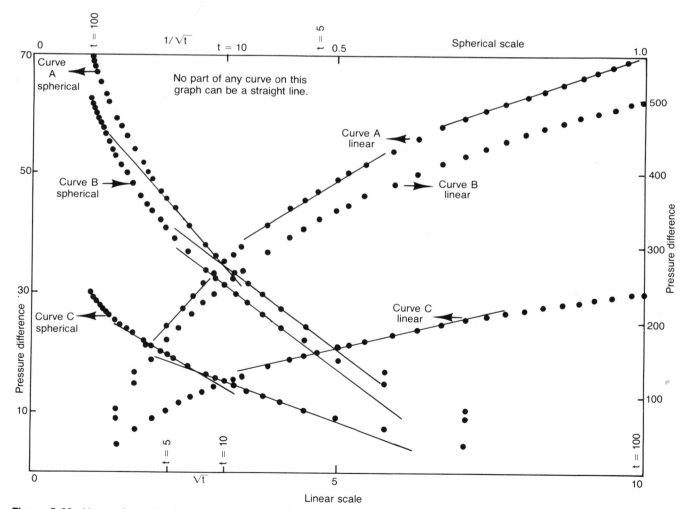

Figure 5–30. *Linear plot and spherical plot. Since semilog curves are all straight lines, no part of any curve on this graph can be a perfectly straight line.*

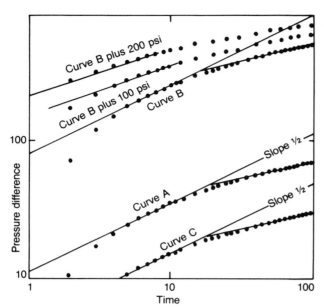

Figure 5–31. *Log-log plot. These curves may have slopes of one and one-half even when well is not fractured and no afterflow occurs. Errors in initial pressure difference-time relations influence location and the shape of these curves.*

from an infinite reservoir under rather ideal conditions. Such values are reached quickly when permeability is good. If permeability is poor and the well has been fractured so that r'_w is large, days and even years may be required for t_D to reach values permitting use of the log function. Graphs prepared by Russell and Truitt or Ramey et al. must then be used.

Use of $r'_w = r_w e^{-s}$, which equals ½ fracture radius, also can be useful. The dimensionless equations can be expressed in log terms, and plots of log t versus log $(p_i - p_{wf})$ have shapes comparable to that of the dimensionless plots when expressed as log-log plots. When the flow or build-up data can be used in the log format, t_D is greater than 50 when r'_w is used.

Raghavan-Ramey et al. have shown that the afterflow period on the log pressure difference versus log time (or Δt) plot is related with a slope of one, while the slope during linear flow (into the fracture) is ½. Radial flow starts about two cycles after the linear straightline ends or about when the radius of investigation = $r_{inv} = \sqrt{0.00105 kt/\phi\mu c_t}$ exceeds the fracture radius by about two (twice the fracture radius). Data

shown on the graphs on this page suggest that slopes of one and ½ are possible even when no afterflow and no fracture is present (Fig. 5–31).

Linear flow is associated with a straight line on a plot of measured pressure (p) versus $\sqrt{t}$ for flow ($\sqrt{t} + \Delta t - \sqrt{\Delta t}$ for build-up); spherical flow is indicated by a straight line when pressure is plotted versus $1/\sqrt{t}$ for flow ($1/\sqrt{t} - 1/\sqrt{t + \Delta t}$ for build-up); and radial flow is associated with a straight line on a plot of pressure versus log t for flow $[(t/\Delta t) + 1$ for build-up]. Δt may be used when t is very large relative to Δt (MDH–Muskat methods) in analysis of build-up tests.

The mathematical relations between curves is of importance. For examples shown here, pressure versus log t is a straight line and other curves are constructed from there. The starting pressure time relationship only shifts the curve but does not change the shape for linear, spherical, and radial curves. The log-log curve uses Δp rather than pressure and the shape of this curve is dependent upon its location on the paper. Thus, a constant must be added to all curves if Δp is not zero at time zero.

All curves should be plotted on scales having accuracy exceeding that of field data. Field data errors can indicate straight lines that are not real. The four sets of data suggest that only one curve can be a straight line at a single time, although some overlap is possible as the flow configuration gradually changes.

Interpretation of field-test data can be complicated because of measurement and plotting problems, which may result in indicated straight lines when none are actually present. Flow configuration must be adequately determined before the proper set of applicable equations can be selected. Curve-matching techniques as well as the linear, spherical, radial, and log-log plots are used for this purpose. The type and vertical location of fractures, presence of multiple pay zones, vertical versus horizontal permeability, crossflow, structural, and fluid barriers, and many other factors can influence the shape of the flow and build-up curves. As in all engineering, when basic data is sparse, all available information relative to the well and reservoir must be combined to obtain the best possible answer.

Complete Analysis of a Drawdown Test*

Transient Analysis

Fig. 5–32 is a plot of measured flowing bottom-hole pressure vs. log t. Pressure data from a flow time of 10 minutes onward during the transient period are linear on this plot. Deviation from the straight line to signal the end of the transient period occurred at a time of about 2 hours. The slope during the transient period is 212 psi/cycle. Calculations are as follows:

$$kh = \frac{162.6\, q\mu B}{m}$$

$$kh = \frac{162.6 \times 800 \times 1.0 \times 1.25}{212}$$

$$kh = 767 \text{ md-ft}$$
$$k = 96 \text{ md}$$

$$s = 1.15 \left[\frac{p_i - p_1\, \text{hr}}{M} - \log \frac{k}{\phi\mu c r_w^2} + 3.23 \right]$$

$$s = 1.15 \left[\frac{1{,}895 - 1690}{212} \right.$$

$$\left. - \log \frac{96}{0.14 \times 1.0 \times 17.7 \times 10^{-6} \times 0.11} + 3.23 \right]$$

$$s = .50$$

Late Transient Analysis

The plot of log $(p_{wf} - \hat{p})$ vs. t, which is the basis of the late transient analysis method, is presented on Fig. 5–34. As was noted previously, the transient period appeared to end at about 2 hours. From the linear plot of p_{wf} vs. t (see Fig. 5–33), it appeared that semi-steady state might have been reached at $t \cong 10$ to 15 hours. Pressure data for the period 2 to 7 hours were included in the late transient analysis. By trial and error, it was established that $p = 1{,}460$ psig gave a reasonable straight-line plot of the data. The intercept and slope values are, respectively,

$$b = 320 \text{ psig and } B = \frac{\log 320 - \log 32}{7.4} = \frac{1}{7.4}\,\text{hr}^{-1}$$

$$kh = \frac{118.6\, q\mu B}{b}$$

$$kh = \frac{118.6 \times 800 \times 1.0 \times 1.25}{320}$$

$$kh = 371 \text{ md-ft}$$
$$k = 46.4 \text{ md}$$
$$V_p = 0.1115 \frac{qB}{\beta bc}$$

$$V_p = 0.1115 \times \frac{800 \times 1.25}{\dfrac{1}{7.4} \times 320 \times 17.7 \times 10^{-6}}$$

$$V_p = 0.146 \times 10^6 \text{ reservoir bbl}$$

This reservoir volume amounts to an equivalent drainage radius of 482 ft, or $\sim$ 17 acres.

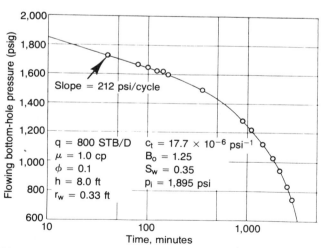

Figure 5–32. *Flowing pressure vs. logarithm of flowing time, extended pressure drawdown test (SPE Monograph 1, 1967).*

* Modified from Matthews and Russell, SPE Monograph 1, 1967.

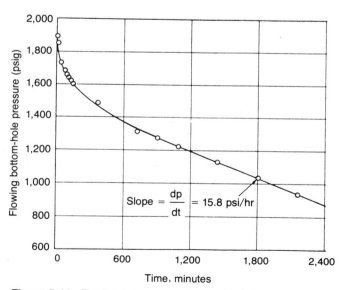

Figure 5–33. *Flowing pressure vs. time, extended pressure drawdown test (SPE Monograph 1, 1967).*

From an equation,

$$s = 0.84\left[\frac{\bar{p} - \hat{p}}{b}\right] - \ln\frac{r_e}{r_w} + \frac{3}{4}$$

$$s = 0.84\left[\frac{1{,}895 - 1{,}460}{320}\right] - \ln\frac{482}{0.33} + 0.75$$

$$s = -5.4$$

The kh and s values obtained here (371 md-ft and −5.4) are not in close agreement with those obtained by the transient analysis (767 md-ft and −0.5). This probably is caused by the fact that the well was given a hydraulic fracture treatment on completion. In fractured wells one expects kh and s values obtained from pressure drawdown tests to be dependent on the range of flow time from which the basic pressure data are derived. Flow at early times is essentially linear in nature (from the zone near the fracture).

Since the theory for transient analysis assumes radial flow, the kh value derived from a transient analysis will be anomalously high. As flow time proceeds, the radial flow in the region away from the fracture becomes dominant and late-transient analysis, which is also based on radial flow theory, more nearly represents the true values of the reservoir parameters. Thus, in the case of a fractured well we believe the late transient results are probably more representative of the true values.

Semisteady State Analysis (Reservoir Limit Test)

The linear plot of p_{wf} vs t is found on Fig. 5–33. This plot appears to be linear for times greater than 15 hours. From the slope of the plot and an equation,

$$V_p = 0.0418\frac{qB}{\beta_L c}$$

$$V_p = 0.0418 \times \frac{800 \times 1.25}{15.8 \times 17.7 \times 10^{-6}}$$

$$V_p = 0.149 \times 10^6 \text{ reservoir bbl}$$

The equivalent drainage area is approximately 17 acres.

The suite of data analyzed here is typical of the type that the engineer must analyze whenever a reservoir is of limited extent. Actually, it is seldom that all the separate facets of the transient, late transient, and reservoir limit analyses are in perfect agreement. In the case we have seen here, there was disagreement between the transient and late transient analyses, which was readily explained by fracturing. The belief that the late transient results are more representative is further supported by the almost-exact agreement in reservoir size calculated from the late transient and reservoir limit analyses.

There are instances in which unique interpretation of pressure drawdown data is not possible. More than one set of conclusions may be feasible. In such cases, the economic consequences of each answer must be considered.

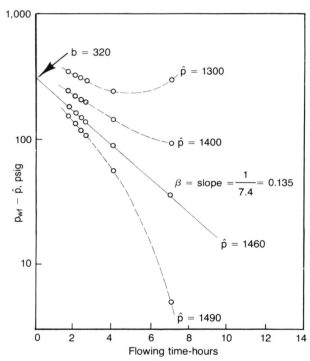

Figure 5–34. *Late transient analysis plot, extended pressure drawdown test (from SPE Monograph 1, 1967).*

Horner–Miller Dyes Hutchinson Build-up Test Analysis

Procedures are the same except that MDH plots time on log scale and a graph is used to obtain average reservoir pressure.

The *cgs* system is identical for first term. The second term must be adjusted to proper log function and p_{1sec} must relate with proper time in equation.

Note: The $\Delta p/m$ term controls the skin value determined and skin determined varies with the value used for permeability.

The analysis may be continued for p_{skin}, flow efficiency, and other terms.

Date of test	4/3/52	6/16/52
Producing interval–well was reperforated 5/3/52	6,258–70	6,258–70
Q = cumulative production, stb	3956	5500
q = producing rate before test, stb/d	96	60
cc/sec @ surface	176.6	110.4
t = pseudoproduction life, Q/q, days	41.2	91.7
10^6 seconds	3.56	7.92
m = slope of straight line, psi/cycle	4.2	2.0
atm/cycle	0.286	0.136
B = bbl @ reservoir/bbl @ surface	1.25	1.25
μ = viscosity, cp	0.65	0.65
ϕ = porosity, fraction	0.219	0.219
c_t = total compressibility, vol/vol/atm	0.00017	0.00017
psi^{-1}	11.6×10^6	11.6×10^6
r_w = well radius, cm	6.3	6.3
ft	0.207	0.207
h = pay thickness, ft	18	18
cm	548	548
p_{1hr}, psi	2,519	2,552
p_{wf} = flowing sand-face pressure before test, psi	2,060	2,522
atm	171.36	170.88
$kh = 162\ q\mu B/m$ in practical units	3,020	3,963
$= 2.303\ q\mu B/12.577m$ in cgs units	92.4	120.8
k = permeability, md (same from both equations)	168	220

Note: Reperforation may cause permeability to change if pay is lenticular. In this example, pay is not lenticular and the difference results because the test design did not allow for sufficient pressure change and slope to allow for accurate determination of values.

Skin in practical units:
$s = 1.51[(p_{1hr} - p_{wf})/m - \log (k/\phi\mu c_t r_w^2) + 3.23]$ when flow is at constant pressure

$(p_{1hr} - p_{wf})m$ (Note: This term is dimensionless)	112.9	18.5
$\log(k/\phi\mu c_t r_w^2)$ (Note: This term is rather constant)	9.37	8.49
Skin	123	15

These data conform with Fig. 5–35.

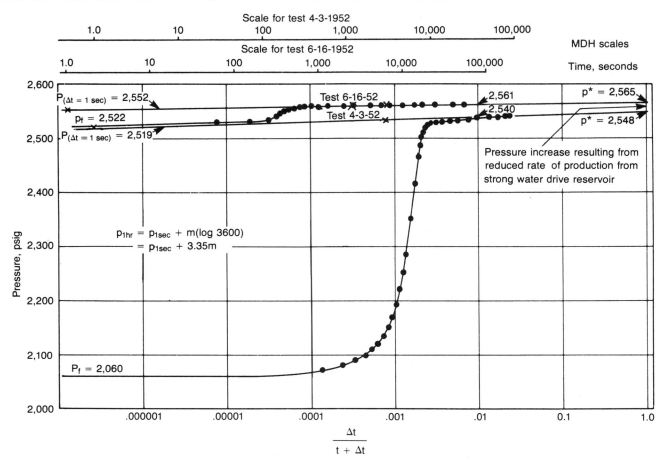

Figure 5–35. *Horner scale for all tests; time scale is a dimensionless ratio.*

Average Reservoir Pressure

The usual analysis of pressure build-up information begins with a plot of measured sand-face pressure versus either $\log \Delta t$ or $\log(t + \Delta t)/\Delta t$. Usually, semilog paper is used rather than the log values on coordinate paper. The first plot is known as the MDH or Miller, Dyes, and Hutchinson plot and the second is called the Horner plot. The Δt plot, MDH, should gradually curve as Δt increases, while the Horner plot should be a straight line if flow is radial in an infinite reservoir that exhibits ideal conditions. Boundary conditions, changing fluids, and other parameters influence the shape of the curves when conditions are not ideal. The MDH method assumes semisteady state flow prior to the shutin for build-up measurements so that r_e, the drainage radius, is known. Both methods relate pressure with Δt, and the radius of investigation relates Δt with $0.000264\, k\Delta t/\phi\mu c r_e^2 = t_{De} =$ dimensionless time. The constant 0.000264 allows one to use practical units such as hours, md, cp, ft, b/d and Mcfd.

Also, $\log(t + \Delta t)/\Delta t = [\log(t + \Delta t) - \log\Delta t]$. When t is very large compared to Δt, the equations become $[\log C - \log\Delta t]$ so that the plot of $-\log\Delta t$ is identical to the plot of $\log(t + \Delta t)/\Delta t$ when $\log(t + \Delta t)$ is a constant. Actually, the log functions have characteristics that increase the time period where the two plots, MDH and Horner, are identical for all practical purposes. But the useable range of the Horner plot often is greater than MDH.

The MDH correction shown on the following curves are easy to use. Curve A is for the case of a depleting solution gas reservoir since no fluids flow across the boundary. For curve B, the pressure is constant at the boundary; no depletion occurs. Since flow is at steady state, the drainage radius can be approximated (radius of spacing if all adjacent wells have been flowing at the same rate and proportional to rate if rates vary between wells). The value of t_{De} can be determined. A related value for the pressure correction can be read from the curves, and $\bar{p}$ is calculated from the equation $p_D = 1.15\,(\bar{p} - p_{ws})/m$. An ideal radial system is assumed.

The MDH correction becomes zero when curve A, which is a straight line as extrapolated, reaches a t_{De} value of 0.1. Use of this value of t_{De} in the above equation for the time scale permits one to calculate a value of Δt, which relates with the semisteady state drainage boundary, r_e. This value of Δt can be placed in the $(t + \Delta t)/\Delta t$ to calculate the corresponding value for $(t + \Delta t)/\Delta t$. The pressure that corresponds with this value of $(t + \Delta t)/\Delta t$ using the Horner plot is a

good approximation of the average reservoir pressure, $\bar{p}$. Thus, the average reservoir pressure can be easily obtained for both MDH and Horner plots when a circular or square boundary is present, as illustrated in Figs. 5–36 and 37.

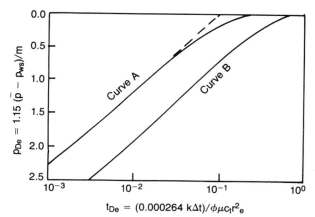

Figure 5–36. *MDH curves for a cylindrical reservoir (from* Trans. AIME 9, 1950). A) No influx of fluid over drainage boundary; B) Constant pressure at drainage boundary.

Matthews, Brons, and Hazebroek prepared a set of curves that enable one to determine the average reservoir pressure, $\bar{p}$, from build-up data. In this method, p^* is obtained by reading the pressure on the straight line of the build-up curve at $(t + \Delta t)/\Delta t = 1$. Also, assume that the slope of the build-up curve is 70 psi/cycle and that the drainage area is 160 acres or $(2640)^2$ sq ft (Figs. 5–38 to 41):

$$(0.000264\,kt/\phi\mu cA) = (0.000264 \times 7.65 \times 13630)/(0.039$$
$$\times .80 \times 17 \times 10^{-6} \times 2640^2 = 7.45$$

If the shape is a square, the appropriate MBH curve can be read at t_{pDA} and the related $p_{DMBH} = 5.45 = (p^* - \bar{p})/(70.6\,q\mu B/kH)$ where:

$$70.6\,q\mu B/kh = m/2.303 = 70/2.303 = 30.4$$
$$\text{then, } p^* - \bar{p} = 5.45 \times 30.4 = 166$$
$$\text{if } p^* = 4585, \ \bar{p} = 4585 - 166 = 4419 \text{ psig} = \text{average reservoir pressure}$$

The time t in the above equations relates with the time to reach the 160-acre drainage boundary using the radius of investigation equation:

$$t = 0.25\,\phi\mu c_t r_e^2/0.000264 \ k = 947\,\phi\mu c_t r_w^2/k$$

	Time obtained from past production/last production			50,800 hours			
	Time required to investigate 640 acres			998 hours			
	Time required to reach end of transient flow			399 hours			

	Curve flow time	Slope	kh	p*	p̄ in psig		
	hours	psi/cycle	md. ft.	psia	Horner	Miller et al	Timmerman
A	50,000	17	581	2895	2859	2860	2859
B	998	17	581	2869			2859
C	399	17	581	2861			2859

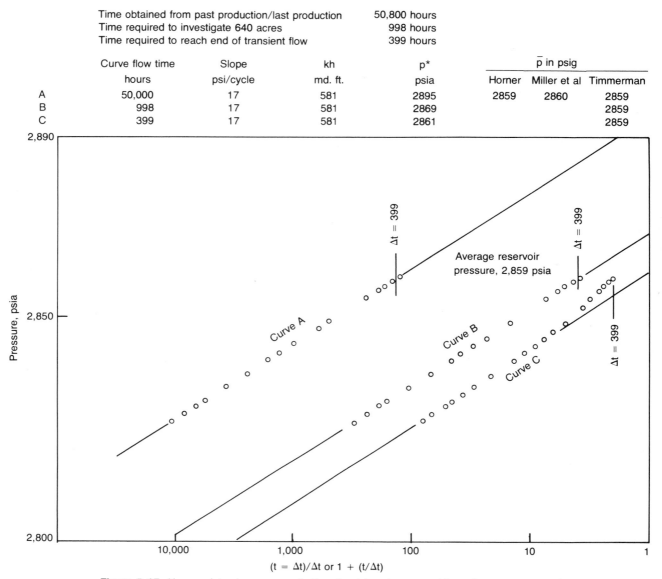

Figure 5–37. *Horner plot using a gas well. Flow time t is unknown and three times are assumed.*

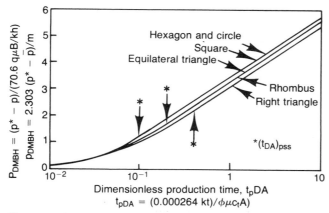

Figure 5–38. *Average reservoir pressure (after Matthews, Brons, and Hazebroek, courtesy SPE-AIME).*

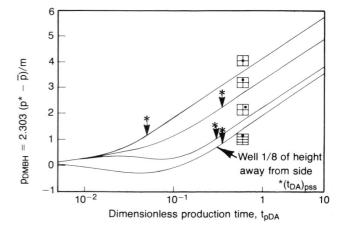

Figure 5–39. *Average reservoir pressure (after Matthews, Brons, and Hazebroek, courtesy SPE-AIME).*

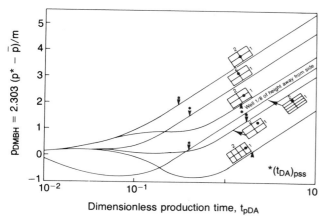

Figure 5–40. *Average reservoir pressure (after Matthews, Brons, and Hazebroek, courtesy SPE-AIME).*

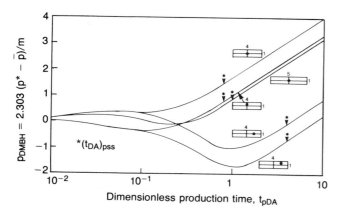

Figure 5–41. *Average reservoir pressure (after Matthews, Brons, and Hazebroek, courtesy SPE-AIME).*

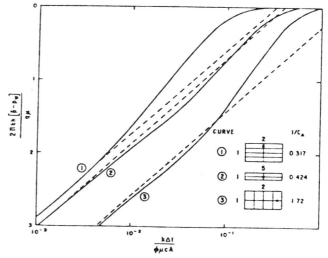

Figure 5–43. *MDH pressure build-up curves for rectangular shapes (from SPE Monograph 5, © SPE-AIME).*

Similar curves using the MDH approach are shown as Figs. 5–42 and 43.

Dietz and other authors have shown that the following equation is applicable at the time when the well is flowing at semisteady state (Fig. 5–44).

$$\bar{p} - p_{wf} = q\mu/(4\pi kh)[\ln (A/C_A r_w^2) + 0.809 + 2s]$$

A is the area of drainage and C_A is a shape-dependent constant read from the following tabulation.

The relationship between various plots of identical data (radial, linear, spherical, and log-log) are shown as Figs. 5–45 through 49.

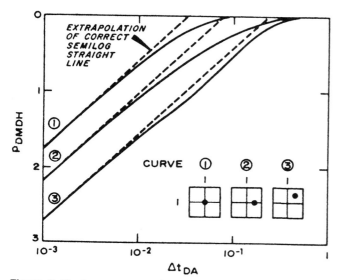

Figure 5–42. *Average reservoir pressure using the MDH method (from SPE Monograph 5, © SPE-AIME).*

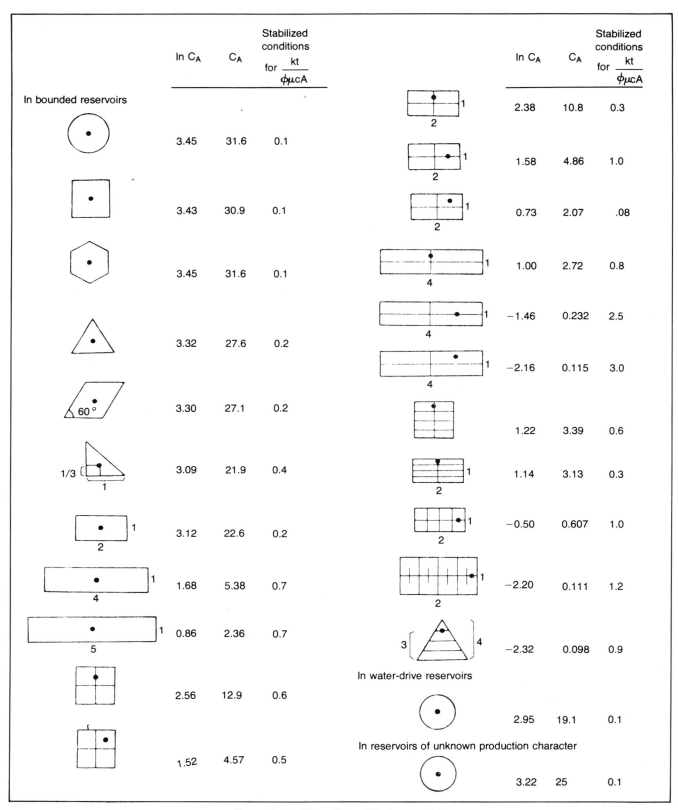

Figure 5-44. (*From Dietz*, JPT, *August 1965*).

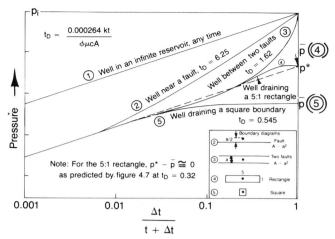

Figure 5–45. *Horner plot showing influence of well boundaries on curves (from SPE Monograph 1, 1967).*

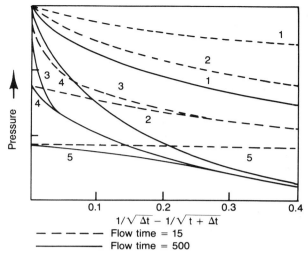

Figure 5–47. *Spherical plot of data from Fig. 5–45.*

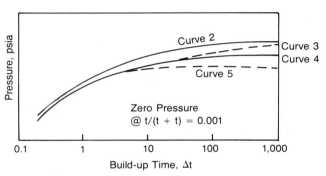

Figure 5–48. *MDH plot of data from Fig. 5–45.*

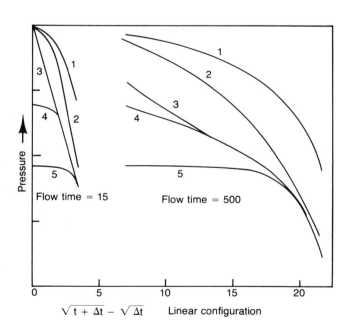

Figure 5–46. *Linear plot of data shown in Fig. 5–45.*

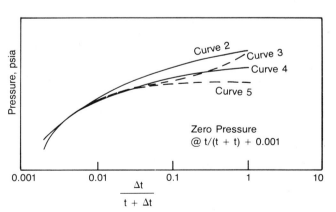

Figure 5–49. *Horner plot of data from Fig. 5–45.*

Technique Used When Oil is Below Bubble Point

The calculations are identical with those for a well above the bubble point, but the input terms must be adjusted to allow for two-phase flow. This implies that permeability must be an average obtained from relative permeability concepts and PVT terms must also be averaged to allow for the presence of two or more fluids. Matthews and Russell in SPE Monograph, 1967, suggested the following procedure for making these adjustments:

Obtaining Total Mobility and Total Compressibility

$$\left[\frac{k}{\mu}\right]_t = \frac{k_o}{\mu_o} + \frac{k_g}{\mu_g} + \frac{k_w}{\mu_w}$$

$$\left[\frac{k}{\mu}\right]_t = \frac{162.6}{mh}\left[B_o q_o + B_g(q_{gt} - q_o R_s) + B_w q_w\right]$$

$$= \frac{162.6}{(135)(20)}[1.227\,(924) + 12.9 \times 10^{-3}$$

$$(2.740 \times 10^6 - 924 \times 53.1) + 0]$$

$$= 2,159$$

$$c_o = \frac{B_g}{B_o}\frac{dR_s}{dp} - \frac{1}{B_o}\frac{dB_o}{dp}$$

$$= \frac{12.9 \times 10^{-3}}{1.227}\,(0.0455) - \frac{1}{1.227}\,(0.0001425)$$

$$= 0.0003622\ \text{psi}^{-1}$$

Values of (dR_s/dp) and (dB_o/dp) are obtained as the slopes of laboratory-determined curves of R_s and B_o vs p; the slope is drawn at the estimated average pressure or 1,315 psia.

Since we do not have a gas analysis, we estimate $T_c = 464°R$ and $p_c = 655$ psia from the gas gravity of 0.93. Then $T_r = 720/464 = 1.55$, $p_r = 1,315/655 = 2.01$. From data, $c_r = 0.56$. Then $c_g = c_r/p_c = 0.56/655 = 0.000854$ psia^{-1}.

We estimate $c_w = 3 \times 10^{-6}$ psi^{-1}, and $c_f = 4 \times 10^{-6}$ psi^{-1} at $\phi = 0.15$. Then $c_t = S_o c_o + S_g c_g + S_w c_w + c_f$, and for $S_o = 0.546$, $S_g = 0.204$ and $S_w = 0.25$, $c_t = 0.546\,(0.0003622) + 0.204\,(0.000854) + 0.25\,(3 \times 10^{-6}) + 4 \times 10^{-6} = 0.000376$.

Analysis of Tight Rocks with Fractures

Finding the proper time interval for determination of slopes used in the equations can be most difficult. Obviously, the wrong slope results in improper results since kh and skin both are controlled by the slope chosen. The time interval from which proper slope should be chosen can be selected by proper use of the afterflow equations, tubing pressure minus bottom-hole pressure plots, the log-log plot slopes, twice the fracture length for both sides of a well, and the proper use of the linear, spherical, and radial configuration plots. These techniques can be illustrated by an analysis of a tight gas well:

Step 1. Assemble all available data concerning the well and reservoir.
Step 2. Design the test properly so that useable information is collected.
    a. Use equations and graphical techniques with best available values for each of the respective terms to determine best values for items such as flow and build-up time, required pressure differences, and flow rates.
    b. Accurately measure all pressures including initial values for reservoir, tubing, casing,

and gradients in the wellbore. Measure flow rates at the time pressures are recorded.
    c. Design the test to approach either constant rate at the sand face or constant pressure at the sand face. Attempt to avoid complications from use of superposition correction.
    d. Consider safety, disposal of produced fluids, pay damage, and environment.
Step 3. Execute actual test in field according to the design. Carefully supervise test in field.
Step 4. Convert all data to the form and units required by the applicable equations:

Transient flow period

$$m_{(P_b)} - m_{(P_{wf})} = 1,637\,q_{sc}T/kh$$
$$\left[\log\frac{kt}{\phi\mu c_t r_w^2} - 3.23 + 0.87s + 0.87\,Dq_{sc}\right]$$

$$P_b^2 - P_w^2 = (1,637\,q_{sc}\mu_{\bar{p}}\bar{z}_{\bar{p}}T/kh)$$
$$\left[\log\frac{kt}{\phi\mu c_t r_w^2} - 3.23 + 0.87s + 0.87\,Dq_{sc}\right]$$

$$P_b - P_{wf} = (818\,q_{sc}\mu_{\bar{p}}\bar{z}_{\bar{p}}T/\tilde{p}kh)$$
$$\left[\log\frac{kt}{\phi\mu c_t r_w^2} - 3.23 + 0.87s + 0.87\,Dq_{sc}\right]$$

where $\tilde{p} = (\bar{p} + p_{wf})/2$ and in bracket μ and c_t are both at $\tilde{p}$.

All pressures must be converted to psia when values are small. Also, when all pressures are above 2,500 psi, the simple pressure may be used. If all pressures are less than 1,000 psi, use of the p^2 approach is reasonable. When some pressure values are between 1,000 and 2,500 psi, the $m(p)$ method or the averaging method proposed in SPE Monograph 1 must be used. A large amount of calculation time can be eliminated if the pressures measured in the field are read at time values shown on the table on "Flow Time for Use in Build-up Testing," p. 252.

Step 5. Determine the best possible values for each term in the various equations to be used. For this example: $h = 50$ feet, $\phi = 0.07$, $\mu_g = 0.0155$ cp, $T = 680°R$ or $460 + 220°F$, $c_t = S_g c_g = 0.0007$ psia^{-1}, $r_w^2 = 0.0352$ sq ft, $S_g = 0.40$ or $(1 - S_w)$.

Step 6. Calculate a range of values for radius of investigation using assumed values for permeability, production times, etc. Plot these data on log-log paper.

method is not so well known and is easy to use. Theory is based on flow at either constant rate or at constant pressure at the sand face. The data suggest that three possible straight lines could be drawn. Experience in analyzing many thousands of wells suggests that the last straight line is most likely to be desired for the expected conditions believed to exist in the vicinity of the well. Other graphs will be plotted to confirm this hunch.

The theoretical p_D versus t_{Dw} curve is overlain by a log-log plot of the field data in Fig. 5–52, as discussed in detail in SPE Monograph 5. As shown by this example, the usual problem of matching two curves when they have little change in slope makes curve match impossible. A curve match is possible if the straight line on Fig. 5–51 is extrapolated in both directions, and thereafter points from this straight line are plotted on Fig. 5–52. The theoretical log function has character and the field data taken from Fig. 5–51 as extrapolated can be matched. Unfortunately, this procedure is equivalent to having the answer and using it to make a confirmation on the log-log plots. When using the log-log plot, care must be taken to obtain an exact relationship between pressure and time at reservoir conditions for initial start-

Probable Range of Values Applicable to Example Well

$r_{inv} = \sqrt{0.00105\ kt/\phi\mu c_t} = \sqrt{0.00105\ kt/0.07 \times 0.0155 \times 0.0007 \times 0.4} = \sqrt{3,456\ kt}$

k	0.02 md	0.03 md	0.04 md	0.05 md	0.06 md	Test condition
30 hr	45 ft	56 ft	64 ft	72 ft	79 ft	
60 hr	64	79	91	102	112	
90 hr	79	97	111	124	137	
146 hr	100	123	142	157	174	End of flow
200 hr	118	144	166	180	204	
350	156	190	220	245	269	End build-up

Step 7. Prepare desired graphs, make necessary calculations, and study-evaluate.

The pressure in the gas well in Fig. 5–50 declined from about 1,725 psia to zero at the well sand face during test period (flow time, 146 hours). The graph shows the recorded pressures, converted to $m(p)$ format, versus time. The recorded flow rate also declined substantially. The data collected during test are not of best quality, but no other data are available. Production at constant pressure was not possible because the operator objected; he suspected mechanical problems.

The superposition technique probably should be used in Fig. 5–51, but an approximation method using a plot of $\Delta p/q$ has been substituted because the

ing conditions. The data from the field indicate that unloading problems were experienced during the first 90 hours of the flow test.

A slope of one on the log-log plot under ideal conditions represent well afterflow effects. A slope of one-half suggests linear flow. Radial flow usually begins one to two cycles after the end of the one-half slope period. Study of isopressure flowstream lines for fractured wells suggests that the flow configuration in a fractured well should be approaching radial after the radius being investigated as calculated by radius of investigation equation equals twice the fracture radius or equals the total fracture length on both sides of the well using $r_w' = r_w e^{-2} = \frac{1}{2} x_f$. The log-log plots and Horner plot should be used very carefully, if at all, until t_{Dw} has a value in excess of 25. The log func-

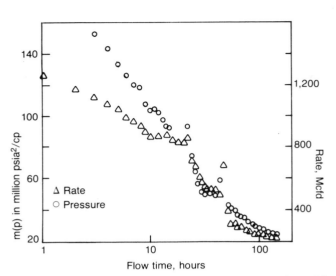

Figure 5–50. *Pressure-rate-time history, pressure expressed as m(p). Although the well had been cleaned up during a test six months earlier, the well produced liquids during the early part of this test.*

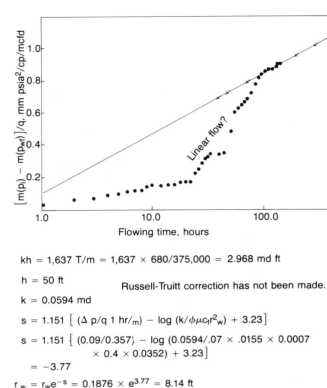

$kh = 1,637 \ T/m = 1,637 \times 680/375,000 = 2.968$ md ft

$h = 50$ ft   Russell-Truitt correction has not been made.

$k = 0.0594$ md

$s = 1.151 \left[(\Delta p/q \ 1 \ hr/m) - \log (k/\phi \mu c_t r^2_w) + 3.23 \right]$

$s = 1.151 \left[(0.09/0.357) - \log (0.0594/.07 \times .0155 \times 0.0007 \right.$
$\left. \times 0.4 \times 0.0352) + 3.23 \right]$

$= -3.77$

$r_w = r_w e^{-s} = 0.1876 \times e^{3.77} = 8.14$ ft

$x_{fr} = 8.14 \times 2 = 16$ ft

Figure 5–51. *Pressure vs. log flow time. Radius investigated is small and R-T correction would reduce k and increase x_{fr}.*

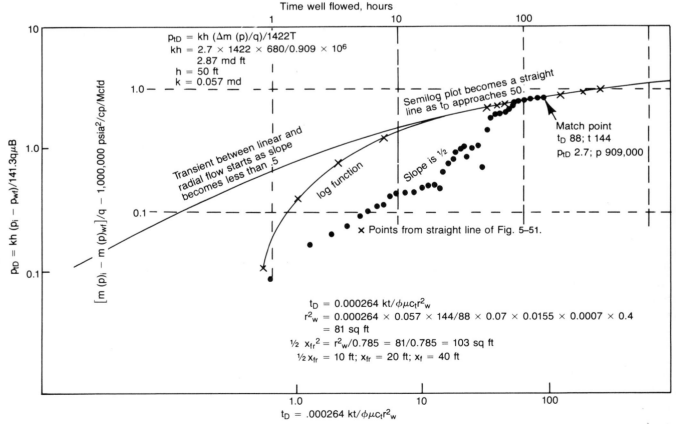

Figure 5–52. *Curve match for flow period.*

tion actually equals the p_{tD} function after tDw has a value of 75.

The linear flow equation contains two unknowns, both A and k (Fig. 5–53). Determination of the area of the two faces of the fracture is most difficult. Permeability usually can be determined with reasonable accuracy by use of a flow and build-up test prior to a fracture job. The early part of the test data are influenced by many factors related to well conditions. It is very important that the pressure element in the bomb be on bottom for a time sufficient to allow for adjustment to subsurface temperature and pressure. If the element is very accurate, measurements are influenced by other factors discussed in a prior section. Also, if permeability is high, interference from pressure transients is a major consideration throughout the tests.

It is easy to show that only one straight line can exist for any specified time period in Figs. 5–51, 53, and 54. Spherical flow usually should not be expected for most fracture wells. Yet a straight line is noted for most wells near the bend in the curve during the early life of the test. The slope of the apparent straight line usually results in values for permeability that are in the ball park when compared with the values calculated using techniques associated with other flow con-

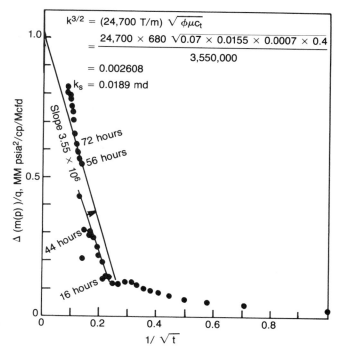

Figure 5–54. *Spherical plot for flow period. If flow is linear to about 100 hours, spherical should not end before 100 hours.*

figurations. Spherical flow contains no pay thickness term, but the calculated value for permeability must be adjusted since $k_s = k_v\,k_h\,k_h$. When permeability is large, k in vertical direction usually approaches that in horizontal direction. In tight pays, the vertical permeability may be much less than the horizontal. The use of all plots helps in selecting the applicable time intervals used to determine slopes and can also be useful in selection of pay thickness, shale barriers, etc. The accuracy of the analysis can be improved if proper techniques are used by experienced analysts.

Other calculations and plots were made to analyze the early part of this flow test. The results were not conclusive and are not shown in this discussion. The analysis of the build-up test that followed the flow test uses theory similar to that for the flow test.

The data shown in Figs. 5–55 and 5–56 employ the curve match technique using two possible time scales, Δt and $t/(t + \Delta t)$, respectively. The technique is similar to that used for the curve match of the flow test, and SPE Monograph 5 should be consulted if the method is unfamilar.

The flow investigates about 150 feet if the permeability is as low as suggested by the test results. Thus the expanding radius of the flow period is being chased by an influence of the shut-in period. It is probable that corrections suggested by Prats, Russell,

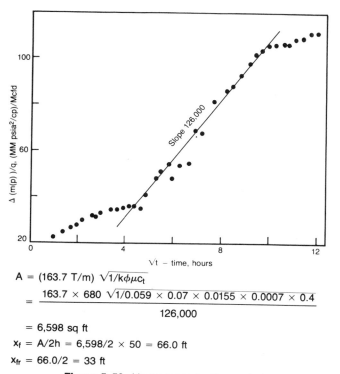

$A = (163.7\ T/m)\ \sqrt{1/k\phi\mu c_t}$

$$= \frac{163.7 \times 680\ \sqrt{1/0.059 \times 0.07 \times 0.0155 \times 0.0007 \times 0.4}}{126,000}$$

$= 6,598$ sq ft

$x_f = A/2h = 6,598/2 \times 50 = 66.0$ ft

$x_{fr} = 66.0/2 = 33$ ft

Figure 5–53. *Linear curve for flow period.*

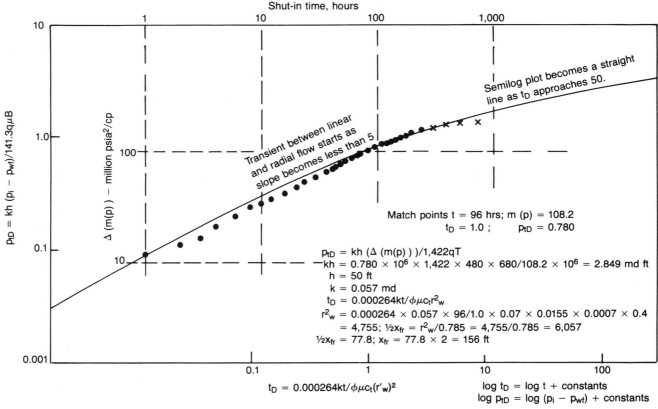

Figure 5–55. *Curve match for build-up using Δt.*

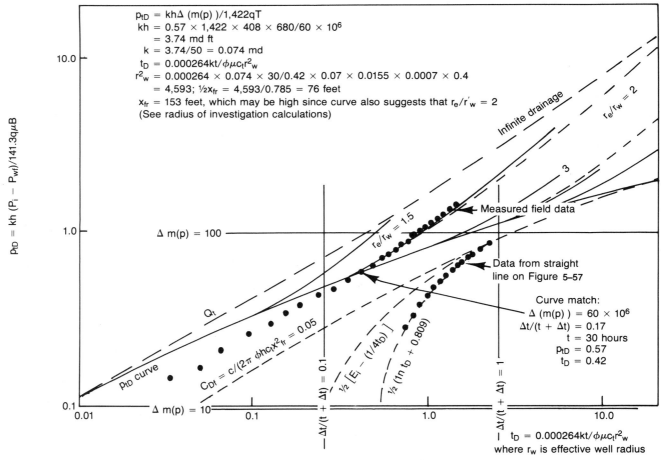

Figure 5–56. *Build-up curve match using Δt/(t + Δt).*

and Truitt would improve results, but the quality of the data hardly support such effort.

Fig. 5–57 is the normal Horner plot for the build-up period. The flow rate prior to the build-up test was not constant. The analysis is based on the ideas of working between limits and the curves are plotted using flow times based on both initial and final flow rates.

Some analysts prefer to use a Miller, Dyes, and Hutchinson plot rather than the Horner plot. The useable range of the MDH plot should be smaller, but both curves should have the same slope when t is very large compared with Δt (Fig. 5–58). For this well, the MDH slope is 111×10^6, compared with a value between 168 to 255×10^6 for the Horner limits.

The analysis results could be improved if the slopes of the various graphs were selected using the overall knowledge gained from this preliminary analysis. However, a more consistent set of results may not improve knowledge as to permeability, pay thickness, skin in the fractured well, etc. The accuracy of the results seemed to meet the present need, and a second analysis was not made. The results of the preliminary analysis are summarized below:

the vertical rather than the horizontal fracture theory is sound.

Since the radius investigated by the flow test is rather short compared with the probable fracture length, refinements such as suggested by Russell and Truitt (see a later section of this chapter) should be made. Also, the curve match techniques should be refined using the base graphs proposed by Earlougher and Kersch as included in SPE Monograph 5. The well is a wildcat, and no interference from other wells should occur. The pay is very tight, and the in-situ water is relatively high. The techniques available—including log—strain the analysis methods used to determine values for porosity and in-situ water. The difficulties are many, but experience, hard work, and sound thinking can convert an analysis of poor quality data into useable conclusions. In this example, the permeability is probably about 0.04 md and the skin is in the range of minus 4 to 5. The fracture length calculated from this skin is less than that forecast from the fracture design, and it is probable that the fracture either healed after the job or the residual permeability of the fracture is not infinite. Also, the indicated short fracture length may result from parallel fractures. The

Summary of Analysis

Curve type	kh, md ft	h, ft	k, md	s	x_{fr}, ft	Applicable time ranges	
Flowing well							
Curve match	2.55	50	0.051		60	Radial after 90 hrs	
Radial flow	2.97		0.059	−3.77	16	90—146 hours	
Spherical flow	—		0.019	—	—	56—80 hours	Refinements such as proposed by Russell-Truitt have not been made.
Linear flow	—		—		32.9	30—80 hours	
Shut-in well build-up							
Curve match							
$t/(t+\Delta t)$	5.88?		0.118?		84?	No good curve match	
Δt	2.85?		0.057?		156?	No good curve match	
Radial Horner	1.50		0.036	−4.76	44		
			0.030	−5.19	67	90–196 hours	
Spherical			0.036			35–108 hours	
Linear					71+	35–100 hours	
MDH			0.074?			Not applicable	

This poor gas well had been shut in for six months prior to this test and the reservoir pressure was measured at the start of the test series. The pressure so measured was 1,725 psi. This value was used in drawing the straight lines for some of the figures. Also, the pay depth was below 2,000 feet so that use of

well unloaded during the early test period, and some 100 *Mcf* of the total production of 2,483 *Mcf* came from the wellbore.

The fractures may penetrate deeply as per the design, but the net result, regardless of cause, usually is a negative skin of about 5.

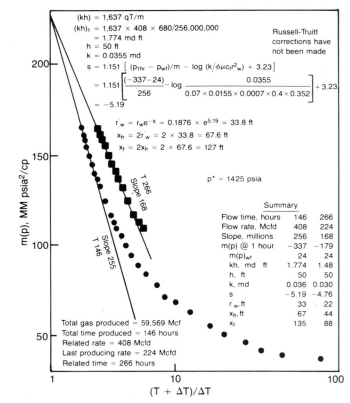

Figure 5–57. *Horner plot for radial flow build-up. For this test, p is known as 1,725 psia.*

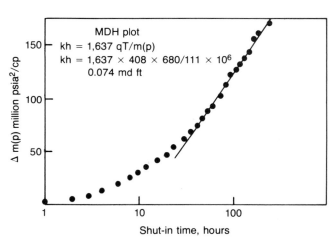

Figure 5–58. *Miller, Dyes, & Hutchinson plot for build-up. This method is not valid since t + Δ t is not large relative to Δt.*

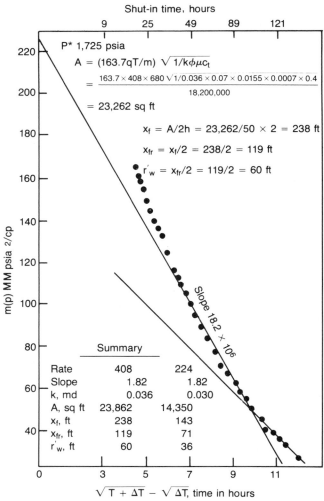

Figure 5–59. *Linear flow plot for build-up.*

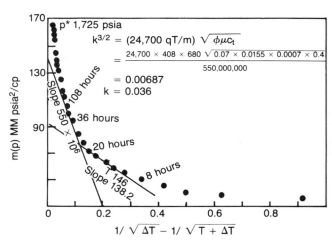

Figure 5–60. *Spherical flow plot for build-up. The straight line occurs during early linear flow rather than between linear and radial flow; do not use these results.*

Two-Rate Flow Test

A transient flow occurs each time the rate of a well is changed. If rate is decreased, the change results in a pressure build-up; if the rate is increased, the change will have drawdown characteristics. Opening the well from shut-in conditions and closing in a flowing well are extreme cases. The two-rate test is simply an intermediate condition. For a two-rate test, the flow rate of the well is stabilized for a long time period, say several days, and the flow rate is changed (either an increase or a decrease) by an amount that will result in an accurate or measurable transient pressure change. The first and second flow rates should both be constant. In wells where stabilization time is large, the two-rate test has the advantage of no shut-in period and total test time is shorter.

$$kh = \frac{162.6 q_1 \mu B}{m} = (162.6 \times 107 \times 0.6 \times 1.5)/90$$

$$= 174; \; k = 174/59 = 3.0 \quad \text{(Fig. 5–61)}$$

$$s = 1.151 \left[\left(\frac{q_1}{q_1 - q_2} \right) \left(\frac{p_{1hr} - p_w}{m} \right) \right.$$
$$\left. - \log \frac{k}{\phi \mu c r_w^2} + 3.23 \right]$$

$$= 1.151[\{(107/(107 - 46)\}\{(3,169 - 3,118)/90\}$$
$$- \log(3/0.06 \times 0.6 \times 0.0000932$$
$$\times 0.04) + 3.23]$$

$$= -3.6$$

$\Delta p \text{ (skin)} = 0.87 \, ms \text{ (at rate } q_1) = 0.87 \times 90 \times -3.6$
$$= 282 \text{ psig}$$

$$p^* = p_w + m[\log(kt/\phi \mu c_t r_w^2) - 3.23 + 0.87s]$$
$$= 3,118 + 90[\log\{(3 \times 5,922)/(0.06 \times 0.6 \times 9.32$$
$$\times 10^5 \times 0.04)\} - 3.23 + 0.87 \times -3.6]$$
$$= 3,548 \text{ psig}$$

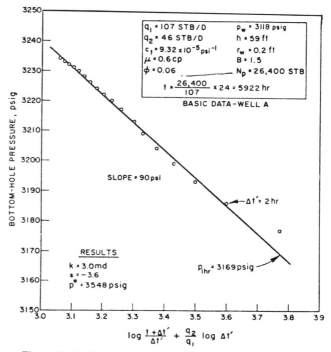

Figure 5–61. *Two-rate test (from SPE Monograph 1, 1967).*

Four-Point Test

The empirical method proposed by Rawlins and Schellhardt during the 1930s is valid for good reservoirs. The gas well is flowed at four rates after stabilization during shut-in between the flow periods. A build-up to say 90% of original pressure often is stabilization. The average reservoir pressure squared minus the flowing pressure squared for each flow rate is plotted versus the corresponding flow rate, as measured on log-log paper. Such points often fall on a straight line having a slope approaching one, as illustrated by Fig. 5–62. Data for the graphs are:

$r_w = 0.23$ ft	$\phi = 0.16$
$S_w = 0.20$	ϕ gas $= 0.128$
$h = 40$ ft	$\mu_g = 0.017 \, cp$
$c_g = 8.61 \times 10^{-4} \text{psi}^{-1}$	$B_g = 8.28 \times 10^{-3}$

Gas gravity $= 0.7$

Total variation in pressure during the test was 82 psi. Thus, gas properties were considered to be constant and were evaluated at 1,691 psia.

The open-flow potential is the value for the flow rate at which the pressure difference denoted as squares is equal to the average reservoir pressure minus the lowest permissible flowing pressure, as indicated on the straight line. The method is not applicable to poor gas reservoirs. Russell applied theoretical concepts to the four-point test to obtain an analysis as shown in the following table (©SPE-AIME):

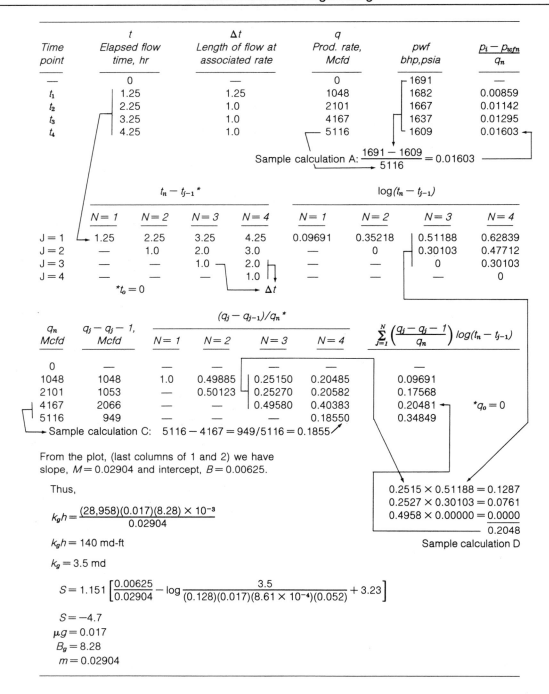

Time point	t Elapsed flow time, hr	Δt Length of flow at associated rate	q Prod. rate, Mcfd	p_{wf} bhp, psia	$\dfrac{p_i - p_{wfn}}{q_n}$
—	0	—	0	1691	—
t_1	1.25	1.25	1048	1682	0.00859
t_2	2.25	1.0	2101	1667	0.01142
t_3	3.25	1.0	4167	1637	0.01295
t_4	4.25	1.0	5116	1609	0.01603

Sample calculation A: $\dfrac{1691 - 1609}{5116} = 0.01603$

	$t_n - t_{j-1}$*				$\log(t_n - t_{j-1})$			
	$N=1$	$N=2$	$N=3$	$N=4$	$N=1$	$N=2$	$N=3$	$N=4$
$J=1$	1.25	2.25	3.25	4.25	0.09691	0.35218	0.51188	0.62839
$J=2$	—	1.0	2.0	3.0	—	0	0.30103	0.47712
$J=3$	—	—	1.0	2.0	—	—	0	0.30103
$J=4$	—	—	—	1.0	—	—	—	0

*$t_o = 0$ Δt

q_n Mcfd	$q_j - q_{j-1}$, Mcfd	$(q_j - q_{j-1})/q_n$*				$\displaystyle\sum_{J=1}^{N}\left(\dfrac{q_j - q_{j-1}}{q_n}\right)\log(t_n - t_{j-1})$
		$N=1$	$N=2$	$N=3$	$N=4$	
0	—	—	—	—	—	—
1048	1048	1.0	0.49885	0.25150	0.20485	0.09691
2101	1053	—	0.50123	0.25270	0.20582	0.17568
4167	2066	—	—	0.49580	0.40383	0.20481
5116	949	—	—	—	0.18550	0.34849

*$q_o = 0$

Sample calculation C: $5116 - 4167 = 949/5116 = 0.1855$

From the plot, (last columns of 1 and 2) we have slope, $M = 0.02904$ and intercept, $B = 0.00625$.

Thus,

$$k_g h = \frac{(28{,}958)(0.017)(8.28)\times 10^{-3}}{0.02904}$$

$k_g h = 140$ md-ft

$k_g = 3.5$ md

$$S = 1.151\left[\frac{0.00625}{0.02904} - \log\frac{3.5}{(0.128)(0.017)(8.61\times 10^{-4})(0.052)} + 3.23\right]$$

$S = -4.7$

$\mu g = 0.017$

$B_g = 8.28$

$m = 0.02904$

$$
\begin{aligned}
0.2515 \times 0.51188 &= 0.1287\\
0.2527 \times 0.30103 &= 0.0761\\
0.4958 \times 0.00000 &= \underline{0.0000}\\
&\ \ 0.2048
\end{aligned}
$$

Sample calculation D

The isochronal test and the flow build-up analysis procedure discussed in an earlier section should be used for intermediate and poor-quality gas reservoirs, respectively. The one-point method, which uses one flow rate and an estimated slope, is worthless unless a prior four-point test is available for the well. A decreasing sequence of four-point flow rates will give the highest open flow potential and will also result in better quality data for analysis purposes.

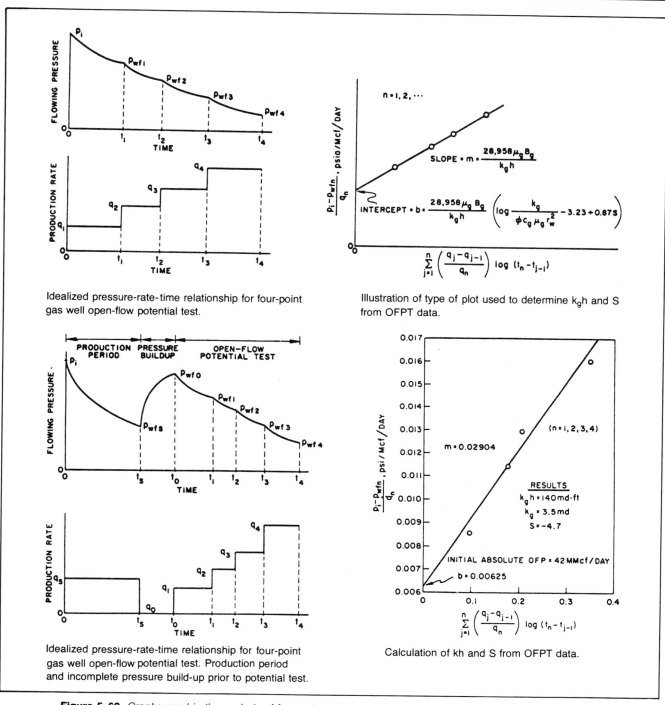

Idealized pressure-rate-time relationship for four-point gas well open-flow potential test.

Illustration of type of plot used to determine $k_g h$ and S from OFPT data.

Idealized pressure-rate-time relationship for four-point gas well open-flow potential test. Production period and incomplete pressure build-up prior to potential test.

Calculation of kh and S from OFPT data.

Figure 5–62. *Graphs used in the analysis of four-point tests (after Russell, JPT, December 1973, © SPE-AIME).*

Isochronal Test

The isochronal method should be used when permeability in gas wells is less than about 50 md. The four-point tests for the well do not adequately stabilize during the test time, and a curved rather than a straight line often is indicated by the four-point test.

The isochronal test is similar to the four-point test, but the flowing sand-face pressures for each flow rate are recorded at the same time intervals for each flow rate. Thus, a plot of $p^2 - p_{wf}^2$ versus flow rate can be made on log-log paper for each of the times that pressures are measured, as shown in Fig. 5–63. The method can be used similar to the four-point method if the curve projected to semisteady state flow is used. In other words, the open flow potential is equal to the maximum available pressure differential—square of differences in values—indicated by the stabilized flow representing semisteady state flow for the wells' drainage area. Since drainage area is related to relative rates of the wells, determining drainage area or radius requires careful analysis. The data can be analyzed as shown below (material from Shell Oil Co.)

Isochronal Test—Determination of Stabilized Performance Coefficient

Data required
Plot of rate vs. $\bar{p}^2 - p_{wf}^2$ as obtained from field data with isochronal test procedure. It is suggested that the performance coefficients C pertaining to the longest practical flow times be used in the calculations to provide increased accuracy.

$$Q = C(\bar{p}^2 - p_{wf}^2)^n$$

Steps preformed
1) Determine characteristic slope n from graph.
   stabilized time = 840 hr
   $n = 1.0283$

2) Determine C_1 and C_2, coefficients corresponding to times t_1 and t_2,
   $t_1 = 2.56$ hr. $C_1 = 150.1$ Mcfd/psi² (millions)
   $t_2 = 25.6$ hr. $C_2 = 123.6$ Mcfd/psi² (millions)
3) Determine time of stabilization t.

   $t = \dfrac{re^2 \phi_g \mu_g \beta_g}{0.2816K}$ or use graphical solution based on ½ spacing

4) Calculate stabilized performance coefficient C from following equations:

$C_1^{1/n} = 150.1^{1/1.0283} = 130.75$
$C_2^{1/n} = 123.6^{1/1.0283} = 108.25$

$$x_2 = \left[\frac{C_2^{1/n}}{2(C_1^{1/n} - C_2^{1/n})} \right]$$

$$\frac{C_2^{1/n}}{2(C_1^{1/n} - C_2^{1/n})} = 2.40555$$

$$x_1 = \left[\frac{C_1^{1/n}}{2(C_1^{1/n} - C_2^{1/n})} \right]$$

$$\frac{\alpha}{r} = \frac{(t_2)^{x_2}}{(t_1)^{x_1}} = \frac{t_2^{2.40555}}{t_1^{2.90555}} = \frac{(25.6)^{2.40555}}{(2.56)^{2.90555}} = 159.02$$

$$C = \frac{C_2 \left[\ln \dfrac{\alpha t_2^{1/2}}{r_w} \right]^n}{\left[\ln \dfrac{\alpha t^{1/2}}{r_w} \right]^n}$$

$$= \frac{123.6 \, [\ln (159.02 \times 25.6^{1/2})]^{1.0283}}{[\ln (159.02 \times 840^{1/2})]^{1.0283}}$$

$$= 97.39 \text{ Mcfd/psi}^2 \text{ (millions)}$$

For low flow rates, acceptable data are collected. At higher flow rates, the performance predictions are increasingly pessimistic. Fig. 5–64 shows stabilization time graphically.

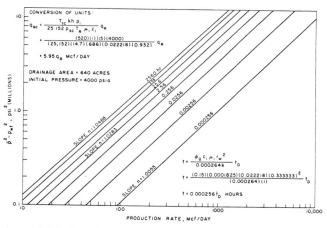

Figure 5–63. *Open-flow potential test isochronal method (after Poettman, JPT, September 1959).*

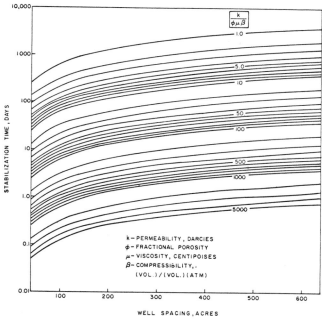

Figure 5–64. *Stabilization time (after Poettman, JPT, September 1959).*

Tests for Gas and Oil Wells—Determining *kh* and Skin

Theory indicates that pressure and $1/q$ should be plotted versus log of time. However, when pay is very tight and flow rate even with maximum drawdown of pressure is rather small, tests in the field may be very long and use of pressure gauges at surface may be impossible. The early life period should be amplified and log-log can be useful. Base graphs can be prepared using the normal equations for linear and radial flow. The interrelationships of these curves can be understood by a study of the following figures, which have been prepared using the log-log curve as a basis.

For the data set in Fig. 5–65, the log-log plots are straight lines. Other graphs use this basic information and vary accordingly because of conditions used in their construction. At lower flow rates, all curves seem to be straight lines, and this fact is used in evaluation of tests.

Base curves have been prepared for a number of assumed reservoir conditions and many are presented in a later chapter, entitled "Analysis and Forecast for Gas Wells." Both the log-log and $1/q$ techniques are useful when analyzing tight oil and gas wells under less than desired test conditions.

All wells produce a build-up or flow transient each time the rate of flow is changed. Theoretical curves showing production rate versus flow time can be constructed for the linear and transient flow period if skins and *kh* are assumed. The actual rate of flow can be plotted or overlain on this same graph. The relative location and slope of the actual production rate may be compared with the theoretical curve to obtain a good approximation of the pay interval *kh* and skin. An example is in Fig. 5–66. Additional theoretical curves are presented in the section related to gas reservoirs. This approach is particularly useful when the well is tested over a long interval but no pressure data can be accumulated. The method is limited to pays having low permeability, so that the transient flow period is lengthy and flow rates are small.

Similar theoretical graphs can be constructed relating q/kh, average reservoir pressure and well-face pressure when flow is at constant sand face pressure. (See Timmerman, *JPT*, January 1966; *JPT*, October 1971; *Petroleum Engineer,* January 1971.) The actual measured field data can also be plotted on these graphs. Methods also are used to prepare forecasts.

Similar curves can be constructed for oil wells using relative permeability when applicable. Also, $1/q$ may be used. Decline curves also are valuable.

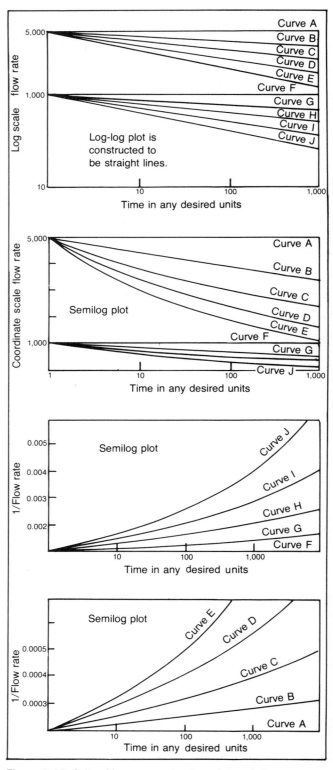

Figure 5–65. *Suite of four curves showing relations between curves.*

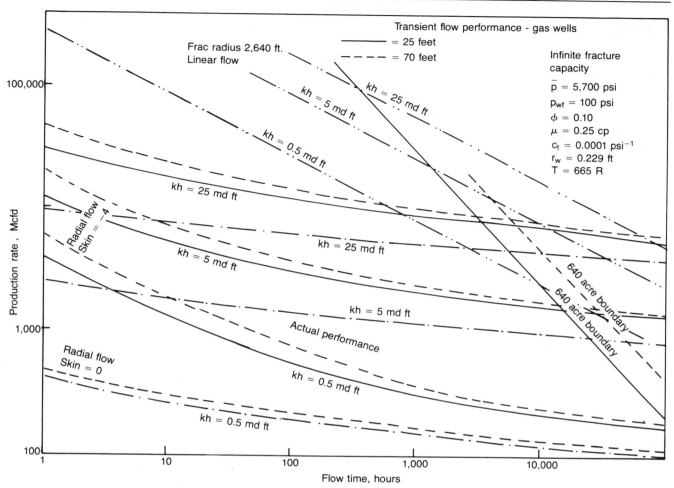

Figure 5–66. *Theoretical flow curves for gas wells during early life.*

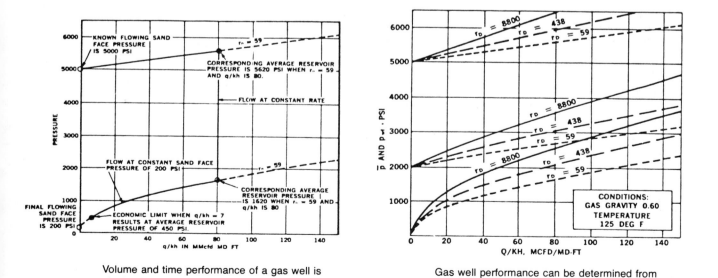

Volume and time performance of a gas well is obtained easily using basic graph.

Gas well performance can be determined from flowing pressure relations.

Figure 5–67. *Long-term behavior of gas wells (courtesy Shell Oil Co.)*

Decline Curves

Classification of Production Decline Curves (after Arps)

Decline type	I. Constant-percentage decline	II. Hyperbolic decline	III. Harmonic decline
Basic Characteristic	Decline is constant $n = 0$	Decline is proportional to a fractional power (n) of the production rate $0 < n < 1$	Decline is proportional to production rate $n = 1$
	$D = K \cdot q^0 = -\dfrac{dq/dt}{q}$	$D = K \cdot q^n = -\dfrac{dq/dt}{q}$ for initial conditions: $K = \dfrac{D_i}{q_i^n}$	$D = K \cdot q^1 = -\dfrac{dq/dt}{q}$ for initial conditions: $K = \dfrac{D_i}{q_i}$
	$\displaystyle\int_0^t D\,dt = -\int_{q_i}^{q_t} \dfrac{dq}{q}$ $-Dt = \log_e \dfrac{q_t}{q_i}$	$\displaystyle\int_0^t \dfrac{D_i}{q_i^n} \cdot dt = -\int_{q_i}^{q_t} \dfrac{dq}{q^{n+1}}$ $\dfrac{n\,D_i t}{q_i^n} = q_t^{-n} - q_i^{-n}$	$\displaystyle\int_0^t \dfrac{D_i}{q_i}\,dt = -\int_{q_i}^{q_t} \dfrac{dq}{q^2}$ $\dfrac{D_i t}{q_i} = \dfrac{1}{q_i} - \dfrac{1}{q_t}$
Rate-Time Relationship	$q_t = q_i \cdot e^{-Dt}$	$q_t = q_i\,(1 + nD_i t)^{-\frac{1}{n}}$	$q_t = q_i\,(1 + D_i t)^{-1}$
	$Q_t = \displaystyle\int_0^t q_t \cdot dt = \int_0^t q_i \cdot e^{-Dt} \cdot dt$ $Q_t = \dfrac{q_i - q_i \cdot e^{-DT}}{D}$ Substitute From Rate-time Equation: $q_i \cdot e^{-DT} = q_t$ To Find:	$Q_t = \displaystyle\int_0^t q_t \cdot dt = \int_0^t (1 + nD_i t)^{-\frac{1}{n}} \cdot dt$ $Q_t = \dfrac{q_i}{(n-1)\,D_i}\left[(1 + n\,D_i t)^{\frac{n-1}{n}} - 1\right]$ Substitute From Rate-time Equation: $(1 + n\,D_i t) = \left(\dfrac{q_i}{q_t}\right)^n$ To Find:	$Q_t = \displaystyle\int_0^t q_t \cdot dt = \int_0^t q_i\,(1 + D_i t)^{-1} \cdot dt$ $Q_t = \dfrac{q_i}{D_i}\left[\log_e (1 + D_i t)\right]$ Substitute From Rate-time Equation: $(1 + D_i\,t) = \dfrac{q_i}{q_t}$ To Find:
Rate-Cumulative Relationship	$Q_t = \dfrac{q_i - q_t}{D}$	$Q_t = \dfrac{q_i^n}{(1-n)D_i}\,(q_i^{1-n} - q_t^{1-n})$	$Q_t = \dfrac{q_i}{D_i} \log_e \dfrac{q_i}{q_t}$

D = Decline as a fraction of production rate
D_i = Initial decline
q_i = Initial production rate
t = Time

q_t = Production rate at time t
Q_t = Cumulative oil production at time t
K = Constant
n = Exponent

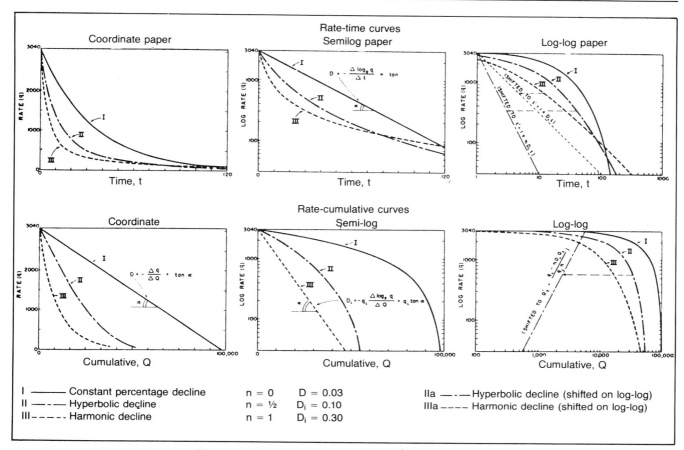

I ——————— Constant percentage decline   n = 0   D = 0.03   IIa —·—·— Hyperbolic decline (shifted on log-log)
II —·—·— Hyperbolic decline   n = ½   D_i = 0.10   IIIa ——— Harmonic decline (shifted on log-log)
III ———— Harmonic decline   n = 1   D_i = 0.30

Figure 5–68. *Decline curves (after Arps,* Trans. AIME, 1956).

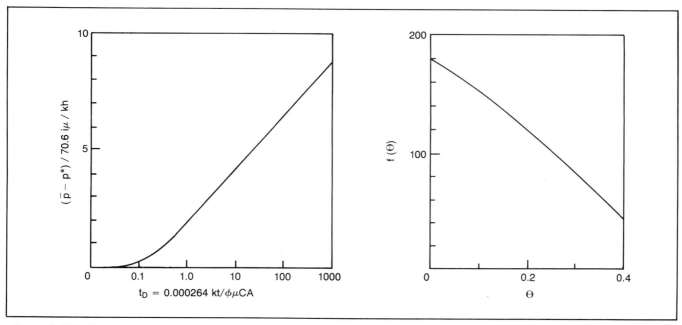

Figure 5–69A. *Function for computing average waterflood pressure.*

Figure 5–69B. *Function for calculating kh.*

Injection Wells (Modified from SPE Monograph 1)*

Example 1A: Pressure Fall-Off Analysis, Liquid-Filled Case, Unit Mobility Ratio

Determining $\bar{p}$ for Fig. 5–69—The first step is to obtain k from the slope of the fall-off curve. Using this k and other data given in this example, we calculate dimensionless flowing time for a 40-acre pattern flood (injection area A of 20 acres):

$$\frac{0.000264\ kt}{\phi\mu cA} = \frac{0.000264\ (21.8)\ 40,100}{0.16\ (0.6)\ 7\times10^{-6}\ (871,200)} = 393$$

From Fig 5–69A, $(\bar{p} - p^*)/(70.6\ i\mu/kh) = 7.91$
Since $70.6\ i\mu/kh = m/2.303$

$$\begin{aligned}\text{then } \bar{p} - p^* &= 7.91\ (m/2.303)\\ &= 7.91\ (130/2.303)\\ &= 447\ \text{psi}\end{aligned}$$

and, obtaining p^* from Fig. 5–69, we find

$$\bar{p} = -322 + 447 = 125\ \text{psig}$$

This result limited to slide-rule accuracy.

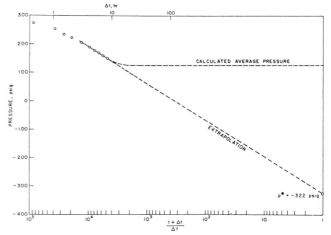

Figure 5–69. *Example pressure fall-off curve (from SPE Monograph 1, 1967).*

1: Pressure Fall-Off Analysis, Liquid-Filled Case, Unit Mobility Ratio

Data:
Date _____ October 30, 1964 _____
Producing Formation ____ Sandstone ____
Hole Size (inches) _____ 8.5 _____
Cum. Inj., W_i (bbl) _____ 2,380,000 _____
Stabilized Daily Inj., i (bbl) ____ 1,426 ____
Effective Prod. Life t (hr) = 24 W_i/i ____ 40,100 ____

Company _____ Shell _____
Lease _____ Zipper _____
Well No. _____ 4 _____
Field _____ Bent _____
State _____ Illinois _____

Calculation of kh (md-ft) and k (md); k is permeability to water, k_w:

$$kh = \frac{162.6\ i\mu B}{m} \qquad k = \frac{kh}{h}$$

h ____ 49 ____ ft
i ____ 1,426 ____ B/D

μ ____ 0.6 ____ cp
B ____ 1.0 ____
m ____ 130 ____ psi/cycle

$$kh = \frac{162.6\times(1,426)\times(0.6)\times(1.0)}{(130)}$$

$$= 1,070\ \text{md-ft}; \quad k = \frac{(1,070)}{(49)} = 21.8\ \text{md} = k_w.$$

Calculation of Skin Effect, s; and Pressure Loss Due to Skin, Δp_{skin} (psi):

$$s = 1.151\left[\frac{p_w - p_{1\,\text{hr}}}{m} - \log\left(\frac{k}{\phi\mu cr_w^2}\right) + 3.23\right].$$

* From Matthews and Russell, © SPE-AIME, 1967

$\Delta p_{skin} = m \times 0.87s.$

k ____21.8____ md

ϕ ____0.16____

μ ____0.6____ cp

c ____7.0×10^{-8}____ psi^{-1}

r_w ____4.25/12____ ft

$p_{1\,hr}$ ____273____ psig

p_w ____525____ psig

m ____130____ psi/cycle

$$s = 1.151 \left[\frac{(525) - (273)}{(130)} \right.$$

$$\left. - \log \frac{(21.8)(144)}{(0.16)(0.6)(7.0 \times 10^{-8})(18.1)} + 3.23 \right] = \underline{-3.73}.$$

$\Delta p_{skin} = (130) \times 0.87\,(-3.73) = -421$ psi (well had been fractured).

Calculation of Injectivity Index (b/d-psi) and Flow Efficiency:

$$I_{(actual)} = \frac{i}{p_w - \bar{p}}$$

$$I_{(ideal)} = \frac{i}{(p_w - \bar{p}) - \Delta p_{skin}}$$

Δp_{skin} ____-421____ psi

i ____1,426____ B/D

$\bar{p}$ ____125____ psig

p_w ____525____ psig

$$I_{(actual)} = \frac{(1,426)}{(525) - (125)} = \underline{3.56}\ \text{B/D-psi.}$$

$$I_{(ideal)} = \frac{(1,426)}{(400) - (-421)} = \underline{1.73}\ \text{B/D-psi.}$$

$$\text{Flow Efficiency} = \frac{I_{(actual)}}{I_{(ideal)}} = \frac{3.56}{1.73} = \underline{2.06}.$$

Assuming $S_o = 0.20$, $S_g = 0$ in the swept zone, we have

$$c = c_t = S_o c_o + S_w c_w + c_f,$$
$$= 0.20\,(3 \times 10^{-6}) + 0.80\,(3 \times 10^{-6}) + 4.0 \times 10^{-6},$$
$$= 7.0 \times 10^{-6}\ \text{psi}^{-1}.$$

Example 2: Pressure Fall-Off Analysis Prior to Reservoir Fill-up, Unit Mobility Ratio

We have the following data on this well: $i = 1,020$ b/d, $h = 45$ ft, $d_t = 6.366$ in. (casing diameter, since no tubing in this well), $p_w = 598$ psi, $p_t = 0$ (zero wellhead injection pressure), $W_i = 6,097$ bbl, $\mu = 0.9$ cp (at 88°F and 8 percent NaCl), $\rho = 1$ gm/cc, $r_w = 1$ ft (estimated on basis of sand removed during swabbing), $\phi = 0.3$, $S_w = 0.32$, $S_g = 0.12$, $S_{gr} = 0$, $S_o = 0.56$, and $S_{or} = 0.2$.

From Fig. 5–70, we find $b_1 = 347$ psi and $\beta_1 = 2.303 \cdot (\log 347 - \log 122)/20 = 0.0522$ hours^{-1}. We have for the case where the pressure drops to zero shortly after closing in,

$$C_1 = \frac{0.0538\ d_t^2 \beta_1 b_1}{\rho i}$$

$$= \frac{(0.0538\,(6.366)^2 (0.0522)\,(347)}{1(1,020)}$$

$$= 0.0386$$

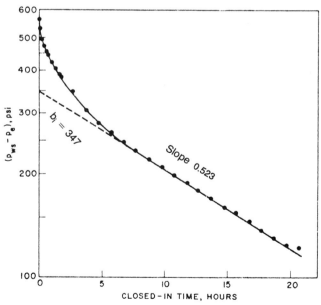

Figure 5–70. Pressure fall-off curve; p_e is known from a lengthy shut-in period at 32 psi (from SPE Monograph 1, 1967).

$C_2 = 0$, since $p_t = 0$

$$C_3 = C_1 \cdot \frac{p_w - p_c}{b_1} = 0.0386 \left(\frac{566}{347}\right) = 0.0628$$

Then,

$$\theta = \frac{C_1(1 - C_3)}{2(1 - C_1 - C_2)} = \frac{0.0386(0.937)}{2(0.961)} = 0.0188$$

$f(\theta) = 176.5$. From Fig. 5–69B,

$$kh = \frac{i\mu}{b_1} \cdot \frac{(1 - C_1 - C_2)}{(1 - C_3)^2} \cdot f(\theta) = \frac{1,020(0.9)}{347}$$

$$\frac{(0.961)}{(0.937)^2} \cdot 176.5 = 511 \text{ md-ft}$$

and

$$s + \ln \frac{r_e}{r_w} = \frac{0.00708(p_w - p_e)}{i_\mu / kh}$$

$$= \frac{0.00708(566)}{1,020(0.9)/511}$$

$$= 2.23$$

$$r_e = \sqrt{\frac{W_i(5.615)}{\pi \phi (S_g - S_{gr})h}}$$

$$= \sqrt{\frac{6,097(5.615)}{\pi (0.3)(0.12)(45)}}$$

$$= 82 \text{ ft}$$

Therefore,

$$s = 2.23 - 2.303 \log (82/1) = -2.18$$

Example 3: Pressure Fall-Off Analysis, Non-Unit Mobility Ratio

From the information given for the preceding example, for $\mu_o = 12$ and for $k_w/k_o = 0.3$, we have

$$\frac{V_o}{V_w} = \frac{S_o - S_{or}}{S_g - S_{gr}} = 3 \text{ and } M = \frac{k_w \mu_o}{k_o \mu_w} = \frac{0.3(12)}{0.9} = 4$$

The ratio k_w/k_o is obtained by measuring core permeability to water at the saturation in the water bank and core permeability to oil at the saturation in oil bank.

$$r_{oD} = \frac{1}{\sqrt{\dfrac{V_o}{V_w} + 1}} = \frac{1}{\sqrt{4}} = 0.5$$

Further, $M = 4$ and $\gamma = 1$ since $c_o \cong c_w$ for this dead oil. Therefore, reading F (Fig. 5–71), we obtain:

$$k_w h = \frac{i_w \mu_w}{b_1} \cdot 2F$$

$$= \frac{1,020(0.9)}{347} \cdot 2(220)$$

$$= 1,170 \text{ md-ft}$$

This value of $k_w h$ is 2.28 times as large as obtained for the single-fluid case.

The skin effect is found from:

$$s = \frac{0.00708(p_w - p_e) k_w h}{i_w \mu_w} - \frac{M - 1}{2} \ln \left(\frac{V_o}{V_w} + 1\right) - \ln \frac{r_e}{r_w}$$

$$= \frac{0.00708(566) \, 1,170}{1,020(0.9)} - \frac{4 - 1}{2} \ln 4 - \ln \frac{82}{1}$$

$$= 5.10 - 2.08 - 4.41$$
$$= -1.39$$

This value of s is less negative (indicating a smaller effective wellbore radius) than the value obtained in the single-fluid case. Thus, use of the single-fluid case has given too large a value for effective wellbore radius and, as noted above, too small a value for kh. This is the result one finds when the water mobility is greater than the oil mobility ($M > 1$). By obtaining too large an effective wellbore radius from use of the single-fluid case, the engineer may incorrectly decide that there is little possibility of injectivity improvement by well stimulation. Use of the proper mobility ratio would lead to a proper recommendation.

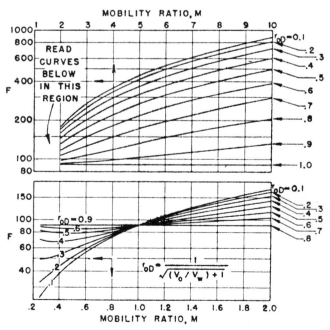

Figure 5–71. *Function for calculating kh, γ = 1 (from SPE Monograph 1, 1967).*

Example 4: Two-Rate Injection Test

An example two-rate injection test is shown in Fig. 5–72 for a 10,000-ft well. In preparation for the test, the well was stabilized at an injection rate of 2,563 b/d. To obtain the transient pressure data the rate was reduced to 742 b/d.

By trial and error as shown on the plot, the average pressure in the region around the well was found to be 3,600 psig. The average pressure is chosen as the highest value for which the plot in Fig. 5–72 is linear at large time. Note that the curves bend down at large time for greater assumed values of $\bar{p}$. Further note that an injection time of 48 hours after the rate change was required to obtain this value of static pressure.

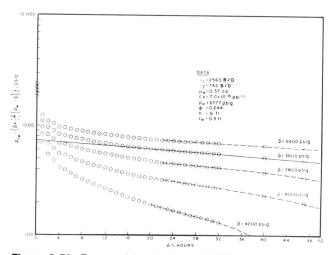

Figure 5–72. *Two-rate injection (from SPE Monograph 1, 1967).*

At 32 hours' injection time, the plot is linear for values of $\bar{p} < 4,200$ psig, as shown. Thus one might have estimated the average pressure as some 600-psi higher at the shorter time.

Data pertinent to the analysis of this test are shown in Fig. 5–72. From the solid line on this figure we find

$$b = 740 \text{ psig (intercept)}$$

and β (absolute value of slope) as

$$\beta = (\log 740 - \log 460)/50 = 0.00413 \text{ hours}^{-1}$$

To determine the formation permeability, we use

$$kh = \frac{181.2\,(i_1 - i_2)\mu}{b}$$

$$= \frac{181.2\,(2,563 - 742)\,0.37}{740}$$

$$= 165 \text{ md-ft}$$

$$k = 5.32 \text{ md}$$

To find the skin factor.

$$s = 1.283 \left(\frac{p_w - \bar{p}}{b}\right) \cdot \left(\frac{i_1 - i_2}{i_1}\right) - 1.151 \log \frac{0.000664\,k}{\beta\phi\mu c r_w^2}$$

$$= 1.283 \left(\frac{6,777 - 3,600}{740}\right) \cdot \left(\frac{2,563 - 742}{2,563}\right) - 1.151 \log$$

$$\frac{0.000664\,(5.32)}{4.13 \times 10^{-3}\,(0.244)\,0.37\,(7.0 \times 10^{-6})\,0.3^2}$$

$$s = -4.36$$

These values agree quite well with those determined by other methods for wells in this reservoir.

Using Early and Late Parts of Transient Data

Use of Earlier Part of Build-up Curve

This method was illustrated by Russell, *JPT* (December 1966), and involves plots of the normal MDH curves and a plot of $(p_{ws} - p_{wf})/(1 - 1/C_2\Delta t)$ versus log time as shown in Figs. 5–73 and 5–74. The use of the extended data is based on the following equations:

$$kh = 162.6\,q_{sc}\mu B/(\text{slope of line } B)$$
$$s = 1.151\{[(p_{ws} - p_{wf})/(1 - 1/C_2\Delta t)]_{\Delta t=1hr}/\text{slope}$$
$$- \log k/\phi\mu c_t r_w^2 + 3.23\}$$
$$p^* = p_{wf} + \text{slope}\,(\log kt/\phi\mu c_t r_w^2 - 3.23 + 0.87s)$$

The curve based on these data has not reached a straight line for the MDH method, and kh values calculated from there are too high. The slope should have been read at later times and weight given last points was too great. The results as presented by Russell follow:

	Conventional analysis	Extended data analysis
kh, md ft	78	92.8
k, md	19.5	23.2
s	2.94	4.39
p*, psig	3455	3300

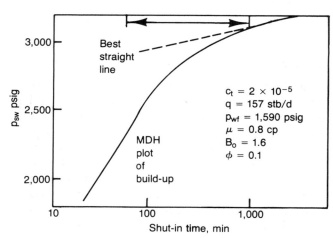

Figure 5–73. *Normal MDH plot (after Russell, JPT, December 1966).* Data used for extended analysis.

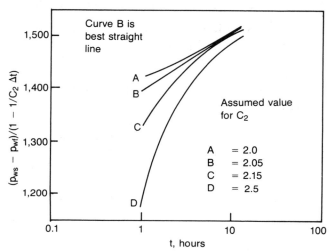

Figure 5–74. *Use of extended data (after Russell, JPT, December 1966).*

It is probable that the MDH had not reached the desired values useable in equations as the proper slope and results for extended data are more useful.

Extended Muskat Method

This method extends the useable portion to the later parts of the build-up curve. The basic equation is:

$$\log(\bar{p}-p_{ws}) = \log(118.6\ q_{sc}\ \mu B/kh) - 0.00168\ k\Delta t/\phi\mu c_t r_e^2$$

The theory implies that the well flowed at semisteady state prior to the build-up test and that the exponential terms associated with transient flow are insignificant. Some drainage boundary effects should be exhibited by the data.

A plot showing $(\bar{p}-p_{ws})$ versus shut-in time should be a straight line, as shown in Fig. 5–76.

The slope of this line is $\beta = 0.00168\ k/\phi\mu c_t r_e^2$; the intercept of of the line with the zero value of shut-in time is $b = 118.6\ q_{sc}\ \mu B/kh$.

$$s = 0.84\left[(\bar{p}-p_{ws})/b\right]$$
$$\qquad - 1.151 \log\left[(0.00168k/\phi\mu c_t r_w^2)(1/\beta)\right] + 0.75\right]$$
$$p_{skin} = bs/0.84$$
$$V_p = 0.1115\ qB_obc_t\beta$$

The routine MDH plot and the extended data plot of Muskat are shown in Figs. 5–75 and 76. Russell presents the following results in *JPT* (December 1966):

	Conventional analysis	Extended Muskat method
k, md	5.14	4.82
s	+0.13	−0.18
p*	2,680	
V_p		7.229×10^4

Results are based on Figs. 5–75 and 76.

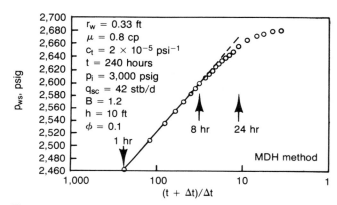

Figure 5–75. *Pressure build-up curve (after Russell, JPT, December 1966).*

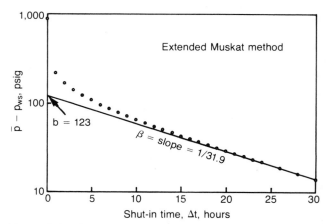

Figure 5–76. *Muskat plot (after Russell, JPT, December 1966).*

Drill-stem Tests

The normal Horner and/or Miller, Dyes, and Hutch-inson methods are applicable to drill-stem tests. At times, shortcut methods as used by service companies in field analysis are reliable.

The drill-stem test often uses two bombs, and one or more flow and shut-in sequences are recorded, as illustrated in Figs. 5–77 and 78.

The following section, taken primarily from the *Journal of Canadian Petroleum Technology,* April–June, 1972, illustrates some of the drill-stem tests that have been published.

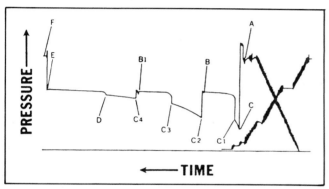

Figure 5–77. *Typical pressure chart; Johnston gauge; triple flow and shut-in.*

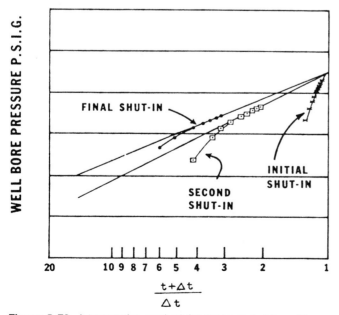

Figure 5–78. *Interpretation method for pressure build-up; Horner plot with minimum afterflow obtainable with packer.*

Example of a Normal Routine Drill-Stem Test

The first flow is very short and is designed to remove any excess pressure which may have resulted from setting the packers. The first build-up is rather long since reliable value for the initial reservoir pressure is desired. The second flow is somewhat longer and is designed to evaluate the formation for some distance from the well. The second shut-in is used to calculate transmissibility and other characteristics of the reservoir. If the second extrapolated pressure is less than the pressure of the first shut-in, depletion of the small reservoir should be suspected.

Drill-Stem Test Involving Single Flow and Build-Up

A. Starting in hole with tool and gauge. Jagged line caused by drag of packer.
B. Tool on bottom. Pressure measures weight of mud column.
C. Packer set. Setting packer usually causes small increase in pressure.
D. Tool opened. Gauge exposed to well bore opposite formation. Flow is at highest rate initially and decreases as fluid column rises inside pipe.
E. Tool shut in. Start of pressure buildup. Buildup equations applicable.
F. Equalizing valve opened. Pressure measures weight of mud column.
G. Starting out of hole.
H. Tool out of hole. Pressure on gauge is atmospheric.

Interpretation of Drill-Stem Test Charts

Many variations of drill-stem test charts are possible. The flow and build-up equations should be studied to obtain an understanding of how fluid and reservoir characteristics influence the shape of drill-stem test curves and how they determine the location of the curves on the charts (see SPE Monograph 1). Also, field information such as flow rates, back pressures, pressure differentials, fluids and amounts recovered, rates of recovery, time of change of operations at the surface all should be carefully reviewed before reaching a decision as to the cause of the specific shape and location of the drill-stem curve on the chart. Two gauges are usually run and their differences may explain what actually happened at the subsurface. Chart interpretation is an art rather than a science, and experience and technological knowledge are both important.

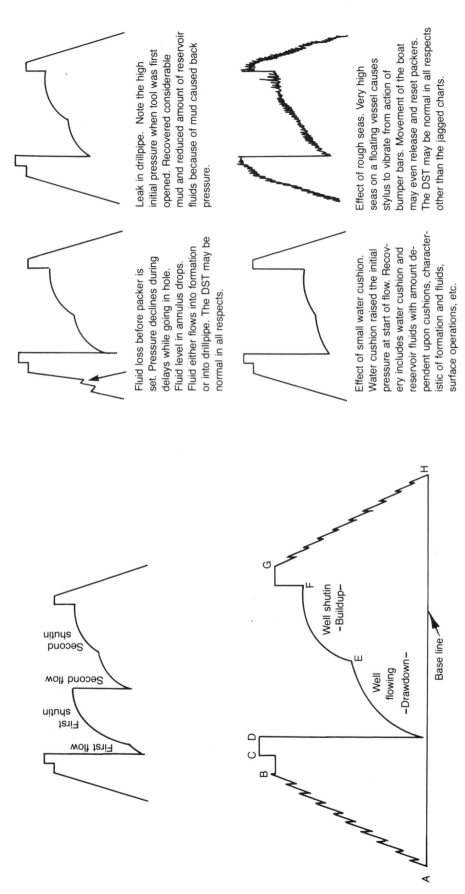

Leak in drillpipe. Note the high initial pressure when tool was first opened. Recovered considerable mud and reduced amount of reservoir fluids because of mud caused back pressure.

Effect of rough seas. Very high seas on a floating vessel causes stylus to vibrate from action of bumper bars. Movement of the boat may even release and reset packers. The DST may be normal in all respects other than the jagged charts.

Fluid loss before packer is set. Pressure declines during delays while going in hole. Fluid level in annulus drops. Fluid either flows into formation or into drillpipe. The DST may be normal in all respects.

Effect of small water cushion. Water cushion raised the initial pressure at start of flow. Recovery includes water cushion and reservoir fluids with amount dependent upon cushions, characteristic of formation and fluids, surface operations, etc.

First flow
First shutin
Second flow
Second shutin

Well shutin
—Buildup—

Well flowing
—Drawdown—

Base line

A
B
C D
E
F
G
H

294

Flow by heads, swabbing, etc. There are many variations caused by swabbing, flow by heads, choke plugging, etc. Recovery includes reservoir fluids. Swabbing may cause a pressure decline if fluid levels are lowered.

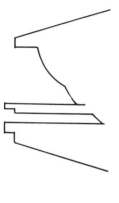

The S-shaped curve. Shut-in at surface, return of gas into solution, small permeable zone in tight liquid filled formation, etc., give such curve during build-up. Recovery includes reservoir fluids in normal amounts.

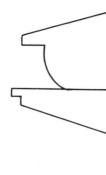

Run away clock. Clock spring released when tool was opened. Recovery included formation fluids in normal amounts.

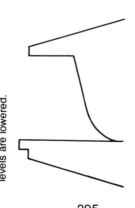

Tool failed to close. No build-up is obtained. Recovery includes formation fluids.

Tool failed to open. No fluid except maybe a little mud is recovered. Measured pressure may tend to decline to that in the reservoir.

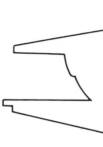

Clock did not run. Recovered reservoir fluids in normal amount.

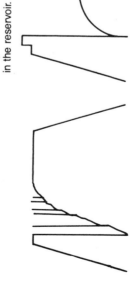

Packer failed and could not be reset. Recovered mud with very little formation fluid.

Packer failed at time of shutin. Recovered reservoir fluids in normal amounts and some mud as required to equalize pressure.

Packer failed, reset, and held. Recovered some mud and reservoir fluids in reduced amounts.

Uniform slippage in clock mechanism. Time scale is in error. Recovered reservoir fluids in normal amounts.

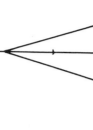

Clock stopped on opening tool and started again when tool was shut in. Recovered reservoir fluids in normal quantities.

Clock stopped upon reaching bottom and started when started coming out of hole. Recovered reservoir fluids in normal quantities.

295

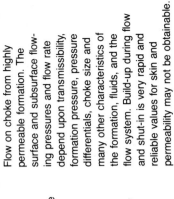

Water produced and well died. Weight of water and a little mud in pipe exceeded reservoir pressure and well then died.

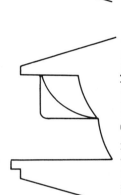

Low formation permeability. Recovered very little mud and some formation fluid.

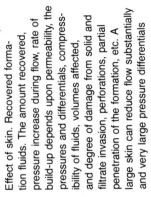

Flow on choke from highly permeable formation. The surface and subsurface flowing pressures and flow rate depend upon transmissibility, formation pressure, pressure differentials, choke size and many other characteristics of the formation, fluids, and the flow system. Build-up during flow and shut-in is very rapid and reliable values for skin and permeability may not be obtainable.

Flow string plugs immediately above gauge. Recovered little fluid, mostly mud. Pressure builds quickly to that of reservoir.

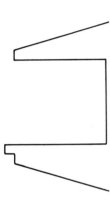

No formation permeability. Small amount of mud may be recovered with very little formation fluid.

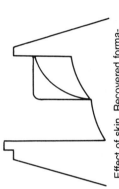

Effect of skin. Recovered formation fluids. The amount recovered, pressure increase during flow, rate of build-up depends upon permeability, the pressures and differentials, compressibility of fluids, volumes affected, and degree of damage from solid and filtrate invasion, perforations, partial penetration of the formation, etc. A large skin can reduce flow substantially and very large pressure differentials can result from skins.

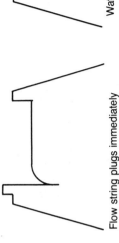

Stylus dragging. Pressure chart has stair step character. Stylus needs adjustment. Drillstem test normal in all respects.

Re-solution of gas in drillpipe when well is shut-in at surface. Test is probably normal in all respects.

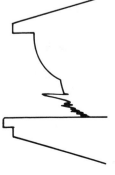

Gradual plugging of flow stream below gauge. Pressure declines as flow rate decreases to a value equal to weight of fluids above gauge. Recovered a little mud and small amount of reservoir fluids.

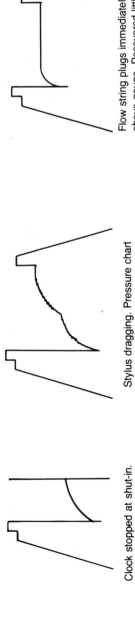

Clock stopped at shut-in. Drillstem normal. Recovered reservoir fluids in normal quantities.

Effect of large super-pressure. Pressure build-up during flow and build-up period is more rapid than usual. Mud recovery also is probably greater. Highest pressure may or may not exceed normal reservoir pressure.

Plugging and unplugging of flow stream at some location above the gauge. Recovered reservoir fluids, possibly in reduced amounts.

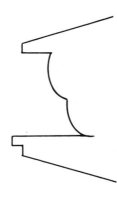

Gauge plugged after packer was set but before tool was opened. Unplugged at lower pressure while coming out of hole. Reservoir fluids recovered in normal amounts.

Gauge plugged during flow and unplugged late in build-up. Reservoir fluids recovered in normal amounts.

Well interference. The DST is usually too short to notice interference with today's well spacing. Pressure declines at long times. Reservoir fluids may be recovered in about normal quantities.

Gauge plugged with tool on bottom before packer was set. Unplugged at same pressure when coming out of hole. Reservoir fluids recovered in normal amount.

Gauge gradually plugged during flow period. Unplugged when coming out of hole. Reservoir fluids recovered in normal quantities.

Hole in chart

Stylus tore chart and could not move further. Reservoir fluids recovered in normal amounts.

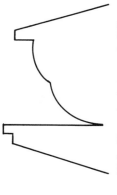

Well with moderate transmissibility. Formation fluids flow at surface. The shape of curves and time fluids reach surface depend upon permeability, viscosity, volume and height of drillpipe, friction, pressures, fluid densities, etc.

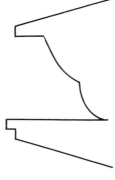

Two layer effects. Caused by two producing zones, poorly connected lenses, fault system, discontinuity, fluid boundaries, etc. Curves change slope — often during flow and build-up and such change may be in two directions depending upon conditions.

Gauge plugged while going into hole. Plugging occurred when fluid weight equalled recorded pressure and unplugged at lower pressure when coming out of hole. Normal recovery of reservoir fluids.

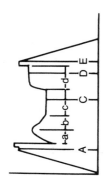

Unloading water cushion in gas well.
a. Water cushion rising to surface.
b. Water cushion being produced
c. Flowing dry gas with choke

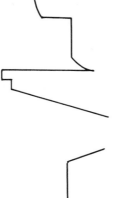
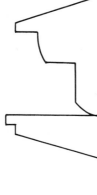

Excessive fluid head inside pipe. Flowing pressure upstream of the surface or subsurface choke remained constant until back pressure due to liquid accumulation became large, resulting in a lower flow rate.

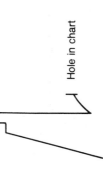

Change in size of pipe string. The change in slope during flow may be in either direction depending upon where larger pipe is located in string. Normal recovery of reservoir fluids.

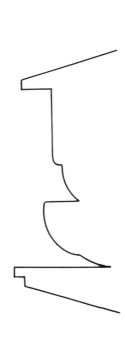

Single test with two gauges. Left gauge suggests highly permeable formation with little or no skin — maybe negative — while right curve indicates gauge which plugged upon reaching bottom and unplugged when starting out of hole. If right gauge was at bottom, it was probably stuck in debris at bottom of hole.

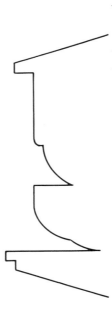

Single test with two gauges. Left gauge shows gradual plugging below gauge while right gauge shows actual behavior of the reservoir. Left gauge measures weight of fluids above gauge rather than reservoir characteristics. The two are not identical because a plug in the flow string has removed the connection between the two pressure sources which are in balance in the normal test.

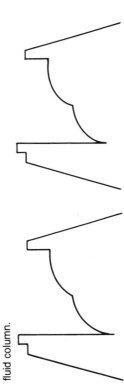

Single test with two gauges. Left gauge measures reservoir characteristics while right gauge plugged while going into the well and did not unplug until inspected at surface.

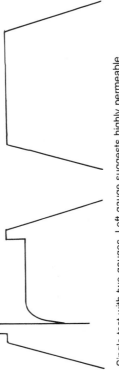

Two tests with the same gauge. Second build-up extrapolated pressure is lower than that of first build-up. Depletion of a small reservoir might be suggested.

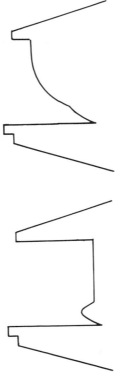

Two tests with the same gauge. Character of the curves for the second test differs appreciably from that of the first test. Skin or other parameter in the flow and build-up equations which might be sensitive to flow or shutin has changed between the two tests. Since some fluid entered drillstring during first test, initial pressure of second test will be higher than that of first test by the weight of this fluid column.

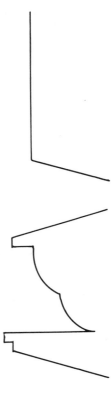

Single test with two gauges. The two sets of curves should be identical except for the small difference in pressure related to location of the gauges within the flow string. This is the normal condition to be expected if both gauges are operating properly and no plugging, etc., is occurring.

298

Productivity (PI) Index

During the early life of the oil business, the PI was defined as rate/pressure drop. Later, the specific production index, SPI, was defined as:

$$SPI = (3.073\ k)/\mu B \log\ (r_e/r_w) = q/h(p_e - p_{wf})$$

When water was produced, the input was corrected by a term (1-water).

As the theory for oil production and flow was developed, the more complicated flow equation,

$$q_o = (2\pi kh/\ln\ (r_e/r_w) \int_{p_w}^{p_e} (k_{or}/\mu_o B_o)\ dp$$

was accepted by engineers. Two-phase flow adjustments may be required.

Unfortunately, everyone finds it difficult to adjust to advancement in technology, and the emperically derived PI index has been adjusted rather than use basic flow equations. By definition the relative permeability, oil viscosity, and formation volume factor change with pressure and pressure changes with time in most reservoirs with production. At higher production rates, the values also change with distance at any specified time. Average values then must be used in all methods of analyses. The overall effect of these changes are illustrated by Fig. 5–79.

These graphs show that the flow rate is not proportional to pressure decline during any single test and that substantial changes in flow rate occur when the pressure in the reservoir also declines during the producing life of a well and reservoir. The curves are not straight lines on coordinate paper, but straight lines often are noted when data are plotted on log-log paper. The equation of the log-log plot straight lines is $q = k(\Delta p)^n$ where n is the slope determined from the log-log plot and k is an intercept. The productivity index for any single test—at a single time—is a constant only if n is one or the slope of log-log plot is one. Such slope is possible in very good permeable pays, which were prevalent when the empirical relations were first noted in well tests.

Vogel developed a method which handles the problem for a single test when no skin is present. Standing added a correction factor when skin was present, as shown in Figs. 5–80, 81, and 82.

Actual well performance and theoretical study suggest that the flow capacity of a well producing under solution-gas drive mechanism during a single test time period can be approximated by the equation:

$$q_o/q_m = 1.0 - 0.20\ (p_{wf}/\bar{p}) - 0.08\ (p_{wf}/\bar{p})^2$$

where q = producing rate at well intake pressure (p_{wf})

q_m = producing rate at maximum intake pressure

$\bar{p}$ = average reservoir pressure

The use of this graph can be understood by studying the example. The Standing correction is simply a correction to Vogel results.

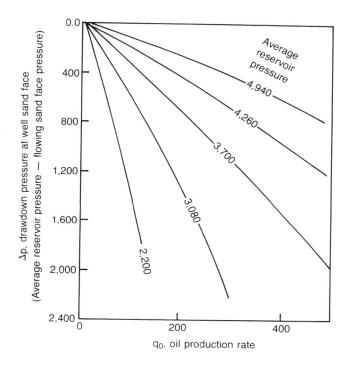

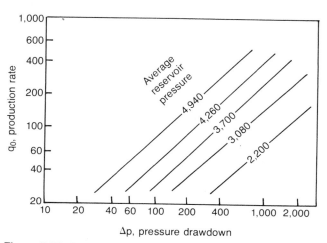

Figure 5–79. *Production characteristics of a well producing under a gas displacement mechanism.*

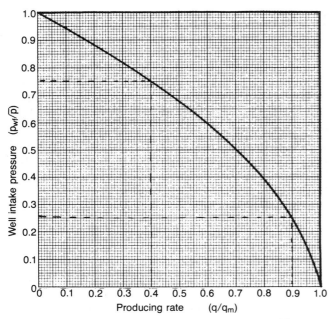

Figure 5–80. *Vogel curve (from* Trans. *AIME, 1968).*

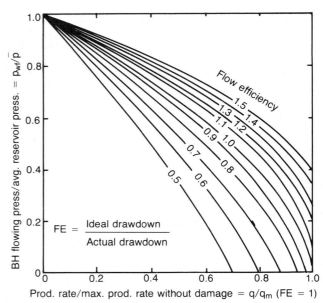

Figure 5–82. *Standing correction (after* JPT, *November 1970).*

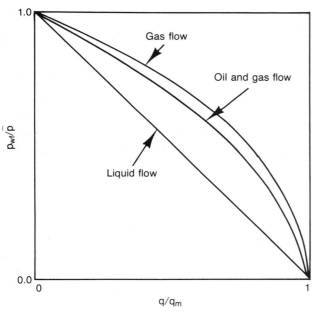

Figure 5–81. *Typical Vogel curves.*

The productivity index relationship is

$$J_o = q_o/(\bar{p} - p_{wf})$$

where

$$\bar{p} > p_{wfd} > p_b$$

For cases where $\bar{p} > p_b > p_{wf}$, Fetkovich proposed an equation which combines single-phase and two-phase flow:

$$q_o = J_o' \, (p_b^2 - p_{wf}^2)^n + J_o \, (\bar{p} - p_b)$$

Fetkovitch presented a method to determine n and J_o' from individual well multirate and pressure tests, or isochronal tests. L. D. Patton (*Petroleum Engineer,* September 1980) presented a method for shifting the axes of the Vogel plot for use when data required by Fetkovitch are not available.

Example—Vogel Method Only

A well tests 65 barrels of tank oil per day with a flowing bottom-hole pressure of 1,500 pisa. The shut-in in bottom hole pressure is 2000 psi, and this value is assumed equal to the average pressure in the reservoir at the time.

1. The maximum theoretical producing rate is:

When $p_{wf} = 1500$ psia:
$$p_{wf}/\bar{p} = 1500/2000 = 0.75$$

Corresponding

$$q/q_m = 0.40 \text{ and}$$
$$q_m = 65/0.40 = 161 \text{ bo/d}$$

2. What is the flow rate when sand-face pressure is reduced to 500 psia?

When $p_{wf} = 500$ psia,
$$p_{wf}/\bar{p} = 500/2000 = 0.25$$
$$q/q_m = 0.90$$
$$q = 0.90 \times 161 = 145 \text{ bo/d}$$

The PI below the bubble point is strongly influenced by the change in relative permeability, but expected PVT functions cause a rather small change in PI with time (Fig. 5–83).

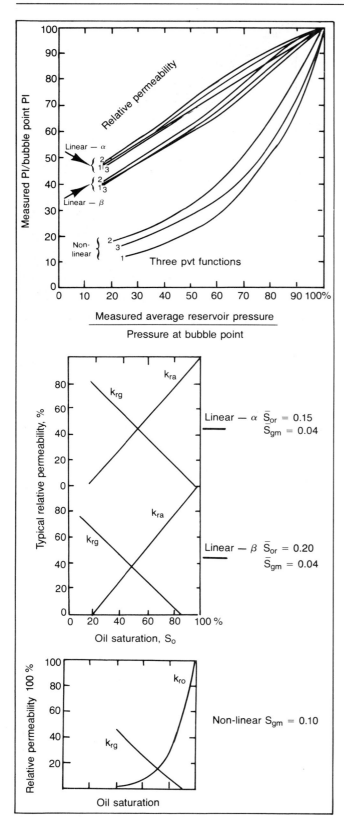

Figure 5–83. *Influence of pvt and relative permeability on PI (courtesy Shell Oil Company).*

Curve-Match Techniques

Curve match techniques should be used only when more accurate techniques are not useable. Always check results with other methods, such as the MDH and Horner semilog techniques when possible. Earlougher and Kersch (*JPT,* July 1974) recommend using Fig. 5–85 for curve match when test time is short. Use of the extended semilog methods also should be considered.

The basic curve uses a dimensionless wellbore storage coefficient defined as

$$C_D = 0.89359 \ C/\phi \ c_t \, hr_w^2$$

where C is wellbore storage coefficient, RB/psi. When skin is present, the equation is changed to

$$C_D \, e^{2s} = 0.89359 \ C \, e^{2s}/\phi \ c_t hr_w^2$$

The basic problem with curve matching is that the base curves all have similar shape. Selection of a unique match usually is difficult—one can only select a wide range for the match of the field data with the base or theoretical curves.

The procedure for using the curve match follows:

1. Plot observed test data as $\Delta p/\Delta t$ on the ordinate and Δt on the abscissa on log-log paper, having exact scales as shown in Fig. 5–84. Place tracing paper over the figure and plot on the tracing paper after tracing the major grid lines for reference and matching purposes. The field data are plotted on the tracing paper using the grid so that scales are exactly the same; p should be in psi and time in hours.
2. Estimate the wellbore storage coefficient from other data sources using $C = V_w c_w$ where V and c are volume of the well and compressibility of well fluids if the well is completely filled. For a well with a rising or falling liquid level, use $C = V_u/(pg/144g_c)$ where V_u is wellbore volume per unit length, B/ft, and g_c is a units conversion factor, 32.17 (g is in ft/sec and also is 32.174).
3. Calculate the location of the horizontal asymptote on the data plot. $(\Delta p/\Delta t)_{1.0} = qB/24C$, and the quantity on the left-hand side of the equation is the value of $\Delta p/\Delta t$ observed on the data plot when $(\Delta p/\Delta t)(24C/qB)$ Fig. 5-84 $= 1.0$.
4. Place the tracing paper over Fig. 5–84 so that the asymptote calculated above overlies the value of one on the ordinate. This implies that $(\Delta p/\Delta t)(24C/qB) = 1.0$.

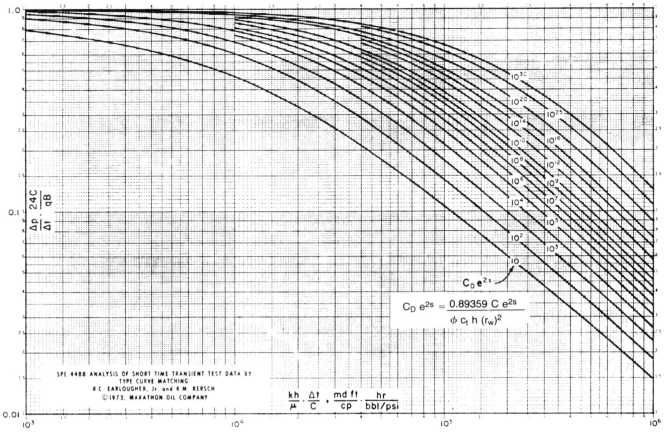

Figure 5–84. *Master type curve (after Earlougher, reprinted by permission of Marathon Oil Co.)*

5. Slide the tracing paper horizontally until the best match is obtained with one of the curves. A slight vertical adjustment also may be required.

6. Sketch the matched curve on the tracing paper and read the value of $C_D e^{2s}$ for the curve so matched. Pick any convenient matched point—any point on graph—and read values for $(\Delta p/\Delta t)$ and Δt from the tracing-paper curve. Likewise, read the corresponding points from Fig. 5–84 to obtain values for $(\Delta p/\Delta t)(24C/qB)$ and $(kh/\mu)(\Delta t/C)$ for the match point.

7. If any vertical movement was required to obtain a match, recalculate the wellbore storage coefficient for new values using match-point data.

$$C = qB[(\Delta p/\Delta t)(24C/qB)_{\text{Fig. 5-84 M}}]/24(\Delta p/\Delta t)_M$$

8. Calculate transmissibility

$$kh/\mu = [C(kh/\mu)(\Delta t/C)_{\text{Fig. 5-84 M}}]/(\Delta t)_M$$

9. Calculate skin factor

$$s = \tfrac{1}{2} \ln \{[\phi c_t h r_w^2 (C_D e^{2s})_M]/0.89359\,C\}$$

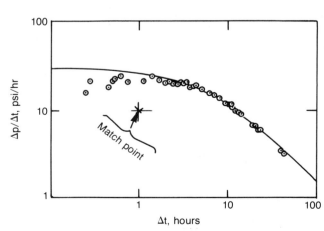

Figure 5–85. *Curve match techniques (after Earlougher, SPE Monograph 5, 1977 © SPE).*

Earlougher and Kersch present an example for a pumping well where exact correlation of pressure with time can cause scatter at early times. The scatter may also be explained for other reasons. In any event, such effects die out in later data of long tests. For such a well, the wellbore fills slowly and C can be calculated from the pressure data to have a value of $C = 0.095$ RB/psi. We also estimate $h = 25$ ft, $\phi = 20\%$. $r_w = 4.5$ in. for shot hole, $B = 1$ RB/stb, $c_t = 1 \times 10^{-5}$ psi^{-1}, $q = 66$ stb/d and $V_\mu = 0.0411$ bbl/ft.

The value calculated for C, the wellbore storage coefficient, is accurate and is used to select a point on a vertical scale for the field data plotted on the tracing paper. This value aids in matching the two sets of curves, and $C e^{2s}$ is 10^4 for the curve so selected. Since the value is the best possible, recalculation of C is unnecessary. The match points read from the two graphs—tracing and Fig. 4-84—are recorded as $(\Delta p/\Delta t) = 10$ psi/hour, Δt is one hour, $(\Delta p/\Delta t)(24C/qB) = 0.345$ and $(kh/\mu)(\Delta t/C) = 4200$ (md ft/cp)(hr/bbl/psi).

$$(\Delta p/\Delta t) = qB/24C = 66 \times 1/24 \times 0.095 = 29 \text{ psi/hr}$$

$$kg/\mu = 0.095 \times 4200/1 = 400 \text{ md ft/cp}$$

$$s = \frac{1}{2} \ln (0.2 \times 10^{-5} \times 25 \times 4.5 \times 4.5 \times 10^4/12 \times$$
$$12 \times 0.89359 \times 0.095)$$
$$= -0.1$$

Use of Curve-Matching Techniques when Build-up Time is Long*

The curve-matching technique is based on a match of (1) a plot of field-measured pressure differentials versus flow time, or $(\Delta t)/(t + \Delta t)$, plotted on log log paper with (2) the theoretical p_D versus t_p curve plotted on identical log log paper. For build-up, Δt may be used when the flow time to reach the drainage boundary is large compared with the shut-in time. The slope of these curves changes slowly and a reliable overlay is difficult.

The problem may be reduced by extrapolating the straight-line portion of the Horner plot of pressure versus $(\Delta t)/(t + \Delta t)$ on semilog paper (see Graph 1, Fig. 5-86) in a backward direction. The data so obtained is plotted on log log paper (see Graph 2). The log-log plot is overlayed on the p_D versus $\frac{1}{2}(\ln t_D + 0.809)$ plot. After a reasonable match is achieved, values of pressure, time, p_D and t_D are read for a match

point selected so that t_D is greater than 25. These values are used to calculate values for kh, k effective well radius r'_w, etc. Such values will be comparable to the values calculated using the Horner procedures as shown in the table.

Comparison of Results

	Field curves on Graphs 2 & 1		Values calculated using Graphs 2 & 3		Values calculated using Graph 1		
Code	$P_{1\,hr}$	p_{wf}	kh	r'_w	kh	s	r'_w
A	2797	2700	4297	40.5	4297	−5.3	45.0
B	2810	2700	4670	37.8	4743	−5.1	36.3
C	2821	2700	5896	26.1	5837	−4.7	25.2
D	2797	2700	4193	51.3	4295	−5.3	45.0
E	2810	2700	4865	40.0	4743	−5.1	36.3
F	2821	2700	5831	26.4	5837	−4.7	25.2
G	2797	2200	4188	.2	4295	− .1	.3
H	2810	2200	4663	.1	4743	− .8	.5

kh is in md ft, r'_w is in ft and p in psia

The shapes of the various graphs, such as linear, spherical, Horner, Miller Dyes and Hutchinson, and log-log plots, are most useful in the interpretation of reservoir characteristics, recognizing errors in data, etc.

Reference

"Advances in Well Test Analysis," SPE Monograph 5.

---

* Material in this section is from "Continuous Tables," *Petroleum Engineer.*

Graph 1. Horner and MDH plot of field buildup.

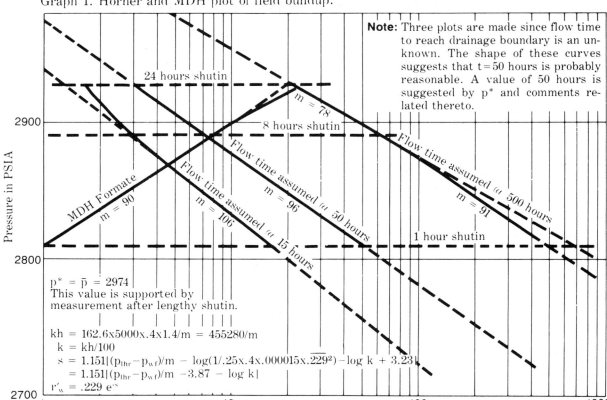

Note: Three plots are made since flow time to reach drainage boundary is an unknown. The shape of these curves suggests that t = 50 hours is probably reasonable. A value of 50 hours is suggested by p* and comments related thereto.

24 hours shutin

m = 78

2900

8 hours shutin

Pressure in PSIA

Flow time assumed @ 500 hours

MDH Formate
m = 90

Flow time assumed @ 50 hours

m = 91

Flow time assumed @ 96

m = 106

Flow time assumed @ 15 hours

1 hour shutin

2800

p* = p̄ = 2974
This value is supported by measurement after lengthy shutin.

kh = 162.6x5000x.4x1.4/m = 455280/m
k = kh/100
s = 1.151|(p_{1hr} − p_{wf})/m − log(1/.25x.4x.000015x.$\overline{229}^2$) − log k + 3.23|
  = 1.151|(p_{1hr} − p_{wf})/m − 3.87 − log k|
r'_w = .229 e^s

2700
1   10   100   1000
(t/Δt) + 1 and Δt

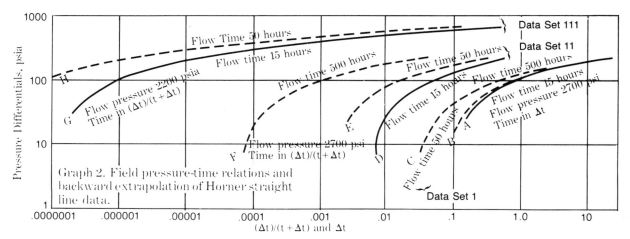

1000

Data Set 111

Flow Time 50 hours

Data Set 11

Flow time 15 hours

Pressure Differentials, psia

100

Flow time 500 hours

Flow time 50 hours

H

Flow pressure 2200 psia
Time in (Δt)/(t + Δt)

Flow time 500 hours

G

Flow time 15 hours

Flow time 50 hours

Flow time 500 hours

E

Flow time 15 hours
Flow pressure 2700 psi
Time in Δt

A

10

Flow pressure 2700 psi
Time in (Δt)/(t + Δt)

F

D

C

B

Flow time 50 hours

Graph 2. Field pressure-time relations and
backward extrapolation of Horner straight
line data.

Data Set 1

1
.0000001   .000001   .00001   .0001   .001   .01   .1   1.0   10
(Δt)/(t + Δt) and Δt

Figure 5–86. *Other curve-matching techniques* (*courtesy* Petroleum Engineer).

Conversion of Bottom-hole Pressure at Instrument Depth to Reservoir Datum

	Measured depth, ft	Subsea depth, ft

A. Instrument depth, feet below _____   _____   _____

B. Top of perforations; top of interval   _____   _____

C. Well gradient between instrument depth
and top of perforated interval, _____psi/ft

D. p at instrument depth as measured, _____psi

E. Gradient in reservoir oil gradient, _____psi/ft

F. Reservoir datum used   _____   _____

G. Top of perforations (interval) minus
instrument depth. B _____ − A _____   _____   _____

H. Top of pay interval minus reservoir
datum. B _____ − F _____   _____   _____

I. G _____ × C _____ = _____ psi

J. D _____ + I _____ = _____ psi

K. H _____ × E _____ = _____ psi

L. J _____ − K _____ = _____ psi

M. Reservoir pressure at datum

        L _____ + 15 psi = _____ psi

A schematic diagram useful in determining the average depth of a reservoir is shown in Fig. 5–87.

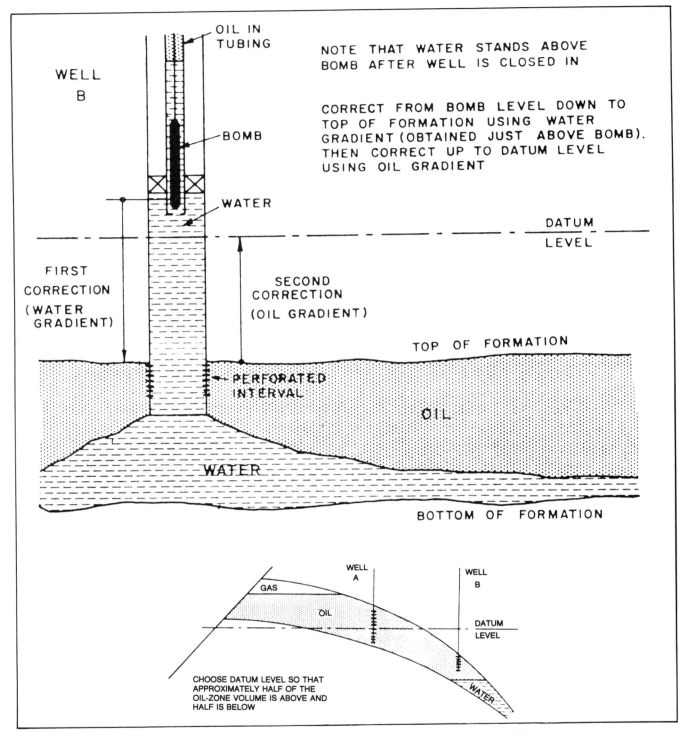

Figure 5–87. *Selection of reservoir datum (from SPE Monograph 1, 1967).*

Form for Reporting Analysis

Well Test date Type of test				

Time, hours
   Flow
   Shut-in				

Linear Flow
   Time interval
   Slope
     Horner
     MDH
     Log-log				

Spherical Flow
   Time interval
   Slope
     Horner
     MDH				

Radial flow
   Time interval
   Slope
     Horner
     MDH
     Log-log				

Pressure, psi
   Flowing
   Initial
   Last measured
   p^*
   At one hour				

Flow rate
   Oil, b/d
   Gas, Mcfd
   Water, b/d
   Used in analysis				

q/m
kh, md ft
   Linear
   Spherical
   Radial
   Used				

h, ft k, md—best value				
$(p_{1hr} - p_{wf})/m$ $\log(k/\phi\mu c_t r_w^2)$				

Skin
   Linear
   Spherical
   Radial
   Used
Δp_{skin}				

Well elevation, ft Bomb depth, ft Datum, ft				
Average reservoir pressure, psi Pressure @ datum, psi Remarks				

Time Scales for Flow and Build-up

Well _____
Test date _____

Flow rates
Oil _____ b/d
Water _____ b/d
Gas _____ Mcfd

Datum _____ ft   Temp. _____ °F
p_{wf} _____ psi
p_{1hr} _____ psi

z _____ c_t 0.000
μ _____ cp ϕ _____
r_w _____ ft S_w _____ S_g _____

$\Delta(m(p))/q$	q	$p_i - p_{wf}$	$m(p)$	Pressure, psi or psia	Δt	$\sqrt{\Delta t}$	$1/\sqrt{\Delta t}$	Radial $(t+\Delta t)/\Delta t$ $t=15$	$t=50$	$t=500$	Linear $\sqrt{t+\Delta t}-\sqrt{\Delta t}$ $t=15$	$t=50$	$t=500$	Spherical $1/\sqrt{\Delta t}-1/\sqrt{t+\Delta t}$ $t=15$	$t=50$	$t=500$
					0.25	0.500	2.000	61.00	201.00	2001.00	3.41	6.59	21.87	1.74	1.859	1.955
					0.50	0.707	1.414	31.00	101.00	1001.00	3.23	6.40	21.67	1.16	1.273	1.369
					0.75	0.866	1.154	21.00	67.67	667.00	3.10	6.26	21.51	0.90	1.014	1.109
					1.00	1.000	1.000	16.00	51.00	501.00	3.00	6.14	21.38	0.75	0.860	0.955
					1.25	1.118	0.894	13.00	41.00	401.00	2.91	6.04	21.27	0.63	0.724	0.859
					1.50	1.224	0.816	11.00	34.33	334.00	2.84	5.95	21.17	0.57	0.677	0.771
					1.75	1.323	0.756	9.50	29.57	286.00	2.77	5.87	21.08	0.52	0.617	0.711
					2.00	1.414	0.707	8.50	26.00	251.00	2.71	5.80	20.99	0.47	0.568	0.662
					2.50	1.581	0.632	7.00	21.00	201.00	2.60	5.67	20.84	0.39	0.494	0.587
					3.00	1.732	0.577	6.000	17.67	168.00	2.51	5.55	20.70	0.34	0.440	0.532
					3.50	1.871	0.534	5.29	15.28	143.00	2.43	5.44	20.57	0.30	0.397	0.489
					4.00	2.000	0.500	4.75	13.50	126.00	2.36	5.35	20.45	0.27	0.364	0.455
					4.50	2.121	0.471	4.33	12.11	112.00	2.30	5.26	20.34	0.24	0.336	0.426
					5.00	2.236	0.447	4.00	11.00	101.00	2.23	5.18	20.24	0.23	0.312	0.402
					6.00	2.449	0.408	3.50	9.33	84.33	2.13	5.03	20.05	0.19	0.274	0.363
					7.00	2.646	0.378	3.14	8.14	72.43	2.04	4.90	19.87	0.17	0.246	0.334
					8.00	2.828	0.354	2.88	7.25	63.50	1.97	4.79	19.71	0.14	0.223	0.310
					9.00	3.000	0.333	2.67	6.56	56.55	1.90	4.68	19.56	0.13	0.203	0.289
					10.00	3.162	0.316	2.50	6.00	51.00	1.84	4.58	19.42	0.12	0.187	0.272
					12.00	3.464	0.289	2.25	5.17	42.67	1.74	4.41	19.16	0.10	0.162	0.245
					14.00	3.742	0.267	2.07	4.57	36.71	1.64	4.26	18.93	0.08	0.142	0.233
					16.00	4.000	0.250	1.94	4.13	32.25	1.57	4.12	18.72	0.07	0.127	0.206
					18.00	4.243	0.236	1.83	3.78	28.78	1.50	4.00	18.52	0.07	0.115	0.192
					20.00	4.472	0.224	1.75	3.50	26.00	1.45	3.90	18.33	0.05	0.104	0.180
					22.	4.690	0.213	1.68	3.27	23.73	1.39	3.79	18.16	0.050	0.095	0.161
					25	5.000	0.200	1.60	3.00	21.00	1.32	3.66	17.91	0.042	0.085	0.156
					30	5.477	0.183	1.50	2.67	17.67	1.25	3.47	17.55	0.034	0.071	0.140

35	0.126	0.061	0.028	17.21	3.30	1.16	15.29	2.43	1.43	5.916	0.169
40	0.115	0.053	0.022	16.91	3.16	1.09	13.50	2.25	1.38	6.324	0.158
45	0.106	0.046	0.020	16.63	3.04	1.04	12.11	2.11	1.33	6.708	0.149
50	0.098	0.041	0.017	16.38	2.93	0.99	11.00	2.00	1.30	7.071	0.141
55	0.092	0.037	0.014	16.14	2.83	0.95	10.09	1.91	1.27	7.416	0.135
60	0.086	0.034	0.014	15.92	2.74	0.92	9.33	1.83	1.25	7.745	0.129
65	0.082	0.031	0.013	15.71	2.66	0.88	8.69	1.77	1.23	8.062	0.124
70	0.078	0.029	0.012	15.50	2.59	0.85	8.14	1.71	1.21	8.367	0.120
75	0.073	0.026	0.010	15.32	2.52	0.83	7.66	1.67	1.20	8.660	0.115
80	0.070	0.024	0.009	15.14	2.46	0.80	7.25	1.63	1.18	8.944	0.112
85	0.067	0.022	0.008	14.97	2.40	0.78	6.88	1.59	1.18	9.220	0.108
90	0.064	0.020	0.007	14.80	2.35	0.76	6.55	1.56	1.17	9.486	0.105
95	0.062	0.020	0.007	14.65	2.29	0.74	6.26	1.53	1.16	9.747	0.103
100	0.059	0.018	0.007	14.50	2.25	0.72	6.00	1.50	1.15	10.000	0.100
110	0.055	0.016	0.006	14.21	2.16	0.69	5.55	1.45	1.14	10.488	0.095
120	0.051	0.014	0.006	13.94	2.08	0.66	5.17	1.42	1.13	10.954	0.091
130	0.048	0.013	0.005	13.70	2.01	0.64	4.85	1.38	1.12	11.402	0.088
140	0.045	0.012	0.005	13.47	1.95	0.62	4.57	1.36	1.11	11.832	0.085
150	0.043	0.011	0.004	13.25	1.90	0.60	4.33	1.33	1.10	12.247	0.082
175	0.038	0.009	0.004	12.75	1.77	0.56	3.86	1.29	1.09	13.268	0.076
200	0.033	0.008	0.003	12.32	1.67	0.52	3.50	1.25	1.08	14.142	0.071
250	0.026	0.005	0.002	11.58	1.51	0.47	3.00	1.20	1.06	15.810	0.063
300	0.023	0.004	—	10.96	1.39	0.43	2.67	1.17	1.05	17.321	0.058
350	0.019	0.003	—	10.45	1.29	0.40	2.43	1.14	1.04	18.708	0.053
400	0.017	0.003	—	10.00	1.21	0.37	2.25	1.13	1.04	20.000	0.050
450	0.015	0.002	—	9.61	1.15	0.35	2.11	1.11	1.03	21.213	0.047
500	0.013	0.002	0.001	9.26	1.09	0.33	2.00	1.10	1.03	22.36	0.045
600	0.011	—	—	8.67	1.00	0.30	1.83	1.08	1.03	24.495	0.041
700	0.009	—	—	8.18	0.93	0.28	1.71	1.07	1.02	26.457	0.038
800	0.007	—	—	7.77	0.87	0.26	1.63	1.06	1.02	28.284	0.035
900	0.006	0.001		7.42	0.82	0.25	1.56	1.06	1.02	30.000	0.033
1000	0.006			7.11	0.78	0.24	1.50	1.05	1.02	31.62	0.32

Nonroutine Techniques—Correction When Rate is not Constant

The basic applicable equation may be written as

$$\Delta p_{(t)} = (q_1\mu/2\pi kh)\Delta p_D(t) + (\mu(q_2 - q_1)/2\pi kh)\Delta p_D(t - t_1)$$
$$+ (\mu(q_3 - q_2)/2\pi kh)\Delta p_D(t - t_2) + \dots$$

$$= (q_1\,\mu/2\pi kh)\Big\{\Delta p_D(t)$$
$$+ \sum_{i=2}^{n}[(q_i - q_{i-1})/q_i]\Delta p_D\,(t - t_{i-1})\Big\}$$

The following example involves a well that served as an injector. Thereafter, it was shut in and then placed on production.

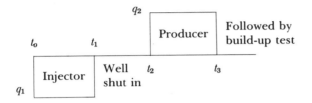

Here $q_i = q_0$ is a negative number

The basic equation is

$$p = p_e - (\mu/4\pi kh)[q_0\ln(t_3 + \Delta t)/(t_3 - t_1 + \Delta t)$$
$$+ q_2\ln(\Delta t)/(t_3 - t_2 + \Delta t)]$$
$$= p_e - (\mu/4\pi kh)[(q_0/q_2)\ln(t_3 - t_1 + \Delta t)/(t_3 + \Delta t)$$
$$+ \ln(t_3 - t_2 + \Delta t)/\Delta t]$$

The term for the shut-in period is omitted since the rate is zero. Also, the difference in the above two equations is the direction of the build-up curve plot. q_0/q_2 is a constant and does not affect slope.

$Q_{(t_3-t_2)} = 4,025,360 =$ cumulative gas, Mcf, produced during period of production
$Q_{(t_1-t_0)} = 2,130,322 =$ cumulative gas, Mcf, injected during injection period
$q_0 = 5,329 =$ average rate during injection period
$q_2 = 6,895 =$ average daily rate during production period
$q_0/q_2 = 5,329/6,895 = 0.7729$
$t_3 = 27,421.99 =$ total time, hours, that well was used for injection, shut-in, and production purposes

Basic Data and Calculations

Measured build-up pressure	Build-up time, hours	$t_3 - t_1 + \Delta t$	$t_3 + \Delta t$	$\dfrac{t_3 - t_1 + \Delta t}{t_3 + \Delta t}$	$\boxed{A}$ $\dfrac{q_0}{q_2}\ln\dfrac{t_3 - t_1 + \Delta t}{t_3 + \Delta t}$	$t_3 - t_2 + \Delta t$	$\dfrac{t_3 - t_2 + \Delta t}{\Delta t}$	$\boxed{B}$ $\ln\dfrac{t_3 - t_2 + \Delta t}{\Delta t}$	$A + B$
3625	4.00	17831.75	27425.99	0.6502	−0.322	14015.75	3503.94	8.16	7.83
3646	7.16	17894.91	27429.15	0.6502	"	14018.91	1957.95	7.58	7.25
3664	22.03	17849.78	27444.02	0.6504	"	14033.78	637.03	6.46	6.13
3679	46.85	17875.60	27469.84	0.6507	"	14059.60	293.83	5.68	5.35
3677	65.73	17893.48	27487.72	0.6510	"	14077.40	214.17	5.37	5.04
3677	67.08	17894.83	27489.07	0.6510	"	14078.83	209.88	5.35	5.02
3677	70.33	17898.08	27492.32	0.6507	"	14082.08	200.23	5.30	4.97
3677	89.47	17917.22	27511.46	0.6513	"	14101.22	157.61	5.06	4.73
3677	94.97	17922.72	27516.96	0.6513	"	14106.72	148.54	5.00	4.67
3684	115.25	17943.00	27537.24	0.6516	−0.331	14127.00	122.58	4.81	4.48

$\ln a = -\ln(1/a)$ or $\ln 4.0 = +1.386$, while $\ln 1/4 = -1.386$ so that the value of $A + B$ does not change except for sign.

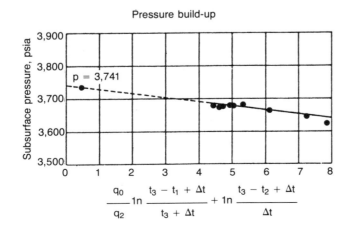

Pressure build-up

$t_1 = 2,130,322 \times 24/5329 =$ time that well was used as an injector, hours
$t_3 - t_2 = 4,025,460 \times 24/6895 =$ pseudotime that well was used as a producer, hours $= 9594.24$
$t_2 - t_1 = 3816 =$ time, hours, that well was shut in
$t_2 = = 9,594.24 + 3816 = 13,410.24$ hours
$t_3 - t_1 = 27,421.99 - 9,595.24 = 17,827.75$ hours

The effect of the large injection period has been essentially eliminated, and it affects the build-up data as a very small constant. If the flow configuration is both infinite and radial, the final build-up pressure is 3,741 psia. The slope of the line is $q\mu/4\pi kh$, and the usual build-up calculations apply.

Techniques from the Ground-Water Industry

Hurst, in *Reservoir Engineering and Conformal Mapping of Oil and Gas Reservoirs* (1979, PennWell Books), should be consulted. Lohman, *in Ground Water Hydraulics* (USGS, 1970), assembled papers presented at a symposium dealing with ground-water studies. Many ground-water aquifers are recharged from rain, rivers, and lakes rather than relying upon compressibility of rocks and reservoir fluids. Ground-water hydraulics is important since some of the techniques are adaptable to reservoir performance and use of the natural filters obtainable by taking water for water floods from sand rather than from lakes and rivers directly can prove economical. Two of the techniques not found in most oil literature are shown.

Flow Net Analysis

A graphical presentation of the isopressure contours and the streamlines that are perpendicular to them are on a scaled map, such as Fig. 5–88. The presentation is similar to that used in horizontal sweep studies of Muskat and others.

The scaled map may represent any water aquifer. The solid lines represent equipotential lines—usually lines of equal pressure if at datum—and the dashed lines are the streamlines that are always perpendicular to the equipotential lines. The streamlines are actually the path followed by a drop of water as it flows toward the well. A grid having a scale or spacing conveniently fitting the data on the map is overlain on the basic map. The equipotentials are equally divided and similarly the flow lines are selected so that the total flow is equally divided between adjacent pairs of flow lines. The theory is based on the observation that movement of each particle of water between adjacent equipotential lines will be along flow paths involving the least work. Hence, it follows that, in isotropic aquifers, such flow paths will be normal to the equipotential lines, and the paths are drawn orthogonal to the latter. The net is constructed so that the two sets of lines form a system of squares. Some of the isopotential lines on the map are curvilinear. Nevertheless, the squares of the overlay are constructed so that the sum of the lengths of each line in one system is closely equal

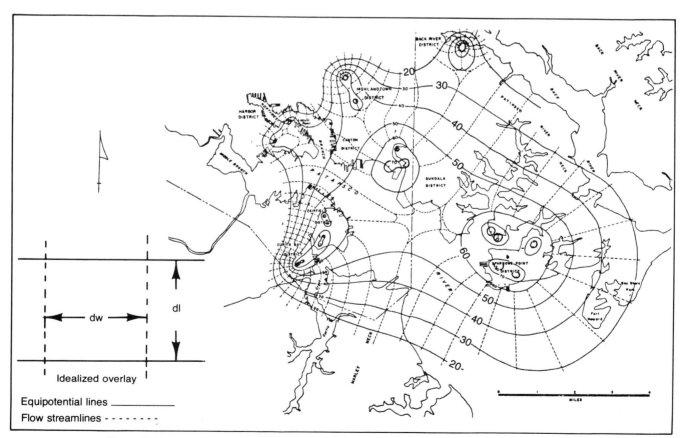

Figure 5–88. *Ground-water pressure map (after Lohman, "Groundwater Hydraulics," USGS, 1970).*

to the sum of the lengths in the other system. Fig. 5–88 illustrates one idealized square of the map whose dimensions are *dw* and *dl*.

Darcy's law may be written as a finite difference equation to obtain:

$$\Delta Q = -kb\Delta w(\Delta h/\Delta l) = -T\Delta w(\Delta h/\Delta l)$$

by construction, $\Delta w = \Delta l$ and $\Delta Q = -T\Delta h$

The above symbols are from ground-water literature where ΔQ is the flow through an element defined by the overlay as a square of thickness *b; h* is the potential in feet or cm as desired.

If we define: n_f = number of flow channels
n_d = number of potential drops
Q = total flow
Then: $Q = n_f\Delta Q$
$\Delta Q = q/n_f$
$h = n_d\Delta h$
$\Delta h = h/n_d$
$Q = -T(n_f/n_d)h$
$T = -q(n_f/n_d)h$

The units of this system are l^2/t such as cm²/sec. The units used can be understood by study of following tables:

Hydraulic conductivity		Field coefficient of permeability
feet/day	m/day	gal day⁻¹ feet⁻²
1.00	0.305	7.48
3.28	1.000	24.5
0.134	0.041	1.00

Transmissivity		
feet²/day	meters²/day	gal day⁻¹ ft⁻¹
1.000	0.0929	7.48
10.76	1.000	80.50
0.134	0.0124	1.00

Intrinsic permeability	Permeability, Darcy	Coefficient of permeability
$k = -q\mu/(d\Phi/dl)$	$k = -q\mu/(dp/dl + \rho g dz/dl)$	$p = q_{60F}/(dh/dl)$
(um²) = 10⁻⁸ cm²	0.987 × 10⁻⁸ cm²	gal day⁻¹ feet⁻² at 60°F
1.000	1.01	18.4
0.987	1.00	18.2
0.054	0.055	1.0

Lohman presents the following example:

The average discharge from the formation during a year was 1,000,000 ft³day⁻¹. The map shows 15 flow channels so that n_f = 15. The number of equipotential drops between the 30 and 60-foot contours is 3 and n_d = 3. The potential drop between the 30 and 60 foot contours is 30 feet and h = 30 feet. Then

$$T = -Q/(n_f/n_d)h = -1000000/(15/3)(-30) = 6670 \text{ ft}^2/d$$

The value of *T* so determined is for a much larger sample of the reservoir than that obtainable from several individual well tests.

Closed Contour Method

The water level contour map containing closed contours around one or more wells of known discharge rates may be used to determine an estimate of transmissivity of an aquifer under steady-state conditions, since $Q = -KA\Delta h/\Delta r = -TL\Delta h/\Delta r$. If any two concentric closed contours have lengths L_i and L_2, $T = -2Q/(L_1 + L_2)(\Delta h/\Delta r)$, where Δh is the contour interval and Δr is distance between two closed contours. If we can determine that two irregularly shaped closed contours have measured lengths—as measured by a wheel-type device—of 27,600 and 44,000 feet respectively, that the contour interval is 10 feet, that the distance between two contours is (1800 + 2200 + 2100 + 1700)/4 = 1950 feet, and that the rate of withdrawal from field within the lowest closed contour is 1,000,000 gal/day:

$$T = -2 \times 1000000/(71600 \times (-10))/(1950 \times 7.48)$$
$$= 730 \text{ ft}^2 d$$

Analysis of Linear Flow

Millhelm and Cichowicz in *JPT* (February 1968) presented an analysis using three flow rates. Ramey and Wattenbarger in *JPT* (May 1969) were critical of time periods selected for slope determination. Using Millhelm data, we find:[*]

The slope of $(p_d^2 - p_w^2)/q$ vs. ln *t* plots is $1/(2B)$ = 670 psia²/Mscfd or $B = 0.000744$ (Fig. 5–89).

[*] This material courtesy SPE-AIME.

The intercept of $(p_d^2 - p_w^2)/q$ vs. q at $q = 0$ is $1/(2B)$ ln $A = 490$ psia²/Mscfd (Fig. 5–90).

The slope of $(p_d^2 - p_w^2)/q$ vs. q is $D/B = 0.270$ psia²/Mscf²/d².

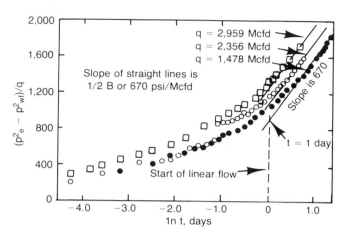

Figure 5–89. Drawdown test using three flow rates (after Millhelm and Cichowitz, JPT, February 1968).

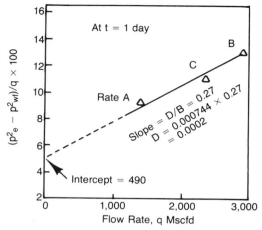

Figure 5–90. Three-rate plot (after Millhelm and Cichowitz, JPT, February 1968).

Effective Permeability Calculation

$B = $ constant, $\dfrac{19.87 \times 10^{-6} kh\, T_{sc}}{z\mu p_{sc} T_f}$·Mscfd/psi²

$B = 1/(2 \times$ slope$)$ of the $(p_d^2 - p_w^2)/q$ vs ln t plot.

$k = \left(\dfrac{\mu z p_{sc} T_f B}{19.87 \times 10^6 T_{sc} h}\right) =$

$\left(\dfrac{(0.016)(0.901)(15.025)(650)1}{(19.87 \times 10^{-6})(520)(1,340)(67)}\right) = 0.152$ md

Effective Wellbore Calculation

$A = $ constant, $\dfrac{1.39 \times 10^{-2}\, k p_d}{\phi_{HC} r_w^2 \mu}$ · dimensionless

$A = 2.08$

$r_w = \sqrt{\dfrac{1.39 \times 10^{-2} p_i k}{\phi_{HC} A \mu}}$

$= \sqrt{\dfrac{(1.39 \times 10^2)(2.684 \times 10^2)(0.152)}{(5.46 \times 10^{-2})(2.08)(1.6 \times 10^{-2})}} = 55.9$ ft

Turbulence Coefficient Calculation

$D = $ constant, $\dfrac{2.715 \times 10^{-15} k p_{sc} M \beta}{h\mu\, T_{sc} r_w}$, (Mscfd)$^{-1}$

$D = 2.01 \times 10^4$ (Mscfd)$^{-1}$

$\beta = \dfrac{\mu\, T_{sc} h r_w D}{2.715 \times 10^{-15} p_{sc} M k}$

$= \dfrac{(1.6 \times 10^{-2})(520)(67)(55.9)(2.01 \times 10^{-4})}{(2.715 \times 10^{-15})(15.025)(17.9)(0.152)}$

$\beta = 5.64 \times 10^{13}$ ft^{-1}

Initial reservoir pressure (psia)	2,684
Pressure at base conditions (psia)	15.025
Formation temperature (°F)	190
Temperature at base conditions (°F)	60
Average reservoir porosity (percent)	8.4
Water saturation (percent)	35
Net pay (ft)	67
Specific gravity of gas	0.62
Casing size (in.)	4½
Tubing size (in.)	2⅜
Total depth (ft)	7,520
Perforation interval (ft)	7,303 to 7,490

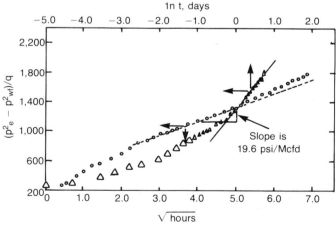

Figure 5–91. Pressure drawdown plots for Well A-1 during testing ($q = 2,955$ Mscfd, after Ramey and Wattenbarger, JPT, May 1969).

To calculate the effective fracture flow area, the plot of $(p_d^2 - p_w^2)/q$ vs $\sqrt{t'}$ is used. The slope of the $(p_d^2 - p_w^2)/q$ vs $\sqrt{t'}$ plot is 194 psia²/Mscfd. For this case, the average reservoir pressure p_d is equal to the initial reservoir pressure p_i.

$$A' = \frac{5.80 \times 10^3 p_{sc} T_f z}{T_{sc} m'} \sqrt{\frac{p_d \mu}{\phi_{HC} K}}$$

$$A' = \frac{(5.80 \times 10^3)(15.025)(650)(0.901)}{(520)(194)}$$

$$\sqrt{\frac{(1.6 \times 10^2)(2.684 \times 10^3)}{(5.46 \times 10^2)(0.152)}} = 36.410 \text{ sq ft}$$

For an effective wellbore radius of 55.9 ft and an effective fracture flow area of 36,410 sq ft, the vertical fracture efficiency is

$$E_f = \frac{A' \, 100}{(8) r_w h} = \frac{(36,410) \times 100}{(8)(55.9)(67)} = 122 \text{ percent}$$

Typical linear flow curves are in Fig. 5–92.

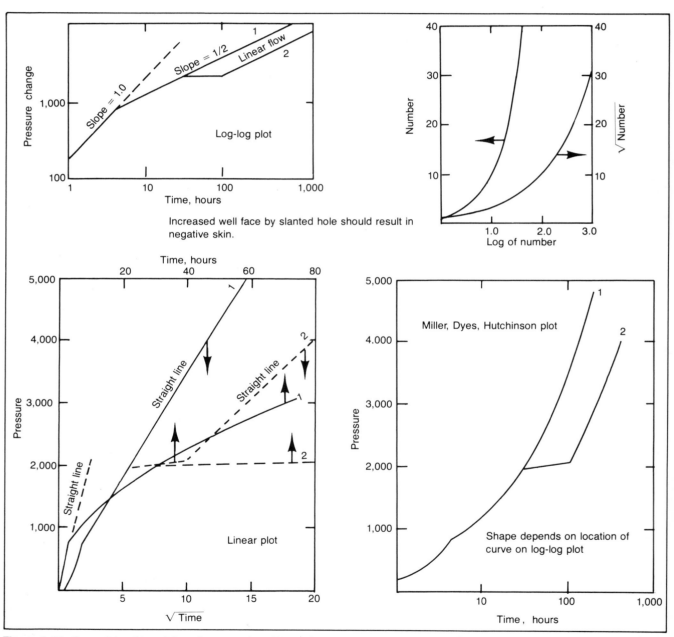

Figure 5–92. *Curves showing relations between plots. Theory indicates that on the log-log plot, a slope of one relates with the afterflow period-wellbore effects. A slope of one-half associates with linear flow. Exact initial pressure-time relation is essential.*

Spherical Flow

When perforations are concentrated in a small interval of a thick section of pay, spherical geometry must be considered. In a homogeneous reservoir containing a single fluid, the early flow is cylindrical or radial in nature, possibly complicated by afterflow effects. Later, the flow becomes spherical until the overlying and underlying boundaries are reached when the flow again becomes radial in nature. Shale stringers penetrated by the well may be barriers and may complicate flow in a manner similar to the problems associated with coning. Also, if a gas cap is present or if water lies below a very heavy oil, coning will take place and relative permeability, etc., must be considered during certain stages of the flow period. During the last radial flow period associated with long tests, coning may be severe, so that characteristics of the gas and water and their pay thickness may control characteristics of the flow. Application of theory to field problems, particularly when shale stringers cause wide variations in permeability, etc., is complicated and requires good judgment.

The spherical flow period may be analyzed by plotting the measured pressures or the related $m(p)$ gas values, if required, versus $1/\sqrt{\Delta t}$ for the flow period and $1/\sqrt{\Delta t} - 1/\sqrt{t + \Delta t}$ for the build-up. The slope is read in the usual manner for the straight-line portions, which must be selected carefully by using time and radius of investigation concepts. The permeability can be computed for spherical flow as follows for gas:*

$$k^{3/2} = \frac{A\sqrt{B}\ q\mu B_o\sqrt{\phi\mu c}}{4\pi^{3/2}m} = \frac{2453\ q\mu B_o\sqrt{\phi\mu c}}{m \text{ when in psi/hr}}$$

$$= \frac{19,002\ q\mu B_o\sqrt{\phi\mu c}}{m \text{ when in psi/min}}$$

The 4 changes to 2 when perforations are near the top or bottom of the interval. The related skin may be computed from the following equation:

$$s = 269.14\ \sqrt{\frac{r_{ew}^2}{k/\phi\mu c}}\left[\frac{p_{\Delta t} - p_{wf}}{7.746\ m} + \frac{1}{\sqrt{60\ t}}\right] - 1.0$$

where Δt may be any time on the straight line which gives wide difference in $p_{\Delta t} - p_{wf}$ in hours.

The equivalent wellbore radius is:

$$r_{ew} = b/2 \ln (b/r_{wr})$$

where r_{wf} is the hole radius of the well.

* The following equations are adapted from Culham, *SPEJ*, December 1974.

The minimum time to reach spherical flow may approximate:

$$6.25\ (b/\ln [\bar{b}/r_{wr}])^2 = 2.637 \times 10^{-4}\ kt/\phi\mu cr_{wr}^2$$

where t is in hours.

The minimum pressure drop should be:

$$p_{min} = (p_o - p_{wf})/8.8623$$

The upper time applicable (start of second radial flow) is:

$$\frac{\partial \psi_H}{\partial \lambda} = -\frac{1}{2}\left[\tau_o\gamma + \frac{2h\gamma}{\pi_b}\sum_{n=0}^{\infty} R_n\left\{\Delta\tau\right.\right.$$

$$exp\left[-\left(\frac{n\pi r_{wr}}{h}\right)^2 (\tau_o + \Delta\tau)\right] - (\tau_o + \Delta\tau)$$

$$\left.\left. exp\left[-\left(\frac{n\pi r_{wr}}{h}\right)^2 \Delta\tau\right]\right\}\right]$$

where

$$R_n = \frac{1}{n}\left[\sin\frac{n\pi l}{h} - \sin\frac{n\pi d}{h}\right]\cos\frac{n\pi z}{h}$$

$$\cdots \cdots \cdots \cdots \cdots \cdots$$

and

$$\gamma = \frac{2\sqrt{\Delta\tau}\sqrt{\tau_o + \Delta\tau}}{\Delta\tau\sqrt{\Delta\tau} - (\tau_o + \Delta\tau)\sqrt{\tau_o + \Delta\tau}}.$$

The radius of investigation is:

$$r_{inv} = \sqrt{0.002637\ kt/\phi\mu c}$$

where t is in hours.

Several examples of an analysis for an oil well are included in Figs. 5–93, 94, and 95.

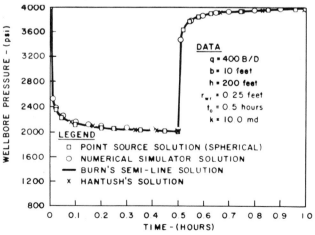

Figure 5–93. *Typical spherical flow and MDH build-up curve (after Culham, SPEJ 6, December 1974).*

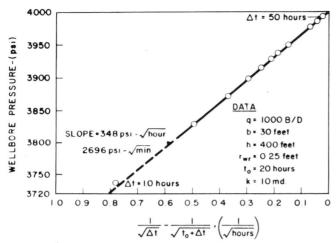

Figure 5–94. *Spherical build-up plot (after Culham, SPEJ 6, December 1974).*

where s_r = skin factor due to restricted entry
h_t = total sand thickness, ft (m)
h_p = length of perforated interval, ft (m)
k_H/k_V = ratio of horizontal to vertical permeability,
r_{wc} = corrected wellbore radius, ft (m), and is given by
$$r_{wc} = r_w e^{0.2126(z_m/h_t+2.753)}$$
for $y > 0$, and $r_{wc} = r_w$ for $y = 0$,
y = distance between the top of the sand and the top of the open interval, and
z_m = distance between the top of the sand and the middle of the open interval.

The relation between y, h_p, and z_m is

$$z_m = y + h_p/2.$$

Because of symmetry, $z_m/h_t \leq 0.5$.

A radial, spherical and thereafter radial flow configuration was assumed in these theoretical calculations. In the field, the first radial is possibly obscured by well and linear flow.

Odeh in *JPT* (June 1980) presents the following equation for calculating skin due to restricted entry from a thick formation:

$$s_r = 1.35 \left(\left(\frac{h_t}{h_p} - 1\right)^{0.825} \left\{ \ln\left(h_t\sqrt{\frac{k_H}{k_V}} + 7\right) - \left[0.49 + 0.1\ln\left(h_t\sqrt{\frac{k_H}{k_V}}\right)\right] \cdot \ln r_{wc} - 1.95 \right\} \right)$$

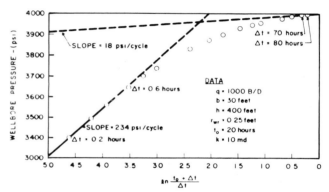

Figure 5–95. *Horner plot of spherical data (after Culham, SPEJ 6, December 1974).*

Vertical Permeability

The spherical flow configuration should be present if a well is only partly perforated. The related time period is during the early life of a test and may be influenced by afterflow and other conditions existing during the early part of a transient test. Prats in *JPT* (May 1970) presented a method for obtaining vertical permeability from a special test.

In this method the well is perforated in a short interval above and below a packer that is set between the two sets of perforations. Fluid is injected in the upper interval—maybe lower if convenient—and pressure is measured at the other set of perforations. As an alternative, the well may simply be produced. Prats shows that if the observed pressure, p_{ws}, is plotted

versus the logarithm of time (Fig. 5–97) from the beginning of injection or production, a straight line should be obtained. The slope is m as used in the following equations, and the intercept at t_{1hr} is a value for p_{1hr}. The horizontal permeability is calculated using this slope and the usual equation, $k_r = 162.6qBu/mh$. The vertical permeability is calculated using this slope and the intercept at t_{1hr} from the following equation:

$$k_v = k_z = (\phi\mu c_t h^2/0.002637) \text{ antilog}$$
$$[(p_{1hr} - p_i)/m - G^* + \{h/(Z_{wf} - Z_{ws})\}/2.3025]$$

The curves in Figs. 5–96 and 97 are used in an example as follows:

$$h = 50 \text{ ft}$$
$$\Delta Z_{wf} = 45 \text{ ft}$$
$$\Delta Z_{ws} = 10 \text{ ft}$$
$$q = -50 \text{ stb/d}$$
$$B = 1.0 \text{ RB/stb}$$
$$\mu = 1.0 \text{ cp}$$
$$c_t = 2.0 \times 10^{-5} \text{ psi}^{-1}$$
$$\phi = 0.10$$
$$p_i = 3{,}015 \text{ psi}$$

The horizontal permeability using slope, m, of 22.5 psi/cycle is

$$k_r = \frac{(-162.6)(-50)(1.0)(1.0)}{(22.5)(50)} = 7.2 \text{ md}$$

Earlougher in Monograph 5 calculates the vertical permeability as follows:

$$\Delta Z_{wf}/h = 45/50 = 0.9$$

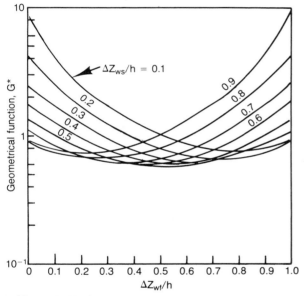

Figure 5–96. *Correction factors (after Prats, May 1970).*

and

$$\Delta Z_{ws}/h = 10/50 = 0.2$$
$$G^* = 0.76$$

$$k_z = \frac{(0.10)(1.0)(2.0 \times 10^{-5})(50)^2}{0.0002637}$$
$$\text{antilog}\left(\frac{3{,}022 - 3{,}015}{22.5} - \frac{0.76 + 50/|45 - 10|}{2.3025}\right)$$
$$= 4.3 \text{ md}$$

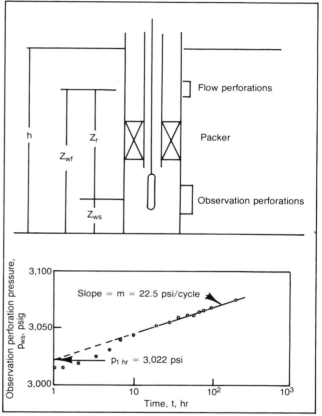

Figure 5–97. *MDH plot (after Prats, JPT, May 1970).*

Pulse Tests*

The three figures shown are one data set separated for purposes of display. The Shell method makes use of the fact that time-dependent terms of the line source solution can be represented by only two curves, which are monotonic functions of a dimensionless time, $(t_{(\Delta p \, max)} - \Delta t)/\Delta t$. If the reservoir is liquid filled or filled with gas at high pressure (small compressibility), the solution proposed is practical.

The pulse test devised by Johnson et al., *JPT* (December 1966) has been reviewed by Earlougher in Monograph 5. Sensitive pressure elements often are used, and instrument stabilization (often several hours) and possible influence of moon and atmosphere must be recognized.

In the Shell method, pulse design involves the determination of the pulse duration, Δt, and the flow rate change, Δq, necessary to give a measureable pressure disturbance, Δp_{max}, at a convenient time, $t_{(\Delta p \, max)}$. The distance between wells, r, is usually known and kh/μ and $k/\phi\mu c_t$ are unknowns that must be assumed for test design purposes.

To use the curves of Prats for design purposes, one must calculate:

1. Value for $(kh/\mu)(\Delta p_{max}/\Delta q)$ from the estimated transmissibility, flow rate, and maximum pressure disturbance desired during the test.
2. Using this value, read the dimensionless time $(t_{\Delta p \, max} - \Delta t)/\Delta t$ from Fig. 5–98.
3. Read value of $(k/\phi\mu c_t)(\Delta t/r^2)$ at the same dimensionless time from Fig. 5–98.
4. Determine the necessary pulse duration, Δt, from value read in step 3 from Fig. 5–98.
5. Use this value for pulse duration, Δt, with values read from the chart in step 2 to obtain a value for the time delay to the maximum pressure disturbance; determine $t_{\Delta p \, max}$.

If test design values are unattainable, redesign. The method also may be used for a quick analysis of the results of a pulse test. The background pressure behavior must be known for accurate analysis.

As an example of the use for design purposes, assume that $k = 300$ md, $h = 50$ ft, $r = 660$ ft, $\phi = 0.20$, $q = 500$ b/d, $\mu = 0.80$ cp, $c_t = 10 \times 10^{-6}$, and desired $p_{max} = 1$ psi. Calculate $(kh/\mu)(\Delta p_{max}/\Delta q) = (300 \times 50 \times 1)/(0.8 \times 500) = 37.5$. From the chart, the related $(t_{\Delta p \, max} - \Delta t)/\Delta t = 0.35$ and $(k/\phi\mu c_t)(\Delta t/r^2) = 1480$. Then, $\Delta t = 1480 \times 660 \times 660 \times 0.2 \times 0.8 \times 10 \times 10^{-6}/300 = 3.4$ hours. Also, $t_{\Delta p \, max} = 3.4(1 + .35) = 4.6$ hours.

* Material in this section courtesy of Shell Oil Co.

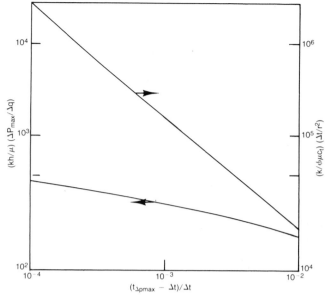

Figure 5–98a. *Curve 1*

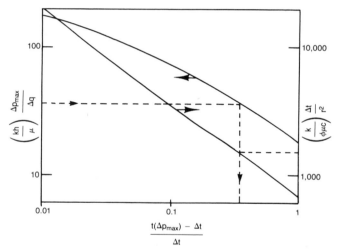

Figure 5–98b. *Curve 2*

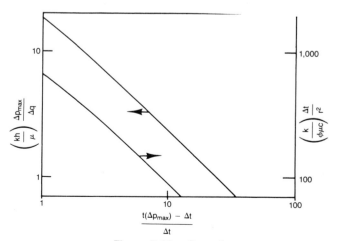

Figure 5–98c. *Curve 3*

Reservoir Limit Tests

The flow test can be divided into three sections: 1. The transient flow period ending when $t = \phi \mu c_t r_e^2 / 0.00264\ k$ or $t_D = 0.1$; 2. The late transient flow period that ends at about $t_D = 0.3$ or when $t = \phi \mu c_t r_e^2 / 0.00088\ k$; and 3. The semisteady-state flow period when the pressure is declining rather uniformly throughout the reservoir. Park Jones, *Oil and Gas Journal* (June 18, 1956), used a plot of the derivative of pressure versus time to amplify this characteristic at start of semisteady-state flow. Boundaries other than complete reservoir boundaries can be reflected by the field data, and an erroneous conclusion as to reservoir size can be drawn. The reservoir limit test should be used with extreme caution. The complete analysis for the drawdown test is desired instead.

Using practical units, the semisteady-state flow equation can be written as:

$$\Delta p = \bar{p} - p_{wf} = (141.2\ q_{sc}\ B\mu/kh)[\ln(r_e/r_w^2) - 0.75 + s + Dq_{sc}]$$

when depletion to $\bar{p}$ is omitted. Dimensionless log-log plot of the pressure change versus flow time has a slope of 45 degrees. The first differential of the pressure change versus time, $d(\Delta p)/dt$, is determined from a plot of pressure change versus time on coordinate paper by obtaining the tangents to this curve at various times and/or pressures. Jones then defines $Y = (d[\Delta p]/dt)/q_{sc}B = 1/(cN)$ where dp/dt is in psi/day, q is in b/d, B is the formation volume factor, c is in psi^{-1} and N is the connected pore volume in bbl. The Park Jones and more routine analysis methods can be compared by studying Figs. 5–99, 100.

The rate or sand-face pressure must be constant as assumed in the theory, or methods such as plotting $(p_e - p_{wf})/q$, superposition, etc., must be used. (Example calculation is included under analysis of drawdown tests in this text.) The example included by Earlougher in SPE Monograph 5 includes a method for determination of a shape factor using constants shown under average reservoir pressure by Deitz. Earlougher proposes the equation

$$C_A = 5.456(m/m^*) \exp [2.303(p_{ihr} - p_{int})/m]$$

where $m^* = 0.23395qB/\phi c_t hA$ and is the slope on coordinate paper—plot pressure versus flow time—effective at the well site prior to placing the well on production. Monograph 5 should be consulted before calculating C_A.

In several instances, apparently reliable limit tests

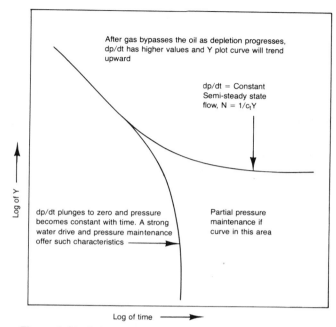

Figure 5–99. *Schematic representation of the Jones plot.*

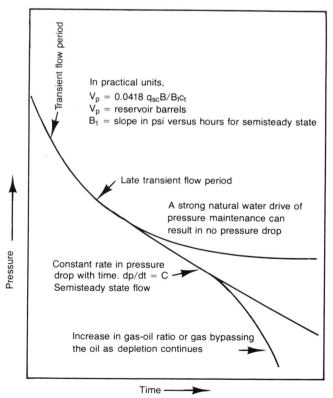

Figure 5–100. *Normal drawdown plot.*

were in error since the wildcat well was located in a poorly connected extension of the main oil accumulation. It is very important that the results of the limit tests be checked against other data using techniques such as the volumetric equation, the material-balance equation, known water and gas-oil contacts, if available, changes in fluid contacts, gravity, seismos, and the geology of the accumulation and reservoir type.

Skins — Positive and Negative Causes

Negative skins usually are related to an enlargement of the wellbore. Such enlargement may be caused by natural fractures, lengthy voids, and by shooting, acidizing, and sandpacking fractures performed by man. Under ideal conditions, the fracture can be considered to have infinite capacity when compared to that of the pay rock. Most theory is based on this assumption, and corrections such as proposed by Cinco, Scott, and Prats (see SPE Index) should be consulted. Also, fractures may penetrate lenses and pay stringers having different characteristics. Fractures at depths less than about 1500 ft tend to lift the overburden, and such fractures usually are horizontal. At greater depth, man-made fractures usually are vertical and have an elliptical shape from point of entry at wellbore. Fracture direction was determined by Dr. King to be perpendicular to the least principal stress at point of fracture for both natural and man-made fractures. At great depth, high pressures are required to initiate fractures and the zero, five-minute, and later time pressure fall-off pressures used to analyze fracture jobs have high values. Fractures tend to heal with time as forces created by fracture injection adjust with time. The literature relative to fracturing should be reviewed. Acidization at high rate is a fracture.

Positive skins are created by many factors, including drilling muds, partial penetration (part spherical flow), limited perforations, and other completion conditions. Shooting, acidization, fracturing, and other well clean-up operations tend to remove positive skins. Under some well conditions, precipitation of solids and movement of solid fines create positive skins during production. The relative magnitude of some of the causes of positive skins is given by Figs. 5–101 through 104.

Some precautions which may be taken to reduce positive skins.*

Don't start any field job without a thorough diagnosis

Don't do any job without fully communicating the total job requirements with everyone who has anything to do with it.

Don't run a job without thorough and proper logistics, down to the last sub to match special tool threads.

Don't start a job without qualified supervision and without an announcement, well in advance, of the total quality control of manpower, materials, tools, and equipment you are going to enforce.

Don't try to "save" by using poor muds in poor condition while penetrating any future production interval.

Don't squeeze cement in production zones. If necessary, squeeze well below, or above, the producing interval. The best approach is a good primary cement job—no squeezing required—money saved.

Don't run casing without adequate centralizers in places where they count.

Don't run primary cement without moving the casing.

Don't use guar gum; try hydroxy ethyl cellulose. It's less damaging and more compatible with other chemicals.

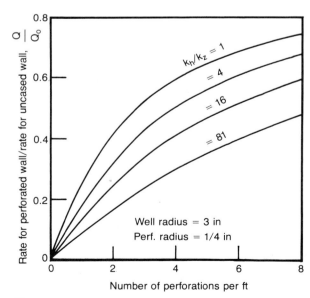

Figure 5–101. *Damage from perforations (after Muskat,* Trans. AIME, 1943).

---

* These precautions courtesy of the *Oil & Gas Journal.*

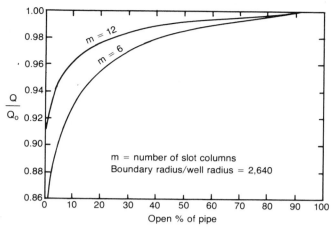

Figure 5-102. *Damage from slotted liners (after Muskat,* Physical Principles of Oil Production, © *SPE).*

Don't "kill" wells with oxidized lease crude oil and dirty lease brine, seawater, or bay water. You really kill them with such fluids.

Don't pull any well without taking into account precisely the pipe volume. Even an old well may blow out, and the subsequent emergency kill procedure can cause irreparable damage.

Don't switch over from mud to completion fluids by cleaning only the tanks. The whole system must be thoroughly cleaned, including pumps, manifolds, flow lines, well bore, etc.

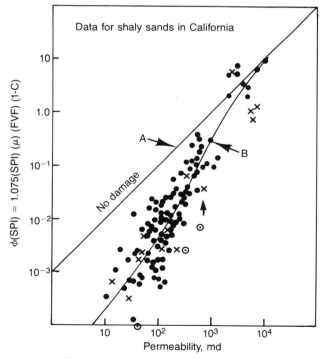

Figure 5-103. *Damage from completions.*

Do a good housecleaning on top of tanks and throughout surface facilities. Loose paper sacks, twine, coffee and water cups, cigarette butts, rags, rotten sand line—all of these make good plugs for perforations.

Don't run any painted tools during completion. Blistered paint is an excellent fluid-loss material, and usually it finds its way to the most prolific perforations.

Don't run down-hole tools without making adequate plans in advance for all steps necessary in case of tool failures.

Don't use pipe dope in boxes in tubing runs. Set the pin in the box, and lightly and sparingly smear the pipe dope on the threads.

Don't sample mud for solids content on the top of the mud tank. Secure samples from the pump discharge for proper and true fluid weight and solid-content determination.

Don't inject any acid into the formation unless both the tubing and the well bore are precleaned with acid.

Don't inject any acid unless you sample and test it for proper density with hydrometers. All good service companies have them on the job.

Don't run slotted or perforated liners with burrs.

Don't use any gravel, sacked, or in bulk, if it contains excessive fines.

Don't use No. 2 diesel for perforation or well completion. Use diesel No. 1 turbine fuel, or jet fuel A. Most No. 2 diesels contain formation-plugging materials.

Don't use expendable guns indiscriminately. In general, expendable guns are good only for a few highly specialized conditions. Under normal conditions they can cause more damage with much less penetration than you think.

Don't perforate in dirty fluids with pressure into the formation.

Don't use perforating guns without proper centralizers (casing guns) or proper positioning devices (tubing guns).

Don't use tubing guns with 90°, 120°, or 180° phasing. Use 0° phasing with proper positioning.

Don't inject into a well any sea or bay water. Both waters can contain large quantities of organic matter such as plankton, bacteria, etc., which, due to their size and nature, are some of the best plugging materials. Seawater and most bay waters also contain sulfates (2–3,000 ppm) capable of forming sulfate scale for which we have no economical remedy. Transport proper water or refined oil to the location in clean containers and filter it. In the long run, this practice pays in every case.

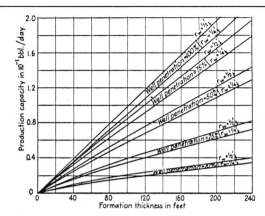

A. The calculated steady-state homogeneous-fluid production capacities of partially penetrating wells, as functions of the formation thickness. r_w = well radius; external-boundary radius = 660 ft; remaining conditions and notation as in C.

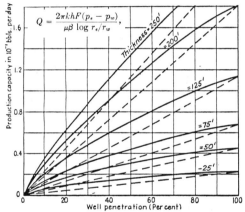

C. The calculated steady-state homogeneous-fluid production capacities of partially penetrating wells as functions of the well penetration, for various thicknesses of the producing stratum. Straight dashed lines given the production capacities for strict radial flow into the exposed well section. $k\Delta p/\mu\beta$ is taken as unity, with k in millidarcys, Δp in psi, and μ in centipoises; well radius = 1/4 ft; external-boundary radius = 660 ft.

E. The calculated effect of casing perforations on the steady-state homogeneous-fluid well productivity, as a function of the perforation density, in anisotropic formations. Q/Q_o = (production capacity of cased and perforated well) / (production capacity of completely penetrating uncased well). k_h/k_z = (horizontal permeability) / (vertical permeability); well radius = 3 in.; perforation radius = 1/4 in. (From AIME Trans., 1943)

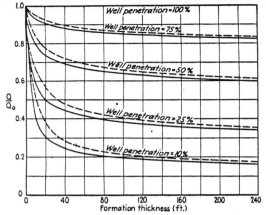

B. The calculated variation of the relative steady-state homogeneous-fluid production capacities of partially penetrating wells with the formation thickness. Q/Q_o = (production capacity of partially penetrating well) / (production capacity of completely penetrating well). For solid curves, well radius = 1/4 ft. For dashed curves, well radius = 1/2 ft. External-boundary radius = 660 ft in all cases.

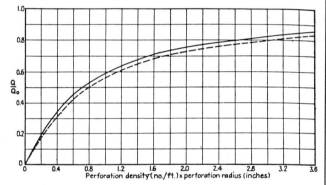

D. The calculated effect of casing perforations on the steady-state homogeneous-fluid well productivity. Q/Q_o = (production capacity of cased and perforated well) / (production capacity of completely penetrating uncased well). For solid curve, casing radius = 3 in. For dashed curve, casing radius = 6 in.

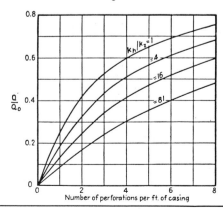

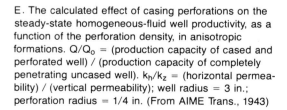

Figure 5–104. *Some causes for well damage (after Muskat,* Physical Principles of Oil Production, © *SPE).*

Don't inject any fluids into a formation unless laboratory compatibility tests are performed with all the fluids to be used.

Check the compatibility on the job for possible changes in each batch of fluid used.

Don't inject prepared brines into a well unless they are checked for solid content. Most sacked salts contain plugging impurities such as diatomaceous or Fullers Earth. Use filters in any case. It is cheaper and safer to plug your surface filters than to plug your well.

While you are using surface filters, don't forget to check and frequently recheck their performance by taking and visually comparing the upstream with the downstream samples.

Don't use mud hoppers for mixing viscosifiers. You'll get many plugging lumps and fisheyes that way. Check with an expert service company, and be prepared to strain your fluids if necessary.

Don't pump high-viscosity fluid into a wellhead before checking it with about a 30-mesh screen for fisheyes and lumps. The test sample should be secured at the wellhead, and the 30-mesh screen can be substituted with a tea strainer

Don't apply excess plastic (epoxy) dope to shoe and float collar. Excess plastic flash will sooner or later find its way into the one of the most-prolific perforations, and in some operations may cause plugging of perforations.

Don't run cement right after an invert mud, without properly weighted invert spaces, which prevent invert mud clobbering and serious formation damage.

Don't run cement after invert without fluid loss-control additives.

Don't swab the hole and fracture the formation by extra-fast trips in and out of the hole. With fast trips you save some "rig-time paper money," but you may lose the prime objective—to obtain a commercial producer or injector having maximum possible rates.

Don't forget density swapping. A few barrels of diesel slug "spotted" in mud, for perforation in diesel, won't be there when your gun is ready to fire.

The harmful effect of formation damage often is removed by fracture treatments, and it often is foolish to spend large amounts of money on mud control, etc., if fracture is planned. However, good housekeeping is beneficial and is essential during and after fracture or acid treatment. Also, any damage-creating materials will move into the fracture if they are in the wellbore at fracture time and may do so if they are in the formation after the fracture job.

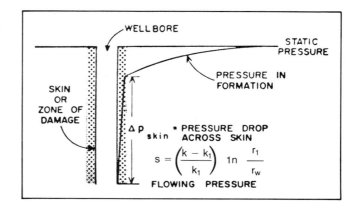

$$r'_w = r_w e^{-s} \cong 0.25 \, x_f \cong L/2$$

$$r_D = \frac{r_e}{r'_w}$$

$$s' = s + Dq = 1.151 \left[\frac{m(p_i) - m(p_{1hr})}{-b} - \log_{10}\left(\frac{k}{\phi(\mu c_t)_i r_w^2}\right) + 3.23 \right]$$

$$s = 1.151 \left[\frac{p_{1hr} - p_{wf}}{m} - \log\left(\frac{k}{\phi \mu c r_w^2}\right) + 3.23 \right]$$

$$\Delta p_{skin} = 0.87 \, m \, s \text{ where } s = \text{skin}$$

Skin is assumed to occur near wellbore.

Improvement Possibilities from Fracturing and Acidization

The benefit in rate of production obtained from deep fractures can be very substantial during the transient flow period, even when no positive skin is present initially when the natural permeability of the pay is small. Unfortunately, the rate of production declines rapidly and substantially with production time, as shown in Fig. 5–105, for the transient flow period–early life of the flow period.

Transient Flow: Assumes that p_{tD} is not required.

Liquids:
$$p_i - p_w = 162.6 \, q_{sc} \, B\mu / kh \, [\log(kt/\phi\mu c r_w^2) - 3.23 + 0.87s]$$

Gas:
$$m(p_i) - m(p_w) = 1637 \, q_{sc} T / kh \, [\log(kt/\phi\mu c r_w^2) - 3.23 + 0.87s + 0.87Dq_{sc}]$$

Semisteady-state flow:

Liquids:
$$p - p_w = 141.2 \, q_{sc} \, B\mu/kh \, [\ln(r_e/r_w) - 0.75 + s]$$

Gases:
$$m(\bar{p}) - m(p_w) = 1423 \, q_{sc} T/kh \, [\ln(r_e/r_w) - 0.75 + s + Dq_{sc}]$$

When r_w is 0.229164 feet: $r'_w = r_w \, e^{-s}$							
Skin	10	2	0	−2	−3	−4	−5	−6
e^{-s}	0.00004	0.14	1	7.389	20.09	54.60	148	403
r'_w	0.000010	0.031	0.229	1.69	4.60	12.51	34.0	92.5
Fracture length			0	6.8	18.4	50.04	136	370
If r_w is 0.3542; r'_w				2.62	7.12	19.34	52.4	142.7
Fracture length			0	10.4	28.5	77.4	209.7	571.0

(r'_w and skin are identical; use one only)

As expected, good pays show little improvement as a result of fracture treatment unless a positive skin was present before treatment. Good pays normally are not fractured or acidized and they usually clean up naturally from drilling—completion damage. Fracturing wells in poor pays where transient flow time is long (low values for $k/\phi\mu c_t r_w^2$) can be very beneficial. But even here the rate of production declines with time and the permanent improvement often approaches a final value with time of around twofold.

A negative value over five for skin is difficult to obtain when normal analysis techniques are used in the analysis and fracture permeability is assumed to be infinite, etc. Nonturbulent flow is assumed (Fig. 5–107).

The improvement during semisteady-state flow is best when spacing is small, as shown in Fig. 5–106.

The log-log plots shown in Fig. 5–108 show the benefits from fracturing and may also be used to determine kh and skin for conditions indicated.

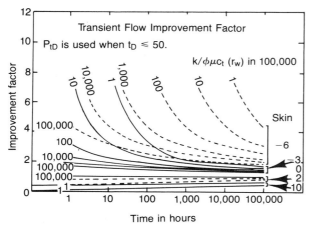

Figure 5–105. *Benefits from fractures; nonturbulent transient flow.*

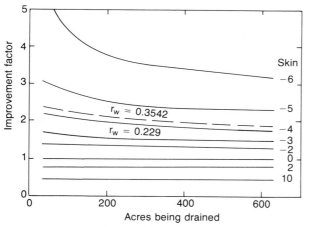

Figure 5–106. *Benefits from fractures; nonturbulent semisteady-state flow when $r_w = 0.229166$ ft.*

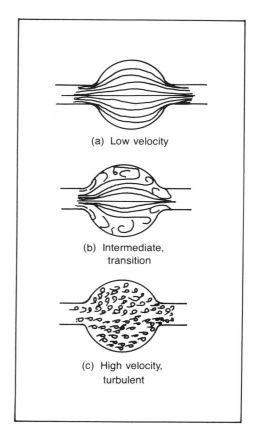

Figure 5–107. *Types of flow can determine the benefits from packed fractures since turbulence can occur.*

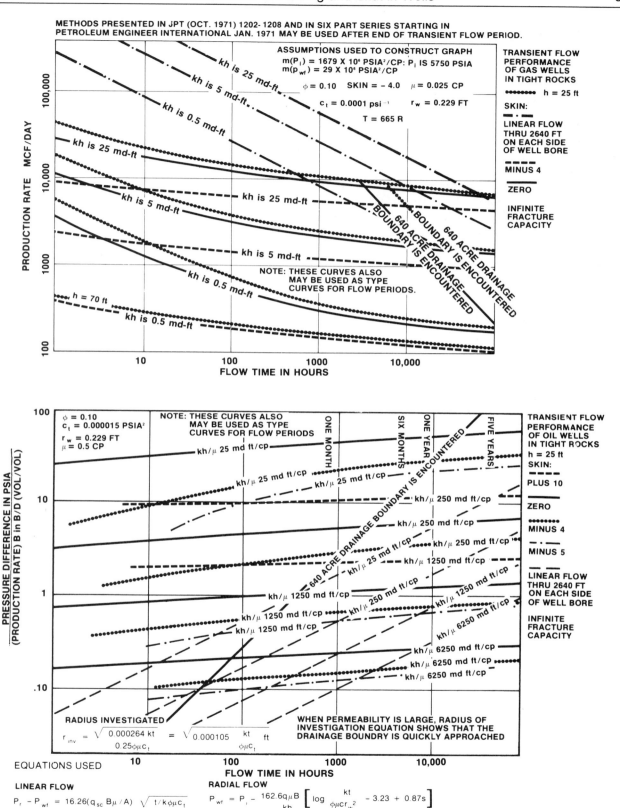

Figure 5–108. *Benefits from fracturing using field test flow data (A–gas; B–oil; courtesy* Petroleum Engineer).

Directionally Drilled Wells

Cinco, Miller, and Ramey in *JPT* (November 1975) conclude that methods for analysis of radial transient flow are applicable since the slanted hole creates a negative skin that is reflected as part of the skin normally determined. The complicated function applicable to slanted holes can be presented as tables that relate t_D, p_D, and h_D as shown.

In an example, these authors assume that normal analysis shows the skin to be 0.8. The semilog slope is 125 psi/cycle and the apparent skin-caused pressure drop is 87 psi. The average angle of inclination is 45 degrees, pay thickness is 100 ft, $r_w = 0.3$ ft, and $k_h/k_z = 5$. The part of total skin caused by hole slant is:

$$h_D = \frac{h}{r_w} \sqrt{\frac{k_r}{k_z}}$$

$$h_D = \frac{100 \text{ ft}}{0.3 \text{ ft}} 5^{1/2} = 745.4$$

$$\theta'_w = \tan^{-1}\left(\sqrt{\frac{k_z}{k_r}}\tan\theta_w\right)$$

$$\theta'_w = \tan^{-1}\left(\frac{\tan 45°}{5^{1/2}}\right) = 24.1°$$

$$\text{skin} = s\theta(\theta'_w, h_D) = -(\theta'_w/41)^{2.06} - [(\theta'_w/56)^{1.865}] \log (h_D/100)$$
$$s_\theta = -(24.1/41)^{2.06} - (24.1/56)^{1.865}$$
$$\times \log_{10} (745.4/100) = -0.66.$$
$$s_d = 0.8 - (-0.66) = +1.46.$$

Dimensionless Wellbore Pressure vs Dimensionless Time for a Fully Penetrating, Slanted Well in an Infinite Slab Reservoir

t_D	$h_D = 100$ 15°	30°	θ_w 45°	60°	75°
1×10^{-1}	0.0120	0.0108	0.0088	0.0062	0.0032
2×10^{-1}	0.0707	0.0634	0.0518	0.0366	0.0189
5×10^{-1}	0.2703	0.2424	0.1979	0.1399	0.0724
7×10^{-1}	0.3767	0.3378	0.2758	0.1950	0.1009
1	0.5043	0.4522	0.3602	0.2611	0.1351
2	0.7841	0.7030	0.5740	0.4059	0.2101
5	1.1919	1.0686	0.8725	0.6170	0.3194
7	1.3477	1.2063	0.9866	0.6976	0.3611
1×10	1.5148	1.3581	1.1089	0.7841	0.4059
2×10	1.8436	1.6529	1.3496	0.9543	0.4940
5×10	2.2829	2.0472	1.6714	1.1816	0.6116
7×10	2.4452	2.1937	1.7914	1.2659	0.6551
1×10^2	2.6179	2.3507	1.9211	1.3571	0.7017
2×10^2	2.9551	2.6612	2.1841	1.5473	0.7981
5×10^2	3.4032	3.0796	2.5508	1.8319	0.9528
7×10^2	3.5681	3.2344	2.6879	1.9421	1.0183
1×10^3	3.7431	3.3991	2.8338	2.0597	1.0911
2×10^3	$°p_D(1,t_D) - 0.123$	3.7227	3.1228	2.2912	1.2375
5×10^3		4.1617	3.5291	2.6291	1.4506
7×10^3		4.3259	3.6859	2.7678	1.5432
1×10^4		4.5012	3.8555	2.9225	1.6528
2×10^4		$°p_D(1,t_D) - 0.516$	4.1916	3.2397	1.9005
5×10^4			4.6434	3.6793	2.2855
7×10^4			$°p_D(1,t_D) - 1.175$	3.8439	2.4382
1×10^5				4.0195	2.6046
2×10^5				$°p_D(1,t_D) - 2.148$	2.9368
5×10^5					3.3861
7×10^5					3.5526
1×10^6					$°p_D(1,t_D) - 3.586$

$°p_D(1,t_D)$ = line-source solution at the wellbore

For this simple case, equations rather than the tables were used. The tables relate with:

$$r_D = \frac{r}{r_w}$$

$$t_D = \frac{0.000264\, k_r t}{\phi\, \mu c_t r_w^2}$$

$$h_{wD} = \frac{h_w}{r_w} \sqrt{\cos^2\theta_w \frac{k_r}{k_z} + \sin^2\theta_w}$$

$$p_D(r_D, \theta, z_D, t_D, \theta'_w, z_{wD}, h_{wD}, h_D)$$

$$= \frac{k_r \sqrt{k_r/k_z}\, h \Delta p(r, \theta, z, t, \theta'_w, z_w, h_w, h)}{141.2\, q_{sc}\, B\mu}$$

Dimensionless Wellbore Pressure vs Dimensionless Time for a Fully Penetrating, Slanted Well in an Infinite Slab Reservoir

t_D	$h_D = 200$		θ_w		
	15°	30°	45°	60°	75°
1×10^{-1}	0.0120	0.0108	0.0088	0.0062	0.0032
2×10^{-1}	0.0707	0.0634	0.0518	0.0366	0.0189
5×10^{-1}	0.2703	0.2424	0.1979	0.1399	0.0724
7×10^{-1}	0.3767	0.3378	0.2758	0.1950	0.1009
1	0.5043	0.4522	0.3692	0.2611	0.1351
2	0.7841	0.7030	0.5740	0.4059	0.2101
5	1.1919	1.0686	0.8725	0.6170	0.3194
7	1.3477	1.2083	0.9866	0.6976	0.3611
1×10	1.5148	1.3581	1.1089	0.7841	0.4059
2×10	1.8436	1.6529	1.3496	0.9543	0.4940
5×10	2.2825	2.0465	1.6709	1.1815	0.6116
7×10	2.4443	2.1915	1.7894	1.2653	0.6550
1×10^2	2.6161	2.3455	1.9151	1.3542	0.7010
2×10^2	2.9506	2.6459	2.1602	1.5273	0.7905
5×10^2	3.4950	3.0495	2.4937	1.7623	0.9107
7×10^2	3.5590	3.2007	2.6221	1.8553	0.9578
1×10^3	3.7332	3.3625	2.6618	1.9599	1.0117
2×10^3	4.0726	3.6798	3.0409	2.1785	1.1322
5×10^3	4.5222	4.1029	3.4159	2.4801	1.3171
7×10^3	4.6879	4.2601	3.5567	2.5925	1.3883
1×10^4	4.8640	4.4286	3.7090	2.7153	1.4655
2×10^4	$°p_D(1,t_D) - 0.152$	4.7620	4.0193	2.9757	1.6301
5×10^4		5.2117	4.4531	3.3692	1.9071
7×10^4		5.3783	4.6166	3.5239	2.0289
1×10^5		$°p_D(1,t_D) - 0.608$	4.7913	3.6919	2.1685
2×10^5			5.1336	4.0260	2.4649
5×10^5			$°p_D(1,t_D) - 1.378$	4.4765	2.8900
7×10^5				4.4765	3.0516
1×10^6				$°p_D(1,t_D) - 2.494$	3.2249
2×10^6					3.5656
5×10^6					$°p_D(1,t_D) - 4.099$

$°p_D(1,t_D)$ = line-source solution at the wellbore

Dimensionless Wellbore Pressure vs Dimensionless Time for a Fully Penetrating, Slanted Well in an Infinite Slab Reservoir

t_D	$h_D = 1,000$		θ_w		
	15°	30°	45°	60°	75°
1×10^{-1}	0.0120	0.0108	0.0088	0.0062	0.0032
2×10^{-1}	0.0707	0.0634	0.0518	0.0366	0.0189
5×10^{-1}	0.2703	0.2424	0.1979	0.1399	0.0724
7×10^{-1}	0.3767	0.3378	0.2758	0.1950	0.1009
1	0.5043	0.4522	0.3692	0.2611	0.1351
2	0.7841	0.7030	0.5740	0.4059	0.2101
5	1.1919	1.0686	0.8725	0.6170	0.3194
7	1.3477	1.2083	0.9866	0.6976	0.3611
1×10	1.5148	1.3581	1.1089	0.7841	0.4059
2×10	1.8436	1.6529	1.3496	0.9543	0.4940
5×10	2.2825	2.0465	1.6709	1.1815	0.6116
7×10	2.4443	2.1915	1.7894	1.2653	0.6550
1×10^2	2.6161	2.3455	1.9151	1.3542	0.7010
2×10^2	2.9503	2.6451	2.1597	1.5272	0.7905
5×10^2	3.3924	3.0416	2.4834	1.7560	0.9090
7×10^2	3.5549	3.1872	2.6023	1.8401	0.9525
1×10^3	3.7271	3.3416	2.7284	1.9293	0.9987
2×10^3	4.0618	3.6417	2.9734	2.1025	1.0884
5×10^3	4.5046	4.0392	3.2979	2.3317	1.2069
7×10^3	4.6676	4.1863	3.4183	2.4163	1.2506
1×10^4	4.8408	4.3437	3.5485	2.5078	1.2973
2×10^4	5.1787	4.6548	3.8119	2.6984	1.3939
5×10^4	5.6271	5.0736	4.1789	2.9832	1.5487
7×10^4	5.7921	5.2284	4.3160	3.0934	1.6143
1×10^5	5.9671	5.3931	4.4620	3.2111	1.6871
2×10^5	6.3084	5.7168	4.7511	3.4426	1.8335
5×10^5	$^\circ p_D(1,t_D) - 0.207$	6.1559	5.1574	3.7805	2.0467
7×10^5		6.3201	5.3142	3.9192	2.1392
1×10^6		6.4954	5.4838	4.0739	2.2488
2×10^6		$^\circ p_D(1,t_D) - 0.824$	5.8199	4.3911	2.4965
5×10^6			6.2717	4.8307	2.8815
7×10^6			$^\circ p_D(1,t_D) - 1.850$	4.9953	3.0343
1×10^7				5.1709	3.2006
2×10^7				$^\circ p_D(1,t_D) - 3.299$	3.5329
5×10^7					3.9821
7×10^7					4.1487
1×10^8					$^\circ p_D(1,t_D) - 5.292$

$^\circ p_D(1,t_D)$ = line-source solution at the wellbore

Build-up Analysis of Deeply Fractured Wells

Fractures and vugs occur in many hard rocks, but even major faults tend to seal in shale intervals in softer rocks such as found in the Texas–Louisiana area of the Gulf Coast. Vugs are found in more permeable limestones or calcareous materials in areas such as West Texas. Fractures and vugs often are associated with rocks having a relatively low porosity and a rather high permeability. Vugs often exhibit a lower permeability, which may be associated with a random poorly connected set of flow channels having higher permeability in localized areas such as found in a specific core. Other heterogeneous characteristics often are found in pays such as directional permeability and permeability differences resulting from differing rocks in stream beds, bars, etc., caused by winnowing due to stream and ocean velocity. Rocks formed by natural processes have variable characteristics that are often not recognized from the limited samples available to the geologist and engineer.

Theory shows that fractures occur in the direction perpendicular to the least principle stress. In most cases, the fracture will be vertical at depths below 1,500 feet due to the overburden forces compared with the lesser forces in other directions. There are at least two major types of fractures. The first type, commonly known as the Pollard type, is similar to the man-made fractures in that the fractures all tend to be in one direction and the long individual fractures may not be adequately connected with each other. In nature, these fractures may be rather wide, and drilling often is difficult since mud is lost to the extent that the well must be drilled without circulation at the surface. Such fractures may be created under tension faulting conditions and when fluids are injected. Recovery from such fractured rocks often is small and waterflooding may not be effective unless flow is between the parallel fractures. Another major fracture type is the Warren-Root type where the fractures are connected in all directions and flow may react as a homogeneous system restricted to the fracture system. The matrix blocks may be rather small, and water injected into the fracture system is imbibed into the matrix so that oil recovery is high under the influence of gravity and capillary forces. When matrix blocks are small, recovery under solution-gas drive and under water drive may approach that of a normal homogeneous pay. Since the reservoir volume or porosity of fractured rocks is often low, water from surrounding shales, shale stringers within the sand, and limited water entry from an aquifer can result in water-drive recovery efficiencies that are not recognized since the water is imbibed rather than produced. Warren-Root type fractures should be expected when compressive forces are present and when the fracturing is related with dome formation caused by moving salt and other natural materials. In many cases, these fractures are limited to the disturbed area and are not associated with regional water movement.

The man-made fracture system may be analyzed using techniques discussed in prior sections of this text. Similar techniques may apply to some Pollard-type fractures. The Pollard example involved a very tight matrix, and a function such as $(1 - e^{-cx})$ was used to estimate average matrix pressure. Determination of the time to reach the drainage boundary using more normal methods was impossible with desired accuracy. This approach was first used for these reservoirs by van Everdingen and Boots in studies made prior to the work of Pollard. Pollard carried the ideas

to involve a determination of matrix porosity, fracture porosity, and skin.

A number or cluster of adjacent and parallel fractures produce a flow improvement equivalent to a single infinite permeability fracture, but a deep fracture is more beneficial than a short fracture. These observations should be recognized when designing man-made fractures, and the single-point fracture procedure should be considered when pay interval is not large. Substantial changes in rock characteristics or compressibility such as a sand-shale interface can at times adjust for the displacement of the fracture so that fracture extension is impossible vertically.

Pollard in *JPT* (March 1959) related the build-up pressure in a well as a function of the closed-in-time as follows:

$$\bar{p} - p_{ws} = C\,e^{-at} + D\,e^{-bt} + (\bar{p} - p_{ws} - C - D)\,e^{-ct}$$

where p_{ws} is measured shut-in pressures. Each of the three terms represents a transient pressure versus time function. The term $C\,e^{-at}$ represents flow from the fine pore system—matrix—and is the difference between the static reservoir pressure and the pressure in the fissure system at any time. The term $D\,e^{-bt}$ is the transient of pressure caused by flow within the coarse fissure system. The remaining term, $(\bar{p} - p_{ws} - C - D)\,e^{-ct}$, is the pressure differential between the coarse fissures near the well and the well bore. The values for C, D and $(\bar{p} - p_{ws} - C - D)$ can be determined when flow time is zero. Since each of the three terms in the equation is an exponential function, plotting the pressure differential associated with any one of the terms against time gives a straight line when Δp is on log scale and time is on coordinate scale. Curve A is the actual measured build-up data plotted as $(\bar{p} - p_{ws})$ versus shut-in time on semilog paper. The second curve is the difference between the actual build-up data and the straight-line extrapolation of this curve. The analysis is shown in Fig. 5–109.

The data can be used to calculate volumes as follows and illustrated by Fig. 5–110: For this example:

$$q_o = 82.23 \text{ m}^3/\text{D just before shut-in}$$
$$h = \text{formation thickness} = 46.8 \text{ meters}$$
$$c_{oe} = c_o + 1/(1 - S_w)[c_w S_w + c_r(1 - \phi)/\phi]$$
$$= 1.614 \times 10^{-4} \text{ atm}^{-1}$$
$$\phi = \text{porosity} = 0.07$$
$$S_w = \text{water saturation} = 10\%$$

The transient slopes read from the curves are: $1/a = 322$, $1/b = 35$, and $1/c = 5.4$ hours/cycle. The values of the intercepts shown are $C = 9.2$ atm, $D = 28.8$ atm, $E = 31.5$ where E is the second differ-

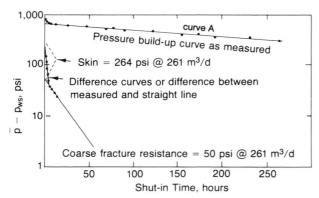

Figure 5–109. *Build-up in Pollard fracture* (JPT, *March 1959*).

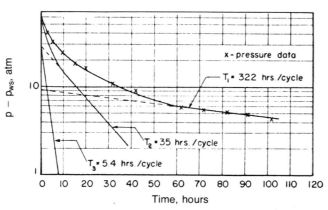

Figure 5–110. *Build-up in Pollard fracture plotted on expanded scales* (JPT, *March 1959*).

ence or difference between the plotted data and straight line of the first difference curve (Fig. 5–110).

The volume of the coarse fissures =

$$V_c = q_o/b(p_{co} - p'_{co})\, c_{eo} \text{ where } p_{co} - p'_{co} = D$$
$$= 82.23 \times 35/ \times 28.8 \times 0.0001614 \times 24 \text{ hr/D}$$
$$= 25{,}800 \text{ m}^3$$

where q_o is at subsurface conditions.

The volume of the fine matrix pores =

$$V_f = q_o/a(D + C)\, c_{oe}$$
$$= 82.23 \times 322/(28.8 + 9.2)0.0001614 \times 24 \text{ hrs/D}$$
$$= 179{,}800 \text{ m}^3$$

Also,

$$V_t = V_c + V_f = r_e^2\, h\, \phi\, (1 - S_w)$$
$$r_e = 488 \text{ feet}$$

As discussed earlier, Pollard-type fractures are useful and can convert a nonproductive reservoir into

an economical situation because of increased flow capacity when fractures are long. However, such fractures do not adequately connect all of the reservoir and recovery is often low.

The general shape of the linear, spherical, and radial flow configuration plots associated with the Pollard-type fractures should be of interest and are included below (Figs. 5–111 to 115):

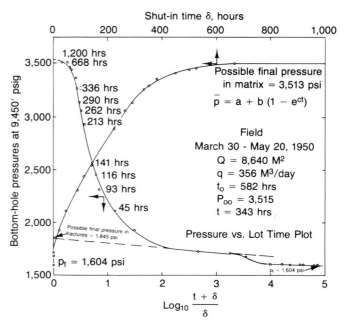

Figure 5–111. *Typical build-up curves for wells producing from fractures and matrix (after Boots and Van Everdingen, courtesy Shell Oil Co.)*

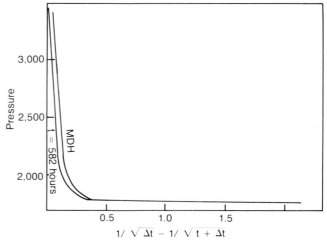

Figure 5–112. *Spherical plot.*

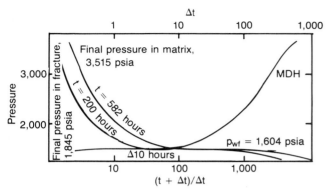

Figure 5–113. *Horner plot.*

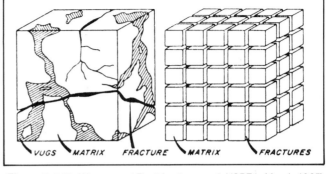

Figure 5–116. *Warren and Root fracture model* (SPEJ, *March 1965*).

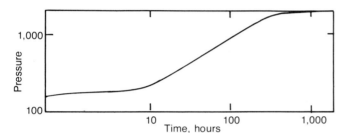

Figure 5–114. *Log-log plot.*

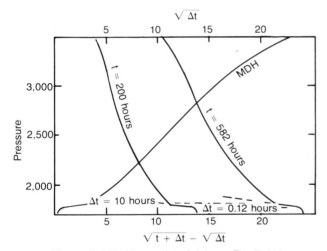

Figure 5–115. *Linear plot of data on Fig. 5–111.*

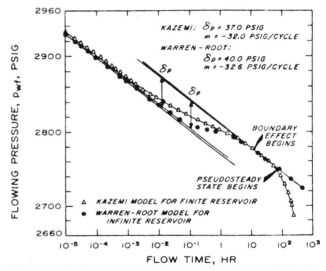

Figure 5–117. *Pressure drawdown case* (after Kazemi, SPEJ, *December 1979*).

The Warren-Root model approaches a homogeneous reservoir performance when the matrix blocks are small, say three feet to the side, and the matrix has say 0.01 md permeability. As blocks are larger and permeability approaches zero, the model approaches the Pollard situation (Fig. 5–116).

The early time period of any transient period should exhibit characteristics of linear flow when the well is fractured by man and naturally when Pollard-type fractures exist. In other cases, as a limit, the fractures behave as a homogeneous reservoir. Fractures, if open, do strongly influence permeability since permeability is a direct function of fracture width.

$$k \text{ in darcy} = 0.544 \times 10^8 \ w^2$$

If fracture width is 0.01 inch wide, the permeability is 5,440 darcys.

Reference to Aguilera's *Naturally Fractured Reservoirs* (Pennwell Books) may prove worthwhile.

Semisteady-state begins when

$$(t_{DA})_{pss} = \frac{0.0002637 \ (kh)_t \ t_{pss}}{[(\phi c_t)_f + (\phi c_t)_{ma}] h \mu A}$$

$$\simeq 0.13.$$

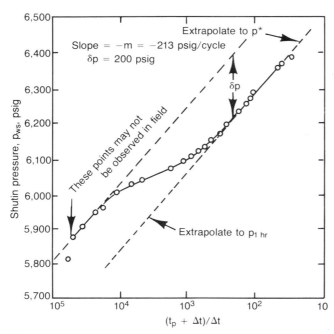

Figure 5–118. *Build-up case (after Warren and Root, SPEJ, March 1965).*

All the results shown are for the infinite reservoir case and are described by two basic parameters:

$$\omega = \phi_2 c_2 / (\phi_1 c_1 - \phi_2 c_2)$$

and

$$\lambda = \alpha k_1 r_w^2 / \bar{k}_2$$

where c_1 = total compressibility, primary system
       c_2 = total compressibility, secondary system
       k_1 = matrix permeability
       $\bar{k}_2$ = effective permeability, fractures
       α = shape factor controlling flow between two systems

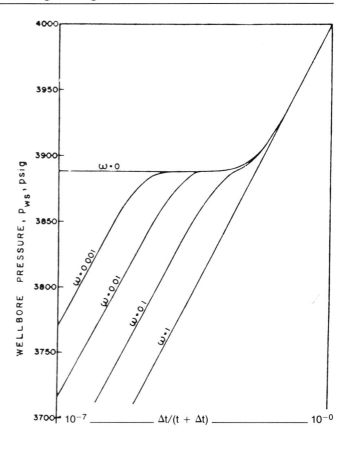

INFINITE RESERVOIR

$\lambda = 5 \times 10^{-8}$ FOR ALL CASES

q = 115 STB/D

t = 21 DAYS

Figure 5–119. *Shift in curve occurs early and may be obscured by well-bore effects (after Warren and Root, SPEJ, September 1963).*

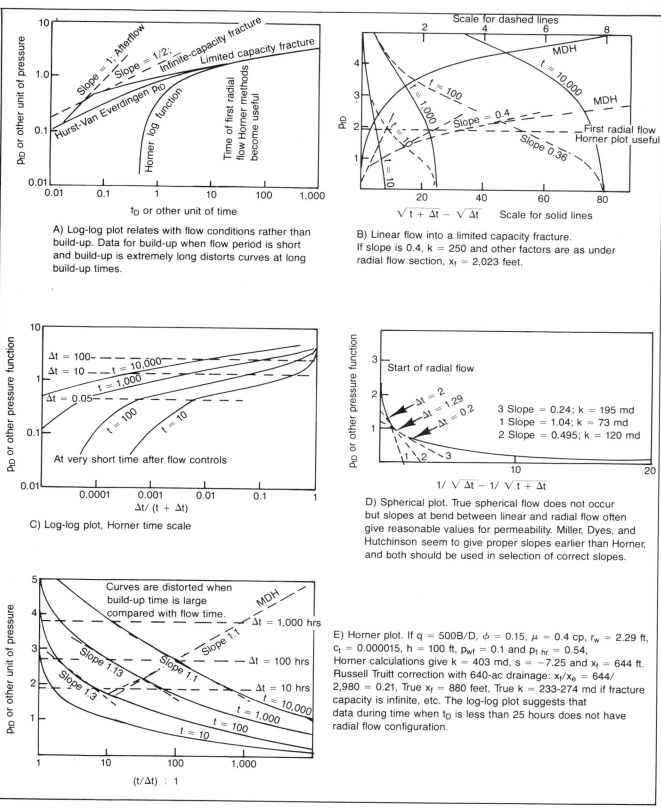

A) Log-log plot relates with flow conditions rather than build-up. Data for build-up when flow period is short and build-up is extremely long distorts curves at long build-up times.

B) Linear flow into a limited capacity fracture. If slope is 0.4, k = 250 and other factors are as under radial flow section, x_f = 2,023 feet.

C) Log-log plot, Horner time scale

D) Spherical plot. True spherical flow does not occur but slopes at bend between linear and radial flow often give reasonable values for permeability. Miller, Dyes, and Hutchinson seem to give proper slopes earlier than Horner, and both should be used in selection of correct slopes.

E) Horner plot. If q = 500B/D, ϕ = 0.15, μ = 0.4 cp, r_w = 2.29 ft, c_t = 0.000015, h = 100 ft, p_{wf} = 0.1 and $p_{1\ hr}$ = 0.54, Horner calculations give k = 403 md, s = −7.25 and x_f = 644 ft. Russell Truitt correction with 640-ac drainage: x_f/x_e = 644/2,980 = 0.21, True x_f = 880 feet, True k = 233-274 md if fracture capacity is infinite, etc. The log-log plot suggests that data during time when t_D is less than 25 hours does not have radial flow configuration.

Figure 5–120. *Sequence of curves for limited-capacity fracture showing relations for various types of build-up curves.*

Corrections when Skins are Negative

The normal calculation of kh, skin, etc., give a combination that is reasonably reliable. However, the values so obtained are not true values, and a fractured well can calculate a value for kh that is in excess of the true value obtained before the well was deeply fractured, (Russell and Truitt, *JPT*, October 1964). In a deeply fractured well, the flow and build-up test record an average for a combination of linear, spherical, and radial flow configurations, and analysis assuming radial flow cannot be expected to give true values without correction. The true value for kh is less than that calculated using normal techniques, and the negative skin also is smaller than calculated as given by the following graphs. In later forecasts, the two data sets should not be mixed. The following graphs are used to obtain true values from the usual calculated values.

When using Figs. 5–121, 122, 123, apparent and calculated values are from normal build-up test procedures. Value x_e is the semisteady-state boundary influencing well performance, and $x_f/2 = r'_w = r_w\, e^{-s}$ where r_w is unfractured well radius and s is calculated skin.

Study of the graphs indicates that few problems and little correction is required when fracture penetration, x_f/x_e, is less than 0.1. The correction is less when spacing is large for the same fracture length. This correction method is applicable when skin is -5 with 40-acre spacing and -6 to -7 when spacing is 640 acres. The usual assumptions are involved in these graphs, and the permeability of the fracture is assumed to be infinite, which is not always correct when fractures are deep and overpacked with nonuniform material. Fractures also may heal nonuniformly. Lenses offer another complication.

The first step in making these corrections is the calculation of kh and skin using the normal Horner and MDH methods. The skin is then converted to x_f using the equations. The graphs then are entered and true values for skin and kh can be read from the curves using trial and error to obtain true x_f. The average reservoir pressure is calculated in the same manner as used for pressure drawdown, or three values are assumed for average reservoir pressure. The $\bar{p} - p_{wf}$ is plotted versus shut-in, time and the straightest line is extrapolated to obtain best value for average reservoir pressure (Fig. 5–124).

Data from the two-rate test require a similar correction as, shown in Fig. 5–125.

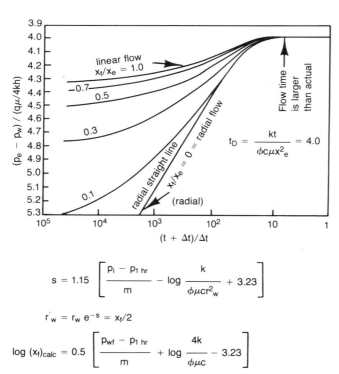

$$s = 1.15 \left[\frac{p_i - p_{1\,hr}}{m} - \log \frac{k}{\phi\mu c r_w^2} + 3.23 \right]$$

$$r'_w = r_w\, e^{-s} = x_f/2$$

$$\log (x_f)_{calc} = 0.5 \left[\frac{p_{wf} - p_{1\,hr}}{m} + \log \frac{4k}{\phi\mu c} - 3.23 \right]$$

Figure 5–121. *Pressure build-up in vertically fractured reservoirs (after Russell and Truitt,* JPT, *October 1963).*

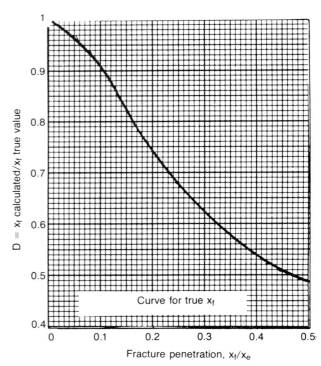

Figure 5–122. *Fracture length correction (after Russell and Truitt,* JPT, *October 1963).*

Sample calculations useful in determining ranges for test design

Skin	-3	-4	-5	-6	-7	-8
e^{-s}	20	55	148	403	1097	2980
r_w, Ft. −6″ id.	0.25	0.25	0.25	0.25	0.25	0.25
r'_w, ft	5	14	37	101	274	745
Calc. x_f, ft.	10	28	74	202	548	1490

Well spacing, ac./well is 40 and 640 and flow before build-up was at semisteady state.

x_e, ft	660	2640	660	2640	660	2640	660	2640	660	2640	660	2640
True x_f, ft.	10	10	29	28	86	76	—	220	—	979	—	—
x_f/x_e	0.015	0.004	0.04	0.01	0.13	0.03	—	0.083	—	0.37	—	—
R, kh_t/kh_a	0.98	1.0	0.92	0.99	0.78	0.94	—	0.84	—	0.5	—	—

Note: A dash implies that flow is almost linear throughout life. The conventional kh from build-up is multiplied by R to obtain the true value for kh of test.

k, md	1		1		0.01		0.01	
ϕ, fraction	0.15		0.15		0.05		0.05	
Sg, fraction	0.70		0.70		0.40		0.40	
Pressure, psi	500		7000		500		7000	
μ_g, cp	0.013		0.030		0.013		0.030	
c_g, psi^{-1}	0.002		0.00006		0.002		0.00006	
$\phi \, Sg\mu_g c_g$	2.73×10^{-6}		0.189×10^{-6}		0.52×10^{-6}		0.036×10^{-6}	
r'_w, ft. (assumed)	25	250	25	250	25	250	25	250
r'^2_w	625	62500	625	62500	625	62500	625	62500
t_{Dw}	0.155t	0.00155t	2.23t	0.0223t	0.0081t	0.000081t	0.117t	0.00117t

If the required t_{Dw} for use of the log function is 50, corresponding real time would be:

t, hours*	323	32258	22	2242	6173	617284	427	42735
days	13	1344	1	93	257	25719	18	1780

Useful data are obtained only if x_e is large compared with x_f, for instance if we assume

$x_f = 2r'_w$, ft	50	500	50	500	50	500	50	500
$r_{inv.} = 2x_f$, ft	100	1000	100	1000	100	1000	100	1000
t for t_{inv}, hrs.**	26	2600	2	180	495	49528	34	3428
, days	1	108	.1	8	21	2063	1	143

Note: * These times are required before the log function should be used in calculations and forecasts work. The P_t functions are applicable and production is less than that obtained from use of log function per same pressure drop and time.
** Time required before useful data are obtained from a build-up or flow test if $x_e = 2x_f = 4r'_w$. These values do not represent true radial flow and are about 10% of values required for log function to be straight line.

Wells With Very Deep Fractures

Russell and Truitt in *JPT* (October 1964) presented a table showing dimensionless pressure drop functions versus a dimensionless time function based on drainage area. A plot of these data is shown in Fig. 5–126.

In the table, values of the dimensionless pressure drop function P are listed as a function of dimensionless time t_D for fracture penetration values of 0.1, 0.2, 0.3, 0.5, 0.7 and 1.

For engineering usage of these tables with the usual system of practical units (i.e., psi, cp, ft, B/D, md, hr), the following equations are valid:

$$p_i - p_{wf} = \frac{221.8 q\mu B}{kh} P\left(t_D, \frac{x_f}{x_e}\right)$$

$$t_D = \frac{0.000264\,kt}{\phi c\mu x_e^2}$$

Values of P for fracture penetrations not listed can be found at any particular t_D value by a simple cross-plot of P vs x_f/x_e.

Advantage of flow when x_f/x_e is one compared with flow at lesser fracture length:

t_{Dxe}	$x_f/x_e = 0.1$	$x_f/x_e = 0.2$	$x_f/x_e = 0.3$	$x_f/x_e = 0.5$
0.00001	10.0	5.0	3.3	2.0
0.001	9.1	4.8	3.3	2.0
0.1	3.8	2.9	2.2	1.5
1.0	1.8	1.5	1.4	1.2

The benefits decrease with time of production.

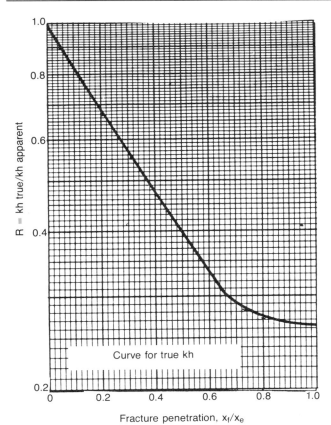

Figure 5–123. *kh correction (after Russell and Truitt*, JPT, *October 1963).*

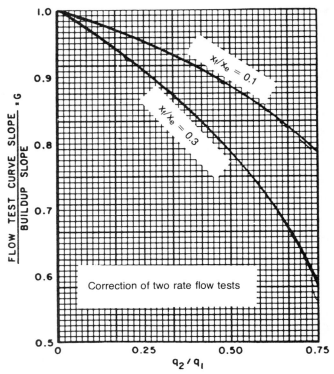

Figure 5–125. *Effect of rate ratio on flow test curve slope in a vertically fractured reservoir (after Russell and Truitt*, JPT, *October 1963).*

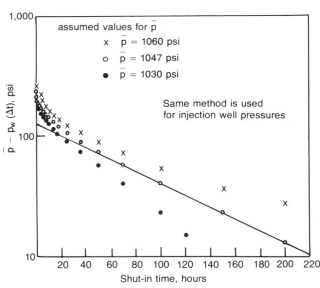

Figure 5–124. *Determination of average pressure in a vertically fractured reservoir,* $[\bar{p} - p_w (\Delta t)]$ *vs* Δt *(after Russell and Truitt*, JPT, *October 1963).*

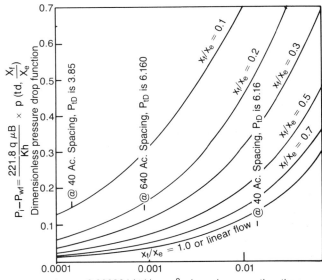

$t_{Dxe} = 0.000264\, kt/\phi\mu c_t\, x^2_e$, based on x_e rather than r_w

The P_t values must be adjusted to allow for constants in equations 141 vs. 221

Figure 5–126. *Pressure in fractured wells. Use these curves instead of the* $r' = r_w\, e^{-s}$ *when the skin is greater than* −5.0.

Horizontal Fractures at Shallow Depths

The first flow naturally involves well bore problems such as unloading. Also, the first flow is linear and measures vertical direction. Thereafter, the flow is a combination of vertical plus horizontal (spherical flow configuration) and finally the radial component predominates when horizontal fracture length is small compared to drainage radius. The performance of the horizontal fracture is compared with that of a vertical fracture expressed in dimensionless terms in Figs. 5–127 and 128.

For the horizontal fracture,

$$t_{Drf} = 0.000264\ k\ t/\phi\mu c_t r_f^2$$

so that the radial-fractured well radius relates as

$$t_{Drf} = t_D(r_w^2/r_f^2)$$

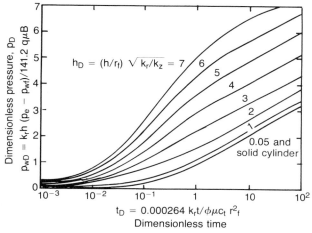

Figure 5–127. *Performance of a flowing well with an infinite conductivity horizontal fracture located in the center of the pay interval (after Ramey et al., SPE 4051, October 1972).*

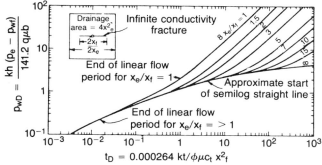

Figure 5–128. *Performance of a flowing well with an infinite-conductivity vertical fracture throughout the vertical pay section (after Ramey et al., SPE 4051, October 1972).*

where r_f is radius of horizontal fracture. Also, the normalized pay thickness,

$$h_D = h/r_f\sqrt{k_r/k_z}$$

where k_r is radial permeability and k_z is vertical permeability. At short flow times, there is a period of linear flow from formation to the horizontal fracture where $p_{Dw} = 2h_D\sqrt{t_{Drf}/\pi}$ and a plot of pressure versus $\sqrt{t}$ should have an early time portion, which is a straight line with slope, m_{Hf}, and an intercept at time zero equal to the pressure in the reservoir, p_e.

and
$$m_{Hf} = (-2.587\ qB/r_f^2)\sqrt{\mu/k_z\phi c_t}$$
$$p_{wf} = p_e + m_{Hf}\sqrt{t}$$

The log-log plot slope is one-half.

Gringarten, Ramey, and Raghavan in *JPT* (July 1975) show that the long-time pressure behavior of an infinite-acting, horizontally fractured well is the same as for an unfractured well having a negative skin. If h_D is less than one, the long-time pressure relation is:

$$p_e - p_{wf} = (70.6\ q\mu B/k_r h)[\ln(.000264\ k_r t/\phi\mu c_t r_f^2)$$
$$+ 1.80907 + h^2 k_r/6 r_f^2 k_z]$$

The short-time relation is:

$$[(k_z)^{1/2}\ r_f^2(p_e - p_{wf})]/141.2 q\mu B = 2(.000264\ t/\pi\phi\mu c)^{1/2}$$

To calculate k_r, k_z and r_f, assuming values for other terms can be obtained from other sources, three independent relations are required. The first can be obtained by using any single point of the pressure difference and time from the log-log plot during the period that slope is one-half to compute the product $(k_z^{1/2}\ r_f^2)$ using the above equation for short times. The products $(\sqrt{k_r k_z}\ r_f)$ and (k_r/r_f^2) can be obtained from the curve match using the pressure and time-scale calculations. These yield two independent relationships. Under some conditions, h_D may also be obtained from the curve-matching technique. The long-time equation indicates that a semilog plot will contain a straight line whose slope may be used to calculate k_r at a distance from the well. At other conditions, an average value for $\sqrt{k_r k_z}$ is only attainable.

Fractures are in the direction perpendicular to least principal forces and horizontal fractures created by man can exist only at shallow depths, say above about 1,500 feet.

The pressure match equation is:

$$\sqrt{k_r k_z}\ r_f = [141.2\ q\mu B\ (p_D/h_D)_M]/(\Delta p)_M$$

The time-scale match equation is:

$$k_r/r_f^2 = [\phi\mu c_t(t_D)_M]/0.000264(t)_M$$

The h_D used to calculate vertical permeability is:

$$k_z = (k_r/r_f^2)[h/(h_D)_M]^2$$

From the pressure match (Fig. 5–129):

$$h_D = 1.5$$
$$p_{wD}/h_D = 0.186$$
$$\Delta p = 100 \text{ psi}$$

$$\frac{p_{wD}}{h_D} = 0.185 = \frac{(\sqrt{k_r k_z}\, r_f)(100)}{141.2 \times 275 \times 1.76 \times 0.23}$$

$$\sqrt{k_r k_z}\, r_f = 29.2 \text{ md ft}$$

From the curve match, $t_D = 0.36$ and $\Delta t = 100$ min

$$t_D = 0.36 = .000264 \times 100\ k_r/$$
$$60 \times .3 \times .23 \times 30 \times 10^{-6}(r_f^2)$$
$$k_r/r_f^2 = 0.00169$$
$$\sqrt{k_r}/r_f = 0.041 \text{ md}^{1/2}/\text{ft}$$
$$h_D = (h/r_f)\sqrt{k_r k_z} = 1.5 \text{ from curve match}$$
$$k_r = 29.2 \times 0.125 = 3.66 \text{ md}$$
$$r_f = \sqrt{3.66}/0.041 = 46.7 \text{ feet}$$
$$\sqrt{k_z} = 29.2/46.7\sqrt{3.66} = .3269 \text{ and } k_z = 0.11 \text{ md}$$

The routine semilog analysis of the radial flow period can also give k_r if h_D is relatively small.

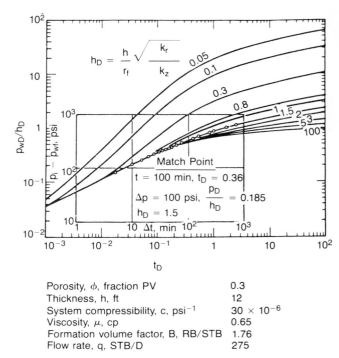

Porosity, ϕ, fraction PV	0.3
Thickness, h, ft	12
System compressibility, c, psi^{-1}	30×10^{-6}
Viscosity, μ, cp	0.65
Formation volume factor, B, RB/STB	1.76
Flow rate, q, STB/D	275

Figure 5–129. *Curve matching for horizontal fracture with uniform influx (after Gringarten et al., JPT, July 1975).*

Turbulent Flow

Engineers must be concerned with the possibility of turbulent flow in flow strings, fractures, and the pore spaces of the pay. Turbulence in pipe was related with roughness and Reynolds number many years ago as shown in Figs. 5–130, 131. The change in flow streamlines from laminar to a semiturbulent pattern occurs at a Reynolds number around 2000. The influence of Reynolds number, pipe size, and fluid viscosity on laminar flow can be illustrated, as shown in Fig. 5–130, 133.

Muskat used data from Fancher et al. (Fig. 5–132) to conclude that Darcy's law applies to most practical reservoir conditions. Muskat states that a well producing 1000 b/d of 30°API oil with 1 cp viscosity will have a velocity of only 0.126 cm/sec at sand face if pay is 10 feet. The Reynolds number is 0.55. The Reynolds number decreases inversely as the radial distance from well, and at 10 feet will have a value of 0.014. A gas well producing 500,000 cf/d from a similar pay will have a Reynolds number of 2.69 at well face and 0.067 at 10 feet. The Fancher data suggest that laminar flow ends at Reynolds numbers of 1 to 10, with the smaller number applicable to pays with lower porosity.

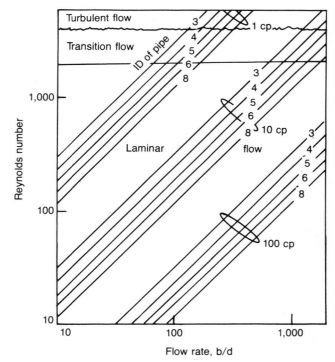

Figure 5–130. *Flow in pipe (courtesy Schlumberger Production Dog Manual).*

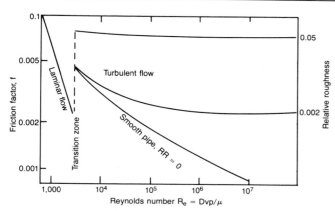

Figure 5–131. *Reynolds number (after Frick,* Petroleum Production Handbook).

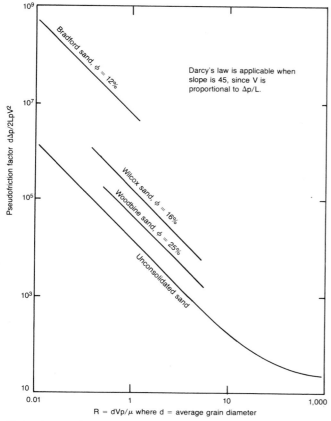

Figure 5–132. *Flowing the gas well at two rates allows one to relate skin with rate and determine skin caused by turbulence (after Fincher et al.,* Muskat's *Physical Principles of Oil Production).*

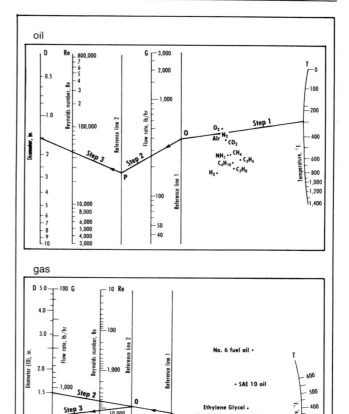

Figure 5–133. *Reynolds number for liquids and gases. Data are based on flow of liquids through tubes (after Ganapathy,* Oil and Gas Journal, *January, 1979).*

Non-Darcy Flow in Gas Wells*

Back-pressure Curves in Groningen Gas Wells (highly permeable thick pay)

Back-pressure test data obtained from the Groningen gas wells indicate that the plot of log $(p_e^2 - p_{wf}^2)$ versus log q_{sc} invariably yields a curved line. Some carefully conducted isochronal tests showed that unsteady state effects are virtually absent. Nor can the curvature be explained by the small variations in the viscosity or the gas deviation factor at the changing average pressures for the different rates. Hence it was surmised that the presence of non-Darcy flow around the wellbore would be responsible for the observed phenomenon. In the following, pertinent equations are derived and thereafter the corresponding analysis of the back pressure curves is shown.

Basic Flow Equation—Steady-State Darcy Flow

Darcy's law may be expressed as:

$$\frac{dp}{dr} = \frac{\mu}{k} v \qquad (1)$$

Where: p = pressure
r = radial distance from center of well
μ = absolute viscosity of the gas
k = permeability of porous medium
v = superficial velocity at reservoir conditions in the direction of the well bore

For radial flow of nonideal gas

$$v = \frac{q_{sc} \, p_{sc} \, T_r \, z}{2 \pi h_d \, T_{sc}} \frac{1}{p \, r} \qquad (2)$$

Where: q_{sc} = volumetric flow rate at standard conditions
h_d = net drained sand thickness
T_r = reservoir temperature
T_{sc} = temperature at standard conditions
P_{sc} = pressure at standard conditions
z = gas deviation factor at p and T_r

If the right-hand side of equation (2) is substituted for v in equation (1) and if constant average values of $\bar{T}_r$, $\bar{z}$ and $\bar{\mu}$ are assumed, equation (1) may be integrated between drainage radius r_e (pressure p_e) and well radius r_w (pressure p_{wf}) to give

$$p_e^2 - p_{wf}^2 = \frac{p_{sc} \, \bar{T}_r \, \bar{z} \, \bar{\mu}}{\pi T_{sc} \, k \, h_d} \ln\left(\frac{r_e}{r_w}\right) q_{sc}$$

* This section courtesy of Shell Oil Co., Netherlands.

After introduction of a (real) skin effect in the well (skin factor S_r), the following simple well inflow performance equation is obtained

$$p_e^2 - p_{wf}^2 = A \, q_{sc} \qquad (3)$$

where

$$A = \frac{p_{sc} \, \bar{T}_r \, \bar{z} \, \bar{\mu}}{\pi T_{sc} \, k \, h_d}\left[\ln \frac{r_e}{r_w} + S_r\right] = \text{Darcy flow coefficient} \qquad (4)$$

Deviation From the Darcy Flow Concept

By equation (3) a log-log plot with $(p_e^2 - r_{wf}^2)$ on the ordinate and q_{sc} on the abscissa should yield a straight line with slope 1. However, in many gas wells a curve is obtained instead, with the slope increasing as the rate increases, and the inflow performance equation is often empirically modified into:

$$(p_e^2 - p_{wf}^2)^n = A \, q_{sc} \qquad (5)$$

where n is the reciprocal slope on the log-log plot.

Modified Flow Equation—Steady-state non-Darcy Flow

The addition of the exponent n in equation (5) can be justified if it is assumed that in the neighborhood of high-capacity gas wells flow is no longer laminar (as assumed by Darcy's law) but is subject to some turbulence. This can be accounted for if Forchheimer's equation is used instead of the Darcy formula:

$$\frac{dp}{dr} = \frac{\mu}{k} v + \beta \, \rho \, v^2 \qquad (6)$$

Where: β = coefficient of inertial resistance

$$\rho = \frac{28.97 \times SG \, p}{z \, R \, T_r} = \begin{array}{l}\text{density of gas at pressure}\\ p \text{ and temperature } T_r\end{array}$$

28.97 = molecular weight of air
SG = specific gravity of gas (air = 1)
R = universal gas constant

Integration of equation (6) yields the new well inflow performance equation:

$$p_e^2 - p_{wf}^2 = A \, q_{sc} + F \, q_{sc}^2 \qquad (7)$$

where: A = Darcy flow coefficient as defined in equation (4)

$$F = \frac{28.97 \times p_{sc}^2 \, \bar{T}_r \, \bar{z} \, SG \, \beta}{2 \pi^2 \, T_{sc}^2 \, R \, h_p^2}\left[\frac{1}{r_w} - \frac{1}{r_e}\right]$$

= turbulent flow coefficient $\qquad (8)$

h_p = perforated thickness (used here as different from h_d previously defined because the turbulence effect is likely to be concentrated near the well bore).

Equation (7) suggests that a plot of $(p_e^2 - p_{wf}^2)/q_{sc}$ versus q_{sc} on linear paper should yield a straight line with the intercept on the ordinate equal to the Darcy flow coefficient (A) and the slope equal to the turbulent flow coefficient (F).

Definition of Exponent *n* from Log-Log Plot in Terms of Turbulence

Combination of equations (5) and (7) yields:

$$n \ln (A q_{sc} + F q_{sc}^2) = \ln (A q_{sc})$$

After differentiation, it follows that

$$n = \frac{A + F q_{sc}}{A + 2 F q_{sc}} \qquad (9)$$

Thus it can be seen that

for $A \gg F$, n approaches 1.0 (low rate, little turbulence)
for $A \ll F$, n approaches 0.5 (high rate, much turbulence)

Turbulence Effect Expressed as a Rate-dependent Apparent Skin

Equation (7) may be written as:

$$p_e^2 - p_{wf}^2 = \frac{p_{sc} \bar{T_r} \bar{z} \bar{\mu}}{\pi T_{sc} k h_d} \left[\ln \frac{r_e}{r_w} + S_r + D q_{sc} \right] q_{sc} \qquad (10)$$

where

$$D = \frac{28.97 \times p_{sc} SG \beta k h_d}{2 \pi T_{sc} R h_p^2 \bar{\mu}} \left[\frac{1}{r_w} - \frac{1}{r_e} \right]$$

= non-Darcy flow coefficient $\qquad$ (11)

The product $D q_{sc}$ can be considered as an apparent, rate-dependent, skin factor.

A Check on the Empirical Flow Coefficients

The Darcy and the turbulent flow coefficients obtained from a plot, as suggested by equation (7), may be checked against the results of pressure build-up analyses as follows:

The factor S_t (total skin) calculated from the pressure build-up should satisfy the relation

$$S_t = S_r + S_a$$

where, from equation (4)

$$S_r = \left(\frac{\pi T_{sc} k h_d}{p_{sc} \bar{T_r} \bar{z} \bar{\mu}} \right) A - \ln \frac{r_e}{r_w} \qquad (12)$$

and from equations (8) and (11)

$$S_a = \left(\frac{\pi T_{sc} k h_d}{p_{sc} \bar{T} \bar{z} \bar{\mu}} \right) F q_{sc} = \frac{F}{A} \left[\ln \frac{r_e}{r_w} + S_r \right] q_{sc} \qquad (13)$$

Conclusion

The results obtained from this line of approach to the analysis of well test data clearly indicated the need to take into account non-Darcy or turbulent flow effects for correctly estimating the inflow performance of high capacity gas wells of the Groningen type.

Effect of Rock and Liquid Variations—Multilayer Reservoirs

Reservoir rocks often are not uniform in either the horizontal or vertical directions. The combinations of the variables are many and include no connection between zones—no crossflow—to complete crossflow where the tighter rocks flow into more permeable channels and into the well.

Commingling of zones when there is no crossflow can be detrimental to recovery efficiency when one of the zones is for practical purposes almost depleted at time of opening of the second zone. The problem can be avoided if the sand face pressure is always below that of the lower pressured zone so that flow from the tighter and higher pressured zone to the more permeable but lower-pressured zone cannot oc-

cur. Since some shut-in periods are often unavoidable, loss of recovery efficiency often must be expected when widely differing zones are commingled under noncrossflow conditions. In many situations, economics dictates accepting possible loss caused by commingling.

A reservoir with crossflow behaves more or less as a single reservoir after adjustment for average values of the various parameters appearing in the equations, as illustrated in Fig. 5–134.

The flow behavior of a well with and without crossflow are compared by Russell and Prats in Fig. 5–135.

Fig. 5–136 extends earlier work by Russell and

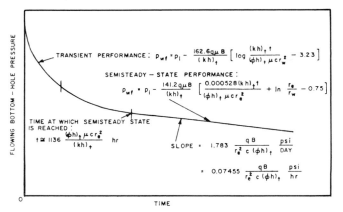

Figure 5–134. *Pressure decline in two-layer reservoir with crossflow (after Russell and Prats, JPT, June 1962).*

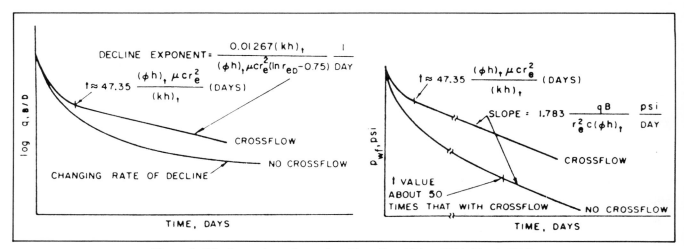

Figure 5–135. *Comparison of crossflow (Russell and Prats, SPEJ, March 1962).*

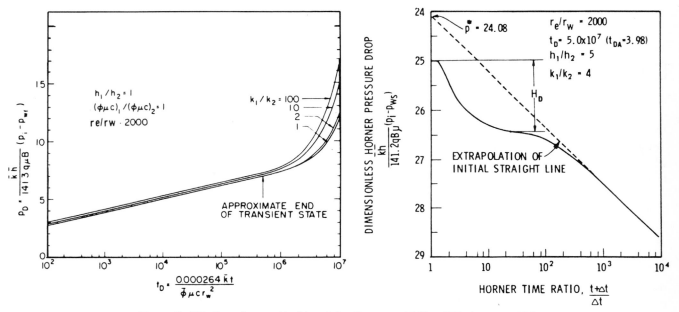

Figure 5–136. *Crossflow and build-up (after Ramey and Miller, JPT, January 1972).*

Prats, which indicates that the dimensionless pressure time function for slow buildup is a straight line having the same slope $2.303/2 = 1.151$.

The relative permeability may differ if fluids are below bubble point, since fluid characteristics in the two zones should differ even when the shut-in period is very long. Also, the data show that the late transient flow period with no crossflow is much longer than for the single layer. *Kh* of a well can be obtained from the *MDH* and Horner plots, and these values are good averages even with no crossflow. Both types of plots show a substantial upward trend at longer shut-in times, which is caused by the tighter, higher-pressured zone flowing into the more permeable lower-pressured zone as the lower-pressured zone approaches stabilization. The early straight line for both flow and build-up plots can be obscured by afterflow and skin effects, and the Muskat analysis method may prove useful when the reservoirs are bounded (Fig. 5–137).

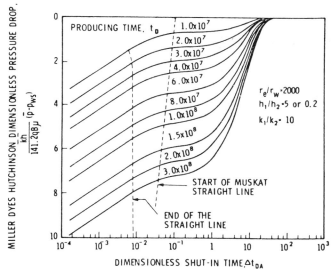

Figure 5–137. *MDH build-up curves (after Raghavan et al., JPT, September 1974).*

The following dimensionless terms are used in analysis of the reservoirs.

$$t_D = \frac{0.000264\,\bar{k}\,t}{\bar{\phi}\mu c r_w^2}$$

$$t_{DA} = 0.000264\,\frac{\bar{k}t}{\bar{\phi}\mu cA}$$

$$p_D = \frac{\bar{k}h}{141.3 q\mu B}\,(p_i - p_{wf})$$

$$\bar{k} = \frac{k_1 h_1 + k_2 h_2}{h_1 + h_2}$$

$$\bar{h} = h_1 + h_2$$

$$\bar{\phi} = \frac{\phi_1 h_1 + \phi_2 h_2}{h_1 + h_2}$$

For the Muskat method, if $(\bar{p} - p_{ws})$ is known or can be evaluated from the Muskat curve for $\Delta t = $ zero from the straight line extrapolation,

$$\bar{kh} = \frac{141.3 q\mu B\,(0.87)}{(\bar{p} - p_{ws})_{\Delta t=0}}$$

Also, the reservoir volume may be obtained from the following equation:

$$\bar{\phi}Ac = 0.000264\,\frac{M\bar{k}}{\mu}\,(\text{slope, log-cycle/hr})^{-1}$$

The paper by Raghavan, Topalough, Cobb, and Ramey published in *JPT* (September 1974) and references included therein should be studied as part of any serious review of flow from two or more reservoirs. Determination of the degree of interconnection between reservoirs at distance is most difficult.

Cobb also discusses the use of the Muskat methods. Matthews, Brons, and Hazebroek in *Trans.* AIME (1954) offer corrections for the two-layer case as shown above. Two poorly connected zones can have characteristics of the Pollard fracture, as illustrated by comparing Figs. 5–138 and 139.

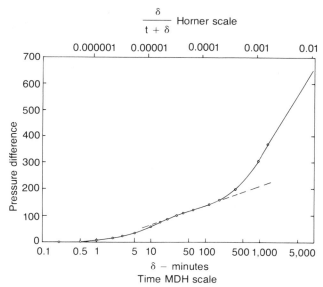

Figure 5–138. *Two-zone example; only top zone perforated.*

Raghavan, Topaloglu, Cobb, and Ramey, *JPT* (September 1974) indicate that at times both the permeability ratio and static pressure can be obtained from the Horner plot.

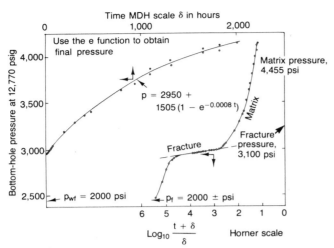

Figure 5-139. *Pollard-type fracture example (after Boots and van Everdingen, courtesy Shell Oil Co.).*

When the time is long and semisteady-state conditions prevail in a two-layer environment,

$$p_{wD}(t_D) = 2\pi t_{DA} + \frac{\left(1 + \frac{k_2}{k_1}\frac{h_2}{h_1}\right)\left(1 + \frac{k_2}{k_1}\frac{h_1}{h_2}\right)\left(\frac{k_1}{k_2}\right)}{\left(1 + \frac{h_2}{h_1}\right)\left(1 + \frac{h_1}{h_2}\right)}$$

$$\cdot \left(\ln\frac{r_e}{r_w} - 0.75\right)$$

where $t_{DA} = t_D(r_w^2/\pi r_e^2) = $ dimensionless time based on drainage area.

Interference Between Wells and Reservoirs

The total pressure drop at Well #2 is the pressure drop caused by Well #1 plus that caused by Well #2 as shown (in Fig. 5-140).

When flow rates are relatively low and permeability is high, pressure throughout the reservoir may be relatively uniform and the model may be represented by water in a bucket. If the rate is high in good reservoirs, the pressure gradients with distance can be substantial and the shut-in or start-up of wells adjacent to the test well may produce interference transients that can severely distort the measured test data. If the reservoir is very tight, interference from Well #1 may not reach Well #2, and infill drilling can be justified.

The pressure at the test well may be declining at a rather uniform rate because of interference from an adjacent well or because the reservoir is small relative to the production rate (rapid depletion of the

reservoir). For these conditions, the starting pressure at time of test must be adjusted for the interference effects in both flow and build-up analysis (Fig. 5-141).

The pressure at any radius from a producing well may be expressed as:

$$p_e - p = (q_w \mu B/2\pi kh)\ p_{DR,TD}$$

$$= (q_w \mu B/2\pi kh)\ \tfrac{1}{2} \int_{R^2}^{\infty} (e^{-x}/x)\ dx$$

$$= (q_w \mu B/2\pi kh)\ \tfrac{1}{2}\ ei\ (R^2/4 t_D)$$

In the wellbore, the equation becomes:

$$p_e - p_{wf} = (q_w \mu B/2\pi kh)\ p_{TD}$$

$$= (q_w \mu B/2\pi kh)\ \tfrac{1}{2} \int_{1/4T_D}^{\infty} (e^{-x}/x)\ dx$$

$$= (q_w \mu B/2\pi kh)\ \tfrac{1}{2}\ ei\ (\tfrac{1}{4}\ T_D)$$

The general equation for use in determining interference among wells may be written as:

$$\Delta p = p_{ext} - p_{obs} = \frac{-m}{2.303}\left[\sum_{j=1}^{NW}\frac{q_j}{q}\left\{Ei\left(\frac{-\phi\mu c a_j^2}{0.00105 k\ (t_j + \Delta t_j)}\right)\right.\right.$$

$$\left.\left. - Ei\left(\frac{-\phi\mu c a_j^2}{0.00105 k t_j}\right)\right\}\right]$$

where Δp represents the pressure drop at the observation well caused by production at other wells. The

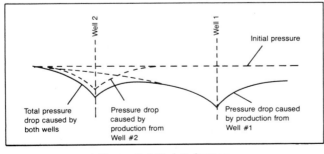

Figure 5-140. *Interference.*

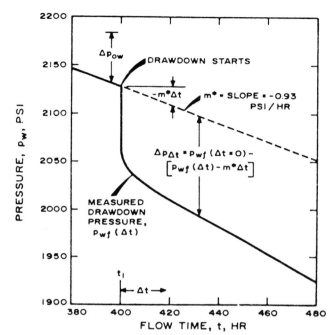

Figure 5–141. *Pressure decline (after Earlougher, SPE Monograph 5, 1977).*

equation may also be written in practical units by inserting applicable conversions:

$$\Delta p = -70.6 \frac{q\mu B}{kh}\left[\sum_{j=1}^{NW}\frac{q_j}{q}\left\{E_i\left(\frac{-\phi\mu ca_j^2}{0.00105k\,(t_j+\Delta t_j)}\right)\right.\right.$$
$$\left.\left.-Ei\left(\frac{-\phi,ca_j^2}{0.00105kt_j}\right)\right\}\right]$$

In a similar manner, the equation for build-up conditions may be written:

$$p_{ws} = p^* - 162.6\frac{q\mu B}{kh}\log\left(\frac{t+\Delta t}{\Delta t}\right)+70.6\frac{q\mu B}{kh}$$

$$\left[\sum_{j=1}^{NW}\frac{q_j}{q}\left\{Ei\left(\frac{-\phi\mu ca_j^2}{0.00105k\,(t_j+\Delta t_j)}\right)-Ei\left(\frac{-\phi\mu ca_j^2}{0.00105kt_j}\right)\right\}\right]$$

In these equations,

$q =$ the production rate at the observation well before shut-in

$q_j =$ rate of production at Well j

$t_j =$ the producing time of the jth well prior to shut-in of the observation well

$\Delta t_j =$ the producing time interval of the jth well subsequent to shut-in of observation well.

$NW =$ number of interfering wells

$a_j =$ distance of jth well from the observation well

When boundaries such as faults are present, image wells are used to represent the fault. In application practice, if some field-wide average pressure is used

as the starting value and field-wide depletion thereafter is recognized by the analyst, the producing and shut-in times can be reasonably approximated by using the equation shown above.

$t =$ Cumulative production at observation well/rate of production prior to shut-in

$t_1 =$ Cumulative production of Well #1/production rate Well #1 during test

$$\Delta t_1 = \frac{\begin{array}{c}\text{Incremental production at Well \#1 subsequent to}\\\text{shut-in of observation well}\end{array}}{\begin{array}{c}\text{Average rate of production } q_1 \text{ during}\\\text{interference test}\end{array}}$$

As an example, calculate the pressure drop at an observation well caused by production at Well #1 at an observation well that has had no production to date. Also available is a prior determination of m at the observation well. There was no production prior to the shut-in test so that $t_1 = 0$. Measurements include:

$Q_1 =$ cumulative production Well #1 subsequent to shut-in of observation well = 23,050 bbl

$q_{obs} =$ production rate at observation well prior to shut-in = 140 b/d

$m_{obs} = (16216\ q\mu B/kh)_{obs} = 270$ psi/cycle as obtained from build-up test of observation well

$B =$ formation volume factor = 1.1 volume in reservoir/volume of tank oil

$q_1 =$ production rate at Well #1 during test at observation well = 180 b/d

$a_1 = 1835 =$ distance between observer and Well #1, feet

$\phi\mu c_t/k = 1\times 10^{-6}$

$t_1 = 23,050/180 = 128$ days = 3070 hours

$p = -270/2,303[(180/140)Ei\,(-1\times 10^{-6}\times 1835\times 1835/0.00105\times 3070)]$

$\quad = -117.2\,[1.285\ Ei\,(-1.042)]$

$\quad = -117.2\times 1.285\times(-0.21)$

$\quad = 32$ psi

The interference of other adjacent wells on the observation well can be calculated in a similar manner and the pressure drop caused by all wells on the observation well can be obtained by adding the individual pressure drops at any selected shut-in time of the observation well. Trial and error techniques may be used to determine values for unknowns if data are accurate and sufficient equations are available using methods such as least squares. The techniques are not included here because of space and detail required is beyond usual field methods. The computer makes such methods practical, and computer input methods should be carefully reviewed when curve match and trial and error techniques are applicable. Refer to SPE

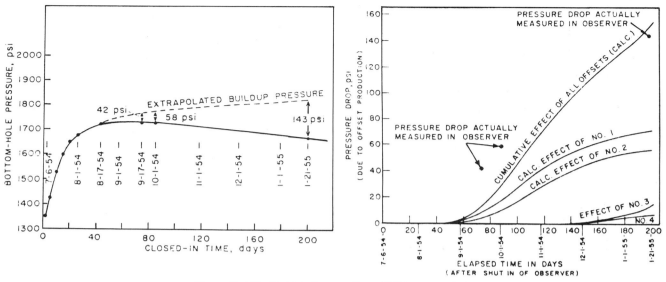

Figure 5–142. *Well interference (from SPE Monograph 1).*

Monograph 1 and SPE Monograph 5 for other examples.

Pressure relations can also be calculated at other times and the entire analysis may be presented on graphs (Fig. 5–142).

Tilted Oil-Water Contacts Under Hydrodynamic Conditions

Water often is moving from higher elevations to lower elevations due to the pressure differentials existing in permeable zones. These pressure differentials across an oil or gas accumulation cause tilted oil-water contacts at time of discovery of the accumulation, and these tilts may exist through the producing life if present initially, as illustrated below.

$$\text{Sin } \alpha = \rho_w / g (\rho_w - \rho_{oil}) \, d\Phi_w / da$$
$$\text{Sin } \alpha = \rho_w / (\rho_w - \rho_{oil}) \, dh_w / da$$

where a is distance and all terms are in *cgs* units such as Darcy, atm, and cm and $g = 987$ cm sec^2

The largest tilts can be expected when oils are heavy since a small difference in fluid densities ($\rho_w - \rho_{oil}$) results in a large value for Sin α. Gas accumulations underlain by flowing water—pressure differential across the accumulation—will show comparatively small tilt. Geological change can cause tilts.

Oil-Water Contacts in Producing Fields

Normal field production practices can cause tilts in water-oil and gas-oil or gas-water contacts as illustrated in an example for oil-water:

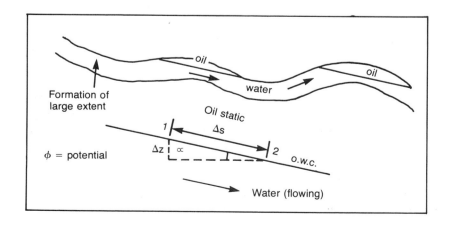

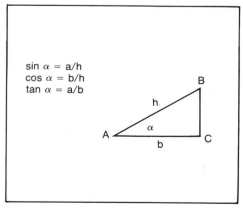

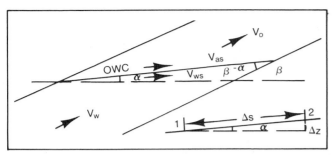

Oil-water contacts in producing fields

$$\frac{\cos(\beta - \alpha)}{\sin \alpha} = \frac{g(\rho_w - \rho_o)}{v\left(\frac{\mu_o}{k_o} - \frac{\mu_w}{k_w}\right)}$$

$$\frac{\cos \beta \cos \alpha + \sin \beta \sin \alpha}{\sin \alpha} = \frac{g(\rho_w - \rho_o)}{v\left(\frac{\mu_o}{k_o} - \frac{\mu_w}{k_w}\right)}$$

$$\frac{\cos \beta}{tg \, \alpha} + \sin \beta = \frac{g(\rho_w - \rho_o)}{v\left(\frac{\mu_o}{k_o} - \frac{\mu_w}{k_w}\right)}$$

If the production rate is increased sufficiently, the angle α can become equal to the angle β, and at that rate the water will bypass the oil without displacing oil ahead of the water. This critical rate has been defined as v_{cr} (often noted as v_s in literature).

Critical Producing Rate

$$\sin \beta = \frac{v_{cr}\left(\frac{\mu_o}{k_o} - \frac{\mu_w}{k_w}\right)}{g(\rho_w - \rho_o)}$$

Also

$$v_{cr} = \frac{g(\rho_w - \rho_o) \sin \beta}{\frac{\mu_o}{k_o} - \frac{\mu_w}{k_w}}$$

$$q_{cr} = \frac{A \, g(\rho_w - \rho_o) \sin \beta}{\frac{\mu_o}{k_o} - \frac{\mu_w}{k_w}}$$

The lower the value of v_{cr}, the better the chance for bypassing of the oil by water. When oils are heavy and viscous and water drive—either natural or injected by man—is present, the critical rate is low. Bypassing or water breakthroughs will invariably occur at an early time during production. In this case, $(\rho_w - \rho_{oil})$ is small and $[(\mu_o/k_o) - (\mu_w/k_w)]$ is large. Here the permeabilities should be recognized as effective values adjusted for relative permeability. Well-known examples of water underrunning the oil on a field-wide as well as individual well-coning basis are Mount Posa (California) and Schoonebeek (Netherlands).

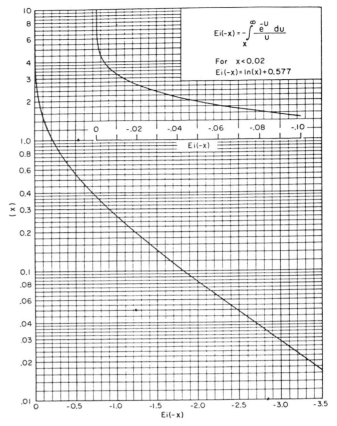

$$Ei(-x) = -\int_{x}^{\infty} \frac{e^{-u}}{u} \, du$$

For $x < 0.02$

$$Ei(-x) = \ln(x) + 0.577$$

Figure 5-143. *Exponential integral values; the Ei function.*

Water usually underruns a field at the base of the pay because of gravity. Gas migrates to top of pay.

In most flow work, the P_T function or the logarithm function should be used, the first for short time and the second for long times with time in dimensionless units. The $E(i)$ function is relatively useless in most field work since log is used (Fig. 5-143).

x	$Ei(-x)$
0.02	-3.335
0.01	-4.028
0.005	-4.721
0.001	-6.331
0.0005	-7.024
0.0001	-8.633

Interference of Wells in Clusters*

In order to calculate the pressure distribution in the drainage areas of the Groningen gas wells, two basic assumptions are made:

* This section courtesy of Shell Oil Co., the Netherlands.

1. In view of the high mobility of the fluid (high permeability and low viscosity) transient effects may be ignored and steady-state considerations applied as relevant to bounded reservoirs.
2. In the light of the observed relatively small pressure drawdowns, equations for the flow of fluids of small and constant compressibility will approximately hold for flow of gas if average values for fluid properties are used.

For the above assumptions it can be shown that if a well is producing at constant rate at the center of a bounded circular drainage area, the relationship between the pressures p_1 and p_2 prevailing in the formation at any distances r_1 and r_2 from the center of the well is given by

$$q = \frac{2\pi kh}{\mu}\left[\frac{(p_2 - p_1)}{\left[\ln\left(\frac{r_2}{r_1}\right) - \frac{r_2^2}{2\,r_e^2} + \frac{r_1^2}{2\,r_e^2}\right]}\right] \qquad (1)$$

Where: q = constant well offtake rate at average reservoir conditions
k = permeability of formation
μ = viscosity of fluid at average reservoir conditions
h = net sand thickness
r_e = radius to the drainage boundary from the center of the well.

If the following substitutions are made:

$r_1 = r_w$ = the radius of the wellbore
$r_2 = r$ = distance to any point in the formation from the well center
$p_1 = p_{wf}$ = the flowing sand-face pressure in the well
$p_2 = p_r$ = pressure at distance r from the well center

and small terms involving r_w^2/r_e^2 are neglected, the following equation is obtained:

$$p_r = p_{wf} + \frac{q\mu}{2\pi kh}\left[\ln\left(\frac{r}{r_w}\right) - \frac{r^2}{2\,r_e^2}\right] \qquad (2)$$

If the average equalized drainage area pressure is denoted by $\bar{p}$,

$$\bar{p} = \frac{2}{r_e^2}\int_{r_w}^{r_e}(p_r\,r)\,dr \qquad (3)$$

If p_r from equation (2) is substituted into equation (3) and the latter integrated, the following relationship is obtained between the average pressure and the well pressure, which is the well-known semisteady-state equation for a single well:

$$\bar{p} = p_{wf} + \frac{q\mu}{2\pi kh}\left[\ln\frac{r_e}{r_w} - \frac{3}{4}\right] \qquad (4)$$

It is assumed that both equations (2) and (4), though basically derived for a well at the center of the circular drainage area, are also valid for a well located somewhat off-center.

Combination of equations (2) and (4) results in:

$$\bar{p} - p_r = \frac{q\mu}{2\pi kh}\left[\ln\frac{r_e}{r} + \frac{r^2}{2\,r_e^2} - \frac{3}{4}\right] \qquad (5)$$

Equation (5) describes the extra pressure drop at any well due to the interference of another well producing at rate q at a distance r.

If there are n wells producing at one cluster, all at identical rates, the total pressure drop at each producing well may be obtained by combining equations (4) and (5). An allowance may be made at this stage for the existence of a skin of dimensionless value S at the sand face. Thus

$$\bar{p} - p_{wf} = \frac{q\mu}{2\pi kh}\left[\ln\frac{r_e}{r_w} + \sum_{i=1}^{i=n-1}\ln\frac{r_e}{r_i}\right.$$
$$\left. + \frac{1}{2\,r_e^2}\sum_{i=1}^{i=n-1}r_i^2 - \frac{3}{4}n + S\right] \qquad (6)$$

Where: r_i = distance of the ith well from the well in which the pressure drop is to be calculated.

For n wells uniformly located in two straight rows with spacing s (with n even), equation (6) may be simplified to:

$$\bar{p} - p_{wf} = \frac{q\mu}{2\pi kh}\left[\ln\frac{r_e}{r_w} + (n-1)\ln\frac{r_e}{s} + C_1\frac{s^2}{r_e^2} - C_2 + S\right] \qquad (7)$$

In equation (7), C_1 and C_2 are two constants determined by the number of wells and the location of the well in which the pressure drop is to be calculated with respect to the other wells in the cluster. However, if the wells at the middle and the ends of a row (which should have the maximum and minimum pressure drops respectively) are considered and their mean values are taken, C_1 and C_2 may be expressed as functions of the number of wells only.

Values of C_1 and C_2

n	4	6	8	10	12	14	16	18	20
C_1	2.0	5.0	12.0	22.5	40.0	63.0	96.0	136.5	190.0
C_2	3.4	4.8	9.1	12.8	17.0	21.5	26.3	31.3	36.6

To judge the relative magnitude of well interference, equations (4) and (7) may be rewritten to express the pressure drops in dimensionless form. Thus, for a single well in the center of the cluster drainage area:

$$\frac{\bar{p}-p_{wf}}{q\mu/2\,\pi\,k\,h}=\left[\ln\frac{r_e}{r_w}-\frac{3}{4}\right]+S \qquad (8)$$

for a cluster well:

$$\frac{\bar{p}-p_{wf}}{q\mu/2\,\pi\,k\,h}=\left[\ln\frac{r_e}{r_w}+(n-1)\ln\frac{r_e}{s}+C_1\frac{s^2}{r_e^2}-C_2\right]+S \qquad (9)$$

Additional Sources of Information

Aguilera, Roberto. *Naturally Fractured Reservoirs.* PennWell Publishing Company, 1980.

Asquith, George. *Log Analysis by Microcomputer.* PennWell Publishing Company, 1980.

————. *Subsurface Carbonate Depositional Models.* PennWell Publishing Company, 1978.

Anderson, Gene. *Coring and Core Analysis.* PennWell Publishing Company, 1978.

Calhoun, John, Jr. *Fundamentals of Reservoir Engineering.* University of Oklahoma Press.

Carlile and Gillett. *Fortran and Computer Mathematics.* PennWell Publishing Company, 1974.

Craft, B. C., and M. F. Hawkins. *Applied Petroleum Reservoir Engineering.* Prentice-Hall Inc.

Campbell, John, et al. *Mineral Property Economics.* Campbell Books, 1978.

Dickey, Parke. *Petroleum Development Geology.* PennWell Publishing Company, 1979.

Frick, Thomas, ed. *Petroleum Production Handbook.* SPE publication.

Katz, D., et al. *Handbook of Natural Gas Engineering.* McGraw-Hill Books, 1959.

Ikoku, Chi. *Natural Gas Engineering.* PennWell Publishing Company, 1980.

Maurer, William. *Advanced Drilling Techniques.* PennWell Publishing Company, 1980.

Megill, Robert. *Introduction to Risk Analysis.* PennWell Publishing Company, 1977.

McCain, William D. *Properties of Petroleum Fluids.* PennWell Publishing Company, 1974.

Moore, Preston, *Drilling Practices Manual.* PennWell Publishing Company, 1974.

Muskat, M. *Flow of Homogeneous Fluids Through Porous Media.* McGraw-Hill Books, 1937.

————. *Principles of Oil Production.* McGraw-Hill Books.

Pirson, S. *Handbook of Well Log Analysis.* Prentice-Hall Inc.

Slider, H. C. *Practical Petroleum Reservoir Engineering Methods.* PennWell Publishing Company, 1978.

Standing, M. B. *Volumetric and Phase Behavior of Oil Field Hydrocarbons.* Reinhold Publishing Company, 1952.

Schlumberger log interpretation manuals. Schlumberger Ltd., Box 2175, Houston 77001.

Dresser-Atlas interpretation and review books. Dresser Industries, Box 1407, Houston 77001.

Welex Services (Halliburton group), Box 2687, Houston 77001.

International Oil and Gas Development Yearbook. International Oil Scouts Assoc., Box 2121, Austin, 78767.

Petroleum Information. 1375 Delaware, Denver 80022.

Handbooks from *World Oil* and Gulf Publishing Co., Box 2608, Houston.

Petroleum Engineers' equations and rules of thumb. *Petroleum Engineer,* Dallas.

Publications sold by the SPE:

Burcik, *Properties of Petroleum Reservoir Fluids*

Cole, *Reservoir Engineering Manual*

Craig, *Reservoir Engineering Aspects of Waterflooding*

Crichlow, *Modern Reservoir Engineering—Stimulation Approach*

Drake, *Fundamentals of Reservoir Engineering*

Earlougher, *Advances in Well Test Analysis*

Farina, *Fortran IV—Self-taught*

Gray and Cole, *Oil-Well Drilling Technology*

Howard and Fast, *Hydraulic Fracturing*

Matthews and Russell, *Pressure Build-up and Flow Tests in Wells*

Nind, *Principles of Oil Well Production*

Nobles, *Using the Computer to Solve Petroleum Engineering Problems*

Peaceman, *Fundamentals of Numerical Reservoir Stimulation*

Williams, *Acidizing Fundamentals*

Zaba and Doherty, *Practical Petroleum Engineer's Handbook*

350

Conversion Factors

To convert	Into	Multiply by*
Length		
centimeters	meters	0.01
centimeters	inches	0.39371
centimeters	feet	0.032808
centimeters	yards	0.010936
chains	inches	792.0
chains	feet	66.0
chains	miles, statute	0.01250
chains	miles, nautical	0.01085
chains	kilometers	0.02012
chains	yards	22.0
chains	meters	20.11684
chains	rods	4.0
chains	links	100.0
fathoms	feet	6.0
fathoms	meters	1.828804
feet	inches	12.0
feet	links	1.5
feet	yards	0.3333
feet	rods	0.06061
feet	miles, statute	0.0001894
feet	miles, nautical British	0.0001529
feet	miles, intern. nautical	0.0001645
feet	kilometers	0.0003048
feet	meters	0.30481
feet	centimeters	30.48
furlongs	meters	201.268
furlongs	yards	220.0
furlongs	rods	40.0
inches	millimeters	25.400
inches	centimeters	2.540
inches	meters	0.0254
inches	feet	0.08333
inches	yards	0.02778
inches	rods	0.005051
kilometers	miles, statute	0.6214
kilometers	miles, nautical British	0.5396
kilometers	miles, intern. nautical	0.5400
kilometers	feet	3,280.83
kilometers	yards	1,093.61
kilometers	inches	39,370.79

* Example: 1 cm × 0.01 = 0.01 meter

To convert	*Into*	*Multiply by*
kilometers	rods	198.838
kilometers	chains	49.7096
links	yards	0.22
links	feet	0.66
links	inches	7.92
links	centimeters	20.12
links	chains	0.01
meters	inches	39.3701
meters	feet	3.2808
meters	yards	1.0936
meters	links	4.9710
meters	rods	0.198838
meters	miles, statute	0.0006214
meters	miles, nautical British	0.0005396
meters	miles, intern. nautical	0.0005400
meters	millimeters	1,000.0
meters	centimeters	100.0
meters	kilometers	0.001
mils	inches	0.001
microns	millimeters	0.001
microns	mils	0.03937
miles, intern. nautical	kilometers	1.85325
miles, intern. nautical	meters	1.853.25
miles, intern. nautical	feet	6,080.204
miles, intern. nautical	yards	2,026.73
miles, nautical British	kilometers	1.85319
miles, nautical British	meters	1,853.19
miles, nautical British	feet	6,080.0
miles, nautical British	yards	2,026.67
miles, statute	kilometers	1.60935
miles, statute	meters	1,609.35
miles, statute	feet	5,280.0
miles, statute	yards	1,760.0
miles, statute	furlongs	8.0
rods	meters	5.0292
rods	yards	5.5
rods	links	25.0
rods	meters	0.05029
rods	feet	16.5
rods	chains	0.25
rods	inches	198.0
yards, U.S.	centimeters	91.4399
yards, U.S.	meters	0.914399
yards, U.S.	feet	3.0
yards, U.S.	inches	36.0
yards, U.S.	rods	0.18182
yards, U.S.	chains	0.04545
yards, U.S.	yards, British	1.0000029
yards, British	yards, U.S.	0.9999971

Area

acres, U.S.	m²	4,046.873
acres, U.S.	sq chains	10.0
acres, U.S.	sq feet	43,560.0
acres, U.S.	sq inches	6,272,640.0
acres, U.S.	km²	0.004047

To convert	*Into*	*Multiply by*
acres, U.S.	sq miles	0.0015625
acres, U.S.	sq rods	160.0
acres, U.S.	sq roods	4.0
acres, U.S.	sq yards	4,840.0
acres, British	sq meters	4,046.849
ares	acres	0.02471
ares	m²	100.0
hectares	sq yards	11,960.0
hectares	m²	10,000.0
hectares	sq miles	0.003861
sq centimeters	sq inches	0.1550
sq chains	sq rods	16.0
sq chains	m²	404.6873
sq chains	acres	0.1
sq feet	sq rods	0.003673
sq feet	acres	2.2957×10^{-5}
sq feet	m²	0.0929
sq feet	sq inches	144.0
sq feet	sq yards	0.1111
sq inches	cm²	6.4516
sq inches	sq feet	0.006944
sq inches	sq yards	0.0007716
sq inches	acres	1.594×10^{-7}
sq kilometers	sq miles (statute)	0.3861
sq kilometers	hectares	100.0
sq kilometers	sq yards	1,195,985.0
sq kilometers	m²	1,000,000.0
sq kilometers	acres	247.1
sq links	sq inches	62.7264
sq links	cm²	404.6873
sq meters	sq inches	1,549.997
sq meters	sq feet	10.764
sq meters	sq yards	1.1960
sq meters	ares	0.01
sq meters	acres	2.4710×10^{-4}
sq miles (statute)	km²	2.590
sq miles (statute)	hectares	259.0
sq miles (statute)	acres	640.0
sq rods	sq feet	272.25
sq rods	acres	0.00625
sq rods	m²	25.29
sq rods	sq inches	39,204.0
sq rods	sq yards	30.25
sq rods	ares	0.2529
sq yards	acres	$2,0661 \times 10^{-4}$
sq yards	sq feet	9.0
sq yards	m²	0.836126
sq yards	sq rods	0.03306
sq yards	sq inches	1,296.0

Volume and capacity

acre-feet, 1 foot-depth	cubic feet	43,560.0
acre-feet, 1 foot-depth	gallons, U.S.	325,851.0
acre-feet, 1 foot-depth	oil barrels	7,758.0
acre-feet, 1 foot-depth	m³	1,233.5
barrels, U.S. liquid	gallons, U.S.	31.5

To convert	*Into*	*Multiply by*
barrels, U.S. liquid	cubic feet	4.21
barrels, U.S. liquid	cubic inches	7,276.5
barrels, U.S. liquid	liters	119.24
barrels, oil	cubic feet	5.6196
barrels, oil	gallons, U.S.	42.0
barrels, oil	liters	158.988
barrels, oil	metric tons, oil API 36	0.1342
bushels, U.S.	bushels, British	0.96945
bushels, U.S.	liters	35.24
bushels, U.S.	cubic feet	1.2445
bushels, U.S.	gallons, dry	8.0
bushels, U.S.	gallons, liquid, U.S.	9.3092
bushels, U.S.	pints	64.0
bushels, U.S.	quarts, dry	32.0
bushels, U.S.	quarts, liquid	37.2368
cubic decimeters, liters	cubic inches	61.024
cubic decimeters, liters	cubic feet	0.035315
cubic decimeters, liters	cubic yards	0.001308
cubic decimeters, liters	gallons, U.S., liquid	0.2642
cubic decimeters, liters	gallons, U.S., dry	0.90808
cubic decimeters, liters	gallons, Imp.	0.2200
cubic decimeters, liters	bushels, U.S.	0.023838
cubic decimeters, liters	barrels oil	0.00629
cubic feet	cubic decimeters, liters	28.317
cubic feet	cubic inches	1,728.0
cubic feet	cubic yards	0.03704
cubic feet	gallons, U.S., liquid	7.48055
cubic feet	gallons, U.S., dry	6.42851
cubic feet	gallons, Imp.	6.2288
cubic feet	m^3	0.028317
cubic feet	oil barrels	0.1781
cubic feet	pints, U.S., liquid	59.8442
cubic inches	cm^3	16.387
cubic inches	cubic decimeters, liters	0.016387
cubic inches	cubic feet	$5,787 \times 10^{-4}$
cubic inches	cubic yards	2.143×10^{-8}
cubic inches	gallons, U.S., liquid	0.01732
cubic inches	gallons, U.S., dry	0.01488
cubic inches	gallons, Imp.	0.003607
cubic meters	cubic decimeters, liters	1,000.0
cubic meters	cubic feet	35.318
cubic meters	cubic yards	1.3080
cubic meters	gallons, U.S., liquid	264.2
cubic meters	gallons, U.S., dry	231.0
cubic meters	gallons, Imp.	220.0
cubic meters	oil barrels	6.2898
cubic yards	cubic decimeters, liters	764.56
cubic yards	cubic inches	46,656.0
cubic yards	cubic feet	27.0
cubic yards	gallons, U.S., liquid	201.974
cubic yards	gallons, U.S., dry	173.570
cubic yards	gallons, Imp.	169.178
cubic yards	bushels, U.S.	21,6962
cubic yards	oil barrels	4.8089
cubic yards	m^3	0.76455
gallons, U.S., dry	gallons, U.S., liquid	1.16365

To convert	*Into*	*Multiply by*
gallons, U.S., dry	bushels	0.125
gallons, Imp.	cubic decimeters, liters	4.537
gallons, Imp.	cubic inches	277.463
gallons, Imp.	cubic feet	0.16046
gallons, Imp.	m³	0.004537
gallons, Imp.	gallons, U.S.	1.2009
gallons, Imp.	oil barrels	0.0286
gallons, U.S., liquid	cubic decimeters, liters	3.7853
gallons, U.S., liquid	cubic inches	231.0
gallons, U.S., liquid	cubic feet	0.1337
gallons, U.S., liquid	cubic yards	0.00495
gallons, U.S., liquid	gallons, U.S., dry	0.85937
gallons, U.S., liquid	bushels, U.S.	0.10742
gallons, U.S., liquid	gallons, Imp.	0.83267
gallons, U.S., liquid	oil barrels	0.02381
metric tons	oil barrels, 42 gallons, 36 API	7.452
pints, U.S., liquid	liters	0.4732
pints, U.S., liquid	cubic inches	28.875
U.S. dry measure	Imp. measure	0.969

Weight

To convert	*Into*	*Multiply by*
drams, avdp.	grains	27.3438
drams, avdp.	ounces, avdp.	0.0625
drams, avdp.	grams	1.7719
grains, avdp.	milligrams	64.7989
grams	ounces, avdp.	0.03527
grams	pounds, avdp.	0.00221
grams	kilograms	0.001
grams	grains, avdp.	15.4324
grams	drams, avdp.	0.5644
grams	ounces, troy	0.03215
hundredweight, long	stones	8.0
hundredweight, long	tons, long	0.05
hundredweight, long	pounds, avdp.	112.0
hundredweight, long	kilograms	50.802352
hundredweight, short	pounds, avdp.	100.0
hundredweight, short	kilograms	45.3592
hundredweight, short	gallons, U.S., water	13.44
hundredweight, short	cubic feet, water	1.8
hundredweight, short	gallons, Imp.	11.2
hundredweight, short	tons, short	0.05
kilograms	tons, metric	0.001
kilograms	tons, long	0.0009842
kilograms	tons, short	0.0011023
kilograms	pounds, avdp.	2.2046
kilograms	pounds, troy	2.6792
kilograms	grams	1,000.0
milligrams	grams	0.001
milligrams	grains	0.015432
ounces, avdp.	grains, avdp.	437.5
ounces, avdp.	pounds, avdp.	0.0625
ounces, avdp.	drams	16.0
ounces, avdp.	grams	28.3495
picul, Netherland Indies	pounds, avdp.	136.0
picul, Netherland Indies	kilograms	61.688
picul, Netherland Indies	tons, long	0.0607

To convert	*Into*	*Multiply by*
picul, Netherland Indies	tons, short	0.0680
pounds, avdp.	hundredweights, long	0.009
pounds, avdp.	ounces, avdp.	16.0
pounds, avdp.	grains	7,000.0
pounds, avdp.	grams	453.5924
pounds, avdp.	pounds, troy	1.21528
pounds, avdp.	kilograms	0.4536
pounds, troy	grains	5.70
pounds, troy	ounces, troy	12.0
pounds, troy	pounds, avdp.	0.82287
pounds, troy	kilograms	0.373242
quarters, avdp.	pounds, avdp.	28.0
quarters, avdp.	kilograms	12.7005
quintals, Netherlands	kilograms	100.0
quintals, Netherlands	pounds, avdp.	220.46
quintals, Netherlands	hundredweight, long	1.9684
stones, avdp.	pounds, avdp.	14.0
stones, avdp.	kilograms	6.350
tons, long	kilograms	1,016.05
tons, long	pounds	2,240.0
tons, long	tons, metric	1.01605
tons, long	tons, short	1.12
tons, metric	kilograms	1,000.0
tons, metric	tons, long	0.98421
tons, metric	tons, short	1.10231
tons, metric	pounds	2,204.6
tons, metric	barrels, water, 60° F.	6.297
tons, metric	barrels, 36 API	7.454
tons, short	tons, metric	0.907185
tons, short	kilograms	907.285
tons, short	tons, long	0.89286
tons, short	pounds	2,000.0

Weight per unit of length

kilograms/meter	pounds/foot	0.6720
kilograms/meter	pounds/yard	2.01591
kilograms/meter	tons, short/mile	1.77400
kilograms/meter	tons, long/mile	1.58393
kilograms/meter	tons, metric/kilometer	1.0
kilograms/kilometer	pounds/mile	3.5480
pounds/foot	kilograms/meter	1.4882
pounds/foot	pounds/yard	3.0
pounds/mile	kilograms/kilometer	0.2818
pounds/yard	pounds/foot	0.3333
pounds/yard	kilograms/meter	0.49605
pounds/yard	tons, short/mile	0.88000
pounds/yard	tons, long/mile	0.78571
pounds/yard	tons, metric/kilometer	0.49605

Flow

barrels, oil/day	gallons, U.S./minute	0.02917
barrels, oil/day	cubic feet/second	6.498×10^{-5}
barrels, oil/day	liters/second	0.00184
barrels, oil/day	m³/day	0.1588
barrels, oil/hour	cubic feet/minute	0.0936
barrels, oil/hour	gallons, U.S./minute	0.7

To convert	Into	Multiply by
barrels, oil/hour	m³/hour	0.1588
barrels, oil/minute	m³/hour	9.539
cubic feet/minute	m³/hour	1.6999
cubic feet/minute	liters/second	0.4720
cubic feet/minute	gallons, U.S./second	0.1247
cubic feet/minute	barrels, oil/hour	10.686
cubic feet/minute	gallons, U.S./minute	7.481
cubic feet/minute	pounds of water/minute	62.428
cubic feet/second	m³/hour	101.94
cubic feet/second	liters/minute	1,699.3
cubic feet/second	gallons, U.S./minute	448.83
cubic feet/second	barrels, oil/day	15,388.0
cubic feet/hour	liters/second	0.007866
cubic feet/hour	liters/minute	0.472
cubic feet/hour	barrels, oil/day	4.2744
m³, 1,000 liters/hour	gallons, U.S./minute	4.403
m³, 1,000 liters/hour	cubic feet/minute	0.5886
m³, 1,000 liters/hour	barrels, oil/day	150.96
gallons, U.S./minute	cubic feet/minute	0.1337
gallons, U.S./minute	cubic feet/hour	8.022
gallons, U.S./minute	barrels, oil/hour	1.429
gallons, U.S./minute	m³/hour	0.2271
gallons, U.S./minute	liters/minute	3.785
liters, cubic decimeters/minute	cubic feet/second	5.885×10^{-4}
liters, cubic decimeters/minute	gallons, U.S./second	0.004403
liters, cubic decimeters/minute	cubic feet/minute	2.12
liters, cubic decimeters/minute	gallons, U.S./minute	15.85
liters, cubic decimeters/minute	barrels, oil/day	543.46

Velocity

degrees/second	revolutions/second	0.002778
degrees/second	revolutions/minute	0.1667
degrees/second	radians/second	0.001745
feet/minute	centimeters/second	0.5080
feet/minute	feet/second	0.01667
feet/minute	meters/minute	0.3048
feet/minute	miles/hour	0.01136
feet/second	meters/second	0.3048
feet/second	feet/hour	3,600.0
feet/second	kilometers/hour	1.097
feet/second	miles/hour	0.68182
kilometers/hour	meters/second	0.2778
kilometers/hour	feet/second	0.9113
kilometers/hour	miles/hour	0.6214
meters/minute	feet/minute	3.281
meters/minute	feet/second	0.05468
meters/minute	meters/hours	60.0
meters/minute	miles/hour	0.03728
meters/second	feet/second	3.2808
meters/second	feet/minute	196.85
meters/second	miles/hour	2.23693
meters/second	kilometers/hour	3.6
miles/hour	meters/second	0.4470
miles/hour	meters/minute	26.82
miles/hour	feet/second	1.4667
miles/hour	feet/minute	88.0

To convert	*Into*	*Multiply by*
miles/hour	kilometers/hour	1.6096
radians/second	revolutions/second	0.1592
radians/second	revolutions/minute	9.549
radians/second	degrees/second	57° 17' 45.8"
revolutions/minute	revolutions/second	0.0167
revolutions/minute	radians/second	0.1047
revolutions/minute	degrees/second	6.0
revolutions/second	revolutions/minute	60.0
revolutions/second	radians/second	6.2832
revolutions/second	degrees/second	360.0

Capacity per unit of length

cubic inches/inch	liters/centimeter	0.006452
cubic feet/foot	liters/meter	92.9
cubic feet/foot	U.S. gallons/foot	7.4805
cubic feet/foot	barrels, oil/foot	0.1781
U.S. gallons/foot	liters/meter	12.419
U.S. gallons/foot	cubic feet/foot	0.1337
U.S. gallons/foot	barrels, oil/foot	0.0238
barrels/foot	liters/meter	521.6
barrels/foot	cubic feet/foot	5.6146
barrels/foot	U.S. gallons/foot	42.0
liters/meter	cubic feet/foot	0.01076
liters/meter	U.S. gallons/foot	0.08052
liters/meter	barrels/foot	0.001917

Weight per unit of volume

pounds/cubic inch	kilograms/centimeter^3	0.0277
pounds/cubic foot	kilograms/centimeter^3	16.0185
pounds/cubic foot	pounds/U.S. gallon	0.1337
pounds/cubic foot	pounds/Imp. gallon	0.1605
pounds/U.S. gallon	kilograms/liter	0.1198
pounds/U.S. gallon	pounds/cubic foot	7.4805
pounds/U.S. gallon	pounds/Imp. gallon	1.2009
pounds/Imp. gallon	kilograms/liter	0.0998
pounds/Imp. gallon	pounds/cubic foot	6.2288
pounds/Imp. gallon	pounds/U.S. gallon	0.8327
kilograms/liter	pounds/cubic foot	62.426
kilograms/liter	pounds/U.S. gallon	8.3455
kilograms/liter	pounds/Imp. gallon	10.0221
kilograms/centimeter^3	pounds/cubic inch	36.127

Specific gravity for fluids lighter than water

specific gravity (G)	degrees Baumé (B)	$\dfrac{1}{G}\left[\dfrac{140}{G}-130\right]$
degrees Baumé (B)	specific gravity (G)	$\dfrac{1}{B}\left[\dfrac{140}{B+130}\right]$
specific gravity (G)	degrees API (A)	$\dfrac{1}{G}\left[\dfrac{141.5}{G}-131.5\right]$
degrees API (A)	specific gravity (G)	$\dfrac{1}{A}\left[\dfrac{141.5}{A+131.5}\right]$
degrees API (A)	degrees Baumé (B)	$\dfrac{1}{A}\left[\dfrac{140\,A+15}{141.5}\right]$

To convert	Into	Multiply by
degrees Baumé (B)	degrees API (A)	$\dfrac{1}{B}\left[\dfrac{141.5\,B - 15}{140}\right]$
specific gravity (G)	degrees Baumé (B)	$\dfrac{1}{G}\left[145 - \dfrac{145}{G}\right]$
degrees Baumé (B)	specific gravity (G)	$\dfrac{1}{B}\left[\dfrac{145}{145 - B}\right]$

Acceleration

feet per second/second	meters per second/second	0.3048
feet per second/second	miles per hour/second	0.6818
gravity, standard value	feet per second/second	32.174
gravity, standard value	cm per second/second	980.665
gravity, at 45° lateral, sea level	feet per second/second	32.172
gravity, at 45° lateral, sea level	cm per second/second	980.616
miles per hour/second	meters per second/second	0.4470
miles per hour/second	feet per second/second	1.4670
miles per hour/second	kilometers per hour/second	1.609
miles per hour/second	cm per second/second	44.704
cm per second/second	miles per hour/second	0.02237
cm per second/second	feet per second/second	0.0328
cm per second/second	kilometers per hour/second	0.0360
meters per second/second	miles per hour/second	2.237
radians per second/second	revs per minute/second	9.5490
revs per second/second	revs per minute/second	60.0

Force conversion factors based on value of
$g = 32.174$ feet/second^2 = 980.665 centimeters/second^2

dynes, gr cm per second^2	pounds weight	2.2481×10^{-6}
dynes, gr cm per second^2	grams weight	1.0197×10^{-3}
grams weight	dynes	980.665
pounds weight	dynes	4.4482×10^{-5}
kilograms weight	dynes	9.8067×10^{-5}

Torque

foot-pounds	kilogram-meters	0.1383
inch-pounds	kilogram-meters	0.0115
kilogram-meters	foot-pounds	7.2330
kilogram-meters	inch-pounds	86.7960

Pressure

pounds/square inch, psi	kilograms/centimeters^2	0.0703
pounds/square inch, psi	atmospheres	0.0680
pounds/square inch, psi	meters head of water	0.7031
pounds/square inch, psi	feet head of water	2.3090
pounds/square inch, psi	inches head of water	27.670
pounds/square inch, psi	inches of mercury	2.0360
pounds/square inch, psi	millimeters of mercury	51.715
pounds/square foot	kilograms/meter^2	4.883
kilograms/centimeter^2, at	psi	14.223
kilograms/centimeter^2, at	pounds/square foot	0.0020482
kilograms/centimeter^2, at	atmospheres	0.9678
kilograms/centimeter^2, at	meters head of water	10.0
kilograms/centimeter^2, at	feet head of water	32.808
kilograms/centimeter^2, at	inches of mercury	28.958
kilograms/centimeter^2, at	millimeters of mercury	735.56

To convert	*Into*	*Multiply by*
kilograms/meter²	pounds/square foot	0.2048
atmospheres	kilograms/centimeter²	1.0333
atmospheres	psi	14.696
atmospheres	meters head of water	10.333
atmospheres	feet head of water	33.899
atmospheres	inches of mercury	29.921
atmospheres	millimeters of mercury	760.0
feet head of water, at 40° C. or 39.1° F.	psi	0.4335
feet head of water	kilograms/centimeter²	0.03048
feet head of water	atmospheres	0.0295
feet head of water	meters head of water	0.3048
feet head of water	inches of mercury	0.8826
feet head of water	millimeters of mercury	22.418
meters head of water, at 4° C. or 39.1° F.	psi	1.4223
meters head of water	kilograms/centimeter²	0.10
meters head of water	atmospheres	0.09678
meters head of water	feet head of water	3.2808
meters head of water	inchs of mercury	2.8958
meters head of water	millimeters of mercury	73.556
inches head of water	psi	0.0361
inches of mercury, at 0° C. or 32° F.	psi	0.4912
inches of mercury	kilograms/centimeter²	0.0345
inches of mercury	atmospheres	0.0334
inches of mercury	meters head of water	0.3453
inches of mercury	feet head of water	1.133
inches of mercury	millimeters of mercury	25.400
millimeters of mercury, at 0° C. or 32° F.	psi	0.019337
millimeters of mercury	kilograms/centimeter²	0.0013596
millimeters of mercury	atmospheres	0.001316
millimeters of mercury	meters head of water	0.013596
millimeters of mercury	feet head of water	0.044606
millimeters of mercury	inches of mercury	0.03937

Work and energy

British thermal units (BTU), mean	kilogram calories, mean	0.252
BTU, mean	foot-pounds	777.97
BTU, mean	kilogram-meters	107.56
BTU, mean	kilowatt-hours, absolute	2.930×10^{-4}
BTU, mean	horsepower-hours	3.929×10^{-4}
BTU, mean	liter-atmospheres	10.41
BTU, mean	cubic foot-atmospheres	3,676.0
BTU, mean	joules, absolute	1,054.8

Thermal units

BTUs, 60° F./°F.	gram calories/°C.	453.59
geal, 15° C./°C.	BTUs, 60° F./°F.	2.2046×10^{-3}
BTUs/pound, mean	geal/gram, mean	0.5556
geal/gram, mean	BTUs/pound, mean	1.8
BTUs, mean/pound °F.	geal per gram/°C., mean	1.0
BTUs/inch	kilogram calorie/meter	9.9213
kilogram-calories/meter	BTUs/inch	0.1008

To convert	Into	Multiply by
BTUs per square foot/°F. per hour	kilogram-calorie per meter²/°C. per hour	4.8824
kilogram-calorie per meter²/°C. per hour	BTUs per square foot/°F. per hour	0.2048

Miscellaneous conversion factors

To convert	Into	Multiply by
Cubic feet of gas/barrel, oil	meters³ gas/meters³ oil	0.1781
meters³ gas/meters³ oil	cubic feet of gas/barrel, oil	5.6146
Imp. gallons/1,000 cubic feet	liters, 1,000 cm³/meters³	0.1605
liters, 1,000 cm³/meters³	Imp. gallons/1,000 cubic feet	6.229
liters, 1,000 cm³/meters³	U.S. gallons/1,000 cubic feet	0.1336
U.S. gallon/1,000 cubic feet	liters, 1,000 cm³/meters³	7.4808
Imp. gallon water, at 4° C. or 39.2° F.	pounds weight	10.0221
U.S. gallon water, at 4° C. or 39.2° F.	pounds weight	8.3454
barrels water, at 4° C. or 39.2° F.	pounds weight	350.5105
cubic foot of water, at 4° C. or 39.2° F.	pounds weight	62.4283
pounds weight	Imp. gallon water, at 4° C.	0.0998
pounds weight	U.S. gallon water, at 4° C.	0.1198
pounds weight	barrels of water, at 4° C.	0.002853
pounds weight	cubic foot of water, at 4° C.	0.016018
meters³/C.V.	cubic feet/horsepower	35.804
meters²/C.V.	square feet/horsepower	10.913
kilograms/C.V.	cubic feet/horsepower	2.235
cubic feet/horsepower	meters³/C.V.	0.0279
square feet/horsepower	meters²/C.V.	0.0916
pounds/horsepower	kilograms/C.V.	0.447
cubic feet of air, at 32° F., 14.7 psi	pounds weight	0.08071
pounds weight	cubic feet of air, at 32° F., 14.7 psi	12.386
cubic feet of air, at 62° F., 14.7 psi	pounds weight	0.07608
pounds weight	cubic feet of air, at 62° F., 14.7 psi	13.144
grams weight	cm² of air at 0° C., 760 millimeters	773.39
cm² of air, at 0° C., 760 millimeters	grams weight	0.001293
grams weight	cm² of mercury, at 0° C.	0.073556
cm² of mercury, at 0° C.	grams weight	13.5951
percentage volumetric recovery	m²/hectare-meter	100.0
percentage volumetric recovery	barrels/acre-foot	77.58
m²/hectare-meter	barrels/acre-foot	0.7758
barrels/acre-foot	m²/hectare-meter	1.289
liters of water, at maximum density, 3.98° C., exact value	cubic decimeters	1.000027
cubic centimeters	liters of water, exact value	0.999973
density	pounds/cubic foot	62.4
degree	radians	0.1745329

Gasoline content of natural gas

To convert	Into	Multiply by
U.S. gallons/1,000 cubic feet	grams/meters³	±90.0
grams/meters³	U.S. gallons/1,000 cubic feet	±0.011

To convert	*Into*	*Multiply by*
Temperature		
degrees Fahrenheit	degrees Celcius	$\left(\dfrac{F-32}{1.8}\right)$
degrees Fahrenheit	degrees absolute	$(F+459.4)$
degrees centigrade	degrees Fahrenheit	$(1.8C+32)$
degrees centigrade	degrees absolute	$(C+273)$
degrees absolute	degrees Fahrenheit	$(T-459.4)$
degrees absolute	degrees centigrade	$(T-273)$

Index